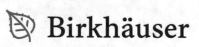

Carlos S. Kubrusly

Spectral Theory of Bounded Linear Operators

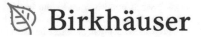 Birkhäuser

Carlos S. Kubrusly
Institute of Mathematics
Federal University of Rio de Janeiro
Rio de Janeiro, Brazil

ISBN 978-3-030-33151-1   ISBN 978-3-030-33149-8   (eBook)
https://doi.org/10.1007/978-3-030-33149-8

Mathematics Subject Classification (2010): 47-00, 47-01, 47-02, 47A10, 47A12, 47A16, 47A25, 47A53, 47A60, 47B15, 47B20, 47B48, 47C05

This book is published under the imprint Birkhäuser, www.birkhauser-science.com by the registered company Springer Nature Switzerland AG
The registered company address is: Gewerbestrasse 11, 6330 Cham, Switzerland

To Alan and Jessica

The positivists have a simple solution: the world must be divided into that which we can say clearly and the rest, which we had better pass over in silence. But can anyone conceive of a more pointless philosophy, seeing that what we can say clearly amounts to next to nothing? If we omitted all that is unclear, we would probably be left with completely uninteresting and trivial tautologies.

Werner Heisenberg [62, Chapter 17, p. 213]

Mathematics is not a language, it's an adventure. [...] Most mathematics is done with a friend over a cup of coffee, with a diagram scribbled on a napkin. Mathematics is and always has been about ideas, and a valuable idea transcends the symbols with which you choose to represent it. As Carl Friedrich Gauss once remarked, "What we need are notions, not notations."

Paul Lockhart [92, Part I, p. 53]

Preface

The book introduces spectral theory for bounded linear operators, giving a modern text for a graduate course focusing on two basic aspects. On the one hand, the spectral theory for normal operators acting on Hilbert spaces is comprehensively investigated, emphasizing recent aspects of the theory with detailed proofs, including Fredholm and multiplicity theories. On the other hand, Riesz–Dunford functional calculus for Banach-space operators is also meticulously presented, as well as Fredholm theory in Banach spaces (which is compared and clearly distinguished from its Hilbert-space counterpart).

Seven portions somewhat uniformly distributed constitute the book contents: five chapters and two chapter-size appendices. The first five chapters consist of a revised, corrected, enlarged, updated, and thoroughly rewritten text based on the author's 2012 book [80], which was heavily focused on Hilbert-space operators. The subsequent appendix draws a parallel between the Hilbert-space and the Banach-space ways of handling Fredholm theory. The final appendix gives a unified approach on multiplicity theory; precisely, on spectral multiplicity.

Prerequisites for a smooth reading and for efficient learning are to some extent modest. A first course in elementary functional analysis, in particular in operator theory, will naturally be helpful. This encompasses a formal acquaintance with an introduction to analysis, including some measure theory and functions of a complex variable. An effort, however, has been made to introduce some essential results in order to make the book as self-contained as possible. Along this line, Chapter 1 supplies some basic concepts from operator theory which will be necessary in subsequent chapters. It does not intend to replace an introductory course on operator theory, but it summarizes notions and fundamental results needed for further chapters, and also establishes notations and terminologies used throughout the book.

Chapter 2 deals with standard spectral results on the Banach algebra of bounded linear operators acting on Banach spaces, including the classical partition of the spectrum and spectral properties for classes of operators. Chapter 3 is central. It is entirely dedicated to the Spectral Theorem for normal operators on Hilbert spaces. After an initial consideration of the compact case, the general case is investigated in detail. Proofs of both versions of the Spectral Theorem for the general case are fully developed. The chapter closes with the Fuglede Theorems. Chapter 4 splits the functional calculus into two distinct but related parts. The functional calculus for normal operators is presented in the first part, which depends of course on the Spectral Theorem. The second part considers the analytic functional calculus (Riesz–Dunford functional calculus) for Banach-space operators. Chapter 5 focuses on Fredholm theory exclusively for Hilbert-space operators. This concentrates on compact perturbations of the spectrum, where a finer analysis is worked out leading to the notions of essential (Fredholm), Weyl, and Browder spectra and, consequently, getting to Weyl's and Browder's Theorems, including recent results.

Each chapter has a final section on *Supplementary Propositions* consisting of (i) auxiliary results required in the sequel and (ii) additional results extending the theory beyond the bounds of the main text. These are followed by a set of *Notes* where each proposition is briefly discussed and references are provided indicating proofs for all of them. Such Supplementary Propositions can also be thought of as a section of proposed problems, and their respective Notes can be viewed as hints for solving them. This in turn is followed by a list of *Suggested Readings* indicating a collection of books on the subject of the chapter where different approaches, proofs, and further results can be found.

Appendix A is the counterpart of Chapter 5 for operators acting on Banach spaces. This certainly is not a mere generalization: unlike Hilbert spaces, Banach spaces are not complemented, and this forces a rather different attitude for approaching Fredholm theory on Banach spaces. The appendix carries out a comparison between these two approaches, referred to as left-right and upper-lower ways of handling semi-Fredholm operators, in the course of which the origins and consequences of the lack of complementation in Banach spaces are carefully analyzed. Appendix B extends Chapter 3 on the Spectral Theorem by presenting an introductory view of multiplicity theory as a key factor for establishing a complete set of unitary invariants for normal operators. This closes the Spectral Theorem saga by stating that multiplicity function and spectral measure form a complete set of unitary invariants for normal operators on separable Hilbert spaces. The same level of detail devoted to proofs of theorems in the main chapters is maintained throughout the appendices. The idea behind such highly detailed proofs is to discuss and explain delicate points, stressing some usually hidden features.

The subject matter of the book has been prepared to be covered in a one-semester graduate course (but the pace of lectures may depend on the boundary conditions). The whole book is the result of an effort to meet the needs of a contemporary course on spectral theory for mathematicians that will also be accessible to a wider audience of students in mathematics, statistics, economics, engineering, and physics, while requiring only a modest prerequisite. It may also be a useful resource for working mathematicians dealing with operator theory in general, and for scientists wishing to apply the theory to their field. I have tried to answer a continual request from students who ask for a text containing complete and detailed proofs, including discussions on a number of questions they have raised throughout the years. The logical dependence of the various sections and chapters is roughly linear, reflecting approximately a minimum amount of material needed to proceed further. There is an updated bibliography at the end of the book, including only references cited within the text, which contains most of the classics (of course) as well as some recent texts reveling modern approaches and present-day techniques.

Since I have been lecturing on this subject for a long time, I have benefited from the help of many friends among students and colleagues, and I am grateful to all of them. In particular, I wish to thank Gustavo C. Amaral and El Hassan Benabdi, who helped with the tracing down of typos and imprecisions. Special thanks are due to Joaquim D. Garcia for his conscientious reading of the whole manuscript, correcting many typos and inaccuracies, and offering hints for sensible improvements. Special thanks are also due the copyeditor Brian Treadway who did an impressive job, including relevant mathematical corrections. I am also grateful to a pair of anonymous reviewers who raised reasonable suggestions and pointed out some necessary modifications.

Rio de Janeiro, June 2019

Carlos S. Kubrusly

Contents

Appendices

1

Introductory Results

This chapter summarizes the essential background required for the rest of the book. Its purpose is notation, terminology, and basics. As for basics, we mean well-known theorems from operator theory needed in the sequel.

1.1 Background: Notation and Terminology

Let $\varnothing$ be the empty set. Denote the sets of nonnegative, positive, and all integers by $\mathbb{N}_0$, $\mathbb{N}$, $\mathbb{Z}$; the fields of rational, real, and complex numbers by $\mathbb{Q}$, $\mathbb{R}$, $\mathbb{C}$; and an arbitrary field by $\mathbb{F}$. Use the same notation for a sequence and its range (e.g., $\{x_n\}$ means $x(\cdot)\colon \mathbb{N} \to X$ and also $\{x \in X\colon x = x(n) \text{ for some } n \in \mathbb{N}\}$).

We assume the reader is familiar with the notions of *linear space* (or vector space), *normed space*, *Banach space*, *inner product space*, *Hilbert space*, as well as *algebra* and *normed algebra*. All spaces considered in this book are complex (i.e., over the complex field $\mathbb{C}$). Given a inner product space $\mathcal{X}$, the sesquilinear form (linear in the first argument) $\langle \cdot \,;\, \cdot \rangle\colon \mathcal{X} \times \mathcal{X} \to \mathbb{C}$ on the Cartesian product $\mathcal{X} \times \mathcal{X}$ stands for the inner product in $\mathcal{X}$. We do not distinguish notation for norms. Thus $\|\cdot\|$ denotes the norm on a normed space $\mathcal{X}$ (in particular, the norm generated by the inner product in an inner product space $\mathcal{X}$ — i.e., $\|x\|^2 = \langle x \,;\, x \rangle$ for all x in $\mathcal{X}$) and also the (induced uniform) operator norm in $\mathcal{B}[\mathcal{X}, \mathcal{Y}]$ (i.e., $\|T\| = \sup_{x \neq 0} \frac{\|Tx\|}{\|x\|}$ for all T in $\mathcal{B}[\mathcal{X}, \mathcal{Y}]$), where $\mathcal{B}[\mathcal{X}, \mathcal{Y}]$ stands for the normed space of all *bounded linear* transformations of a normed space $\mathcal{X}$ into a normed space $\mathcal{Y}$. The induced uniform norm has the *operator norm property*; that is, if $\mathcal{X}$, $\mathcal{Y}$, and $\mathcal{Z}$ are normed spaces over the same scalar field, and if $T \in \mathcal{B}[\mathcal{X}, \mathcal{Y}]$ and $S \in \mathcal{B}[\mathcal{Y}, \mathcal{Z}]$, then $ST \in \mathcal{B}[\mathcal{X}, \mathcal{Z}]$ and $\|ST\| \leq \|S\|\,\|T\|$.

Between normed spaces, *continuous linear transformation* and *bounded linear transformation* are synonyms. A transformation $T\colon \mathcal{X} \to \mathcal{Y}$ between normed spaces $\mathcal{X}$ and $\mathcal{Y}$ is *bounded* if there exists a constant $\beta \geq 0$ for which $\|Tx\| \leq \beta \|x\|$ for every x in $\mathcal{X}$. It is said to be *bounded below* if there is a constant $\alpha > 0$ for which $\alpha \|x\| \leq \|Tx\|$ for every x in $\mathcal{X}$. An *operator* on a normed space $\mathcal{X}$ is precisely a bounded linear (i.e., a continuous linear) transformation of $\mathcal{X}$ into itself. Set $\mathcal{B}[\mathcal{X}] = \mathcal{B}[\mathcal{X}, \mathcal{X}]$ for short: the normed algebra

© Springer Nature Switzerland AG 2020
C. S. Kubrusly, *Spectral Theory of Bounded Linear Operators*,
https://doi.org/10.1007/978-3-030-33149-8_1

of all operators on $\mathcal{X}$. If $\mathcal{X} \neq \{0\}$, then $\mathcal{B}[\mathcal{X}]$ contains the identity operator I and $\|I\| = 1$ and so $\mathcal{B}[\mathcal{X}]$ is a *unital normed algebra*. If $\mathcal{X} \neq \{0\}$ is a Banach space, then $\mathcal{B}[\mathcal{X}]$ is a *unital (complex) Banach algebra*. Since the induced uniform norm has the operator norm property, $\|T^n\| \leq \|T\|^n$ for every operator T in $\mathcal{B}[\mathcal{X}]$ on a normed space $\mathcal{X}$ and every integer $n \geq 0$.

A transformation $T \in \mathcal{B}[\mathcal{X}, \mathcal{Y}]$ is a *contraction* if $\|T\| \leq 1$ or, equivalently, if $\|Tx\| \leq \|x\|$ for every $x \in \mathcal{X}$. It is a *strict contraction* if $\|T\| < 1$. If $\mathcal{X} \neq \{0\}$, then a transformation T is a contraction (or a strict contraction) if and only if $\sup_{x \neq 0}(\|Tx\|/\|x\|) \leq 1$ (or $\sup_{x \neq 0}(\|Tx\|/\|x\|) < 1$). If $\mathcal{X} \neq \{0\}$, then

$$\|T\| < 1 \quad \Longrightarrow \quad \|Tx\| < \|x\| \ \text{ for every } 0 \neq x \in \mathcal{X} \quad \Longrightarrow \quad \|T\| \leq 1.$$

If T satisfies the middle inequality, then it is called a *proper contraction*.

Let $\mathcal{X}$ be a Banach space. A sequence $\{T_n\}$ of operators in $\mathcal{B}[\mathcal{X}]$ *converges uniformly* to an operator T in $\mathcal{B}[\mathcal{X}]$ if $\|T_n - T\| \to 0$, and $\{T_n\}$ *converges strongly* to T if $\|(T_n - T)x\| \to 0$ for every x in $\mathcal{X}$. If $\mathcal{X}$ is a Hilbert space, then $\{T_n\}$ *converges weakly* to T if $\langle (T_n - T)x \,; y \rangle \to 0$ for every x, y in $\mathcal{X}$ (or, equivalently, $\langle (T_n - T)x \,; x \rangle \to 0$ for every x in $\mathcal{X}$ if the Hilbert space is complex). These will be denoted by $T_n \xrightarrow{u} T$, $T_n \xrightarrow{s} T$, and $T_n \xrightarrow{w} T$, respectively. The sequence $\{T_n\}$ is *bounded* if $\sup_n \|T_n\| < \infty$. As is readily verified,

$$T_n \xrightarrow{u} T \quad \Longrightarrow \quad T_n \xrightarrow{s} T \quad \Longrightarrow \quad T_n \xrightarrow{w} T \quad \Longrightarrow \quad \sup_n \|T_n\| < \infty$$

(the last implication is a consequence of the Banach–Steinhaus Theorem).

A *linear manifold* $\mathcal{M}$ of a linear space $\mathcal{X}$ over a field $\mathbb{F}$ is a subset of $\mathcal{X}$ for which $u + v \in \mathcal{M}$ and $\alpha u \in \mathcal{M}$ for every $u, v \in \mathcal{M}$ and $\alpha \in \mathbb{F}$. A *subspace* of a normed space $\mathcal{X}$ is a *closed* linear manifold of it. The closure of a linear manifold is a subspace. A subspace of a Banach space is a Banach space. A linear manifold $\mathcal{M}$ of a linear space $\mathcal{X}$ is *nontrivial* if $\{0\} \neq \mathcal{M} \neq \mathcal{X}$. A subset A of $\mathcal{X}$ is *F-invariant* for a function $F \colon \mathcal{X} \to \mathcal{X}$ (or A is an *invariant set* for F) if $F(A) \subseteq A$. An *invariant linear manifold* (*invariant subspace*) for $T \in \mathcal{B}[\mathcal{X}]$ is a linear manifold (subspace) of $\mathcal{X}$ which, as a subset of $\mathcal{X}$, is T-invariant. The zero space $\{0\}$ and the whole space $\mathcal{X}$ (i.e., the *trivial subspaces*) are trivially invariant for every T in $\mathcal{B}[\mathcal{X}]$. If $\mathcal{M}$ is an invariant linear manifold for T, then its closure $\mathcal{M}^-$ is an invariant subspace for T.

Let $\mathcal{X}$ and $\mathcal{Y}$ be linear spaces. The *kernel* (or *null space*) of a linear transformation $T \colon \mathcal{X} \to \mathcal{Y}$ is the inverse image of $\{0\}$ under T,

$$\mathcal{N}(T) = T^{-1}(\{0\}) = \{x \in \mathcal{X} \colon Tx = 0\},$$

which is a linear manifold of $\mathcal{X}$. The *range* of T is the image of $\mathcal{X}$ under T,

$$\mathcal{R}(T) = T(\mathcal{X}) = \{y \in \mathcal{Y} \colon y = Tx \text{ for some } x \in \mathcal{X}\},$$

which is a linear manifold of $\mathcal{Y}$. If $\mathcal{X}$ and $\mathcal{Y}$ are normed spaces and T lies in $\mathcal{B}[\mathcal{X}, \mathcal{Y}]$ (i.e., if the linear transformation T is bounded), then $\mathcal{N}(T)$ is a subspace of $\mathcal{X}$ (i.e., if T is bounded, then $\mathcal{N}(T)$ is closed).

If $\mathcal{M}$ and $\mathcal{N}$ are linear manifolds of a linear space $\mathcal{X}$, then the *ordinary sum* (or simply, the *sum*) $\mathcal{M} + \mathcal{N}$ is the linear manifold of $\mathcal{X}$ consisting of all sums $u + v$ with u in $\mathcal{M}$ and v in $\mathcal{N}$. The *direct sum* $\mathcal{M} \oplus \mathcal{N}$ is the linear space of all ordered pairs (u, v) in $\mathcal{M} \times \mathcal{N}$ (i.e., u in $\mathcal{M}$ and v in $\mathcal{N}$), where vector addition and scalar multiplication are defined coordinatewise. Although ordinary and direct sums are different linear spaces, they are isomorphic if $\mathcal{M} \cap \mathcal{N} = \{0\}$. A pair of linear manifolds $\mathcal{M}$ and $\mathcal{N}$ are *algebraic complements* of each other if $\mathcal{M} + \mathcal{N} = \mathcal{X}$ and $\mathcal{M} \cap \mathcal{N} = \{0\}$. In this case, each vector x in $\mathcal{X}$ can be uniquely written as $x = u + v$ with u in $\mathcal{M}$ and v in $\mathcal{N}$. If two subspaces (linear manifolds) of a normed space are algebraic complements of each other, then they are called *complementary subspaces* (*complementary linear manifolds*).

1.2 Inverse Theorems on Banach Spaces

A fundamental theorem of (linear) functional analysis is the OPEN MAPPING THEOREM: *a continuous linear transformation of a Banach space onto a Banach space is an open mapping* (see, e.g., [78, Section 4.5] — a mapping is *open* if it maps open sets onto open sets). A crucial corollary of it is the next result.

Theorem 1.1. (INVERSE MAPPING THEOREM). *If $\mathcal{X}$ and $\mathcal{Y}$ are Banach spaces and if $T \in \mathcal{B}[\mathcal{X}, \mathcal{Y}]$ is injective and surjective, then $T^{-1} \in \mathcal{B}[\mathcal{Y}, \mathcal{X}]$.*

Proof. An invertible transformation T is precisely an injective and surjective one. Since the inverse T^{-1} of an invertible linear transformation is linear, and since an invertible transformation is open if and only if it has a continuous inverse, the stated result follows from the Open Mapping Theorem. $\qquad \square$

Thus an injective and surjective bounded linear transformation between Banach spaces has a *bounded* (linear) inverse. Let $\mathcal{G}[\mathcal{X}, \mathcal{Y}]$ denote the collection of all invertible (i.e., injective and surjective) elements from $\mathcal{B}[\mathcal{X}, \mathcal{Y}]$ with a bounded inverse. (Recall: inverse of linear transformation is linear, thus $\mathcal{G}[\mathcal{X}, \mathcal{Y}]$ is the set of all *topological isomorphism* of $\mathcal{X}$ onto $\mathcal{Y}$.) By the above theorem, if $\mathcal{X}$ and $\mathcal{Y}$ are Banach spaces, then every invertible transformation in $\mathcal{B}[\mathcal{X}, \mathcal{Y}]$ lies in $\mathcal{G}[\mathcal{X}, \mathcal{Y}]$. Set $\mathcal{G}[\mathcal{X}] = \mathcal{G}[\mathcal{X}, \mathcal{X}]$, the group (under multiplication) of all invertible operators from $\mathcal{B}[\mathcal{X}]$ with a bounded inverse (or of all invertible operators from $\mathcal{B}[\mathcal{X}]$ if $\mathcal{X}$ is a Banach space). If $\mathcal{X}$ and $\mathcal{Y}$ are normed spaces and if there exists $T^{-1} \in \mathcal{B}[\mathcal{R}(T), \mathcal{X}]$ such that $T^{-1}T = I \in \mathcal{B}[\mathcal{X}]$, then $TT^{-1} = I \in \mathcal{B}[\mathcal{R}(T)]$ (i.e., if $y = Tx$ for some $x \in \mathcal{X}$, then $TT^{-1}y = TT^{-1}Tx = Tx = y$). In this case T is said to have *a bounded inverse on its range.*

Theorem 1.2. (BOUNDED INVERSE THEOREM). *Let $\mathcal{X}$ and $\mathcal{Y}$ be Banach spaces and take any $T \in \mathcal{B}[\mathcal{X}, \mathcal{Y}]$. The following assertions are equivalent.*

(a) *T has a bounded inverse on its range.*

(b) *T is bounded below.*

(c) *T is injective and has a closed range.*

Proof. Part (i). The equivalence between (a) and (b) still holds if $\mathcal{X}$ and $\mathcal{Y}$ are just normed spaces. Indeed, if there exists $T^{-1} \in \mathcal{B}[\mathcal{R}(T), \mathcal{X}]$, then there exists a constant $\beta > 0$ for which $\|T^{-1}y\| \le \beta \|y\|$ for every $y \in \mathcal{R}(T)$. Take an arbitrary $x \in \mathcal{X}$ so that $Tx \in \mathcal{R}(T)$. Thus $\|x\| = \|T^{-1}Tx\| \le \beta \|Tx\|$, and so $\frac{1}{\beta}\|x\| \le \|Tx\|$. Hence (a) implies (b). Conversely, if (b) holds true, then $0 < \|Tx\|$ for every nonzero x in $\mathcal{X}$, and so $\mathcal{N}(T) = \{0\}$. Then T has a (linear) inverse on its range — *a linear transformation is injective if and only if it has a null kernel*. Take an arbitrary $y \in \mathcal{R}(T)$ so that $y = Tx$ for some $x \in \mathcal{X}$. Thus $\|T^{-1}y\| = \|T^{-1}Tx\| = \|x\| \le \frac{1}{\alpha}\|Tx\| = \frac{1}{\alpha}\|y\|$ for some constant $\alpha > 0$. Hence T^{-1} is bounded. Thus (b) implies (a).

Part (ii). Take an arbitrary $\mathcal{R}(T)$-valued convergent sequence $\{y_n\}$. Since each y_n lies in $\mathcal{R}(T)$, then there exists an $\mathcal{X}$-valued sequence $\{x_n\}$ for which $y_n = Tx_n$ for each n. Since $\{Tx_n\}$ converges in $\mathcal{Y}$, then it is a Cauchy sequence in $\mathcal{Y}$. Thus if T is bounded below, then there exists $\alpha > 0$ such that

$$0 \le \alpha \|x_m - x_n\| \le \|T(x_m - x_n)\| = \|Tx_m - Tx_n\|$$

for every m, n. Hence $\{x_n\}$ is a Cauchy sequence in $\mathcal{X}$, and so it converges in $\mathcal{X}$ to, say, $x \in \mathcal{X}$ if $\mathcal{X}$ is a Banach space. Since T is continuous, it preserves convergence and hence $y_n = Tx_n \to Tx$. Then the (unique) limit of $\{y_n\}$ lies in $\mathcal{R}(T)$. Conclusion: $\mathcal{R}(T)$ is closed in $\mathcal{Y}$ by the classical Closed Set Theorem. That is, $\mathcal{R}(T)^- = \mathcal{R}(T)$ whenever $\mathcal{X}$ is a Banach space, where $\mathcal{R}(T)^-$ stands for the closure of $\mathcal{R}(T)$. Moreover, since (b) trivially implies $\mathcal{N}(T) = \{0\}$, it follows that (b) implies (c). On the other hand, if $\mathcal{N}(T) = \{0\}$, then T is injective. If in addition the linear manifold $\mathcal{R}(T)$ is closed in the Banach space $\mathcal{Y}$, then it is itself a Banach space and so $T \colon \mathcal{X} \to \mathcal{R}(T)$ is an injective and surjective bounded linear transformation of the Banach space $\mathcal{X}$ onto the Banach space $\mathcal{R}(T)$. Hence its inverse T^{-1} lies in $\mathcal{B}[\mathcal{R}(T), \mathcal{X}]$ by the Inverse Mapping Theorem (Theorem 1.1). Thus (c) implies (a). $\qquad\square$

If one of the Banach spaces $\mathcal{X}$ or $\mathcal{Y}$ is zero, then the results in the previous theorems are either void or hold trivially. Indeed, if $\mathcal{X} \neq \{0\}$ and $\mathcal{Y} = \{0\}$, then the unique operator in $\mathcal{B}[\mathcal{X}, \mathcal{Y}]$ is not injective (thus not bounded below and not invertible); if $\mathcal{X} = \{0\}$ and $\mathcal{Y} \neq \{0\}$, then the unique operator in $\mathcal{B}[\mathcal{X}, \mathcal{Y}]$ is not surjective but is bounded below (thus injective) and has a bounded inverse on its singleton range; if $\mathcal{X} = \{0\}$ and $\mathcal{Y} = \{0\}$ the unique operator in $\mathcal{B}[\mathcal{X}, \mathcal{Y}]$ has an inverse in $\mathcal{B}[\mathcal{X}, \mathcal{Y}]$.

The next theorem is a rather useful result establishing a *power series expansion* for $(\lambda I - T)^{-1}$. This is the *Neumann expansion* (or *Neumann series*) due to C.G. Neumann. The condition $\|T\| < |\lambda|$ will be weakened in Corollary 2.12.

Theorem 1.3. (NEUMANN EXPANSION). *If $T \in \mathcal{B}[\mathcal{X}]$ is an operator on a Banach space $\mathcal{X}$, and if λ is any scalar such that $\|T\| < |\lambda|$, then $\lambda I - T$ has an inverse in $\mathcal{B}[\mathcal{X}]$ given by the following uniformly convergent series:*

$$(\lambda I - T)^{-1} = \frac{1}{\lambda} \sum_{k=0}^{\infty} \left(\frac{T}{\lambda}\right)^k.$$

Proof. Take an operator $T \in \mathcal{B}[\mathcal{X}]$ and a nonzero scalar λ. If $\|T\| < |\lambda|$, then $\sup_n \sum_{k=0}^n \left\|\left(\frac{T}{\lambda}\right)\right\|^k < \infty$. Thus, since $\sum_{k=0}^n \left\|\left(\frac{T}{\lambda}\right)^k\right\| \leq \sum_{k=0}^n \left\|\left(\frac{T}{\lambda}\right)\right\|^k$ for every $n \geq 0$, the increasing sequence of nonnegative numbers $\left\{\sum_{k=0}^n \left\|\left(\frac{T}{\lambda}\right)^k\right\|\right\}$ is bounded, and hence it converges in $\mathbb{R}$. Thus the $\mathcal{B}[\mathcal{X}]$-valued sequence $\left\{\left(\frac{T}{\lambda}\right)^n\right\}$ is absolutely summable, and so it is summable (since $\mathcal{B}[\mathcal{X}]$ is a Banach space — see, e.g., [78, Proposition 4.4]). That is, the series $\sum_{k=0}^\infty \left(\frac{T}{\lambda}\right)^k$ converges in $\mathcal{B}[\mathcal{X}]$. Equivalently, there is an operator in $\mathcal{B}[\mathcal{X}]$, say $\sum_{k=0}^\infty \left(\frac{T}{\lambda}\right)^k$, for which

$$\sum_{k=0}^n \left(\tfrac{T}{\lambda}\right)^k \xrightarrow{\ u\ } \sum_{k=0}^\infty \left(\tfrac{T}{\lambda}\right)^k.$$

Now, as is also readily verified by induction, for every $n \geq 0$

$$(\lambda I - T)\tfrac{1}{\lambda}\sum_{k=0}^n \left(\tfrac{T}{\lambda}\right)^k = \tfrac{1}{\lambda}\sum_{k=0}^n \left(\tfrac{T}{\lambda}\right)^k(\lambda I - T) = I - \left(\tfrac{T}{\lambda}\right)^{n+1}.$$

But $\left(\frac{T}{\lambda}\right)^n \xrightarrow{\ u\ } O$ (since $\left\|\left(\frac{T}{\lambda}\right)^n\right\| \leq \left\|\left(\frac{T}{\lambda}\right)\right\|^n \to 0$ when $\|T\| < |\lambda|$), and so

$$(\lambda I - T)\tfrac{1}{\lambda}\sum_{k=0}^n \left(\tfrac{T}{\lambda}\right)^k \xrightarrow{\ u\ } I \quad \text{and} \quad \tfrac{1}{\lambda}\sum_{k=0}^n \left(\tfrac{T}{\lambda}\right)^k(\lambda I - T) \xrightarrow{\ u\ } I,$$

where O stands for null operator. Hence

$$(\lambda I - T)\tfrac{1}{\lambda}\sum_{k=0}^\infty \left(\tfrac{T}{\lambda}\right)^k = \tfrac{1}{\lambda}\sum_{k=0}^\infty \left(\tfrac{T}{\lambda}\right)^k(\lambda I - T) = I,$$

and so $\frac{1}{\lambda}\sum_{k=0}^\infty \left(\frac{T}{\lambda}\right)^k \in \mathcal{B}[\mathcal{X}]$ is the inverse of $\lambda I - T \in \mathcal{B}[\mathcal{X}]$. $\qquad\square$

Remark. Under the hypothesis of Theorem 1.3,

$$\|(\lambda I - T)^{-1}\| \leq \tfrac{1}{|\lambda|}\sum_{k=0}^\infty \left(\tfrac{\|T\|}{|\lambda|}\right)^k = \left(|\lambda| - \|T\|\right)^{-1}.$$

1.3 Orthogonal Structure in Hilbert Spaces

Two vectors x and y in an inner product space $\mathcal{X}$ are *orthogonal* if $\langle x\,;y\rangle = 0$. In this case we write $x \perp y$. Two subsets A and B of $\mathcal{X}$ are *orthogonal* (notation: $A \perp B$) if every vector in A is orthogonal to every vector in B. The *orthogonal complement* of a set A is the set $A^\perp$ made up of all vectors in $\mathcal{X}$ orthogonal to every vector of A,

$$A^\perp = \{x \in \mathcal{X}\colon \langle x\,;y\rangle = 0 \text{ for every } y \in A\},$$

which is a subspace (i.e., a closed linear manifold) of $\mathcal{X}$, and therefore $A^\perp = (A^\perp)^- = (A^-)^\perp$. If $\mathcal{M}$ and $\mathcal{N}$ are orthogonal ($\mathcal{M} \perp \mathcal{N}$) linear manifolds of an inner product space $\mathcal{X}$, then $\mathcal{M} \cap \mathcal{N} = \{0\}$. In particular, $\mathcal{M} \cap \mathcal{M}^\perp = \{0\}$. If $\mathcal{M} \perp \mathcal{N}$, then $\mathcal{M} \oplus \mathcal{N}$ is an *orthogonal direct sum*. If $\mathcal{X}$ is a Hilbert space, then $\mathcal{M}^{\perp\perp} = \mathcal{M}^-$, and $\mathcal{M}^\perp = \{0\}$ if and only if $\mathcal{M}^- = \mathcal{X}$. Moreover, *if $\mathcal{M}$ and $\mathcal{N}$ are orthogonal subspaces of a Hilbert space $\mathcal{X}$, then $\mathcal{M} + \mathcal{N}$ is a subspace of $\mathcal{X}$.* In this case, $\mathcal{M} + \mathcal{N} \cong \mathcal{M} \oplus \mathcal{N}$: ordinary and direct sums of orthogonal subspaces of a Hilbert space are unitarily equivalent — the inner product in $\mathcal{M} \oplus \mathcal{N}$ is given by $\langle (u_1, v_1) ; (u_2, v_2) \rangle = \langle u_1 ; u_2 \rangle + \langle v_1 ; v_2 \rangle$ for every (u_1, v_1) and (u_2, v_2) in $\mathcal{M} \oplus \mathcal{N}$. Two linear manifolds $\mathcal{M}$ and $\mathcal{N}$ are *orthogonal complementary manifolds* if $\mathcal{M} + \mathcal{N} = \mathcal{X}$ and $\mathcal{M} \perp \mathcal{N}$. If $\mathcal{M}$ and $\mathcal{N}$ are orthogonal complementary subspaces of a Hilbert space $\mathcal{X}$, then we identify the ordinary sum $\mathcal{M} + \mathcal{N}$ with the orthogonal sum $\mathcal{M} \oplus \mathcal{N}$, and write (an abuse of notation, actually) $\mathcal{X} = \mathcal{M} \oplus \mathcal{N}$ (instead of $\mathcal{X} \cong \mathcal{M} \oplus \mathcal{N}$).

A central result of Hilbert-space geometry, the PROJECTION THEOREM, says: *if $\mathcal{H}$ is a Hilbert space and $\mathcal{M}$ is a subspace of it, then* (cf. [78, Theorem 5.20])

$$\mathcal{H} = \mathcal{M} + \mathcal{M}^{\perp}.$$

(Thus every subspace of a Hilbert space is complemented — cf. Remark 5.3(a) and Appendix B.) Ordinary and direct sums of *orthogonal* subspaces are unitarily equivalent. So the Projection Theorem is equivalently stated in terms of *orthogonal* direct sum up to unitary equivalence, $\mathcal{H} \cong \mathcal{M} \oplus \mathcal{M}^{\perp}$, often written

$$\mathcal{H} = \mathcal{M} \oplus \mathcal{M}^{\perp}.$$

This leads to the notation $\mathcal{M}^{\perp} = \mathcal{H} \ominus \mathcal{M}$, if $\mathcal{M}$ is a *subspace* of $\mathcal{H}$.

For a nonempty subset A of a normed space $\mathcal{X}$, let span A denote the *span* (or *linear span*) of A: the linear manifold of $\mathcal{X}$ consisting of all (finite) linear combinations of vectors of A or, equivalently, the smallest linear manifold of $\mathcal{X}$ including A (i.e., the intersection of all linear manifolds of $\mathcal{X}$ that include A). Its closure $(\text{span}\, A)^{-}$, denoted by $\bigvee A$ (and sometimes also called the *span* of A), is a subspace of $\mathcal{X}$. Let $\{\mathcal{M}_{\gamma}\}_{\gamma \in \Gamma}$ be a family of subspaces of $\mathcal{X}$ indexed by an arbitrary (not necessarily countable) nonempty index set Γ. The subspace $\left[\text{span} \left(\bigcup_{\gamma \in \Gamma} \mathcal{M}_{\gamma} \right) \right]^{-}$ of $\mathcal{X}$ is the *topological sum* of $\{\mathcal{M}_{\gamma}\}_{\gamma \in \Gamma}$, usually denoted by $\left(\sum_{\gamma \in \Gamma} \mathcal{M}_{\gamma} \right)^{-}$ or $\bigvee_{\gamma \in \Gamma} \mathcal{M}_{\gamma}$. Thus

$$\left(\sum_{\gamma \in \Gamma} \mathcal{M}_{\gamma} \right)^{-} = \left[\text{span} \left(\bigcup_{\gamma \in \Gamma} \mathcal{M}_{\gamma} \right) \right]^{-} = \bigvee \left(\bigcup_{\gamma \in \Gamma} \mathcal{M}_{\gamma} \right) = \bigvee_{\gamma \in \Gamma} \mathcal{M}_{\gamma}.$$

A pivotal result of Hilbert-space theory is the ORTHOGONAL STRUCTURE THEOREM, which reads as follows. *Let $\{\mathcal{M}_{\gamma}\}_{\gamma \in \Gamma}$ be a family of pairwise orthogonal subspaces of a Hilbert space $\mathcal{H}$. For every $x \in \left(\sum_{\gamma \in \Gamma} \mathcal{M}_{\gamma} \right)^{-}$ there is a unique summable family $\{u_{\gamma}\}_{\gamma \in \Gamma}$ of vectors $u_{\gamma} \in \mathcal{M}_{\gamma}$ such that*

$$x = \sum_{\gamma \in \Gamma} u_{\gamma}.$$

Moreover, $\|x\|^2 = \sum_{\gamma \in \Gamma} \|u_{\gamma}\|^2$. Conversely, if $\{u_{\gamma}\}_{\gamma \in \Gamma}$ is a square-summable family of vectors in $\mathcal{H}$ with $u_{\gamma} \in \mathcal{M}_{\gamma}$ for each $\gamma \in \Gamma$, then $\{u_{\gamma}\}_{\gamma \in \Gamma}$ is summable and $\sum_{\gamma \in \Gamma} u_{\gamma}$ lies in $\left(\sum_{\gamma \in \Gamma} \mathcal{M}_{\gamma} \right)^{-}$. (See, e.g., [78, Section 5.5].)

For the definition of *summable family* and *square-summable family* see [78, Definition 5.26]. A summable family in a normed space has only a countable number of nonzero vectors, and so for each x the sum $x = \sum_{\gamma \in \Gamma} u_{\gamma}$ has only a countable number of nonzero summands u_{γ} (see, e.g., [78, Corollary 5.28]).

Take the direct sum $\bigoplus_{\gamma \in \Gamma} \mathcal{M}_\gamma$ of pairwise orthogonal subspaces $\mathcal{M}_\gamma$ of a Hilbert space $\mathcal{H}$ made up of square-summable families of vectors. That is, $\mathcal{M}_\alpha \perp \mathcal{M}_\beta$ for $\alpha \neq \beta$, and $x = \{x_\gamma\}_{\gamma \in \Gamma} \in \bigoplus_{\gamma \in \Gamma} \mathcal{M}_\gamma$ if and only if each x_γ lies in $\mathcal{M}_\gamma$ and $\sum_{\gamma \in \Gamma} \|x_\gamma\|^2 < \infty$. This is a Hilbert space with inner product

$$\langle x \,; y \rangle = \sum_{\gamma \in \Gamma} \langle x_\gamma \,; y_\gamma \rangle$$

for every $x = \{x_\gamma\}_{\gamma \in \Gamma}$ and $y = \{y_\gamma\}_{\gamma \in \Gamma}$ in $\bigoplus_{\gamma \in \Gamma} \mathcal{M}_\gamma$. A consequence (a restatement, actually) of the Orthogonal Structure Theorem reads as follows. The orthogonal direct sum $\bigoplus_{\gamma \in \Gamma} \mathcal{M}_\gamma$ and the topological sum $\left(\sum_{\gamma \in \Gamma} \mathcal{M}_\gamma\right)^-$ are unitarily equivalent Hilbert spaces. In other words,

$$\bigoplus_{\gamma \in \Gamma} \mathcal{M}_\gamma \cong \left(\sum_{\gamma \in \Gamma} \mathcal{M}_\gamma\right)^-.$$

If, in addition, $\{\mathcal{M}_\gamma\}_{\gamma \in \Gamma}$ spans $\mathcal{H}$ (i.e., $\left(\sum_{\gamma \in \Gamma} \mathcal{M}_\gamma\right)^- = \bigvee_{\gamma \in \Gamma} \mathcal{M}_\gamma = \mathcal{H}$), then it is usual to express the above identification by writing

$$\mathcal{H} = \bigoplus_{\gamma \in \Gamma} \mathcal{M}_\gamma.$$

Similarly, the *external orthogonal direct sum* $\bigoplus_{\gamma \in \Gamma} \mathcal{H}_\gamma$ of a collection of Hilbert spaces $\{\mathcal{H}_\gamma\}_{\gamma \in \Gamma}$ (not necessarily subspaces of a larger Hilbert space) consists of square-summable families $\{x_\gamma\}_{\gamma \in \Gamma}$ (i.e., $\sum_{\gamma \in \Gamma} \|x_\gamma\|^2 < \infty$), with each x_γ in $\mathcal{H}_\gamma$. This is a Hilbert space with inner product given as above, and the Hilbert spaces $\{\mathcal{H}_\gamma\}_{\gamma \in \Gamma}$ are referred to as being *externally orthogonal*.

Let $\mathcal{H}$ and $\mathcal{K}$ be Hilbert spaces. The *direct sum* of $T \in \mathcal{B}[\mathcal{H}]$ and $S \in \mathcal{B}[\mathcal{K}]$ is the operator $T \oplus S \in \mathcal{B}[\mathcal{H} \oplus \mathcal{K}]$ such that $(T \oplus S)(x, y) = (Tx, Sy)$ for every $(x, y) \in \mathcal{H} \oplus \mathcal{K}$. Similarly, if $\{\mathcal{H}_\gamma\}_{\gamma \in \Gamma}$ is a collection of Hilbert spaces and if $\{T_\gamma\}_{\gamma \in \Gamma}$ is a *bounded* family of operators $T_\gamma \in \mathcal{B}[\mathcal{H}_\gamma]$ (i.e., $\sup_{\gamma \in \Gamma} \|T_\gamma\| < \infty$), then their *direct sum* $T \colon \bigoplus_{\gamma \in \Gamma} \mathcal{H}_\gamma \to \bigoplus_{\gamma \in \Gamma} \mathcal{H}_\gamma$ is the mapping

$$T\{x_\gamma\}_{\gamma \in \Gamma} = \{T_\gamma x_\gamma\}_{\gamma \in \Gamma} \quad \text{for every} \quad \{x_\gamma\}_{\gamma \in \Gamma} \in \bigoplus_{\gamma \in \Gamma} \mathcal{H}_\gamma,$$

which is an operator in $\mathcal{B}\big[\bigoplus_{\gamma \in \Gamma} \mathcal{H}_\gamma\big]$ with $\|T\| = \sup_{\gamma \in \Gamma} \|T_\gamma\|$, denoted by

$$T = \bigoplus_{\gamma \in \Gamma} T_\gamma.$$

1.4 Orthogonal Projections on Inner Product Spaces

A function $F \colon X \to X$ of a set X into itself is *idempotent* if $F^2 = F$, where F^2 stands for the composition of F with itself. A *projection* E is an idempotent (i.e., $E = E^2$) linear transformation $E \colon \mathcal{X} \to \mathcal{X}$ of a linear space $\mathcal{X}$ into itself. If E is a projection, then so is $I - E$, which is referred to as the *complementary projection* of E, and their null spaces and ranges are related as follows:

$$\mathcal{R}(E) = \mathcal{N}(I - E) \quad \text{and} \quad \mathcal{N}(E) = \mathcal{R}(I - E).$$

The range of a projection is the set of all its fixed points,

$$\mathcal{R}(E) = \{x \in \mathcal{X} \colon Ex = x\},$$

which forms with the kernel a pair of algebraic complements,

$$\mathcal{R}(E) + \mathcal{N}(E) = \mathcal{X} \quad \text{and} \quad \mathcal{R}(E) \cap \mathcal{N}(E) = \{0\}.$$

An *orthogonal projection* $E \colon \mathcal{X} \to \mathcal{X}$ on an inner product space $\mathcal{X}$ is a projection for which $\mathcal{R}(E) \perp \mathcal{N}(E)$. If E is an orthogonal projection on $\mathcal{X}$, then so is the complementary projection $I - E \colon \mathcal{X} \to \mathcal{X}$. *Every orthogonal projection is bounded (i.e., continuous — $E \in \mathcal{B}[\mathcal{H}]$).* In fact, every orthogonal projection is a *contraction*. Indeed, every $x \in \mathcal{X}$ can be (uniquely) written as $x = u + v$ with $u \in \mathcal{R}(E)$ and $v \in \mathcal{N}(E)$, and so $\|Ex\|^2 = \|Eu + Ev\|^2 = \|Eu\|^2 = \|u\|^2 \le \|x\|^2$ since $\|x\|^2 = \|u\|^2 + \|v\|^2$ by the Pythagorean Theorem (as $u \perp v$). Hence $\mathcal{R}(E)$ is a subspace (i.e., it is closed, because $\mathcal{R}(E) = \mathcal{N}(I - E)$ and $I - E$ is bounded). Moreover, $\mathcal{N}(E) = \mathcal{R}(E)^\perp$; equivalently, $\mathcal{R}(E) = \mathcal{N}(E)^\perp$. Conversely, if any of these properties holds for a projection E, then it is an orthogonal projection.

Two orthogonal projections E_1 and E_2 on $\mathcal{X}$ are said to be *orthogonal to each other* (or *mutually orthogonal*) if $\mathcal{R}(E_1) \perp \mathcal{R}(E_2)$ (equivalently, if $E_1 E_2 = E_2 E_1 = O$). Let Γ be an arbitrary index set (not necessarily countable). If $\{E_\gamma\}_{\gamma \in \Gamma}$ is a family of orthogonal projections on an inner product space $\mathcal{X}$ which are orthogonal to each other ($\mathcal{R}(E_\alpha) \perp \mathcal{R}(E_\beta)$ for $\alpha \ne \beta$), then $\{E_\gamma\}_{\gamma \in \Gamma}$ is an *orthogonal family of orthogonal projections* on $\mathcal{X}$. An *orthogonal sequence of orthogonal projections* $\{E_k\}_{k=0}^\infty$ is similarly defined. If $\{E_\gamma\}_{\gamma \in \Gamma}$ is an orthogonal family of orthogonal projections and

$$\sum_{\gamma \in \Gamma} E_\gamma x = x \quad \text{for every } x \in \mathcal{X},$$

then $\{E_\gamma\}_{\gamma \in \Gamma}$ is called a *resolution of the identity* on $\mathcal{X}$. (Recall: for each x in $\mathcal{X}$ the sum $\sum_{\gamma \in \Gamma} E_\gamma x$ has only a countable number of nonzero vectors.) If $\{E_k\}_{k=0}^\infty$ is an infinite sequence, then the above identity in $\mathcal{X}$ means convergence in the strong operator topology:

$$\sum_{k=0}^n E_k \xrightarrow{s} I \quad \text{as } n \to \infty.$$

If $\{E_k\}_{k=0}^n$ is a finite family, then the above identity in $\mathcal{X}$ obviously coincides with the identity $\sum_{k=0}^n E_k = I$ in $\mathcal{B}[\mathcal{X}]$ (e.g., $\{E_1, E_2\}$ is a resolution of the identity on $\mathcal{X}$ whenever $E_2 = I - E_1$ is the complementary projection of the orthogonal projection E_1 on $\mathcal{X}$).

The Projection Theorem and the Orthogonal Structure Theorem of the previous section can be written in terms of orthogonal projections. Indeed, *for* *every subspace $\mathcal{M}$ of a Hilbert space $\mathcal{H}$ there is a unique orthogonal projection $E \in \mathcal{B}[\mathcal{H}]$ such that $\mathcal{R}(E) = \mathcal{M}$,* referred to as *the orthogonal projection onto $\mathcal{M}$.* This is an equivalent version of the Projection Theorem (see, e.g., [78, Theorem 5.52]). An equivalent version of the Orthogonal Structure Theorem reads as follows. *If $\{E_\gamma\}_{\gamma \in \Gamma}$ is a resolution of the identity on $\mathcal{H}$, then*

$$\mathcal{H} = \left(\sum\nolimits_{\gamma \in \Gamma} \mathcal{R}(E_\gamma) \right)^{-}.$$

Conversely, if $\{\mathcal{M}_\gamma\}_{\gamma \in \Gamma}$ is a family of pairwise orthogonal subspaces of $\mathcal{H}$ for which $\mathcal{H} = \left(\sum_{\gamma \in \Gamma} \mathcal{M}_\gamma \right)^{-}$, then the family of orthogonal projections $\{E_\gamma\}_{\gamma \in \Gamma}$ with each $E_\gamma \in \mathcal{B}[\mathcal{H}]$ onto each $\mathcal{M}_\gamma$ (i.e., with $\mathcal{R}(E_\gamma) = \mathcal{M}_\gamma$) is a resolution of the identity on $\mathcal{H}$ (see, e.g., [78, Theorem 5.59]). Since $\{\mathcal{R}(E_\gamma)\}_{\gamma \in \Gamma}$ is a family of pairwise orthogonal subspaces, the above orthogonal decomposition of $\mathcal{H}$ is unitarily equivalent to the orthogonal direct sum $\bigoplus_{\gamma \in \Gamma} \mathcal{R}(E_\gamma)$, that is, $\left(\sum_{\gamma \in \Gamma} \mathcal{R}(E_\gamma) \right)^{-} \cong \bigoplus_{\gamma \in \Gamma} \mathcal{R}(E_\gamma)$, which is commonly written as

$$\mathcal{H} = \bigoplus\nolimits_{\gamma \in \Gamma} \mathcal{R}(E_\gamma).$$

If $\{E_\gamma\}_{\gamma \in \Gamma}$ is a resolution of the identity made up of nonzero projections on a nonzero Hilbert space $\mathcal{H}$, and if $\{\lambda_\gamma\}_{\gamma \in \Gamma}$ is a similarly indexed family of scalars, then consider the following subset of $\mathcal{H}$:

$$\mathcal{D}(T) = \big\{ x \in \mathcal{H} \colon \{\lambda_\gamma E_\gamma x\}_{\gamma \in \Gamma} \text{ is a summable family of vectors in } \mathcal{H} \big\}.$$

The mapping $T \colon \mathcal{D}(T) \to \mathcal{H}$ defined by

$$Tx = \sum\nolimits_{\gamma \in \Gamma} \lambda_\gamma E_\gamma x \quad \text{for every } x \in \mathcal{D}(T)$$

is called a *weighted sum of projections.* The domain $\mathcal{D}(T)$ of a weighted sum of projections is a linear manifold of $\mathcal{H}$, and T is a linear transformation. Moreover, *T is bounded if and only if $\{\lambda_\gamma\}_{\gamma \in \Gamma}$ is a bounded family of scalars, which happens if and only if $\mathcal{D}(T) = \mathcal{H}$.* In this case $T \in \mathcal{B}[\mathcal{H}]$ and $\|T\| = \sup_{\gamma \in \Gamma} |\lambda_\gamma|$ (see, e.g., [78, Proposition 5.61]).

1.5 The Adjoint Operator on Hilbert Space

Throughout the text $\mathcal{H}$ and $\mathcal{K}$ are Hilbert spaces. Take any $T \in \mathcal{B}[\mathcal{H}, \mathcal{K}]$. The *adjoint T^** of T is the unique mapping of $\mathcal{K}$ into $\mathcal{H}$ for which

$$\langle Tx \,; y \rangle = \langle x \,; T^* y \rangle \quad \text{for every } x \in \mathcal{H} \text{ and } y \in \mathcal{K}.$$

In fact, $T^* \in \mathcal{B}[\mathcal{K}, \mathcal{H}]$ (i.e., T^* is bounded and linear), $T^{**} = T$, and

$$\|T^*\|^2 = \|T^* T\| = \|T T^*\| = \|T\|^2.$$

Moreover, if $\mathcal{Z}$ is a Hilbert space and $S \in \mathcal{B}[\mathcal{K}, \mathcal{Z}]$, then $ST \in \mathcal{B}[\mathcal{H}, \mathcal{Z}]$ is such that $(ST)^* = T^* S^*$ (see, e.g., [78, Proposition 5.65 and Section 5.12]).

Lemma 1.4. *For every* $T \in \mathcal{B}[\mathcal{H}, \mathcal{K}]$,

$$\mathcal{N}(T) = \mathcal{R}(T^*)^\perp = \mathcal{N}(T^*T) \quad and \quad \mathcal{R}(T)^- = \mathcal{N}(T^*)^\perp = \mathcal{R}(TT^*)^-.$$

Proof. By definition $x \in \mathcal{R}(T^*)^\perp$ if and only if $\langle x \,; T^* y \rangle = 0$ or, equivalently, $\langle Tx \,; y \rangle = 0$, for every y in $\mathcal{K}$, which means $Tx = 0$; that is, $x \in \mathcal{N}(T)$. Hence

$$\mathcal{R}(T^*)^\perp = \mathcal{N}(T).$$

Since $\|Tx\|^2 = \langle Tx \,; Tx \rangle = \langle T^*Tx \,; x \rangle$ for every $x \in \mathcal{H}$, we get $\mathcal{N}(T^*T) \subseteq \mathcal{N}(T)$. But $\mathcal{N}(T) \subseteq \mathcal{N}(T^*T)$ trivially and so

$$\mathcal{N}(T) = \mathcal{N}(T^*T).$$

This completes the proof of the first identities. Since these hold for every T in $\mathcal{B}[\mathcal{H}, \mathcal{K}]$, they also hold for T^* in $\mathcal{B}[\mathcal{K}, \mathcal{H}]$ and for TT^* in $\mathcal{B}[\mathcal{K}]$. Thus

$$\begin{aligned}
\mathcal{R}(T)^- &= \mathcal{R}(T^{**})^{\perp\perp} = \mathcal{N}(T^*)^\perp = \mathcal{N}(T^{**}T^*)^\perp \\
&= \mathcal{N}(TT^*)^\perp = \mathcal{R}((TT^*)^*)^{\perp\perp} = \mathcal{R}(T^{**}T^*)^{\perp\perp} = \mathcal{R}(TT^*)^-,
\end{aligned}$$

because $\mathcal{M}^{\perp\perp} = \mathcal{M}^-$ for every linear manifold $\mathcal{M}$ and $T^{**} = T$. $\qquad\square$

Lemma 1.5. $\mathcal{R}(T)$ *is closed in* $\mathcal{K}$ *if and only if* $\mathcal{R}(T^*)$ *is closed in* $\mathcal{H}$.

Proof. Let $T|_{\mathcal{M}}$ denote the restriction of T to any linear manifold $\mathcal{M} \subseteq \mathcal{H}$. Set $T_\perp = T|_{\mathcal{N}(T)^\perp}$ in $\mathcal{B}[\mathcal{N}(T)^\perp, \mathcal{K}]$. Since $\mathcal{H} = \mathcal{N}(T) + \mathcal{N}(T)^\perp$, every $x \in \mathcal{H}$ can be written as $x = u + v$ with $u \in \mathcal{N}(T)$ and $v \in \mathcal{N}(T)^\perp$. If $y \in \mathcal{R}(T)$, then $y = Tx = Tu + Tv = Tv = T|_{\mathcal{N}(T)^\perp} v$ for some $x \in \mathcal{H}$, and so $y \in \mathcal{R}(T|_{\mathcal{N}(T)^\perp})$. Hence $\mathcal{R}(T) \subseteq \mathcal{R}(T|_{\mathcal{N}(T)^\perp})$. Since $\mathcal{R}(T|_{\mathcal{N}(T)^\perp}) \subseteq \mathcal{R}(T)$ we get

$$\mathcal{R}(T_\perp) = \mathcal{R}(T) \quad and \quad \mathcal{N}(T_\perp) = \{0\}.$$

If $\mathcal{R}(T) = \mathcal{R}(T)^-$, then take the inverse $T_\perp^{-1}$ in $\mathcal{B}[\mathcal{R}(T), \mathcal{N}(T)^\perp]$ (Theorem 1.2). For any $w \in \mathcal{N}(T)^\perp$ consider the functional $f_w \colon \mathcal{R}(T) \to \mathbb{C}$ defined by

$$f_w(y) = \langle T_\perp^{-1} y \,; w \rangle$$

for every $y \in \mathcal{R}(T)$. Since $T_\perp^{-1}$ is linear and the inner product is linear in the first argument, then f is linear. Since $|f_w(y)| \le \|T_\perp^{-1}\| \|w\| \|y\|$ for every y in $\mathcal{R}(T)$, then f is also bounded. Moreover, $\mathcal{R}(T)$ is a subspace of the Hilbert space $\mathcal{K}$ (it is closed in $\mathcal{K}$), and so a Hilbert space itself. Then by the Riesz Representation Theorem in Hilbert space (see, e.g., [78, Theorem 5.62]) there exists a unique vector z_w in the Hilbert space $\mathcal{R}(T)$ for which

$$f_w(y) = \langle y \,; z_w \rangle$$

for every $y \in \mathcal{R}(T)$. Thus, since an arbitrary $x \in \mathcal{H}$ is written as $x = u + v$ with $u \in \mathcal{N}(T)$ and $v \in \mathcal{N}(T)^{\perp}$, we get

$$
\begin{aligned}
\langle x \,; T^* z_w \rangle = \langle Tx \,; z_w \rangle &= \langle Tu \,; z_w \rangle + \langle Tv \,; z_w \rangle = \langle Tv \,; z_w \rangle \\
&= f_w(Tv) = \langle T_{\perp}^{-1} Tv \,; w \rangle = \langle T_{\perp}^{-1} T_{\perp} v \,; w \rangle \\
&= \langle v \,; w \rangle = \langle u \,; w \rangle + \langle v \,; w \rangle = \langle x \,; w \rangle.
\end{aligned}
$$

Hence $\langle x \,; T^* z_w - w \rangle = 0$ for every $x \in \mathcal{H}$; that is, $T^* z_w = w$. So $w \in \mathcal{R}(T^*)$. Therefore $\mathcal{N}(T)^{\perp} \subseteq \mathcal{R}(T^*)$. On the other hand, $\mathcal{R}(T^*) \subseteq \mathcal{R}(T^*)^- = \mathcal{N}(T)^{\perp}$ by Lemma 1.4 (since $T^{**} = T$), and hence

$$
\mathcal{R}(T^*) = \mathcal{R}(T^*)^-.
$$

Thus $\mathcal{R}(T) = \mathcal{R}(T)^-$ implies $\mathcal{R}(T^*) = \mathcal{R}(T^*)^-$, and consequently the converse also holds because $T^{**} = T$. So

$$
\mathcal{R}(T) = \mathcal{R}(T)^- \quad \text{if and only if} \quad \mathcal{R}(T^*) = \mathcal{R}(T^*)^-. \qquad \square
$$

Take the *adjoint operator* $T^* \in \mathcal{B}[\mathcal{H}]$ of an operator $T \in \mathcal{B}[\mathcal{H}]$ on a Hilbert space $\mathcal{H}$. Let $\mathcal{M}$ be a subspace of $\mathcal{H}$. If $\mathcal{M}$ and its orthogonal complement $\mathcal{M}^{\perp}$ are both T-invariant (i.e., if $T(\mathcal{M}) \subseteq \mathcal{M}$ and $T(\mathcal{M}^{\perp}) \subseteq \mathcal{M}^{\perp}$), then we say that $\mathcal{M}$ *reduces* T (or $\mathcal{M}$ is a *reducing subspace* for T). An operator is *reducible* if it has a nontrivial reducing subspace (otherwise it is called *irreducible*), and *reductive* if all its invariant subspaces are reducing. Since $\mathcal{M} \neq \{0\}$ if and only if $\mathcal{M}^{\perp} \neq \mathcal{H}$, and $\mathcal{M}^{\perp} \neq \{0\}$ if and only if $\mathcal{M} \neq \mathcal{H}$, then an operator T on a Hilbert space $\mathcal{H}$ is reducible if there exists a subspace $\mathcal{M}$ of $\mathcal{H}$ such that both $\mathcal{M}$ and $\mathcal{M}^{\perp}$ are nonzero and T-invariant or, equivalently, if $\mathcal{M}$ is nontrivial and invariant for both T and T^* as we will see next.

Lemma 1.6. *Take an operator T on a Hilbert space $\mathcal{H}$. A subspace $\mathcal{M}$ of $\mathcal{H}$ is T-invariant if and only if $\mathcal{M}^{\perp}$ is T^*-invariant. $\mathcal{M}$ reduces T if and only if $\mathcal{M}$ is invariant for both T and T^*, which implies $(T|_{\mathcal{M}})^* = T^*|_{\mathcal{M}}$.*

Proof. Take any vector $y \in \mathcal{M}^{\perp}$. If $Tx \in \mathcal{M}$ whenever $x \in \mathcal{M}$, then $\langle x \,; T^* y \rangle = \langle Tx \,; y \rangle = 0$ for every $x \in \mathcal{M}$, and so $T^* y \perp \mathcal{M}$; that is, $T^* y \in \mathcal{M}^{\perp}$. Thus $T(\mathcal{M}) \subseteq \mathcal{M}$ implies $T^*(\mathcal{M}^{\perp}) \subseteq \mathcal{M}^{\perp}$. Since this holds for every operator in $\mathcal{B}[\mathcal{H}]$, $T^*(\mathcal{M}^{\perp}) \subseteq \mathcal{M}^{\perp}$ implies $T^{**}(\mathcal{M}^{\perp\perp}) \subseteq \mathcal{M}^{\perp\perp}$, and hence $T(\mathcal{M}) \subseteq \mathcal{M}$ (because $T^{**} = T$ and $\mathcal{M}^{\perp\perp} = \mathcal{M}^- = \mathcal{M}$). Then $T(\mathcal{M}) \subseteq \mathcal{M}$ if and only if $T^*(\mathcal{M}^{\perp}) \subseteq \mathcal{M}^{\perp}$, and so $T(\mathcal{M}^{\perp}) \subseteq \mathcal{M}^{\perp}$ if and only if $T^*(\mathcal{M}) \subseteq \mathcal{M}$. Therefore $T(\mathcal{M}) \subseteq \mathcal{M}$ and $T(\mathcal{M}^{\perp}) \subseteq \mathcal{M}^{\perp}$ if and only if $T(\mathcal{M}) \subseteq \mathcal{M}$ and $T^*(\mathcal{M}) \subseteq \mathcal{M}$, and so $\langle (T|_{\mathcal{M}})x \,; y \rangle = \langle Tx \,; y \rangle = \langle x \,; T^* y \rangle = \langle x \,; (T^*|_{\mathcal{M}})y \rangle$ for $x, y \in \mathcal{M}$. $\qquad \square$

An operator $T \in \mathcal{B}[\mathcal{H}]$ on a Hilbert space $\mathcal{H}$ is *self-adjoint* (or *Hermitian*) if it coincides with its adjoint (i.e., if $T^* = T$). A characterization of self-adjoint operators reads as follows: *on a complex Hilbert space $\mathcal{H}$, an operator T is self-adjoint if and only if $\langle Tx \,; x \rangle \in \mathbb{R}$ for every vector $x \in \mathcal{H}$. Moreover, if*

$T \in \mathcal{B}[\mathcal{H}]$ *is self-adjoint, then* $T = O$ *if and only if* $\langle Tx\,;x\rangle = 0$ *for all* $x \in \mathcal{H}$. (See, e.g., [78, Proposition 5.79 and Corollary 5.80].) A self-adjoint operator T is *nonnegative* (notation: $O \leq T$) if $0 \leq \langle Tx\,;x\rangle$ for every $x \in \mathcal{H}$, and it is *positive* (notation: $O < T$) if $0 < \langle Tx\,;x\rangle$ for every nonzero $x \in \mathcal{H}$. An invertible positive operator T in $\mathcal{G}[\mathcal{H}]$ is called *strictly positive* (notation: $O \prec T$). For every T in $\mathcal{B}[\mathcal{H}]$, the operators T^*T and TT^* in $\mathcal{B}[\mathcal{H}]$ are always nonnegative (reason: $0 \leq \|Tx\|^2 = \langle Tx\,;Tx\rangle = \langle T^*Tx\,;x\rangle$ for every $x \in \mathcal{H}$). If S and T are operators in $\mathcal{B}[\mathcal{H}]$, then we write $S \leq T$ if $T - S$ is nonnegative (thus self-adjoint). Similarly, we also write $S < T$ and $S \prec T$ if $T - S$ is positive or strictly positive (thus self-adjoint). So *if* $T - S$ *is self-adjoint* (in particular, if both T and S are), *then* $S = T$ *if and only if* $S \leq T$ *and* $T \leq S$.

1.6 Normal and Hyponormal Operators

Let S and T be operators on the same space $\mathcal{X}$. They *commute* if $ST = TS$. An operator $T \in \mathcal{B}[\mathcal{H}]$ on a Hilbert space $\mathcal{H}$ is *normal* if it commutes with its adjoint (i.e., if $TT^* = T^*T$, or $O = T^*T - TT^*$). Observe that $T^*T - TT^*$ is always self-adjoint. An operator $T \in \mathcal{B}[\mathcal{H}]$ is *hyponormal* if $TT^* \leq T^*T$ (i.e., if $O \leq T^*T - TT^*$), which is equivalent to saying that $(\lambda I - T)(\lambda I - T)^* \leq (\lambda I - T)^*(\lambda I - T)$ for every scalar $\lambda \in \mathbb{C}$. An operator $T \in \mathcal{B}[\mathcal{H}]$ is *cohyponormal* if its adjoint is hyponormal (i.e., if $T^*T \leq TT^*$, or $O \leq TT^* - T^*T$). An operator T is normal if and only if it is both hyponormal and cohyponormal. Thus *every normal operator is hyponormal*. If it is either hyponormal or cohyponormal, then it is called *seminormal*.

Theorem 1.7. $T \in \mathcal{B}[\mathcal{H}]$ *is hyponormal if and only if* $\|T^*x\| \leq \|Tx\|$ *for every* $x \in \mathcal{H}$. *Moreover, the following assertions are pairwise equivalent.*

(a) T *is normal.*

(b) $\|T^*x\| = \|Tx\|$ *for every* $x \in \mathcal{H}$.

(c) T^n *is normal for every positive integer* n.

(d) $\|T^{*n}x\| = \|T^nx\|$ *for every* $x \in \mathcal{H}$ *and every* $n \geq 1$.

Proof. Take any $T \in \mathcal{B}[\mathcal{H}]$. The characterization for hyponormal operators goes as follows: $TT^* \leq T^*T$ if and only if $\langle TT^*x\,;x\rangle \leq \langle T^*Tx\,;x\rangle$ or, equivalently, $\|T^*x\|^2 \leq \|Tx\|^2$, for every $x \in \mathcal{H}$. Since T is normal if and only if it is both hyponormal and cohyponormal, assertions (a) and (b) are equivalent. Thus, as $T^{*n} = T^{n*}$ for each $n \geq 1$, (c) and (d) are equivalent. If T^* commutes with T, then it commutes with T^n and, dually, T^n commutes with $T^{*n} = T^{n*}$. Hence (a) implies (c), and (d) implies (b) trivially. $\square$

In spite of the equivalence between (a) and (c) above, the square of a hyponormal operator is not necessarily hyponormal. It is clear that every self-adjoint operator is normal, and so is every nonnegative operator. Moreover, normality distinguishes the orthogonal projections among the projections.

Theorem 1.8. *If $E \in \mathcal{B}[\mathcal{H}]$ is a nonzero projection, then the following assertions are pairwise equivalent.*

(a) E *is an orthogonal projection.*

(b) E *is nonnegative.*

(c) E *is self-adjoint.*

(d) E *is normal.*

(e) $\|E\| = 1$.

(f) $\|E\| \leq 1$.

Proof. Let E be a projection (i.e., a linear idempotent). If E is an orthogonal projection (i.e., $\mathcal{R}(E) \perp \mathcal{N}(E)$), then $I - E$ is again an orthogonal projection, thus bounded, and so $\mathcal{R}(E) = \mathcal{N}(I - E)$ is closed. Hence $\mathcal{R}(E^*)$ is closed by Lemma 1.5. Moreover, $\mathcal{R}(E) = \mathcal{N}(E)^\perp$. But $\mathcal{N}(E)^\perp = \mathcal{R}(E^*)^-$ by Lemma 1.4 (since $E^{**} = E$). Then $\mathcal{R}(E) = \mathcal{R}(E^*)$, and so for every $x \in \mathcal{H}$ there is a $z \in \mathcal{H}$ for which $Ex = E^*z$. Therefore, $E^*Ex = (E^*)^2 z = (E^2)^* z = E^* z = Ex$. That is, $E^*E = E$. Hence E is nonnegative (it is self-adjoint, and E^*E is always nonnegative). Outcome: (a) $\Rightarrow$ (b) $\Rightarrow$ (c) $\Rightarrow$ (d),

$$\mathcal{R}(E) \perp \mathcal{N}(E) \implies E \geq O \implies E^* = E \implies E^*E = EE^*.$$

Conversely, if $E \in \mathcal{B}[\mathcal{H}]$ is normal, then $\|E^*x\| = \|Ex\|$ for every x in $\mathcal{H}$ (Theorem 1.7) and so $\mathcal{N}(E^*) = \mathcal{N}(E)$. Hence $\mathcal{R}(E) = \mathcal{N}(E^*)^\perp = \mathcal{N}(E)^\perp$ by Lemma 1.4 (because $\mathcal{R}(E) = \mathcal{N}(I - E)$ is closed). Therefore $\mathcal{R}(E) \perp \mathcal{N}(E)$. Then

$$E^*E = EE^* \implies \mathcal{R}(E) \perp \mathcal{N}(E).$$

Thus (d) $\Rightarrow$ (a). To show that (c) $\Rightarrow$ (e), which trivially implies (f), proceed as follows. If $O \neq E = E^*$, then $\|E\|^2 = \|E^*E\| = \|E^2\| = \|E\| \neq 0$. So $\|E\| = 1$:

$$E = E^* \implies \|E\| = 1 \implies \|E\| \leq 1.$$

Now take any $v \in \mathcal{N}(E)^\perp$. Then $(I - E)v \in \mathcal{N}(E)$ (since $\mathcal{R}(I - E) = \mathcal{N}(E)$). Hence $(I - E)v \perp v$ and so $0 = \langle (I - E)v \, ; v \rangle = \|v\|^2 - \langle Ev \, ; v \rangle$. If $\|E\| \leq 1$, then $\|v\|^2 = \langle Ev \, ; v \rangle \leq \|Ev\| \|v\| \leq \|E\| \|v\|^2 \leq \|v\|^2$. Thus $\|Ev\| = \|v\| = \langle Ev \, ; v \rangle^{\frac{1}{2}}$. So $\|(I - E)v\|^2 = \|Ev - v\|^2 = \|Ev\|^2 - 2\operatorname{Re}\langle Ev \, ; v \rangle + \|v\|^2 = 0$, and hence $v \in \mathcal{N}(I - E) = \mathcal{R}(E)$. Then $\mathcal{N}(E)^\perp \subseteq \mathcal{R}(E)$, which implies that $\mathcal{R}(E^*) \subseteq \mathcal{N}(E^*)^\perp$ by Lemma 1.4, and therefore $\mathcal{R}(E^*) \perp \mathcal{N}(E^*)$. But E^* is a projection whenever E is ($E = E^2$ implies $E^* = E^{2*} = E^{*2}$). Thus E^* is an orthogonal projection. Then E^* is self-adjoint, and so is E. Hence (f) $\Rightarrow$ (c):

$$\|E\| \leq 1 \implies E = E^*. \qquad \square$$

Corollary 1.9. *Every bounded weighted sum of projections is normal.*

Proof. Let $T \in \mathcal{B}[\mathcal{H}]$ be a weighted sum of projections, which means $Tx = \sum_{\gamma \in \Gamma} \lambda_\gamma E_\gamma x$ for every x in the Hilbert space $\mathcal{H}$, where $\{\lambda_\gamma\}_{\gamma \in \Gamma}$ is a bounded family of scalars and $\{E_\gamma\}_{\gamma \in \Gamma}$ is a resolution of the identity on $\mathcal{H}$. Since $E_\gamma^* = E_\gamma$ for every γ and $\mathcal{R}(E_\alpha) \perp \mathcal{R}(E_\beta)$ whenever $\alpha \neq \beta$, it is readily verified that the adjoint T^* of T is given by $T^*x = \sum_{\gamma \in \Gamma} \overline{\lambda}_\gamma E_\gamma$ for every x in $\mathcal{H}$ (i.e., $\langle Tx \, ; y \rangle = \langle x \, ; T^*y \rangle$ for every x, y in $\mathcal{H}$). Moreover, since $E_\gamma^2 = E_\gamma$ for every γ and $E_\alpha E_\beta = E_\beta E_\alpha = O$ whenever $\alpha \neq \beta$, we also get $TT^*x = T^*Tx$ for every $x \in \mathcal{H}$. Hence T is normal. $\qquad \square$

An *isometry* between metric spaces is a map that preserves distance. Thus every isometry is injective. A linear transformation V between normed spaces is an isometry if and only if it preserves norm ($\|Vx\| = \|x\|$ — hence $\|V\| = 1$) and, between inner product spaces, if and only if it preserves inner product ($\langle Vx \, ; Vy \rangle = \langle x \, ; y \rangle$). An invertible linear isometry (i.e., a surjective linear isometry) is an *isometric isomorphism*. The inverse of an invertible linear isometry is again a linear isometry ($\|V^{-1}x\| = \|VV^{-1}x\| = \|x\|$). If $\mathcal{W}, \mathcal{X}, \mathcal{Y}, \mathcal{Z}$ are normed spaces, $S \in \mathcal{B}[\mathcal{W}, \mathcal{X}]$, $T \in \mathcal{B}[\mathcal{Y}, \mathcal{Z}]$, and $V \in \mathcal{B}[\mathcal{X}, \mathcal{Y}]$ is an isometry, then

$$\|VS\| = \|S\| \quad \text{and} \quad \|TV\| = \|T\| \text{ if } V \text{ is surjective.}$$

A transformation $V \in \mathcal{B}[\mathcal{H}, \mathcal{K}]$ between Hilbert spaces is an *isometry* if and only if $V^*V = I$ (identity on $\mathcal{H}$), and a *coisometry* if its adjoint is an isometry ($VV^* = I$ — identity on $\mathcal{K}$). A *unitary transformation* is an isometry and a coisometry or, equivalently, an invertible transformation with $V^{-1} = V^*$. Thus a *unitary operator* $U \in \mathcal{B}[\mathcal{H}]$ is precisely a normal isometry: $UU^* = U^*U = I$; and so normality distinguishes the unitary operators among the isometries.

Lemma 1.10. *If $\mathcal{X}$ is a normed space and $T \in \mathcal{B}[\mathcal{X}]$, then the real-valued sequence $\{\|T^n\|^{\frac{1}{n}}\}$ converges in $\mathbb{R}$.*

Proof. Take any positive integer m. Every positive integer n can be written as $n = m p_n + q_n$ for some nonnegative integers p_n, q_n with $q_n < m$. Hence

$$\|T^n\| = \|T^{mp_n + q_n}\| = \|T^{mp_n} T^{q_n}\| \leq \|T^{mp_n}\| \|T^{q_n}\| \leq \|T^m\|^{p_n} \|T^{q_n}\|.$$

Suppose $T \neq O$. Set $\alpha = \max_{0 \leq k \leq m-1} \{\|T^k\|\} \neq 0$. (Recall: $q_n \leq m - 1$.) Thus

$$\|T^n\|^{\frac{1}{n}} \leq \|T^m\|^{\frac{p_n}{n}} \alpha^{\frac{1}{n}} = \alpha^{\frac{1}{n}} \|T^m\|^{\frac{1}{m} - \frac{q_n}{mn}}.$$

Since $\alpha^{\frac{1}{n}} \to 1$ and $\|T^m\|^{\frac{1}{m} - \frac{q_n}{mn}} \to \|T^m\|^{\frac{1}{m}}$ as $n \to \infty$, we get

$$\limsup_n \|T^n\|^{\frac{1}{n}} \leq \|T^m\|^{\frac{1}{m}}.$$

Therefore, since m is an arbitrary positive integer,

$$\limsup_n \|T^n\|^{\frac{1}{n}} \leq \liminf_n \|T^n\|^{\frac{1}{n}},$$

and so $\{\|T^n\|^{\frac{1}{n}}\}$ converges in $\mathbb{R}$. $\qquad \square$

For each operator $T \in \mathcal{B}[\mathcal{X}]$ let $r(T)$ denote the limit of $\{\|T^n\|^{\frac{1}{n}}\}$,

$$r(T) = \lim_n \|T^n\|^{\frac{1}{n}}.$$

As we saw before (proof of Lemma 1.10), $r(T) \leq \|T^n\|^{\frac{1}{n}}$ for every $n \geq 1$. In particular, $r(T) \leq \|T\|$. Also $r(T^k)^{\frac{1}{k}} = \lim_n \|T^{kn}\|^{\frac{1}{kn}} = r(T)$ for each $k \geq 1$ (as $\{\|T^{kn}\|^{\frac{1}{kn}}\}$ is a subsequence of the convergent sequence $\{\|T^n\|^{\frac{1}{n}}\}$). Then $r(T^k) = r(T)^k$ for every positive integer k. Thus for every operator $T \in \mathcal{B}[\mathcal{X}]$ on a normed space $\mathcal{X}$ and for each nonnegative integer n,

$$0 \leq r(T)^n = r(T^n) \leq \|T^n\| \leq \|T\|^n.$$

An operator $T \in \mathcal{B}[\mathcal{X}]$ on a normed space $\mathcal{X}$ is *normaloid* if $r(T) = \|T\|$. Here is an alternate definition of a normaloid operator.

Theorem 1.11. *$r(T) = \|T\|$ if and only if $\|T^n\| = \|T\|^n$ for every $n \geq 1$.*

Proof. Immediate by the above inequalities and the definition of $r(T)$. $\qquad \square$

Theorem 1.12. *Every hyponormal operator is normaloid.*

Proof. Let $T \in \mathcal{B}[\mathcal{H}]$ be a hyponormal operator on a Hilbert space $\mathcal{H}$.

Claim 1. $\|T^n\|^2 \leq \|T^{n+1}\|\|T^{n-1}\|$ for every positive integer n.

Proof. Indeed, for any operator $T \in \mathcal{B}[\mathcal{H}]$,

$$\|T^n x\|^2 = \langle T^n x \,; T^n x \rangle = \langle T^* T^n x \,; T^{n-1} x \rangle \leq \|T^* T^n x\|\|T^{n-1} x\|$$

for each $n \geq 1$ and every $x \in \mathcal{H}$. Now if T is hyponormal, then

$$\|T^* T^n x\|\|T^{n-1} x\| \leq \|T^{n+1} x\|\|T^{n-1} x\| \leq \|T^{n+1}\|\|T^{n-1}\|\|x\|^2$$

by Theorem 1.7 and hence, for each $n \geq 1$ and every $x \in \mathcal{H}$,

$$\|T^n x\|^2 \leq \|T^{n+1}\|\|T^{n-1}\|\|x\|^2,$$

which ensures the claimed result. This completes the proof of Claim 1.

Claim 2. $\|T^k\| = \|T\|^k$ for every positive integer $k \leq n$, for all $n \geq 1$.

Proof. The above result holds trivially if $T = O$ and it also holds trivially for $n = 1$ (for all $T \in \mathcal{B}[\mathcal{H}]$). Let $T \neq O$ and suppose the above result holds for some integer $n \geq 1$. By Claim 1 we get

$$\|T\|^{2n} = (\|T\|^n)^2 = \|T^n\|^2 \leq \|T^{n+1}\|\|T^{n-1}\| = \|T^{n+1}\|\|T\|^{n-1}.$$

Therefore, as $\|T^n\| \leq \|T\|^n$ for every $n \geq 1$, and since $T \neq O$,

$$\|T\|^{n+1} = \|T\|^{2n}(\|T\|^{n-1})^{-1} \leq \|T^{n+1}\| \leq \|T\|^{n+1}.$$

Hence $\|T^{n+1}\| = \|T\|^{n+1}$. Then the claimed result holds for $n+1$ whenever it holds for n, which concludes the proof of Claim 2 by induction.

Therefore, $\|T^n\| = \|T\|^n$ for every integer $n \geq 1$ by Claim 2, and so T is normaloid by Theorem 1.11. $\qquad\square$

Since $\|T^{*n}\| = \|T^n\|$ for each $n \geq 1$, then $r(T^*) = r(T)$. Thus an operator T is normaloid if and only if its adjoint T^* is normaloid, and so (by Theorem 1.12) every seminormal operator is normaloid. In particular, *every normal operator is normaloid*. Summing up: an operator T is normal if it commutes with its adjoint (i.e., $TT^* = T^*T$), hyponormal if $TT^* \leq T^*T$, and normaloid if $r(T) = \|T\|$. These classes are related by proper inclusion:

$$\text{Normal} \;\subset\; \text{Hyponormal} \;\subset\; \text{Normaloid}.$$

1.7 Orthogonal Reducing Eigenspaces

Let I denote the identity operator on a Hilbert space $\mathcal{H}$. A *scalar operator* on $\mathcal{H}$ is a scalar multiple of the identity, say λI for any scalar λ in $\mathbb{C}$. (The terms "scalar operator" or "scalar type operator" are also employed with a different meaning — see the paragraph following Proposition 4.J in Section 4.5.)

Take an arbitrary operator T on $\mathcal{H}$. For every scalar λ in $\mathbb{C}$ consider the kernel $\mathcal{N}(\lambda I - T)$, which is a (closed) subspace of $\mathcal{H}$ (since T is bounded). If this kernel is not zero, then it is called an *eigenspace* of T.

Lemma 1.13. *Take any operator $T \in \mathcal{B}[\mathcal{H}]$ and an arbitrary scalar $\lambda \in \mathbb{C}$.*

(a) *If T is hyponormal, then $\mathcal{N}(\lambda I - T) \subseteq \mathcal{N}(\overline{\lambda} I - T^*)$.*

(b) *If T is normal, then $\mathcal{N}(\lambda I - T) = \mathcal{N}(\overline{\lambda} I - T^*)$.*

Proof. For any operator $T \in \mathcal{B}[\mathcal{H}]$ and any scalar $\lambda \in \mathbb{C}$,

$$(\lambda I - T)^*(\lambda I - T) - (\lambda I - T)(\lambda I - T)^* = T^*T - TT^*.$$

Thus $\lambda I - T$ is hyponormal (normal) if and only if T is hyponormal (normal). Hence if T is hyponormal, then so is $\lambda I - T$, and therefore

$$\|(\overline{\lambda} I - T^*)x\| \;\leq\; \|(\lambda I - T)x\| \quad \text{for every } x \in \mathcal{H} \text{ and every } \lambda \in \mathbb{C}$$

by Theorem 1.7, which yields the inclusion in (a). If T is normal, then the preceding inequality becomes an identity yielding the identity in (b). $\qquad\square$

Lemma 1.14. *Take $T \in \mathcal{B}[\mathcal{H}]$ and $\lambda \in \mathbb{C}$. If $\mathcal{N}(\lambda I - T) \subseteq \mathcal{N}(\overline{\lambda} I - T^*)$, then*

(a) *$\mathcal{N}(\lambda I - T) \perp \mathcal{N}(\zeta I - T)$ whenever $\zeta \neq \lambda$,   and*

(b) *$\mathcal{N}(\lambda I - T)$ reduces T.*

Proof. (a) Take any $x \in \mathcal{N}(\lambda I - T)$ and any $y \in \mathcal{N}(\zeta I - T)$. Thus $\lambda x = Tx$ and $\zeta y = Ty$. If $\mathcal{N}(\lambda I - T) \subseteq \mathcal{N}(\bar{\lambda} I - T^*)$, then $x \in \mathcal{N}(\bar{\lambda} I - T^*)$ and so $\bar{\lambda} x = T^* x$. Hence $\langle \zeta y \,;\, x \rangle = \langle Ty \,;\, x \rangle = \langle y \,;\, T^* x \rangle = \langle y \,;\, \bar{\lambda} x \rangle = \langle \lambda y \,;\, x \rangle$ and therefore $(\zeta - \lambda)\langle y \,;\, x \rangle = 0$, which implies $\langle y \,;\, x \rangle = 0$ whenever $\zeta \neq \lambda$.

(b) If $x \in \mathcal{N}(\lambda I - T) \subseteq \mathcal{N}(\bar{\lambda} I - T^*)$, then $Tx = \lambda x$ and $T^* x = \bar{\lambda} x$ and hence $\lambda T^* x = \lambda \bar{\lambda} x = \bar{\lambda} \lambda x = \bar{\lambda} T x = T \bar{\lambda} x = T T^* x$, which implies $T^* x \in \mathcal{N}(\lambda I - T)$. Thus $\mathcal{N}(\lambda I - T)$ is T^*-invariant. But $\mathcal{N}(\lambda I - T)$ is clearly T-invariant (if $Tx = \lambda x$, then $T(Tx) = \lambda(Tx)$). So $\mathcal{N}(\lambda I - T)$ reduces T by Lemma 1.6. $\square$

Lemma 1.15. *If $T \in \mathcal{B}[\mathcal{H}]$ is hyponormal, then*

(a) $\mathcal{N}(\lambda I - T) \perp \mathcal{N}(\zeta I - T)$ *for every $\lambda, \zeta \in \mathbb{C}$ such that $\lambda \neq \zeta$,   and*

(b) $\mathcal{N}(\lambda I - T)$ *reduces T for every $\lambda \in \mathbb{C}$.*

Proof. Apply Lemma 1.13(a) to Lemma 1.14. $\square$

Theorem 1.16. *If $\{\lambda_\gamma\}_{\gamma \in \Gamma}$ is a (nonempty) family of distinct complex numbers, and if $T \in \mathcal{B}[\mathcal{H}]$ is hyponormal, then the topological sum*

$$\mathcal{M} = \left(\sum\nolimits_{\gamma \in \Gamma} \mathcal{N}(\lambda_\gamma I - T) \right)^-$$

reduces T, and the restriction of T to it, $T|_{\mathcal{M}} \in \mathcal{B}[\mathcal{M}]$, is normal.

Proof. For each $\gamma \in \Gamma \neq \varnothing$ write $\mathcal{N}_\gamma = \mathcal{N}(\lambda_\gamma I - T)$, which is a T-invariant subspace of $\mathcal{H}$. Thus $T|_{\mathcal{N}_\gamma}$ lies in $\mathcal{B}[\mathcal{N}_\gamma]$ and coincides with the scalar operator $\lambda_\gamma I$ on $\mathcal{N}_\gamma$. By Lemma 1.15(a), $\{\mathcal{N}_\gamma\}_{\gamma \in \Gamma}$ is a family of pairwise orthogonal subspaces of $\mathcal{H}$. Take an arbitrary $x \in \left(\sum_{\gamma \in \Gamma} \mathcal{N}_\gamma \right)^-$. According to the Orthogonal Structure Theorem, $x = \sum_{\gamma \in \Gamma} u_\gamma$ with each u_γ in $\mathcal{N}_\gamma$. Moreover, Tu_γ and $T^* u_\gamma$ lie in $\mathcal{N}_\gamma$ for every $\gamma \in \Gamma$ because $\mathcal{N}_\gamma$ reduces T (Lemmas 1.6 and 1.15(b)). Thus since T and T^* are linear and continuous we conclude that $Tx = \sum_{\gamma \in \Gamma} Tu_\gamma$ lies in $\left(\sum_{\gamma \in \Gamma} \mathcal{N}_\gamma \right)^-$ and $T^* x = \sum_{\gamma \in \Gamma} T^* u_\gamma$ lies in $\left(\sum_{\gamma \in \Gamma} \mathcal{N}_\gamma \right)^-$. Then $\left(\sum_{\gamma \in \Gamma} \mathcal{N}_\gamma \right)^-$ reduces T (Lemma 1.6). Finally, since $\{\mathcal{N}_\gamma\}_{\gamma \in \Gamma}$ is a family of orthogonal subspaces (Lemma 1.15(a)), we may identify the topological sum $\mathcal{M}$ with the orthogonal direct sum $\bigoplus_{\gamma \in \Gamma} \mathcal{N}_\gamma$ by the Orthogonal Structure Theorem, where each $\mathcal{N}_\gamma$ reduces $T|_{\mathcal{M}}$ (it reduces T by Lemma 1.15(b)), and so $T|_{\mathcal{M}}$ is identified with the direct sum of operators $\bigoplus_{\gamma \in \Gamma} T|_{\mathcal{N}_\gamma}$, which is normal because each $T|_{\mathcal{N}_\gamma}$ is normal. $\square$

1.8 Compact Operators and the Fredholm Alternative

A set A in a topological space is *compact* if every open covering of it has a finite subcovering. It is *sequentially compact* if every A-valued sequence has a subsequence that converges in A. It is *relatively compact* or *conditionally compact* if its closure is compact. A set in a metric space is *totally bounded* if for every $\varepsilon > 0$ there is a finite partition of it into sets of diameter less than ε. The

COMPACTNESS THEOREM says that *a set in a metric space is compact if and only if it is sequentially compact, if and only if it is complete and totally bounded.* In a metric space every compact set is closed and every totally bounded set is bounded. *In a finite-dimensional normed space a set is compact if and only if it is closed and bounded.* This is the HEINE–BOREL THEOREM.

Let $\mathcal{X}$ and $\mathcal{Y}$ be normed spaces. A linear transformation $T\colon \mathcal{X}\to\mathcal{Y}$ is *compact* (or *completely continuous*) if it maps bounded subsets of $\mathcal{X}$ into relatively compact subsets of $\mathcal{Y}$; that is, if $T(A)^-$ is compact in $\mathcal{Y}$ whenever A is bounded in $\mathcal{X}$ or, equivalently, if $T(A)$ lies in a compact subset of $\mathcal{Y}$ whenever A is bounded in $\mathcal{X}$. Every compact linear transformation is bounded, thus continuous. Every linear transformation on a finite-dimensional space is compact. Let $\mathcal{B}_\infty[\mathcal{X},\mathcal{Y}]$ denote the collection of all compact linear transformations of $\mathcal{X}$ into $\mathcal{Y}$. Hence $\mathcal{B}_\infty[\mathcal{X},\mathcal{Y}]\subseteq\mathcal{B}[\mathcal{X},\mathcal{Y}]$ (and $\mathcal{B}_\infty[\mathcal{X},\mathcal{Y}]=\mathcal{B}[\mathcal{X},\mathcal{Y}]$ whenever $\mathcal{X}$ is finite-dimensional). $\mathcal{B}_\infty[\mathcal{X},\mathcal{Y}]$ is a linear manifold of $\mathcal{B}[\mathcal{X},\mathcal{Y}]$ which is closed in $\mathcal{B}[\mathcal{X},\mathcal{Y}]$ if $\mathcal{Y}$ is a Banach space. Set $\mathcal{B}_\infty[\mathcal{X}]=\mathcal{B}_\infty[\mathcal{X},\mathcal{X}]$, the collection of all *compact operators* on $\mathcal{X}$. $\mathcal{B}_\infty[\mathcal{X}]$ is an ideal (i.e., a two-sided ideal) of the normed algebra $\mathcal{B}[\mathcal{X}]$. That is, $\mathcal{B}_\infty[\mathcal{X}]$ is a subalgebra of $\mathcal{B}[\mathcal{X}]$ for which the product of a compact operator with a bounded operator is again compact.

Compact operators are supposed in this section to act on a complex nonzero Hilbert space $\mathcal{H}$, although the theory for compact operators equally applies, and usually is developed, in Banach spaces. (But a Hilbert-space approach considerably simplifies proofs since Hilbert spaces are complemented — a normed space is *complemented* if every subspace has a complementary subspace.)

Theorem 1.17. *If $T\in\mathcal{B}_\infty[\mathcal{H}]$ and $\lambda\in\mathbb{C}\backslash\{0\}$, then $\mathcal{R}(\lambda I-T)$ is closed.*

Proof. Let $\mathcal{M}$ be a subspace of a complex Banach space $\mathcal{X}$. Take any compact transformation K in $\mathcal{B}_\infty[\mathcal{M},\mathcal{X}]$. Let I be the identity on $\mathcal{M}$, let λ be a nonzero complex number, and consider the operator $\lambda I-K$ in $\mathcal{B}[\mathcal{M},\mathcal{X}]$.

Claim. If $\mathcal{N}(\lambda I-K)=\{0\}$, then $\mathcal{R}(\lambda I-K)$ is closed in $\mathcal{X}$.

Proof. Suppose $\mathcal{N}(\lambda I-K)=\{0\}$. If $\mathcal{R}(\lambda I-K)$ is not closed in $\mathcal{X}$, then $\lambda I-K$ is not bounded below (Theorem 1.2). Thus for every $\varepsilon>0$ there is a nonzero vector $x_\varepsilon\in\mathcal{M}$ for which $\|(\lambda I-K)x_\varepsilon\|<\varepsilon\|x_\varepsilon\|$. Then there is a sequence $\{x_n\}$ of unit vectors ($\|x_n\|=1$) in $\mathcal{M}$ for which $\|(\lambda I-K)x_n\|\to 0$. Since K is compact and $\{x_n\}$ is bounded, Proposition 1.S ensures that $\{Kx_n\}$ has a convergent subsequence, say $\{Kx_k\}$, and so $Kx_k\to y\in\mathcal{X}$. However,

$$\|\lambda x_k-y\|=\|\lambda x_k-Kx_k+Kx_k-y\|\leq\|(\lambda I-K)x_k\|+\|Kx_k-y\|\to 0.$$

Thus the $\mathcal{M}$-valued sequence $\{\lambda x_k\}$ converges in $\mathcal{X}$ to y, and so $y\in\mathcal{M}$ (because $\mathcal{M}$ is closed in $\mathcal{X}$). Since y is in the domain of K, then $y\in\mathcal{N}(\lambda I-K)$ (because $Ky=K\lim_k\lambda x_k=\lambda\lim_k Kx_k=\lambda y$ since K is continuous). Moreover, $y\neq 0$ (because $0\neq|\lambda|=\|\lambda x_k\|\to\|y\|$). Hence $\mathcal{N}(\lambda I-K)\neq\{0\}$, which is a contradiction. Therefore $\mathcal{R}(\lambda I-K)$ is closed in $\mathcal{X}$, concluding the proof of the claimed result.

Take $T \in \mathcal{B}[\mathcal{H}]$ and $\lambda \neq 0$. Consider the restriction $(\lambda I - T)|_{\mathcal{N}(\lambda I - T)^{\perp}}$ of $\lambda I - T$ to $\mathcal{N}(\lambda I - T)^{\perp}$ in $\mathcal{B}[\mathcal{N}(\lambda I - T)^{\perp}, \mathcal{H}]$. Clearly $\mathcal{N}\big((\lambda I - T)|_{\mathcal{N}(\lambda I - T)^{\perp}}\big) = \{0\}$. If T is compact, then so is the restriction $T|_{\mathcal{N}(\lambda I - T)^{\perp}}$ in $\mathcal{B}[\mathcal{N}(\lambda I - T)^{\perp}, \mathcal{H}]$ (Proposition 1.V in Section 1.9). Since $(\lambda I - T)|_{\mathcal{N}(\lambda I - T)^{\perp}} = \lambda I - T|_{\mathcal{N}(\lambda I - T)^{\perp}}$ (where the symbol I in the right-hand side denotes the identity on $\mathcal{N}(\lambda I - T)^{\perp}$) we get $\mathcal{N}(\lambda I - T|_{\mathcal{N}(\lambda I - T)^{\perp}}) = \{0\}$. Therefore $\mathcal{R}(\lambda I - T|_{\mathcal{N}(\lambda I - T)^{\perp}})$ is closed by the above claim. But $\mathcal{R}\big((\lambda I - T)|_{\mathcal{N}(\lambda I - T)^{\perp}}\big) = \mathcal{R}(\lambda I - T)$. $\qquad\square$

Theorem 1.18. *If* $T \in \mathcal{B}_{\infty}[\mathcal{H}]$, $\lambda \in \mathbb{C}\backslash\{0\}$, *and* $\mathcal{N}(\lambda I - T) = \{0\}$, *then* $\mathcal{R}(\lambda I - T) = \mathcal{H}$.

Proof. Take any operator $T \in \mathcal{B}[\mathcal{H}]$ and any scalar $\lambda \in \mathbb{C}$. Consider the sequence $\{\mathcal{M}_n\}_{n=0}^{\infty}$ of linear manifolds of $\mathcal{H}$ recursively defined by

$$\mathcal{M}_{n+1} = (\lambda I - T)(\mathcal{M}_n) \quad \text{for every } n \geq 0, \quad \text{with} \quad \mathcal{M}_0 = \mathcal{H}.$$

As it can be verified by induction,

$$\mathcal{M}_{n+1} \subseteq \mathcal{M}_n \quad \text{for every } n \geq 0.$$

In fact, $\mathcal{M}_1 = \mathcal{R}(\lambda I - T) \subseteq \mathcal{H} = \mathcal{M}_0$ and, if the above inclusion holds for some $n \geq 0$, then $\mathcal{M}_{n+2} = (\lambda I - T)(\mathcal{M}_{n+1}) \subseteq (\lambda I - T)(\mathcal{M}_n) = \mathcal{M}_{n+1}$, which concludes the induction. Suppose $T \in \mathcal{B}_{\infty}[\mathcal{H}]$ and $\lambda \neq 0$ so that $\lambda I - T$ has a closed range (Theorem 1.17). If $\mathcal{N}(\lambda I - T) = \{0\}$, then $\lambda I - T$ has a bounded inverse on its range (Theorem 1.2). Since $(\lambda I - T)^{-1} \colon \mathcal{R}(\lambda I - T) \to \mathcal{H}$ is continuous, $\lambda I - T$ sends closed sets into closed sets. (*A map between topological spaces is continuous if and only if the inverse image of closed sets is closed.*) Hence, since $\mathcal{M}_0$ is closed, by another induction we get

$$\{\mathcal{M}_n\}_{n=0}^{\infty} \text{ is a decreasing sequence of subspaces of } \mathcal{H}.$$

Now suppose $\mathcal{R}(\lambda I - T) \neq \mathcal{H}$. If $\mathcal{M}_{n+1} = \mathcal{M}_n$ for some integer n, then (since $\mathcal{M}_0 = \mathcal{H} \neq \mathcal{R}(\lambda I - T) = \mathcal{M}_1$) there exists an integer $k \geq 1$ for which $\mathcal{M}_{k+1} = \mathcal{M}_k \neq \mathcal{M}_{k-1}$, and this leads to a contradiction, namely, if $\mathcal{M}_{k+1} = \mathcal{M}_k$, then $(\lambda I - T)(\mathcal{M}_k) = \mathcal{M}_k$ and so $\mathcal{M}_k = (\lambda I - T)^{-1}(\mathcal{M}_k) = \mathcal{M}_{k-1}$. Outcome: $\mathcal{M}_{n+1}$ is properly included in $\mathcal{M}_n$ for each n; that is,

$$\mathcal{M}_{n+1} \subset \mathcal{M}_n \quad \text{for every } n \geq 0.$$

Hence each $\mathcal{M}_{n+1}$ is a nonzero proper subspace of the Hilbert space $\mathcal{M}_n$. Then each $\mathcal{M}_{n+1}$ has a nonzero complementary subspace in $\mathcal{M}_n$, namely $\mathcal{M}_n \ominus \mathcal{M}_{n+1} \subset \mathcal{M}_n$. Therefore, for each integer $n \geq 0$ there is an $x_n \in \mathcal{M}_n$ with $\|x_n\| = 1$ such that $\frac{1}{2} < \inf_{u \in \mathcal{M}_{n+1}} \|x_n - u\|$. In light of this, take any pair of integers $0 \leq m < n$ and set

$$x = x_n + \lambda^{-1}\big((\lambda I - T)x_m - (\lambda I - T)x_n\big)$$

(recall: $\lambda \neq 0$). So $Tx_n - Tx_m = \lambda(x - x_m)$. Since x lies in $\mathcal{M}_{m+1}$,

$$\tfrac{1}{2}|\lambda| < |\lambda|\|x - x_m\| = \|Tx_n - Tx_m\|,$$

and the sequence $\{Tx_n\}$ has no convergent subsequence (every subsequence of it is not Cauchy). Since $\{x_n\}$ is bounded, T is not compact (cf. Proposition 1.S again), which is a contradiction. Thus $\mathcal{R}(\lambda I - T) = \mathcal{H}$. $\square$

Theorem 1.19. *If* $T \in \mathcal{B}_\infty[\mathcal{H}]$ *and* $\lambda \in \mathbb{C}\backslash\{0\}$ *(i.e., if* T *is compact and* λ *is nonzero), then* $\dim\mathcal{N}(\lambda I - T) = \dim\mathcal{N}(\bar{\lambda}I - T^*) < \infty$.

Proof. Let $\dim\mathcal{X}$ stand for the dimension of a linear space $\mathcal{X}$ as usual. Take an arbitrary compact operator $T \in \mathcal{B}_\infty[\mathcal{H}]$ and an arbitrary nonzero scalar $\lambda \in \mathbb{C}$. By Theorem 1.17 we get $\mathcal{R}(\lambda I - T) = \mathcal{R}(\lambda I - T)^-$ and so $\mathcal{R}(\bar{\lambda}I - T^*) = \mathcal{R}(\bar{\lambda}I - T^*)^-$ by Lemma 1.5. Thus (cf. Lemma 1.4),

$$\mathcal{N}(\lambda I - T) = \{0\} \quad \text{if and only if} \quad \mathcal{R}(\bar{\lambda}I - T^*) = \mathcal{H},$$

$$\mathcal{N}(\bar{\lambda}I - T^*) = \{0\} \quad \text{if and only if} \quad \mathcal{R}(\lambda I - T) = \mathcal{H}.$$

Since $T, T^* \in \mathcal{B}_\infty[\mathcal{H}]$ (Proposition 1.W in Section 1.9), then by Theorem 1.18

$$\mathcal{N}(\lambda I - T) = \{0\} \quad \text{implies} \quad \mathcal{R}(\lambda I - T) = \mathcal{H},$$

$$\mathcal{N}(\bar{\lambda}I - T^*) = \{0\} \quad \text{implies} \quad \mathcal{R}(\bar{\lambda}I - T^*) = \mathcal{H}.$$

Hence, $\mathcal{N}(\lambda I - T) = \{0\}$ if and only if $\mathcal{N}(\bar{\lambda}I - T^*) = \{0\}$. Equivalently,

$$\dim\mathcal{N}(\lambda I - T) = 0 \quad \text{if and only if} \quad \dim\mathcal{N}(\bar{\lambda}I - T^*) = 0.$$

Now suppose $\dim\mathcal{N}(\lambda I - T) \neq 0$, and so $\dim\mathcal{N}(\bar{\lambda}I - T^*) \neq 0$. The subspace $\mathcal{N}(\lambda I - T) \neq \{0\}$ is invariant for T (cf. Proposition 1.E(b) in Section 1.9) and $T|_{\mathcal{N}(\lambda I - T)} = \lambda I$ in $\mathcal{B}[\mathcal{N}(\lambda I - T)]$. If T is compact, then so is $T|_{\mathcal{N}(\lambda I - T)}$ (Proposition 1.V). Thus $\lambda I \neq O$ is compact on $\mathcal{N}(\lambda I - T) \neq \{0\}$, and hence $\dim\mathcal{N}(\lambda I - T) < \infty$ (Proposition 1.Y). Dually, since T^* is compact, we get $\dim\mathcal{N}(\bar{\lambda}I - T^*) < \infty$. Hence there are positive integers m and n for which

$$\dim\mathcal{N}(\lambda I - T) = m \quad \text{and} \quad \dim\mathcal{N}(\bar{\lambda}I - T^*) = n.$$

We show next that $m = n$. Let $\{e_i\}_{i=1}^m$ and $\{f_i\}_{i=1}^n$ be orthonormal bases for the Hilbert spaces $\mathcal{N}(\lambda I - T)$ and $\mathcal{N}(\bar{\lambda}I - T^*)$. Set $k = \min\{m, n\} \geq 1$ and consider the mappings $S\colon \mathcal{H} \to \mathcal{H}$ and $S^*\colon \mathcal{H} \to \mathcal{H}$ defined by

$$Sx = \sum\nolimits_{i=1}^k \langle x\,;e_i\rangle f_i \quad \text{and} \quad S^*x = \sum\nolimits_{i=1}^k \overline{\langle x\,;f_i\rangle}\, e_i$$

for every $x \in \mathcal{H}$. Clearly, S and S^* lie in $\mathcal{B}[\mathcal{H}]$ and S^* is the adjoint of S; that is, $\langle Sx\,;y\rangle = \langle x\,;S^*y\rangle$ for every $x, y \in \mathcal{H}$. Actually,

$$\mathcal{R}(S) \subseteq \bigvee\{f_i\}_{i=1}^k \subseteq \mathcal{N}(\bar{\lambda}I - T^*) \quad \text{and} \quad \mathcal{R}(S^*) \subseteq \bigvee\{e_i\}_{i=1}^k \subseteq \mathcal{N}(\lambda I - T),$$

and so S and S^* in $\mathcal{B}[\mathcal{H}]$ are finite-rank operators, thus compact (i.e., they lie in $\mathcal{B}_\infty[\mathcal{H}]$), and hence $T + S$ and $T^* + S^*$ also lie in $\mathcal{B}_\infty[\mathcal{H}]$ (since $\mathcal{B}_\infty[\mathcal{H}]$

is a subspace of $\mathcal{B}[\mathcal{H}]$). First suppose $m \leq n$ (which implies $k = m$). If x lies in $\mathcal{N}(\lambda I - (T + S))$, then $(\lambda I - T)x = Sx$. But $\mathcal{R}(S) \subseteq \mathcal{N}(\overline{\lambda} I - T^*) = \mathcal{R}(\lambda I - T)^{\perp}$ (Lemma 1.4), and hence $(\lambda I - T)x = Sx = 0$ (since Sx lies in $\mathcal{R}(\lambda I - T)$). Then $x \in \mathcal{N}(\lambda I - T) = \operatorname{span}\{e_i\}_{i=1}^{m}$ and hence $x = \sum_{i=1}^{m} \alpha_i e_i$ (for some family of scalars $\{\alpha_i\}_{i=1}^{m}$). Therefore $0 = Sx = \sum_{j=1}^{m} \alpha_j S e_j = \sum_{j=1}^{m} \alpha_j \sum_{i=1}^{m} \langle e_j ; e_i \rangle f_i = \sum_{i=1}^{m} \alpha_i f_i$, and so $\alpha_i = 0$ for every $i = 1, \ldots, m$ — $\{f_i\}_{i=1}^{m}$ is an orthonormal set, thus linearly independent. That is, $x = 0$. Then $\mathcal{N}(\lambda I - (T + S)) = \{0\}$. Thus, by Theorem 1.18,

$$m \leq n \quad \text{implies} \quad \mathcal{R}(\lambda I - (T + S)) = \mathcal{H}.$$

Dually, using exactly the same argument,

$$n \leq m \quad \text{implies} \quad \mathcal{R}(\overline{\lambda} I - (T^* + S^*)) = \mathcal{H}.$$

If $m < n$, then $k = m < m + 1 \leq n$ and $f_{m+1} \in \mathcal{R}(\lambda I - (T + S)) = \mathcal{H}$ so that there exists $v \in \mathcal{H}$ for which $(\lambda I - (T + S))v = f_{m+1}$. Hence

$$\begin{aligned}
1 = \langle f_{m+1} ; f_{m+1} \rangle &= \langle (\lambda I - (T + S))v ; f_{m+1} \rangle \\
&= \langle (\lambda I - T)v ; f_{m+1} \rangle - \langle Sv ; f_{m+1} \rangle = 0,
\end{aligned}$$

which is a contradiction. Indeed, $\langle (\lambda I - T)v ; f_{m+1} \rangle = \langle Sv ; f_{m+1} \rangle = 0$ since $f_{m+1} \in \mathcal{N}(\overline{\lambda} I - T^*) = \mathcal{R}(\lambda I - T)^{\perp}$ and $Sv \in \mathcal{R}(S) \subseteq \operatorname{span}\{f_i\}_{i=1}^{m}$. If $n < m$, then $k = n < n + 1 \leq m$, and $e_{n+1} \in \mathcal{R}(\overline{\lambda} I - (T^* + S^*)) = \mathcal{H}$ so that there exists $u \in \mathcal{H}$ for which $(\overline{\lambda} I - (T^* + S^*))u = e_{n+1}$. Hence

$$\begin{aligned}
1 = \langle e_{m+1} ; e_{m+1} \rangle &= \langle (\overline{\lambda} I - (T^* + S^*))u ; e_{n+1} \rangle \\
&= \langle (\overline{\lambda} I - T^*)u ; e_{n+1} \rangle - \langle S^* u ; e_{n+1} \rangle = 0,
\end{aligned}$$

which is again a contradiction (because $e_{n+1} \in \mathcal{N}(\lambda I - T) = \mathcal{R}(\overline{\lambda} I - T^*)^{\perp}$ and $S^* u \in \mathcal{R}(S^*) \subseteq \operatorname{span}\{e_i\}_{i=1}^{n}$). Therefore $m = n$. $\qquad\square$

Together, the results of Theorems 1.17 and 1.19 are referred to as the Fredholm Alternative, which is stated below.

Corollary 1.20. (FREDHOLM ALTERNATIVE). *If $T \in \mathcal{B}_{\infty}[\mathcal{H}]$ and $\lambda \in \mathbb{C} \backslash \{0\}$, then $\mathcal{R}(\lambda I - T)$ is closed and $\dim \mathcal{N}(\lambda I - T) = \dim \mathcal{N}(\overline{\lambda} I - T^*) < \infty$.*

1.9 Supplementary Propositions

Let $\mathcal{X}$ be a complex linear space. A map $f \colon \mathcal{X} \times \mathcal{X} \to \mathbb{C}$ is a sesquilinear form if it is additive in both arguments, homogeneous in the first argument, and conjugate homogeneous in the second argument. A Hermitian symmetric sesquilinear form that induces a positive quadratic form is an inner product.

Proposition 1.A. (POLARIZATION IDENTITIES). *If $f \colon \mathcal{X} \times \mathcal{X} \to \mathbb{C}$ is a sesquilinear form on a complex linear space $\mathcal{X}$, then for every $x, y \in \mathcal{X}$*

(a_1)
$$f(x,y) = \tfrac{1}{4}\big(f(x+y,x+y) - f(x-y,x-y)$$
$$+ i\,f(x+iy,x+iy) - i\,f(x-iy,x-iy)\big).$$

If $\mathcal{X}$ is a complex inner product space and $S,T \in \mathcal{B}[\mathcal{X}]$, then for $x,y \in \mathcal{X}$

(a_2)
$$\langle Sx\,;Ty\rangle = \tfrac{1}{4}\big(\langle S(x+y)\,;T(x+y)\rangle - \langle S(x-y)\,;T(x-y)\rangle$$
$$+ i\langle S(x+iy)\,;T(x+iy)\rangle - i\langle S(x-iy)\,;T(x-iy)\rangle\big).$$

(PARALLELOGRAM LAW). *If $\mathcal{X}$ is an inner product space and $\|\cdot\|$ is the norm generated by the inner product, then for every $x,y \in \mathcal{X}$*

(b)
$$\|x+y\|^2 + \|x-y\|^2 = 2\big(\|x\|^2 + \|y\|^2\big).$$

Proposition 1.B. *A linear manifold of a Banach space is a Banach space if and only if it is a subspace. A linear transformation is bounded if and only if it maps bounded sets into bounded sets. If $\mathcal{X}$ and $\mathcal{Y}$ are nonzero normed spaces, then $\mathcal{B}[\mathcal{X},\mathcal{Y}]$ is a Banach space if and only if $\mathcal{Y}$ is a Banach space.*

A sum of two subspaces of a Hilbert space may not be a subspace (it may not be closed if the subspaces are not orthogonal). But if one of them is finite-dimensional, then the sum is closed (see Proposition A.6 in Appendix A).

Proposition 1.C. *If $\mathcal{M}$ and $\mathcal{N}$ are subspaces of a normed space $\mathcal{X}$, and if $\dim\mathcal{N} < \infty$, then the sum $\mathcal{M}+\mathcal{N}$ is a subspace of $\mathcal{X}$.*

Every subspace of a Hilbert space has a complementary subspace (e.g., its orthogonal complement). However, this is not true for every Banach space — see Remarks A.4 and A.5 in Appendix A.

Proposition 1.D. *If $\mathcal{M}$ and $\mathcal{N}$ are complementary subspaces of a Banach space $\mathcal{X}$, then the unique projection $E\colon \mathcal{X}\to\mathcal{X}$ with $\mathcal{R}(E) = \mathcal{M}$ and $\mathcal{N}(E) = \mathcal{N}$ is continuous. Conversely, if $E \in \mathcal{B}[\mathcal{X}]$ is a (continuous) projection, then $\mathcal{R}(E)$ and $\mathcal{N}(E)$ are complementary subspaces of $\mathcal{X}$.*

Proposition 1.E. *Let T and S be arbitrary operators on a normed space $\mathcal{X}$.*

(a) *The subspaces $\mathcal{N}(T)$ and $\mathcal{R}(T)^-$ of $\mathcal{X}$ are invariant for T. If S and T commute, then the subspaces $\mathcal{N}(S)$, $\mathcal{N}(T)$, $\mathcal{R}(S)^-$ and $\mathcal{R}(T)^-$ of $\mathcal{X}$ are invariant for both S and T.*

(b) *If T is an operator on a normed space, then $\mathcal{N}(p(T))$ and $\mathcal{R}(p(T))^-$ are T-invariant subspaces for every polynomial $p(T) = \sum_{i=0}^{n}\alpha_i T^i$ in T — in particular, $\mathcal{N}(\lambda I - T)$ and $\mathcal{R}(\lambda I - T)^-$ are T-invariant for every $\lambda \in \mathbb{C}$.*

Proposition 1.F. *Let $T \in \mathcal{B}[\mathcal{X}]$ be an operator on a Banach space $\mathcal{X}$. If there exists an invertible operator $S \in \mathcal{G}[\mathcal{X}]$ for which $\|S - T\| < \|S^{-1}\|^{-1}$, then T is itself invertible (i.e., $T \in \mathcal{G}[\mathcal{X}]$ and so $T^{-1} \in \mathcal{B}[\mathcal{X}]$).*

Proposition 1.G. *Let $\mathcal{M}$ be a linear manifold of a normed space $\mathcal{X}$, let $\mathcal{Y}$ be a Banach space, and take $T \in \mathcal{B}[\mathcal{M},\mathcal{Y}]$.*

(a) *There exists a unique extension* $\widehat{T} \in \mathcal{B}[\mathcal{M}^-, \mathcal{Y}]$ *of* $T \in \mathcal{B}[\mathcal{M}, \mathcal{Y}]$ *over the closure of* $\mathcal{M}$. *Moreover,* $\|\widehat{T}\| = \|T\|$.

(b) *If* $\mathcal{X}$ *is a Hilbert space, then* $\widehat{T}E \in \mathcal{B}[\mathcal{X}, \mathcal{Y}]$ *is a (bounded linear) extension of* $T \in \mathcal{B}[\mathcal{M}, \mathcal{Y}]$ *over the whole space* $\mathcal{X}$, *where* $E \in \mathcal{B}[\mathcal{X}]$ *is the orthogonal projection onto the subspace* $\mathcal{M}^-$ *of* $\mathcal{X}$, *so that* $\|\widehat{T}E\| \leq \|T\|$.

Proposition 1.H. *Let* $\mathcal{M}$ *and* $\mathcal{N}$ *be subspaces of a Hilbert space.*

(a) $\mathcal{M}^\perp \cap \mathcal{N}^\perp = (\mathcal{M} + \mathcal{N})^\perp$.

(b) $\dim \mathcal{N} < \infty \implies \dim\left(\mathcal{N} \ominus (\mathcal{M} \cap \mathcal{N})\right) = \dim\left(\mathcal{M}^\perp \ominus (\mathcal{M} + \mathcal{N})^\perp\right)$.

Proposition 1.I. *A subspace* $\mathcal{M}$ *of a Hilbert space* $\mathcal{H}$ *is invariant (or reducing) for a nonzero* $T \in \mathcal{B}[\mathcal{H}]$ *if and only if the unique orthogonal projection* $E \in \mathcal{B}[\mathcal{H}]$ *with* $\mathcal{R}(E) = \mathcal{M}$ *is such that* $ETE = TE$ *(or* $TE = ET$).

Take $T \in \mathcal{B}[\mathcal{X}]$ and $S \in \mathcal{B}[\mathcal{Y}]$ on normed spaces $\mathcal{X}$ and $\mathcal{Y}$, and $X \in \mathcal{B}[\mathcal{X}, \mathcal{Y}]$ of $\mathcal{X}$ into $\mathcal{Y}$. If $XT = SX$, then X *intertwines* T to S (or T is *intertwined* to S through X). If there is an X with dense range intertwining T to S, then T is *densely intertwined* to S. If X has dense range and is injective, then it is *quasiinvertible* (or a *quasiaffinity*). If a quasiinvertible X intertwines T to S, then T is a *quasiaffine transform* of S. If T is a quasiaffine transform of S and S is a quasiaffine transform of T, then T and S are *quasisimilar*. If an invertible X (with a bounded inverse) intertwines T to S (and so X^{-1} intertwines S to T), then T and S are *similar* (or *equivalent*). Unitary equivalence is a special case of similarity on inner product spaces through a (surjective) isometry: operators T and S on inner product spaces $\mathcal{X}$ and $\mathcal{Y}$ are *unitarily equivalent* (notation: $T \cong S$) if there is a unitary transformation X intertwining them.

A linear manifold (or a subspace) of a normed space $\mathcal{X}$ is *hyperinvariant* for an operator $T \in \mathcal{B}[\mathcal{X}]$ if it is invariant for every operator in $\mathcal{B}[\mathcal{X}]$ that commutes with T. Obviously, hyperinvariant subspaces are invariant.

Proposition 1.J. *Similarity preserves invariant subspaces (if two operators are similar, and if one has a nontrivial invariant subspace, then so has the other), and quasisimilarity preserves hyperinvariant subspaces (if two operators are quasisimilar, and if one has a nontrivial hyperinvariant subspace, then so has the other).*

Proposition 1.K. *If* $\{E_k\}$ *is an orthogonal sequence of orthogonal projections on a Hilbert space* $\mathcal{H}$, *then* $\sum_{k=1}^n E_k \xrightarrow{s} E$, *where* $E \in \mathcal{B}[\mathcal{H}]$ *is the orthogonal projection with* $\mathcal{R}(E) = \left(\sum_{k \in \mathbb{N}} \mathcal{R}(E_k)\right)^-$.

Proposition 1.L. *A bounded linear transformation between Hilbert spaces is invertible if and only if its adjoint is (i.e.,* $T \in \mathcal{G}[\mathcal{H}, \mathcal{K}] \iff T^* \in \mathcal{G}[\mathcal{K}, \mathcal{H}]$). *Moreover,* $(T^*)^{-1} = (T^{-1})^*$.

Proposition 1.M. *Every nonnegative operator* $Q \in \mathcal{B}[\mathcal{H}]$ *on a Hilbert space* $\mathcal{H}$ *has a unique nonnegative square root* $Q^{\frac{1}{2}} \in \mathcal{B}[\mathcal{H}]$ *(i.e.,* $(Q^{\frac{1}{2}})^2 = Q$*), which commutes with every operator in* $\mathcal{B}[\mathcal{H}]$ *that commutes with* Q.

Proposition 1.N. *Every* $T \in \mathcal{B}[\mathcal{H}, \mathcal{K}]$ *between Hilbert spaces* $\mathcal{H}$ *and* $\mathcal{K}$ *has a unique* polar decomposition $T = W|T|$*, where* $W \in \mathcal{B}[\mathcal{H}, \mathcal{K}]$ *is a* partial isometry *(i.e.,* $W|_{\mathcal{N}(W)^\perp} : \mathcal{N}(W)^\perp \to \mathcal{K}$ *is an isometry) and* $|T| = (T^*T)^{\frac{1}{2}}$.

Proposition 1.O. *Every operator* $T \in \mathcal{B}[\mathcal{H}]$ *on a complex Hilbert space* $\mathcal{H}$ *has a unique* Cartesian decomposition $T = A + iB$*, with* $A = \frac{1}{2}(T + T^*)$ *and* $B = -\frac{i}{2}(T - T^*)$ *are self-adjoint operators on* $\mathcal{H}$*. Moreover, the operator* T *is normal if and only if* A *and* B *commute and, in this case,* $T^*T = A^2 + B^2$ *and* $\max\{\|A\|^2, \|B\|^2\} \le \|T\|^2 \le \|A\|^2 + \|B\|^2$.

Proposition 1.P. *The restriction* $T|_{\mathcal{M}}$ *of a hyponormal operator* T *to an invariant subspace* $\mathcal{M}$ *is hyponormal. If* $T|_{\mathcal{M}}$ *is normal, then* $\mathcal{M}$ *reduces* T.

Proposition 1.Q. *The restriction* $T|_{\mathcal{M}}$ *of a normal operator* T *to an invariant subspace* $\mathcal{M}$ *is normal if and only if* $\mathcal{M}$ *reduces* T.

Recall from Section 1.5: a Hilbert-space operator is *reductive* if every invariant subspace is reducing (i.e., T is reductive if $T(\mathcal{M}) \subseteq \mathcal{M}$ implies $T(\mathcal{M}^\perp) \subseteq \mathcal{M}^\perp$ or, equivalently, if $T(\mathcal{M}) \subseteq \mathcal{M}$ implies $T^*(\mathcal{M}) \subseteq \mathcal{M}$). Therefore Proposition 1.Q can be restated as follows.

A normal operator is reductive if and only if its restriction to every invariant subspace is normal.

Proposition 1.R. *A transformation* U *between Hilbert spaces is unitary if and only if it is an invertible contraction whose inverse also is a contraction* ($\|U\| \le 1$ *and* $\|U^{-1}\| \le 1$*), which happens if and only if* $\|U\| = \|U^{-1}\| = 1$.

Proposition 1.S. *A linear transformation between normed spaces is compact if and only if it maps bounded sequences into sequences that have a convergent subsequence.*

Proposition 1.T. *A linear transformation of a normed space into a Banach space is compact if and only if it maps bounded sets into totally bounded sets.*

Proposition 1.U. *A linear transformation of a Hilbert space into a normed space is compact if and only if it maps weakly convergent sequences into strongly convergent sequences.*

Proposition 1.V. *The restriction of a compact linear transformation to a linear manifold is again a compact linear transformation.*

Proposition 1.W. *A linear transformation between Hilbert spaces is compact if and only if its adjoint is (i.e.,* $T \in \mathcal{B}_\infty[\mathcal{H}, \mathcal{K}] \iff T^* \in \mathcal{B}_\infty[\mathcal{K}, \mathcal{H}]$*).*

Proposition 1.X. *A finite-rank (i.e., one with a finite-dimensional range) bounded linear transformation between normed spaces is compact.*

Proposition 1.Y. *The identity I on a normed space X is compact if and only if $\dim X < \infty$. Hence, if $T \in B_\infty[X]$ is invertible, then $\dim X < \infty$.*

A sequence $\{x_n\}$ of vectors in a normed space X is a *Schauder basis* for X if each vector x in X is uniquely written as countable linear combination $x = \sum_k \alpha_k x_k$ of $\{x_n\}$. Banach spaces with a Schauder basis are *separable* (i.e., have a countable dense subset or, equivalently, are spanned by a countable set — see, e.g., [78, Problem 4.11, Theorem 4.9]).

Proposition 1.Z. (a) *If a sequence of finite-rank bounded linear transformations between Banach spaces converges uniformly, then its limit is compact.*

(b) *Every compact linear transformation between Banach spaces possessing a Schauder basis is the uniform limit of a sequence of finite-rank bounded linear transformations.*

Notes: These are standard results required in the sequel. Proofs can be found in many texts on operator theory (see Suggested Reading). Proposition 1.A can be found in [78, Propositions 5.4 and Problem 5.3], and Proposition 1.B in [78, Propositions 4.7, 4.12, 4.15, and Example 4.P]. For Proposition 1.C see [30, Proposition III.4.3] or [58, Problem 13] — see Proposition A.6 in Appendix A. For Proposition 1.D see [30, Theorem III.13.2] or [78, Problem 4.35] — see Remark A.5 in Appendix A. For Propositions 1.E(a,b) see [78, Problems 4.20 and 4.22]. Proposition 1.F is a consequence of the Neumann expansion (Theorem 1.3) — see, e.g., [58, Solution 100] or [30, Corollary VII.2.39b)]. Proposition 1.G(a) is called *extension by continuity*; see [78, Theorem 4.35]. It implies Proposition 1.G(b): an orthogonal projection is continuous and $\widehat{T} = \widehat{T}E|_{\mathcal{R}(E)}$ in $B[\mathcal{M}^-, \mathcal{H}]$ is a restriction of $\widehat{T}E$ in $B[X, \mathcal{H}]$ to $\mathcal{R}(E) = \mathcal{M}^-$. Proposition 1.H will be needed in Theorem 5.4. For Proposition 1.H(a) see [78, Problems 5.8]. Here is a proof for Proposition 1.H(b).

Proof of Proposition 1.H(b). Since $(\mathcal{M} \cap \mathcal{N}) \perp (\mathcal{M}^\perp \cap \mathcal{N})$, consider the orthogonal projection $E \in B[\mathcal{H}]$ onto $\mathcal{M}^\perp \cap \mathcal{N}$ (i.e., $\mathcal{R}(E) = \mathcal{M}^\perp \cap \mathcal{N}$) along $\mathcal{M} \cap \mathcal{N}$ (i.e., $\mathcal{N}(E) = \mathcal{M} \cap \mathcal{N}$). Let $L: \mathcal{N} \ominus (\mathcal{M} \cap \mathcal{N}) \to \mathcal{M}^\perp \cap \mathcal{N}$ be the restriction of E to $\mathcal{N} \ominus (\mathcal{M} \cap \mathcal{N})$. It is clear that $L = E|_{\mathcal{N} \ominus (\mathcal{M} \cap \mathcal{N})}$ is linear, injective ($\mathcal{N}(L) = \{0\}$), and surjective ($\mathcal{R}(L) = \mathcal{R}(E) = \mathcal{M}^\perp \cap \mathcal{N}$). Therefore $\mathcal{N} \ominus (\mathcal{M} \cap \mathcal{N})$ is isomorphic to $\mathcal{M}^\perp \cap \mathcal{N}$. Since $\mathcal{M}$ is closed and $\mathcal{N}$ is finite-dimensional, $\mathcal{M} + \mathcal{N}$ is closed (even if $\mathcal{M}$ and $\mathcal{N}$ are not orthogonal — cf. Proposition 1.C). Then (Projection Theorem) $\mathcal{H} = (\mathcal{M} + \mathcal{N})^\perp \oplus (\mathcal{M} + \mathcal{N})$ and $\mathcal{M}^\perp = (\mathcal{M} + \mathcal{N})^\perp \oplus [\mathcal{M}^\perp \ominus (\mathcal{M} + \mathcal{N})^\perp]$, and hence $\mathcal{M}^\perp \ominus (\mathcal{M} + \mathcal{N})^\perp$ coincides with $\mathcal{M}^\perp \cap (\mathcal{M} + \mathcal{N}) = \mathcal{M}^\perp \cap \mathcal{N}$. Therefore $\mathcal{N} \ominus (\mathcal{M} \cap \mathcal{N})$ is isomorphic to $\mathcal{M}^\perp \ominus (\mathcal{M} + \mathcal{N})^\perp$, and so they have the same dimension. $\square$

For Proposition 1.I see, e.g., [75, Solutions 4.1 and 4.6], and for Proposition 1.J see, e.g., [74, Corollaries 4.2 and 4.8]. By the way, *does quasisimilarity preserve nontrivial invariant subspaces?* (See, e.g., [98, p. 194].) As for Propositions 1.K and 1.L see [78, Propositions 5.28 and p. 385]). Propositions 1.M, 1.N, and 1.O are the classical SQUARE ROOT THEOREM, the POLAR DECOMPOSITION THEOREM, and the CARTESIAN DECOMPOSITION THEOREM (see, e.g., [78, Theorems 5.85 and 5.89, and Problem 5.46]). Propositions 1.P, 1.Q, 1.R can be found in [78, Problems 6.16 and 6.17, and Proposition 5.73]. Propositions 1.S, 1.T, 1.U, 1.V, 1.W, 1.X, 1.Y, and 1.Z are all related to compact operators (see, e.g., [78, Theorem 4.52(d,e), Problem 4.51, p. 258, Problem 5.41, Proposition 4.50, remark to Theorem 4.52, Corollary 4.55, Problem 4.58]).

Suggested Readings

Akhiezer and Glazman [4, 5]
Beals [13]
Berberian [18, 19]
Brown and Pearcy [23, 24]
Conway [30, 31, 32, 33]
Douglas [38]
Dunford and Schwartz [45]

Fillmore [48]
Halmos [54, 55, 58]
Harte [60]
Istrăţescu [69]
Kubrusly [74, 75, 78]
Radjavi and Rosenthal [98]
Weidmann [113]

2

Spectrum of an Operator

Let $T\colon\mathcal{M}\to\mathcal{X}$ be a linear transformation, where $\mathcal{X}$ is a nonzero normed space and $\mathcal{M}=\mathcal{D}(T)$ is the domain of T which is a linear manifold of $\mathcal{X}$. Let $\mathbb{F}$ denote either the real field $\mathbb{R}$ or the complex field $\mathbb{C}$, and let I be the identity operator on $\mathcal{M}$. The general notion of spectrum applies to bounded or unbounded linear transformations T from $\mathcal{M}\subseteq\mathcal{X}$ to $\mathcal{X}$.

The *resolvent set* $\rho(T)$ of T is the set of all scalars λ in $\mathbb{F}$ for which the linear transformation $\lambda I-T\colon\mathcal{M}\to\mathcal{X}$ has a densely defined continuous inverse:

$$\rho(T)=\big\{\lambda\in\mathbb{F}\colon(\lambda I-T)^{-1}\in\mathcal{B}[\mathcal{R}(\lambda I-T),\mathcal{M}]\ \text{ and }\ \mathcal{R}(\lambda I-T)^{-}=\mathcal{X}\big\}$$

(see, e.g., [9, Definition 18.2] — there are different definitions of resolvent set for unbounded linear transformations, but they all coincide for the bounded case). We will, however, restrict the theory to *bounded* linear transformations of a complex Banach space into itself (i.e., to elements of the complex Banach algebra $\mathcal{B}[\mathcal{X}]$). The *spectrum* $\sigma(T)$ of T is the complement in $\mathbb{C}$ of $\rho(T)$.

2.1 Basic Spectral Properties

Throughout this chapter $T\colon\mathcal{X}\to\mathcal{X}$ will be a *bounded* linear transformation of $\mathcal{X}$ into itself (i.e., an operator on $\mathcal{X}$), and so $\mathcal{D}(T)=\mathcal{X}$ where $\mathcal{X}\neq\{0\}$ is a *complex Banach space*. In other words, $T\in\mathcal{B}[\mathcal{X}]$ where $\mathcal{X}$ is a nonzero complex Banach space. In such a case (i.e., in the *unital complex Banach algebra* $\mathcal{B}[\mathcal{X}]$), the resolvent set $\rho(T)$ is precisely the set of all complex numbers λ for which $\lambda I-T\in\mathcal{B}[\mathcal{X}]$ is invertible (i.e., has a bounded inverse on $\mathcal{X}$), according to Theorem 1.2. Thus (see also Theorem 1.1) the resolvent set is given by

$$\rho(T)=\big\{\lambda\in\mathbb{C}\colon\lambda I-T\in\mathcal{G}[\mathcal{X}]\big\}=\big\{\lambda\in\mathbb{C}\colon\lambda I-T\text{ has an inverse in }\mathcal{B}[\mathcal{X}]\big\}$$
$$=\big\{\lambda\in\mathbb{C}\colon\mathcal{N}(\lambda I-T)=\{0\}\ \text{ and }\ \mathcal{R}(\lambda I-T)=\mathcal{X}\big\},$$

and the spectrum is

$$\sigma(T)=\mathbb{C}\setminus\rho(T)=\big\{\lambda\in\mathbb{C}\colon\lambda I-T\text{ has no inverse in }\mathcal{B}[\mathcal{X}]\big\}$$
$$=\big\{\lambda\in\mathbb{C}\colon\mathcal{N}(\lambda I-T)\neq\{0\}\ \text{ or }\ \mathcal{R}(\lambda I-T)\neq\mathcal{X}\big\}.$$

© Springer Nature Switzerland AG 2020
C. S. Kubrusly, *Spectral Theory of Bounded Linear Operators*,
https://doi.org/10.1007/978-3-030-33149-8_2

Theorem 2.1. $\rho(T)$ *is nonempty and open, and* $\sigma(T)$ *is compact.*

Proof. Take any $T \in \mathcal{B}[\mathcal{X}]$. By the Neumann expansion (Theorem 1.3), if $\|T\| < |\lambda|$, then $\lambda \in \rho(T)$. Equivalently, since $\sigma(T) = \mathbb{C} \backslash \rho(T)$,

$$|\lambda| \le \|T\| \quad \text{for every } \lambda \in \sigma(T).$$

Thus $\sigma(T)$ is bounded, and therefore $\rho(T) \ne \varnothing$.

Claim. If $\lambda \in \rho(T)$, then the open ball $B_\delta(\lambda)$ with center at λ and (positive) radius $\delta = \|(\lambda I - T)^{-1}\|^{-1}$ is included in $\rho(T)$.

Proof. If $\lambda \in \rho(T)$, then $\lambda I - T \in \mathcal{G}[\mathcal{X}]$. Hence $(\lambda I - T)^{-1}$ is nonzero and bounded. Thus $0 < \|(\lambda I - T)^{-1}\|^{-1} < \infty$. Set $\delta = \|(\lambda I - T)^{-1}\|^{-1}$, let $B_\delta(0)$ be the nonempty open ball of radius δ about the origin of the complex plane $\mathbb{C}$, and take any ζ in $B_\delta(0)$. Since $|\zeta| < \|(\lambda I - T)^{-1}\|^{-1}$, then $\|\zeta(\lambda I - T)^{-1}\| < 1$. Therefore $[I - \zeta(\lambda I - T)^{-1}] \in \mathcal{G}[\mathcal{X}]$ by Theorem 1.3, and so $(\lambda - \zeta)I - T = (\lambda I - T)[I - \zeta(\lambda I - T)^{-1}] \in \mathcal{G}[\mathcal{X}]$. Thus $\lambda - \zeta \in \rho(T)$ so that

$$B_\delta(\lambda) = B_\delta(0) + \lambda = \lambda - B_\delta(0) = \{\lambda - \zeta \in \mathbb{C} \colon \text{for some } \zeta \in B_\delta(0)\} \subseteq \rho(T),$$

which completes the proof of the claimed result.

Thus $\rho(T)$ is open (it includes a nonempty open ball centered at each of its points) and so $\sigma(T)$ is closed. Compact in $\mathbb{C}$ means closed and bounded. $\square$

Remark. Since $B_\delta(\lambda) \subseteq \rho(T)$, the distance of any λ in $\rho(T)$ to the spectrum $\sigma(T)$ is greater than or equal to δ; that is (compare with Proposition 2.E),

$$\lambda \in \rho(T) \quad \text{implies} \quad \|(\lambda I - T)^{-1}\|^{-1} \le d(\lambda, \sigma(T)).$$

The *resolvent function* $R_T \colon \rho(T) \to \mathcal{G}[\mathcal{X}]$ of an operator $T \in \mathcal{B}[\mathcal{X}]$ is the mapping of the resolvent set $\rho(T)$ of T into the group $\mathcal{G}[\mathcal{X}]$ of all invertible operators from $\mathcal{B}[\mathcal{X}]$ defined by

$$R_T(\lambda) = (\lambda I - T)^{-1} \quad \text{for every } \lambda \in \rho(T).$$

Since $R_T(\lambda) - R_T(\zeta) = R_T(\lambda)[R_T(\zeta)^{-1} - R_T(\lambda)^{-1}]R_T(\zeta)$, we get

$$R_T(\lambda) - R_T(\zeta) = (\zeta - \lambda)R_T(\lambda)R_T(\zeta)$$

for every $\lambda, \zeta \in \rho(T)$ (because $R_T(\zeta)^{-1} - R_T(\lambda)^{-1} = (\zeta - \lambda)I$). This is the *resolvent identity*. Swapping λ and ζ in the resolvent identity, it follows that $R_T(\lambda)$ and $R_T(\zeta)$ commute for every $\lambda, \zeta \in \rho(T)$. Also, $T R_T(\lambda) = R_T(\lambda)T$ for every $\lambda \in \rho(T)$ (since $R_T(\lambda)^{-1}R_T(\lambda) = R_T(\lambda)R_T(\lambda)^{-1}$ trivially).

Let Λ be a nonempty open subset of the complex plane $\mathbb{C}$. Take a function $f \colon \Lambda \to \mathbb{C}$ and a point $\zeta \in \Lambda$. Suppose there exists a complex number $f'(\zeta)$ with the following property. For every $\varepsilon > 0$ there is a $\delta > 0$ for which

$\left|\frac{f(\lambda)-f(\zeta)}{\lambda-\zeta} - f'(\zeta)\right| < \varepsilon$ for all λ in Λ such that $0 < |\lambda - \zeta| < \delta$. If there exists such an $f'(\zeta) \in \mathbb{C}$, then it is called the *derivative* of f at ζ. If $f'(\zeta)$ exists for every ζ in Λ, then $f \colon \Lambda \to \mathbb{C}$ is *analytic* or *holomorphic* (on its domain Λ). In this context the above terms are synonyms, implying infinite differentiability and power series expansion. A function $f \colon \mathbb{C} \to \mathbb{C}$ is *entire* if it is analytic on the whole complex plane $\mathbb{C}$. To prove the next result we need the LIOUVILLE THEOREM, which says that *every bounded entire function is constant*.

Theorem 2.2. *If $\mathcal{X}$ is nonzero, then the spectrum $\sigma(T)$ is nonempty.*

Proof. (Note: if $\mathcal{X} = \{0\}$, then $T = O = I$, and so $\rho(T) = \mathbb{C}$.) Let $T \in \mathcal{B}[\mathcal{X}]$ be an operator on a *nonzero* complex Banach space $\mathcal{X}$. Let $\mathcal{B}[\mathcal{X}]^* = \mathcal{B}[\mathcal{B}[\mathcal{X}], \mathbb{C}]$ be the Banach space of all bounded linear functionals on $\mathcal{B}[\mathcal{X}]$; that is, the dual of $\mathcal{B}[\mathcal{X}]$. Since $\mathcal{X} \neq \{0\}$, then $\mathcal{B}[\mathcal{X}] \neq \{O\}$, and so $\mathcal{B}[\mathcal{X}]^* \neq \{0\}$ (a consequence of the Hahn–Banach Theorem — see, e.g., [78, Corollary 4.64]). Take any nonzero ξ in $\mathcal{B}[\mathcal{X}]^*$ (i.e., a nonzero bounded linear functional $\xi \colon \mathcal{B}[\mathcal{X}] \to \mathbb{C}$), and consider the composition of it with the resolvent function, $\xi \circ R_T \colon \rho(T) \to \mathbb{C}$. Recall: $\rho(T) = \mathbb{C} \backslash \sigma(T)$ is nonempty and open in $\mathbb{C}$.

Claim 1. If $\sigma(T)$ is empty, then $\xi \circ R_T \colon \rho(T) \to \mathbb{C}$ is bounded.

Proof. The resolvent function $R_T \colon \rho(T) \to \mathcal{G}[\mathcal{X}]$ is continuous (since scalar multiplication and addition are continuous mappings, and inversion is also a continuous mapping; see, e.g., [78, Problem 4.48]). Thus $\|R_T(\cdot)\| \colon \rho(T) \to \mathbb{R}$ is continuous. Then $\sup_{|\lambda| \leq \|T\|} \|R_T(\lambda)\| < \infty$ if $\sigma(T) = \varnothing$. In fact if $\sigma(T)$ is empty, then $\rho(T) \cap B_{\|T\|}[0] = B_{\|T\|}[0] = \{\lambda \in \mathbb{C} \colon |\lambda| \leq \|T\|\}$ is a compact set in $\mathbb{C}$, so that the continuous function $\|R_T(\cdot)\|$ attains its maximum on it by the WEIERSTRASS THEOREM: *a continuous real-valued function attains its maximum and minimum on any compact set in a metric space.* On the other hand, if $\|T\| < |\lambda|$, then $\|R_T(\lambda)\| \leq (|\lambda| - \|T\|)^{-1}$ (see the remark following Theorem 1.3), and so $\|R_T(\lambda)\| \to 0$ as $|\lambda| \to \infty$. Thus, since $\|R_T(\cdot)\| \colon \rho(T) \to \mathbb{R}$ is continuous, $\sup_{\|T\| \leq \lambda} \|R_T(\lambda)\| < \infty$. Hence $\sup_{\lambda \in \rho(T)} \|R_T(\lambda)\| < \infty$. Thus

$$\sup_{\lambda \in \rho(T)} \|(\xi \circ R_T)(\lambda)\| \leq \|\xi\| \sup_{\lambda \in \rho(T)} \|R_T(\lambda)\| < \infty,$$

which completes the proof of Claim 1.

Claim 2. $\xi \circ R_T \colon \rho(T) \to \mathbb{C}$ is analytic.

Proof. If λ and ζ are distinct points in $\rho(T)$, then

$$\frac{R_T(\lambda) - R_T(\zeta)}{\lambda - \zeta} + R_T(\zeta)^2 = (R_T(\zeta) - R_T(\lambda)) R_T(\zeta)$$

by the resolvent identity. Set $f = \xi \circ R_T \colon \rho(T) \to \mathbb{C}$. Let $f' \colon \rho(T) \to \mathbb{C}$ be defined by $f'(\lambda) = -\xi(R_T(\lambda)^2)$ for each $\lambda \in \rho(T)$. Therefore,

$$\left|\frac{f(\lambda) - f(\zeta)}{\lambda - \zeta} - f'(\zeta)\right| = \left|\xi[(R_T(\zeta) - R_T(\lambda)) R_T(\zeta)]\right|$$

$$\leq \|\xi\| \|R_T(\zeta)\| \|R_T(\zeta) - R_T(\lambda)\|$$

so that $f: \rho(T) \to \mathbb{C}$ is analytic because $R_T: \rho(T) \to \mathcal{G}[\mathcal{X}]$ is continuous, which completes the proof of Claim 2.

Thus, by Claims 1 and 2, if $\sigma(T) = \varnothing$ (i.e., if $\rho(T) = \mathbb{C}$), then $\xi \circ R_T: \mathbb{C} \to \mathbb{C}$ is a bounded entire function, and so a constant function by the Liouville Theorem. But (see proof of Claim 1) $\|R_T(\lambda)\| \to 0$ as $|\lambda| \to \infty$, and hence $\xi(R_T(\lambda)) \to 0$ as $|\lambda| \to \infty$ (since ξ is continuous). Then $\xi \circ R_T = 0$ for all ξ in $\mathcal{B}[\mathcal{H}]^* \neq \{0\}$ and so $R_T = O$ (by the Hahn–Banach Theorem). That is, $(\lambda I - T)^{-1} = O$ for $\lambda \in \mathbb{C}$, which is a contradiction. Thus $\sigma(T) \neq \varnothing$.   □

Remark. The spectrum $\sigma(T)$ is compact and nonempty, and so is its boundary $\partial\sigma(T)$. Hence, $\partial\sigma(T) = \partial\rho(T) \neq \varnothing$.

2.2 A Classical Partition of the Spectrum

The spectrum $\sigma(T)$ of an operator T in $\mathcal{B}[\mathcal{X}]$ is the set of all scalars λ in $\mathbb{C}$ for which the operator $\lambda I - T$ fails to be an invertible element of the algebra $\mathcal{B}[\mathcal{X}]$ (i.e., fails to have a bounded inverse on $\mathcal{R}(\lambda I - T) = \mathcal{X}$). According to the nature of such a failure, $\sigma(T)$ can be split into many disjoint parts. A classical partition comprises three parts. The set $\sigma_P(T)$ of those λ for which $\lambda I - T$ has no inverse (i.e., for which the operator $\lambda I - T$ is not injective) is the *point spectrum* of T,

$$\sigma_P(T) = \{\lambda \in \mathbb{C}: \mathcal{N}(\lambda I - T) \neq \{0\}\}.$$

A scalar $\lambda \in \mathbb{C}$ is an *eigenvalue* of T if there exists a nonzero vector x in $\mathcal{X}$ such that $Tx = \lambda x$. Equivalently, λ is an *eigenvalue* of T if $\mathcal{N}(\lambda I - T) \neq \{0\}$. If $\lambda \in \mathbb{C}$ is an eigenvalue of T, then the nonzero vectors in $\mathcal{N}(\lambda I - T)$ are the *eigenvectors* of T, and $\mathcal{N}(\lambda I - T)$ is the *eigenspace* (which is a subspace of $\mathcal{X}$), associated with the eigenvalue λ. The *multiplicity of an eigenvalue* is the dimension of the respective eigenspace. Thus *the point spectrum of T is precisely the set of all eigenvalues* of T. Now consider the set $\sigma_C(T)$ of those λ for which $\lambda I - T$ has a densely defined but unbounded inverse on its range,

$$\sigma_C(T) = \{\lambda \in \mathbb{C}: \mathcal{N}(\lambda I - T) = \{0\}, \ \mathcal{R}(\lambda I - T)^- = \mathcal{X} \ \text{and} \ \mathcal{R}(\lambda I - T) \neq \mathcal{X}\}$$

(see Theorem 1.3), which is referred to as the *continuous spectrum* of T. The *residual spectrum* of T is the set $\sigma_R(T) = \sigma(T) \backslash (\sigma_P(T) \cup \sigma_C(T))$ of all scalars λ such that $\lambda I - T$ has an inverse on its range which is not densely defined:

$$\sigma_R(T) = \{\lambda \in \mathbb{C}: \mathcal{N}(\lambda I - T) = \{0\} \ \text{and} \ \mathcal{R}(\lambda I - T)^- \neq \mathcal{X}\}.$$

The collection $\{\sigma_P(T), \sigma_C(T), \sigma_R(T)\}$ is a partition of $\sigma(T)$:

$$\sigma_P(T) \cap \sigma_C(T) = \sigma_P(T) \cap \sigma_R(T) = \sigma_C(T) \cap \sigma_R(T) = \varnothing,$$

$$\sigma(T) = \sigma_P(T) \cup \sigma_C(T) \cup \sigma_R(T).$$

The diagram below, borrowed from [74], summarizes such a partition of the spectrum. The residual spectrum is split into two disjoint parts, $\sigma_R(T) = \sigma_{R_1}(T) \cup \sigma_{R_2}(T)$, and the point spectrum into four disjoint parts, $\sigma_P(T) = \bigcup_{i=1}^{4} \sigma_{P_i}(T)$. We adopt the following abbreviated notation: $T_\lambda = \lambda I - T$, $\mathcal{N}_\lambda = \mathcal{N}(T_\lambda)$, and $\mathcal{R}_\lambda = \mathcal{R}(T_\lambda)$. Recall that if $\mathcal{N}(T_\lambda) = \{0\}$, then its linear inverse T_λ^{-1} on $\mathcal{R}_\lambda$ is continuous if and only if $\mathcal{R}_\lambda$ is closed (Theorem 1.2).

		$\mathcal{R}_\lambda^- = \mathcal{X}$		$\mathcal{R}_\lambda^- \neq \mathcal{X}$	
		$\mathcal{R}_\lambda^- = \mathcal{R}_\lambda$	$\mathcal{R}_\lambda^- \neq \mathcal{R}_\lambda$	$\mathcal{R}_\lambda^- \neq \mathcal{R}_\lambda$	$\mathcal{R}_\lambda^- = \mathcal{R}_\lambda$
$\mathcal{N}_\lambda = \{0\}$	$T_\lambda^{-1} \in \mathcal{B}[\mathcal{R}_\lambda, \mathcal{X}]$	$\rho(T)$	$\varnothing$	$\varnothing$	$\sigma_{R_1}(T)$
	$T_\lambda^{-1} \notin \mathcal{B}[\mathcal{R}_\lambda, \mathcal{X}]$	$\varnothing$	$\sigma_C(T)$	$\sigma_{R_2}(T)$	$\varnothing$
$\mathcal{N}_\lambda \neq \{0\}$		$\sigma_{P_1}(T)$	$\sigma_{P_2}(T)$	$\sigma_{P_3}(T)$	$\sigma_{P_4}(T)$

$\left.\vphantom{\begin{matrix} a \\ b \\ c \end{matrix}}\right\} \sigma_{AP}(T)$

$\underbrace{\hphantom{xxxxxxxxxxxxxxxx}}_{\sigma_{CP}(T)}$

Fig. § 2.2. A classical partition of the spectrum

According to Theorem 2.2, $\sigma(T) \neq \varnothing$. But any of the above disjoint parts of the spectrum may be empty (Section 2.7). However, if $\sigma_P(T) \neq \varnothing$, then a set of eigenvectors associated with *distinct* eigenvalues is linearly independent.

Theorem 2.3. *Let* $\{\lambda_\gamma\}_{\gamma \in \Gamma}$ *be a family of distinct eigenvalues of* T. *For each* $\gamma \in \Gamma$ *let* x_γ *be an eigenvector associated with* λ_γ. *The set* $\{x_\gamma\}_{\gamma \in \Gamma}$ *is linearly independent.*

Proof. For each $\gamma \in \Gamma$ take $0 \neq x_\gamma \in \mathcal{N}(\lambda_\gamma I - T) \neq \{0\}$, and consider the set $\{x_\gamma\}_{\gamma \in \Gamma}$ (whose existence is ensured by the Axiom of Choice).

Claim. Every finite subset of $\{x_\gamma\}_{\gamma \in \Gamma}$ is linearly independent.

Proof. Take any finite subset $\{x_i\}_{i=1}^{m}$ of $\{x_\gamma\}_{\gamma \in \Gamma}$. The singleton $\{x_1\}$ is trivially linearly independent. Suppose $\{x_i\}_{i=1}^{n}$ is linearly independent for some $1 \leq n < m$. If $\{x_i\}_{i=1}^{n+1}$ is linearly dependent, then $x_{n+1} = \sum_{i=1}^{n} \alpha_i x_i$, where the family $\{\alpha_i\}_{i=1}^{n}$ of complex numbers has at least one nonzero number. Thus

$$\lambda_{n+1} x_{n+1} = T x_{n+1} = \sum_{i=1}^{n} \alpha_i T x_i = \sum_{i=1}^{n} \alpha_i \lambda_i x_i.$$

If $\lambda_{n+1} = 0$, then $\lambda_i \neq 0$ for every $i \neq n+1$ (because the eigenvalues are distinct) and $\sum_{i=1}^{n} \alpha_i \lambda_i x_i = 0$ so that $\{x_i\}_{i=1}^{n}$ is not linearly independent, which is a contradiction. If $\lambda_{n+1} \neq 0$, then $x_{n+1} = \sum_{i=1}^{n} \alpha_i \lambda_{n+1}^{-1} \lambda_i x_i$, and therefore $\sum_{i=1}^{n} \alpha_i (1 - \lambda_{n+1}^{-1} \lambda_i) x_i = 0$ so that $\{x_i\}_{i=1}^{n}$ is not linearly independent (since $\lambda_i \neq \lambda_{n+1}$ for every $i \neq n+1$ and $\alpha_i \neq 0$ for some i), which is again a contradiction. This completes the proof by induction: $\{x_i\}_{i=1}^{n+1}$ is linearly independent.

However, if every finite subset of $\{x_\gamma\}_{\gamma \in \Gamma}$ is linearly independent, then so is the set $\{x_\gamma\}_{\gamma \in \Gamma}$ itself (see, e.g., [78, Proposition 2.3]). $\square$

There are some overlapping parts of the spectrum which are commonly used: for instance, the *compression spectrum* $\sigma_{CP}(T)$ and the *approximate point spectrum* (or *approximation spectrum*) $\sigma_{AP}(T)$, which are defined by

$$\sigma_{CP}(T) = \{\lambda \in \mathbb{C} \colon \mathcal{R}(\lambda I - T) \text{ is not dense in } \mathcal{X}\}$$
$$= \sigma_{P_3}(T) \cup \sigma_{P_4}(T) \cup \sigma_R(T),$$

$$\sigma_{AP}(T) = \{\lambda \in \mathbb{C} \colon \lambda I - T \text{ is not bounded below}\}$$
$$= \sigma_P(T) \cup \sigma_C(T) \cup \sigma_{R_2}(T) = \sigma(T) \backslash \sigma_{R_1}(T).$$

The points of $\sigma_{AP}(T)$ are referred to as the *approximate eigenvalues* of T.

Theorem 2.4. *The following assertions are pairwise equivalent.*

(a) *For every $\varepsilon > 0$ there is a unit vector x_ε in $\mathcal{X}$ such that $\|(\lambda I - T)x_\varepsilon\| < \varepsilon$.*

(b) *There is a sequence $\{x_n\}$ of unit vectors in $\mathcal{X}$ such that $\|(\lambda I - T)x_n\| \to 0$.*

(c) *$\lambda \in \sigma_{AP}(T)$.*

Proof. Clearly (a) implies (b). If (b) holds, then there is no constant $\alpha > 0$ for which $\alpha = \alpha\|x_n\| \le \|(\lambda I - T)x_n\|$ for all n. Thus $\lambda I - T$ is not bounded below, and so (b) implies (c). Conversely, if $\lambda I - T$ is not bounded below, then there is no constant $\alpha > 0$ such that $\alpha\|x\| \le \|(\lambda I - T)x\|$ for all $x \in \mathcal{X}$ or, equivalently, for every $\varepsilon > 0$ there exists a nonzero y_ε in $\mathcal{X}$ for which $\|(\lambda I - T)y_\varepsilon\| < \varepsilon\|y_\varepsilon\|$. Set $x_\varepsilon = \|y_\varepsilon\|^{-1}y_\varepsilon$, and hence (c) implies (a). $\qquad\square$

Theorem 2.5. *The approximate point spectrum $\sigma_{AP}(T)$ is a nonempty closed subset of $\mathbb{C}$ that includes the boundary $\partial\sigma(T)$ of the spectrum $\sigma(T)$.*

Proof. By Theorems 2.1 and 2.2, $\rho(T)$ is nonempty and $\sigma(T)$ is nonempty and compact. Hence $\varnothing \ne \partial\sigma(T) = \partial\rho(T) = \rho(T)^- \cap \sigma(T)$. Take an arbitrary λ in $\partial\sigma(T)$. Then λ is a point of adherence of $\rho(T)$ and so there is a $\rho(T)$-valued sequence $\{\lambda_n\}$ (thus $\lambda_n I - T \in \mathcal{G}[\mathcal{X}]$) such that $\lambda_n \to \lambda$. Moreover, for each n

$$(\lambda_n I - T) - (\lambda I - T) = (\lambda_n - \lambda)I.$$

Claim. $\sup_n \|(\lambda_n I - T)^{-1}\| = \infty$.

Proof. Since $\lambda_n \in \rho(T)$, then $(\lambda_n I - T)^{-1} \in \mathcal{B}[\mathcal{X}]$. So $0 < \|(\lambda_n I - T)^{-1}\| < \infty$ for each n. Suppose $\sup_n \|(\lambda_n I - T)^{-1}\| < \infty$. Thus there is a $\beta > 0$ for which $0 < \|(\lambda_n I - T)^{-1}\| < \beta$ or, equivalently, $\beta^{-1} < \|(\lambda_n I - T)^{-1}\|^{-1} < \infty$, for all n. Since $\lambda_n \in \rho(T) \to \lambda \notin \rho(T)$, take n large enough so that $0 < |\lambda_n - \lambda| < \beta^{-1}$. Hence by the above displayed identity

$$\|(\lambda_n I - T) - (\lambda I - T)\| < \beta^{-1} < \|(\lambda_n I - T)^{-1}\|^{-1}.$$

Thus according to Proposition 1.F in Section 1.9 $\lambda I - T$ is invertible (i.e., $\lambda I - T \in \mathcal{G}[\mathcal{X}]$) and so λ lies in $\rho(T)$, which is a contradiction because λ was taken in $\partial\sigma(T) \subseteq \sigma(T)$. This completes the proof of the claimed result.

Since $\|(\lambda_n I - T)^{-1}\| = \sup_{\|y\|=1} \|(\lambda_n I - T)^{-1}y\|$, then for each positive integer n we may take a unit vector y_n in $\mathcal{X}$ such that

$$\|(\lambda_n I - T)^{-1}\| - \tfrac{1}{n} \leq \|(\lambda_n I - T)^{-1}y_n\| \leq \|(\lambda_n I - T)^{-1}\|.$$

But $\sup_n \|(\lambda_n I - T)^{-1}y_n\| = \infty$ according to the preceding claim, and therefore $\inf_n \|(\lambda_n I - T)^{-1}y_n\|^{-1} = 0$. So there exist subsequences $\{\lambda_k\}$ and $\{y_k\}$ of $\{\lambda_n\}$ and $\{y_n\}$ for which

$$\|(\lambda_k I - T)^{-1}y_k\|^{-1} \to 0.$$

Set $x_k = \|(\lambda_k I - T)^{-1}y_k\|^{-1}(\lambda_k I - T)^{-1}y_k$ and get a sequence $\{x_k\}$ of unit vectors in $\mathcal{X}$ such that $\|(\lambda_k I - T)x_k\| = \|(\lambda_k I - T)^{-1}y_k\|^{-1}$. Hence

$$\|(\lambda I - T)x_k\| = \|(\lambda_k I - T)x_k - (\lambda_k - \lambda)x_k\| \leq \|(\lambda_k I - T)^{-1}y_k\|^{-1} + |\lambda_k - \lambda|.$$

Since $\lambda_k \to \lambda$ and $\|(\lambda_k I - T)^{-1}y_k\|^{-1} \to 0$, then $\|(\lambda I - T)x_k\| \to 0$ and so $\lambda \in \sigma_{AP}(T)$ by Theorem 2.4. Thus

$$\partial\sigma(T) \subseteq \sigma_{AP}(T).$$

Then $\sigma_{AP}(T) \neq \varnothing$. Finally, take an arbitrary $\lambda \in \mathbb{C}\backslash\sigma_{AP}(T)$. By definition $\lambda I - T$ is bounded below. Hence there exists an $\alpha > 0$ for which

$$\alpha\|x\| \leq \|(\lambda I - T)x\| = \|(\lambda I - \zeta I + \zeta I - T)x\| \leq \|(\zeta I - T)x\| + \|(\lambda - \zeta)x\|$$

and so

$$(\alpha - |\lambda - \zeta|)\|x\| \leq \|(\zeta I - T)x\|$$

for every $x \in \mathcal{X}$ and every $\zeta \in \mathbb{C}$. Then $\zeta I - T$ is bounded below for every ζ such that $0 < \alpha - |\lambda - \zeta|$. Equivalently, $\zeta \in \mathbb{C}\backslash\sigma_{AP}(T)$ for every ζ sufficiently close to λ (i.e., if $|\lambda - \zeta| < \alpha$). Thus the nonempty open ball $B_\alpha(\lambda)$ centered at λ is included in $\mathbb{C}\backslash\sigma_{AP}(T)$. So $\mathbb{C}\backslash\sigma_{AP}(T)$ is open, and $\sigma_{AP}(T)$ is closed. $\square$

Remark. Therefore, since $\sigma_{AP}(T) = \sigma(T)\backslash\sigma_{R_1}(T)$ is closed in $\mathbb{C}$ and includes the common boundary $\partial\sigma(T) = \partial\rho(T)$,

$$\mathbb{C}\backslash\sigma_{R_1}(T) = \rho(T) \cup \sigma_{AP}(T) = \rho(T) \cup \partial\rho(T) \cup \sigma_{AP}(T) = \rho(T)^- \cup \sigma_{AP}(T)$$

is closed in $\mathbb{C}$. Outcome:

$$\sigma_{R_1}(T) \text{ is an open subset of } \mathbb{C}.$$

Next we assume T is an operator acting on a *nonzero complex Hilbert space* $\mathcal{H}$. The Hilbert-space structure brings forth important simplifications. A particularly useful example of such simplifications is the formula for the residual spectrum in terms of point spectra in Theorem 2.6 below. This considerably simplifies the notion of residual spectrum for Hilbert-space operators and gives a complete characterization for the spectrum of T^* and its parts in terms of the spectrum of T and its parts. First recall the following piece of notation. If Λ is any subset of $\mathbb{C}$, then set

$$\Lambda^* = \{\bar{\lambda} \in \mathbb{C} : \lambda \in \Lambda\}.$$

Clearly, $\Lambda^{**} = \Lambda$, $(\mathbb{C}\backslash\Lambda)^* = \mathbb{C}\backslash\Lambda^*$, and $(\Lambda_1 \cup \Lambda_2)^* = \Lambda_1^* \cup \Lambda_2^*$.

Theorem 2.6. *If $T^* \in \mathcal{B}[\mathcal{H}]$ is the adjoint of $T \in \mathcal{B}[\mathcal{H}]$, then*

$$\rho(T) = \rho(T^*)^*, \quad \sigma(T) = \sigma(T^*)^*, \quad \sigma_C(T) = \sigma_C(T^*)^*,$$

and the residual spectrum of T is given by the formula

$$\sigma_R(T) = \sigma_P(T^*)^*\backslash\sigma_P(T).$$

As for the subparts of the point and residual spectra,

$$\sigma_{P_1}(T) = \sigma_{R_1}(T^*)^*, \quad \sigma_{P_2}(T) = \sigma_{R_2}(T^*)^*,$$

$$\sigma_{P_3}(T) = \sigma_{P_3}(T^*)^*, \quad \sigma_{P_4}(T) = \sigma_{P_4}(T^*)^*.$$

For the compression and approximate point spectra we get

$$\sigma_{CP}(T) = \sigma_P(T^*)^*, \quad \sigma_{AP}(T^*)^* = \sigma(T)\backslash\sigma_{P_1}(T),$$

$$\partial\sigma(T) \subseteq \sigma_{AP}(T) \cap \sigma_{AP}(T^*)^* = \sigma(T)\backslash(\sigma_{P_1}(T) \cup \sigma_{R_1}(T)).$$

Proof. Since $S \in \mathcal{G}[\mathcal{H}]$ if and only if $S^* \in \mathcal{G}[\mathcal{H}]$, we get $\rho(T) = \rho(T^*)^*$. Hence $\sigma(T)^* = (\mathbb{C}\backslash\rho(T))^* = \mathbb{C}\backslash\rho(T^*) = \sigma(T^*)$. Recall: $\mathcal{R}(S)^- = \mathcal{R}(S)$ if and only if $\mathcal{R}(S^*)^- = \mathcal{R}(S^*)$, and $\mathcal{N}(S) = \{0\}$ if and only if $\mathcal{R}(S^*)^\perp = \{0\}$ (Lemmas 1.4 and 1.5) which means $\mathcal{R}(S^*)^- = \mathcal{H}$. So $\sigma_{P_1}(T) = \sigma_{R_1}(T^*)^*$, $\sigma_{P_2}(T) = \sigma_{R_2}(T^*)^*$, $\sigma_{P_3}(T) = \sigma_{P_3}(T^*)^*$, and $\sigma_{P_4}(T) = \sigma_{P_4}(T^*)^*$. Using the same argument, $\sigma_C(T) = \sigma_C(T^*)^*$ and $\sigma_{CP}(T) = \sigma_P(T^*)^*$. Hence

$$\sigma_R(T) = \sigma_{CP}(T)\backslash\sigma_P(T) \quad \text{implies} \quad \sigma_R(T) = \sigma_P(T^*)^*\backslash\sigma_P(T).$$

Moreover, since $\sigma(T^*)^* = \sigma(T)$ and $\sigma_{R_1}(T^*)^* = \sigma_{P_1}(T)$, and also recalling that $\sigma_{AP}(T) = \sigma(T)\backslash\sigma_{R_1}(T)$, we get

$$\sigma_{AP}(T^*)^* = \sigma(T^*)^*\backslash\sigma_{R_1}(T^*)^* = \sigma(T)\backslash\sigma_{P_1}(T).$$

Thus $\sigma_{AP}(T^*)^* \cap \sigma_{AP}(T) = \sigma(T)\backslash(\sigma_{P_1}(T) \cup \sigma_{R_1}(T))$. But $\sigma(T)$ is closed and $\sigma_{R_1}(T)$ is open (and so is $\sigma_{P_1}(T) = \sigma_{R_1}(T^*)^*$) in $\mathbb{C}$. Then $\sigma_{P_1}(T) \cup \sigma_{R_1}(T) \subseteq \sigma(T)^\circ$, where $\sigma(T)^\circ$ denotes the interior of $\sigma(T)$. Therefore

$$\partial\sigma(T) \subseteq \sigma(T)\backslash(\sigma_{P_1}(T) \cup \sigma_{R_1}(T)). \qquad \square$$

Remark. Since $\sigma_{P_1}(T) = \sigma_{R_1}(T^*)^*$ and since $\sigma_{R_1}(T)$ is open in $\mathbb{C}$ for every operator T (as in the previous remark), then

$$\sigma_{P_1}(T) \text{ is an open subset of } \mathbb{C}.$$

2.3 Spectral Mapping Theorems

Spectral Mapping Theorems are crucial results in spectral theory. Here we focus on the important particular case for polynomials. Further versions will be considered in Chapter 4. Let $p\colon \mathbb{C} \to \mathbb{C}$ be a polynomial with complex coefficients, take any subset Λ of $\mathbb{C}$, and consider its image under p, viz.,

$$p(\Lambda) = \{p(\lambda) \in \mathbb{C} \colon \lambda \in \Lambda\}.$$

Theorem 2.7. (SPECTRAL MAPPING THEOREM FOR POLYNOMIALS). *Take an operator $T \in \mathcal{B}[\mathcal{X}]$ on a complex Banach space $\mathcal{X}$. If p is an arbitrary polynomial with complex coefficients, then*

$$\sigma(p(T)) = p(\sigma(T)).$$

Proof. To avoid trivialities, let $p\colon \mathbb{C} \to \mathbb{C}$ be an arbitrary *nonconstant* polynomial with complex coefficients,

$$p(\lambda) = \sum\nolimits_{i=0}^{n} \alpha_i \lambda^i \ \text{ with } \ n \geq 1 \ \text{ and } \ \alpha_n \neq 0$$

for every $\lambda \in \mathbb{C}$. Take an arbitrary $\zeta \in \mathbb{C}$ and consider the factorization

$$\zeta - p(\lambda) = \beta_n \prod\nolimits_{i=1}^{n} (\lambda_i - \lambda),$$

with $\beta_n = (-1)^{n+1}\alpha_n$ where $\{\lambda_i\}_{i=1}^{n}$ are the roots of $\zeta - p(\lambda)$. Hence

$$\zeta I - p(T) = \zeta I - \sum\nolimits_{i=0}^{n} \alpha_i T^i = \beta_n \prod\nolimits_{i=1}^{n} (\lambda_i I - T).$$

If $\lambda_i \in \rho(T)$ for every $i = 1, \ldots, n$, then $\beta_n \prod_{i=1}^{n}(\lambda_i I - T) \in \mathcal{G}[\mathcal{X}]$ so that $\zeta \in \rho(p(T))$. Thus if $\zeta \in \sigma(p(T))$, then there exists $\lambda_j \in \sigma(T)$ for some $j = 1, \ldots, n$. However, λ_j is a root of $\zeta - p(\lambda)$; that is,

$$\zeta - p(\lambda_j) = \beta_n \prod\nolimits_{i=1}^{n} (\lambda_i - \lambda_j) = 0,$$

and so $p(\lambda_j) = \zeta$. Thus if $\zeta \in \sigma(p(T))$, then

$$\zeta = p(\lambda_j) \in \{p(\lambda) \in \mathbb{C} \colon \lambda \in \sigma(T)\} = p(\sigma(T))$$

because $\lambda_j \in \sigma(T)$. Therefore

$$\sigma(p(T)) \subseteq p(\sigma(T)).$$

Conversely, suppose $\zeta \in p(\sigma(T)) = \{p(\lambda) \in \mathbb{C} \colon \lambda \in \sigma(T)\}$. Then $\zeta = p(\lambda)$, and so $\zeta - p(\lambda) = 0$, for some $\lambda \in \sigma(T)$. Hence there exists λ in $\sigma(T)$ for which $\lambda = \lambda_j$ for some $j = 1, \ldots, n$. Now consider the factorization

$$\zeta I - p(T) = \beta_n \prod_{i=1}^{n} (\lambda_i I - T)$$

$$= (\lambda_j I - T) \beta_n \prod_{j \neq i = 1}^{n} (\lambda_i I - T) = \beta_n \prod_{j \neq i = 1}^{n} (\lambda_i I - T)(\lambda_j I - T),$$

which holds true since $(\lambda_j I - T)$ commutes with $(\lambda_i I - T)$ for every i. If $\zeta \in \rho(p(T))$, then $(\zeta I - p(T)) \in \mathcal{G}[\mathcal{X}]$ so that

$$(\lambda_j I - T) \left(\beta_n \prod_{j \neq i = 1}^{n} (\lambda_i I - T)(\zeta I - p(T))^{-1} \right)$$

$$= (\zeta I - p(T))(\zeta I - p(T))^{-1} = I = (\zeta I - p(T))^{-1}(\zeta I - p(T))$$

$$= \left((\zeta I - p(T))^{-1} \beta_n \prod_{j \neq i = 1}^{n} (\lambda_i I - T) \right)(\lambda_j I - T).$$

Thus $(\lambda_j I - T)$ has a right and a left inverse (i.e., it is invertible), and so $(\lambda_j I - T) \in \mathcal{G}[\mathcal{X}]$ by Theorem 1.1. Then $\lambda = \lambda_j \in \rho(T)$, which contradicts the fact that $\lambda \in \sigma(T)$. Conclusion: if $\zeta \in p(\sigma(T))$, then $\zeta \notin \rho(p(T))$. Equivalently, $\zeta \in \sigma(p(T))$. Therefore

$$p(\sigma(T)) \subseteq \sigma(p(T)). \qquad \square$$

In particular,
$$\sigma(T^n) = \sigma(T)^n \quad \text{for every } n \geq 0,$$
which means: $\zeta \in \sigma(T)^n = \{\lambda^n \in \mathbb{C} : \lambda \in \sigma(T)\}$ if and only if $\zeta \in \sigma(T^n)$. Also

$$\sigma(\alpha T) = \alpha \sigma(T) \quad \text{for every } \alpha \in \mathbb{C},$$

which means: $\zeta \in \alpha \sigma(T) = \{\alpha \lambda \in \mathbb{C} : \lambda \in \sigma(T)\}$ if and only if $\zeta \in \sigma(\alpha T)$. Now set $\Lambda^{-1} = \{\lambda^{-1} \in \mathbb{C} : \lambda \in \Lambda\}$ if $0 \notin \Lambda$. In addition, if $T \in \mathcal{G}[\mathcal{X}]$, then we get

$$\sigma(T^{-1}) = \sigma(T)^{-1}$$

(even though this is not a particular case of the Spectral Mapping Theorem for polynomials), which means: $\zeta \in \sigma(T)^{-1} = \{\lambda^{-1} \in \mathbb{C} : 0 \neq \lambda \in \sigma(T)\}$ if and only if $\zeta \in \sigma(T^{-1})$. Indeed, if $T \in \mathcal{G}[\mathcal{X}]$ (i.e., $0 \in \rho(T)$) and if $\zeta \neq 0$, then $-\zeta T^{-1}(\zeta^{-1} I - T) = \zeta I - T^{-1}$. Thus $\zeta^{-1} \in \rho(T)$ if and only if $\zeta \in \rho(T^{-1})$. Moreover, if $T \in \mathcal{B}[\mathcal{H}]$, then

$$\sigma(T^*) = \sigma(T)^*$$

by Theorem 2.6, where $\mathcal{H}$ is a complex Hilbert space and $\sigma(T)^*$ stands for the set of all complex conjugates of elements of $\sigma(T)$.

The next result is an extension of the Spectral Mapping Theorem for polynomials which holds for normal operators acting on a Hilbert space. With $\Lambda^* = \{\bar{\lambda} \in \mathbb{C} : \lambda \in \Lambda\}$ for an arbitrary subset Λ of $\mathbb{C}$, set

$$p(\Lambda, \Lambda^*) = \{p(\lambda, \overline{\lambda}) \in \mathbb{C}: \lambda \in \Lambda\}$$

where $p(\cdot, \cdot): \Lambda \times \Lambda \to \mathbb{C}$ is any polynomial in λ and $\overline{\lambda}$ of the form $p(\lambda, \overline{\lambda}) = \sum_{i,j=0}^{n,m} \alpha_{i,j} \lambda^i \overline{\lambda}^j$, which can also be viewed as a function $p(\cdot, \overline{\cdot}): \Lambda \to \mathbb{C}$ on Λ.

Theorem 2.8. (SPECTRAL MAPPING THEOREM FOR NORMAL OPERATORS). *If $T \in \mathcal{B}[\mathcal{H}]$ is normal on a complex Hilbert space $\mathcal{H}$ and $p(\cdot, \cdot): \Lambda \times \Lambda \to \mathbb{C}$ is a polynomial in λ and $\overline{\lambda}$, then*

$$\sigma(p(T, T^*)) = p(\sigma(T), \sigma(T^*)) = \{p(\lambda, \overline{\lambda}) \in \mathbb{C}: \lambda \in \sigma(T)\}.$$

Proof. Take any normal operator $T \in \mathcal{B}[\mathcal{H}]$. If $p(\lambda, \overline{\lambda}) = \sum_{i,j=0}^{n,m} \alpha_{i,j} \lambda^i \overline{\lambda}^j$, then set $p(T, T^*) = \sum_{i,j=0}^{n,m} \alpha_{i,j} T^i T^{*j}$. Let $\mathcal{P}(T, T^*)$ be the collection of all polynomials $p(T, T^*)$, which is a commutative subalgebra of the Banach algebra $\mathcal{B}[\mathcal{H}]$ since T commutes with T^*. Consider the collection $\mathcal{T}$ of all commutative subalgebras of $\mathcal{B}[\mathcal{H}]$ containing T and T^*, which is partially ordered (in the inclusion ordering) and nonempty (e.g., $\mathcal{P}(T, T^*) \in \mathcal{T}$). Moreover, every chain in $\mathcal{T}$ has an upper bound in $\mathcal{T}$ (the union of all subalgebras in a given chain of subalgebras in $\mathcal{T}$ is again a subalgebra in $\mathcal{T}$). Thus Zorn's Lemma says that $\mathcal{T}$ has a maximal element, say $\mathcal{A}(T)$. Outcome: If T is normal, then there is a maximal (thus closed) commutative subalgebra $\mathcal{A}(T)$ of $\mathcal{B}[\mathcal{H}]$ containing T and T^*. Since $\mathcal{P}(T, T^*) \subseteq \mathcal{A}(T) \in \mathcal{T}$, and since every $p(T, T^*) \in \mathcal{P}(T, T^*)$ is normal, then $\mathcal{A}(p(T, T^*)) = \mathcal{A}(T)$ for every nonconstant $p(T, T^*)$. Furthermore,

$$\Phi(p(T, T^*)) = p(\Phi(T), \Phi(T^*))$$

for every homomorphism $\Phi: \mathcal{A}(T) \to \mathbb{C}$. Thus, by Proposition 2.Q(b),

$$\sigma(p(T, T^*)) = \{p(\Phi(T), \Phi(T^*)) \in \mathbb{C}: \Phi \in \widehat{\mathcal{A}}(T)\},$$

where $\widehat{\mathcal{A}}(T)$ is the collection of all algebra homomorphisms of $\mathcal{A}(T)$ onto $\mathbb{C}$. Take a surjective homomorphism $\Phi: \mathcal{A}(T) \to \mathbb{C}$ (i.e., take any $\Phi \in \widehat{\mathcal{A}}(T)$). Consider the Cartesian decomposition $T = A + iB$, where $A, B \in \mathcal{B}[\mathcal{H}]$ are self-adjoint, and so $T^* = A - iB$ (Proposition 1.O). Thus $\Phi(T) = \Phi(A) + i\Phi(B)$ and $\Phi(T^*) = \Phi(A) - i\Phi(B)$. Since $A = \frac{1}{2}(T + T^*)$ and $B = -\frac{i}{2}(T - T^*)$ lie in $\mathcal{P}(T, T^*)$, we get $\mathcal{A}(A) = \mathcal{A}(B) = \mathcal{A}(T)$. Also, since A and B are self-adjoint, $\{\Phi(A) \in \mathbb{C}: \Phi \in \widehat{\mathcal{A}}(T)\} = \sigma(A) \subset \mathbb{R}$ and $\{\Phi(B) \in \mathbb{C}: \Phi \in \widehat{\mathcal{A}}(T)\} = \sigma(B) \subset \mathbb{R}$ (Propositions 2.A and 2.Q(b)), and so $\Phi(A) \in \mathbb{R}$ and $\Phi(B) \in \mathbb{R}$. Hence

$$\Phi(T^*) = \overline{\Phi(T)}.$$

($\overline{\Phi(T)}$ is the complex conjugate of $\Phi(T) \in \mathbb{C}$.) Thus since $\sigma(T^*) = \sigma(T)^*$ for every $T \in \mathcal{B}[\mathcal{H}]$ by Theorem 2.6, and recalling Proposition 2.Q(b), we get

$$\sigma(p(T, T^*)) = \{p(\Phi(T), \overline{\Phi(T)}) \in \mathbb{C}: \Phi \in \widehat{\mathcal{A}}(T)\}$$
$$= \{p(\lambda, \overline{\lambda}) \in \mathbb{C}: \lambda \in \{\Phi(T) \in \mathbb{C}: \Phi \in \widehat{\mathcal{A}}(T)\}\}$$
$$= \{p(\lambda, \overline{\lambda}) \in \mathbb{C}: \lambda \in \sigma(T)\}$$
$$= p(\sigma(T), \sigma(T)^*) = p(\sigma(T), \sigma(T^*)). \qquad \square$$

2.4 Spectral Radius and Normaloid Operators

The *spectral radius* of an operator $T \in \mathcal{B}[\mathcal{X}]$ on a nonzero complex Banach space $\mathcal{X}$ is the nonnegative number

$$r_\sigma(T) = \sup_{\lambda \in \sigma(T)} |\lambda| = \max_{\lambda \in \sigma(T)} |\lambda|.$$

The first identity in the above expression defines the spectral radius $r_\sigma(T)$, and the second one is a consequence of the Weierstrass Theorem (cf. proof of Theorem 2.2) since $\sigma(T) \neq \varnothing$ is compact in $\mathbb{C}$ and the function $|\cdot| \colon \mathbb{C} \to \mathbb{R}$ is continuous. A straightforward consequence of the Spectral Mapping Theorem for polynomials reads as follows.

Corollary 2.9. $r_\sigma(T^n) = r_\sigma(T)^n$ *for every* $n \geq 0$.

Proof. Take any nonnegative integer n. By Theorem 2.7, $\sigma(T^n) = \sigma(T)^n$. Thus $\zeta \in \sigma(T^n)$ if and only if $\zeta = \lambda^n$ for some $\lambda \in \sigma(T)$, and so $\sup_{\zeta \in \sigma(T^n)} |\zeta| = \sup_{\lambda \in \sigma(T)} |\lambda^n| = \sup_{\lambda \in \sigma(T)} |\lambda|^n = \left(\sup_{\lambda \in \sigma(T)} |\lambda| \right)^n$. $\square$

If $\lambda \in \sigma(T)$, then $|\lambda| \leq \|T\|$. This follows by the Neumann expansion of Theorem 1.3 (cf. proof of Theorem 2.1). Hence $r_\sigma(T) \leq \|T\|$. Therefore, for every operator $T \in \mathcal{B}[\mathcal{X}]$ and for each nonnegative integer n,

$$0 \leq r_\sigma(T^n) = r_\sigma(T)^n \leq \|T^n\| \leq \|T\|^n.$$

Thus $r_\sigma(T) \leq 1$ if T is *power bounded* (i.e., if $\sup_n \|T^n\| < \infty$). Indeed, in this case, $r_\sigma(T)^n \leq \|T^n\| \leq \sup_k \|T^k\|$ for $n \geq 1$ and $\lim_n (\sup_k \|T^k\|)^{\frac{1}{n}} = 1$. Then

$$\sup_n \|T^n\| < \infty \quad \text{implies} \quad r_\sigma(T) \leq 1.$$

Remark. If T is a *nilpotent* operator (i.e., if $T^n = O$ for some $n \geq 1$), then $r_\sigma(T) = 0$, and so $\sigma(T) = \sigma_P(T) = \{0\}$ (cf. Proposition 2.J). An operator $T \in \mathcal{B}[\mathcal{X}]$ is *quasinilpotent* if $r_\sigma(T) = 0$ (i.e., if $\sigma(T) = \{0\}$). Thus every nilpotent is quasinilpotent. Since $\sigma_P(T)$ may be empty for a quasinilpotent operator (see, e.g., Proposition 2.N), these classes are related by proper inclusion:

$$\text{Nilpotent} \quad \subset \quad \text{Quasinilpotent}.$$

The next result is the well-known Gelfand–Beurling formula for the spectral radius. Its proof requires another piece of elementary complex analysis, namely, *every analytic function has a power series representation*. That is, if $f \colon \Lambda \to \mathbb{C}$ is analytic, and if $B_{\alpha,\beta}(\zeta) = \{\lambda \in \mathbb{C} \colon 0 \leq \alpha < |\lambda - \zeta| < \beta\}$ lies in the open set $\Lambda \subseteq \mathbb{C}$ for some $0 \leq \alpha < \beta$, then f has a *unique Laurent expansion* about the point ζ, viz., $f(\lambda) = \sum_{k=-\infty}^{\infty} \gamma_k (\lambda - \zeta)^k$ for every $\lambda \in B_{\alpha,\beta}(\zeta)$.

Theorem 2.10. (GELFAND–BEURLING FORMULA).

$$r_\sigma(T) = \lim_n \|T^n\|^{\frac{1}{n}}.$$

Proof. Since $r_\sigma(T)^n \leq \|T^n\|$ for every positive integer n, and since the limit of the sequence $\{\|T^n\|^{\frac{1}{n}}\}$ exists by Lemma 1.10, we get

$$r_\sigma(T) \leq \lim_n \|T^n\|^{\frac{1}{n}}.$$

For the reverse inequality proceed as follows. Consider the Neumann expansion (Theorem 1.3) for the resolvent function $R_T \colon \rho(T) \to \mathcal{G}[\mathcal{X}]$,

$$R_T(\lambda) = (\lambda I - T)^{-1} = \lambda^{-1} \sum_{k=0}^{\infty} T^k \lambda^{-k}$$

for every $\lambda \in \rho(T)$ such that $\|T\| < |\lambda|$, where the above series converges in the (uniform) topology of $\mathcal{B}[\mathcal{X}]$. Take an arbitrary bounded linear functional $\xi \colon \mathcal{B}[\mathcal{X}] \to \mathbb{C}$ in $\mathcal{B}[\mathcal{X}]^*$ (cf. proof of Theorem 2.2). Since ξ is linear continuous,

$$\xi(R_T(\lambda)) = \lambda^{-1} \sum_{k=0}^{\infty} \xi(T^k) \lambda^{-k}$$

for every $\lambda \in \rho(T)$ such that $\|T\| < |\lambda|$.

Claim. The above displayed identity holds whenever $r_\sigma(T) < |\lambda|$.

Proof. The series $\lambda^{-1} \sum_{k=0}^{\infty} \xi(T^k) \lambda^{-k}$ is a Laurent expansion of $\xi(R_T(\lambda))$ about the origin for every $\lambda \in \rho(T)$ such that $\|T\| < |\lambda|$. But $\xi \circ R_T$ is analytic on $\rho(T)$ (cf. Claim 2 in Theorem 2.2) and so $\xi(R_T(\lambda))$ has a unique Laurent expansion about the origin for every $\lambda \in \rho(T)$, and hence for every $\lambda \in \mathbb{C}$ such that $r_\sigma(T) < |\lambda|$. Then $\xi(R_T(\lambda)) = \lambda^{-1} \sum_{k=0}^{\infty} \xi(T^k) \lambda^{-k}$, which holds whenever $r_\sigma(T) \leq \|T\| < |\lambda|$, must be the Laurent expansion about the origin for every $\lambda \in \mathbb{C}$ such that $r_\sigma(T) < |\lambda|$. This proves the claimed result.

Thus if $r_\sigma(T) < |\lambda|$, then the series of complex numbers $\sum_{k=0}^{\infty} \xi(T^k) \lambda^{-k}$ converges, and so $\xi((\lambda^{-1}T)^k) = \xi(T^k)\lambda^{-k} \to 0$, for every ξ in the dual space $\mathcal{B}[\mathcal{X}]^*$ of $\mathcal{B}[\mathcal{X}]$. This means that the $\mathcal{B}[\mathcal{X}]$-valued sequence $\{(\lambda^{-1}T)^k\}$ converges weakly. Then it is bounded (in the uniform topology of $\mathcal{B}[\mathcal{X}]$ as a consequence of the Banach–Steinhaus Theorem). Thus the operator $\lambda^{-1}T$ is power bounded. Hence $|\lambda|^{-n}\|T^n\| \leq \sup_k \|(\lambda^{-1}T)^k\| < \infty$, so that

$$|\lambda|^{-1}\|T^n\|^{\frac{1}{n}} \leq \left(\sup_k \|(\lambda^{-1}T)^k\|\right)^{\frac{1}{n}}$$

for every n. Therefore $|\lambda|^{-1} \lim_n \|T^n\|^{\frac{1}{n}} \leq 1$, and so $\lim_n \|T^n\|^{\frac{1}{n}} \leq |\lambda|$, for every $\lambda \in \mathbb{C}$ such that $r_\sigma(T) < |\lambda|$. Then $\lim_n \|T^n\|^{\frac{1}{n}} \leq r_\sigma(T) + \varepsilon$ for every $\varepsilon > 0$. Outcome:

$$\lim_n \|T^n\|^{\frac{1}{n}} \leq r_\sigma(T). \qquad \square$$

What Theorem 2.10 says is $r_\sigma(T) = r(T)$, where $r_\sigma(T)$ is the spectral radius of T and $r(T)$ is the limit of the sequence $\{\|T^n\|^{\frac{1}{n}}\}$ (whose existence was proved in Lemma 1.10). In light of this, we will adopt one and the same notation (the simpler one, of course) for both of them. So from now on we write

$$r(T) = \sup_{\lambda \in \sigma(T)} |\lambda| = \max_{\lambda \in \sigma(T)} |\lambda| = \lim_n \|T^n\|^{\frac{1}{n}}$$

for the spectral radius. A normaloid operator was defined in Section 1.6 as an operator T for which $r(T) = \|T\|$. Thus *a normaloid operator acting on a complex Banach space is precisely an operator whose norm coincides with the spectral radius*, by Theorem 2.10. Moreover, T is normaloid if and only if $\|T^n\| = \|T\|^n$ for every $n \geq 0$, by Theorem 1.11. Furthermore, since on a complex Hilbert space $\mathcal{H}$ every normal operator is normaloid, and so is every nonnegative operator, and since T^*T is always nonnegative, then

$$r(T^*T) = r(TT^*) = \|T^*T\| = \|TT^*\| = \|T\|^2 = \|T^*\|^2$$

for $T \in \mathcal{B}[\mathcal{H}]$. Further useful properties follow from Theorem 2.10. For instance,

$$r(\alpha T) = |\alpha|\, r(T) \quad \text{for every} \ \alpha \in \mathbb{C}.$$

Also $r(ST) = r(TS)$ (see Proposition 2.C). In a Hilbert space, $r(T^*) = r(T)$ and $r(Q^{\frac{1}{2}}) = r(Q)^{\frac{1}{2}}$ if $Q \geq O$ (cf. Theorem 2.6 and Proposition 2.D). Note: although $\sigma(T^{-1}) = \sigma(T)^{-1}$, in general $r(T^{-1}) \neq r(T)^{-1}$ (e.g., $T = \mathrm{diag}(1,2)$).

An important application of the Gelfand–Beurling formula is the characterization of *uniform stability* in terms of the spectral radius. A normed-space operator $T \in \mathcal{B}[\mathcal{X}]$ is *uniformly stable* if the power sequence $\{T^n\}$ converges uniformly to the null operator (notation: $T^n \xrightarrow{u} O$, i.e., if $\|T^n\| \to 0$).

Corollary 2.11. *If $T \in \mathcal{B}[\mathcal{X}]$ is an operator on a complex Banach space $\mathcal{X}$, then the following assertions are pairwise equivalent.*

(a) $T^n \xrightarrow{u} O$.

(b) $r(T) < 1$.

(c) $\|T^n\| \leq \beta \alpha^n$ *for every $n \geq 0$, for some $\beta \geq 1$ and some $\alpha \in (0,1)$.*

Proof. Since $r(T)^n = r(T^n) \leq \|T^n\|$ for each $n \geq 1$, if $\|T^n\| \to 0$ then $r(T) < 1$. Thus (a) $\Rightarrow$ (b). Suppose $r(T) < 1$ and take an arbitrary α in $(r(T), 1)$. Since $r(T) = \lim_n \|T^n\|^{\frac{1}{n}}$ (Gelfand–Beurling formula), there is an integer $n_\alpha \geq 1$ for which $\|T^n\| \leq \alpha^n$ whenever $n \geq n_\alpha$. Thus $\|T^n\| \leq \beta \alpha^n$ for every $n \geq 0$ with $\beta = \max_{0 \leq n \leq n_\alpha} \|T^n\| \alpha^{-n_\alpha}$, and so $\|T^n\| \to 0$. Thus (b) $\Rightarrow$ (c) $\Rightarrow$ (a). $\square$

A normed-space operator $T \in \mathcal{B}[\mathcal{X}]$ is *strongly stable* if the power sequence $\{T^n\}$ converges strongly to the null operator (notation: $T^n \xrightarrow{s} O$, i.e., if $\|T^n x\| \to 0$ for every $x \in \mathcal{X}$), and T is *weakly stable* if $\{T^n\}$ converges weakly to the null operator (notation: $T^n \xrightarrow{w} O$). In a Hilbert space $\mathcal{H}$ this means $\langle T^n x\,;y\rangle \to 0$ for every $x, y \in \mathcal{H}$, or $\langle T^n x\,;x\rangle \to 0$ for every $x \in \mathcal{H}$ if $\mathcal{H}$ is complex (cf. Section 1.1). Thus from what we have seen so far, in a Hilbert space,

$$r(T) < 1 \iff T^n \xrightarrow{u} O \implies$$

$$T^n \xrightarrow{s} O \implies T^n \xrightarrow{w} O \implies \sup_n \|T^n\| < \infty \implies r(T) \leq 1.$$

The converses to the above one-way implications fail in general. The next result applies the characterization of uniform stability in Corollary 2.11 to extend the Neumann expansion of Theorem 1.3.

Corollary 2.12. *Let $T \in \mathcal{B}[\mathcal{X}]$ be an operator on a complex Banach space, and let $\lambda \in \mathbb{C}$ be any nonzero complex number.*

(a) *$r(T) < |\lambda|$ if and only if $\left\{\sum_{k=0}^{n} \left(\frac{T}{\lambda}\right)^k\right\}$ converges uniformly. In this case we get $\lambda \in \rho(T)$ and $R_T(\lambda) = (\lambda I - T)^{-1} = \frac{1}{\lambda} \sum_{k=0}^{\infty} \left(\frac{T}{\lambda}\right)^k$ where $\sum_{k=0}^{\infty} \left(\frac{T}{\lambda}\right)^k$ denotes the uniform limit of $\left\{\sum_{k=0}^{n} \left(\frac{T}{\lambda}\right)^k\right\}$.*

(b) *If $r(T) = |\lambda|$ and $\left\{\sum_{k=0}^{n} \left(\frac{T}{\lambda}\right)^k\right\}$ converges strongly, then $\lambda \in \rho(T)$ and $R_T(\lambda) = (\lambda I - T)^{-1} = \frac{1}{\lambda} \sum_{k=0}^{\infty} \left(\frac{T}{\lambda}\right)^k$ where $\sum_{k=0}^{\infty} \left(\frac{T}{\lambda}\right)^k$ denotes the strong limit of $\left\{\sum_{k=0}^{n} \left(\frac{T}{\lambda}\right)^k\right\}$.*

(c) *If $|\lambda| < r(T)$, then $\left\{\sum_{k=0}^{n} \left(\frac{T}{\lambda}\right)^k\right\}$ does not converge strongly.*

Proof. If $\left\{\sum_{k=0}^{n} \left(\frac{T}{\lambda}\right)^k\right\}$ converges uniformly, then $\left(\frac{T}{\lambda}\right)^n \xrightarrow{u} O$, and therefore $|\lambda|^{-1} r(T) = r\left(\frac{T}{\lambda}\right) < 1$ by Corollary 2.11. On the other hand, if $r(T) < |\lambda|$, then $\lambda \in \rho(T)$ so that $\lambda I - T \in \mathcal{G}[\mathcal{X}]$, and also $r\left(\frac{T}{\lambda}\right) = |\lambda|^{-1} r(T) < 1$. Hence $\left\|\left(\frac{T}{\lambda}\right)^n\right\| \leq \beta \alpha^n$ for every $n \geq 0$, for some $\beta \geq 1$ and $\alpha \in (0, 1)$ according to Corollary 2.11, and so $\sum_{k=0}^{\infty} \left\|\left(\frac{T}{\lambda}\right)^k\right\| < \infty$, which means the $\mathcal{B}[\mathcal{X}]$-valued sequence $\left\{\left(\frac{T}{\lambda}\right)^n\right\}$ is absolutely summable. Now follow the steps in the proof of Theorem 1.3 to conclude the results in (a). If $\left\{\sum_{k=0}^{n} \left(\frac{T}{\lambda}\right)^k\right\}$ converges strongly, then $\left(\frac{T}{\lambda}\right)^n x \to 0$ in $\mathcal{X}$ for every $x \in \mathcal{X}$ and so $\sup_n \left\|\left(\frac{T}{\lambda}\right)^n x\right\| < \infty$ for every $x \in \mathcal{X}$. Hence $\sup_n \left\|\left(\frac{T}{\lambda}\right)^n\right\| < \infty$ (by the Banach–Steinhaus Theorem). Thus $|\lambda|^{-1} r(T) = r\left(\frac{T}{\lambda}\right) \leq 1$, which proves (c). Moreover,

$$(\lambda I - T)\frac{1}{\lambda}\sum_{k=0}^{n} \left(\frac{T}{\lambda}\right)^k = \frac{1}{\lambda}\sum_{k=0}^{n} \left(\frac{T}{\lambda}\right)^k (\lambda I - T) = I - \left(\frac{T}{\lambda}\right)^{n+1} \xrightarrow{s} I.$$

Therefore $(\lambda I - T)^{-1} = \frac{1}{\lambda} \sum_{k=0}^{\infty} \left(\frac{T}{\lambda}\right)^k$, where $\sum_{k=0}^{\infty} \left(\frac{T}{\lambda}\right)^k \in \mathcal{B}[\mathcal{X}]$ is the strong limit of $\left\{\sum_{k=0}^{n} \left(\frac{T}{\lambda}\right)^k\right\}$, which concludes the proof of (b). $\square$

2.5 Numerical Radius and Spectraloid Operators

The *numerical range* $W(T)$ of an operator $T \in \mathcal{B}[\mathcal{H}]$ acting on a nonzero complex Hilbert space $\mathcal{H}$ is the nonempty and convex set consisting of the inner products $\langle Tx\,; x\rangle$ for all unit vectors $x \in \mathcal{H}$:

$$W(T) = \left\{\lambda \in \mathbb{C}\colon \lambda = \langle Tx\,; x\rangle \text{ for some } x \text{ with } \|x\| = 1\right\}.$$

Indeed, the numerical range $W(T)$ is a convex set in $\mathbb{C}$ (see, e.g., [58, Problem 210]) which is nonempty ($W(T) \neq \varnothing$) since $\mathcal{H} \neq \{0\}$, and clearly

$$W(T^*) = W(T)^*.$$

Theorem 2.13. $\sigma_P(T) \cup \sigma_R(T) \subseteq W(T)$ *and* $\sigma(T) \subseteq W(T)^-$.

Proof. Take any operator $T \in \mathcal{B}[\mathcal{H}]$ on a nonzero complex Hilbert space $\mathcal{H}$. If $\lambda \in \sigma_P(T)$, then there is a unit vector $x \in \mathcal{H}$ such that $Tx = \lambda x$. Hence $\langle Tx\,; x \rangle = \lambda\|x\|^2 = \lambda$ and so $\lambda \in W(T)$. If $\lambda \in \sigma_R(T)$, then $\overline{\lambda} \in \sigma_P(T^*)$ by Theorem 2.6, and so $\overline{\lambda} \in W(T^*)$. Thus $\lambda \in W(T)$. Therefore

$$\sigma_P(T) \cup \sigma_R(T) \subseteq W(T).$$

If $\lambda \in \sigma_{AP}(T)$, then there is a sequence $\{x_n\}$ of unit vectors in $\mathcal{H}$ for which $\|(\lambda I - T)x_n\| \to 0$ by Theorem 2.4. Hence

$$0 \leq |\lambda - \langle Tx_n\,; x_n \rangle| = |\langle(\lambda I - T)x_n\,; x_n\rangle| \leq \|(\lambda I - T)x_n\| \to 0$$

and hence $\langle Tx_n\,; x_n \rangle \to \lambda$. Since each $\langle Tx_n\,; x_n \rangle$ lies in $W(T)$, it follows by the classical Closed Set Theorem that $\lambda \in W(T)^-$. Thus $\sigma_{AP}(T) \subseteq W(T)^-$ and so (since $\sigma_R(T) \subseteq W(T)$)

$$\sigma(T) = \sigma_R(T) \cup \sigma_{AP}(T) \subseteq W(T)^-. \qquad \square$$

The *numerical radius* of an operator $T \in \mathcal{B}[\mathcal{H}]$ on a nonzero complex Hilbert space $\mathcal{H}$ is the nonnegative number

$$w(T) = \sup_{\lambda \in W(T)} |\lambda| = \sup_{\|x\|=1} |\langle Tx\,; x \rangle|.$$

As is readily verified,

$$w(T^*) = w(T) \quad \text{and} \quad w(T^*T) = \|T\|^2.$$

Unlike the spectral radius, the numerical radius is a norm on $\mathcal{B}[\mathcal{H}]$. That is, $0 \leq w(T)$ for every $T \in \mathcal{B}[\mathcal{H}]$ and $0 < w(T)$ if $T \neq O$, $w(\alpha T) = |\alpha|w(T)$, and $w(T + S) \leq w(T) + w(S)$ for every $\alpha \in \mathbb{C}$ and every $S, T \in \mathcal{B}[\mathcal{H}]$. However, the numerical radius does not have the operator norm property. In other words, the inequality $w(ST) \leq w(S)w(T)$ is *not* true for all operators $S, T \in \mathcal{B}[\mathcal{H}]$. Nevertheless, the *power inequality* holds: $w(T^n) \leq w(T)^n$ for all $T \in \mathcal{B}[\mathcal{H}]$ and every positive integer n (see, e.g., [58, p. 118 and Problem 221]). Moreover, the numerical radius is a norm equivalent to the (induced uniform) operator norm of $\mathcal{B}[\mathcal{H}]$ and dominates the spectral radius, as in the next theorem.

Theorem 2.14. $0 \leq r(T) \leq w(T) \leq \|T\| \leq 2w(T)$.

Proof. Take any operator T in $\mathcal{B}[\mathcal{H}]$. Since $\sigma(T) \subseteq W(T)^-$ we get $r(T) \leq w(T)$. Moreover, $w(T) = \sup_{\|x\|=1} |\langle Tx\,; x \rangle| \leq \sup_{\|x\|=1} \|Tx\| = \|T\|$. Now, by the polarization identity (cf. Proposition 1.A),

$$\langle Tx\,; y \rangle = \tfrac{1}{4}\big(\langle T(x + y)\,; (x + y)\rangle - \langle T(x - y)\,; (x - y)\rangle$$
$$+ i\langle T(x + iy)\,; (x + iy)\rangle - i\langle T(x - iy)\,; (x - iy)\rangle\big)$$

for every x, y in $\mathcal{H}$. Therefore, since $|\langle Tz\,;z\rangle| \leq \sup_{\|u\|=1} |\langle Tu\,;u\rangle| \|z\|^2 = w(T)\|z\|^2$ for every $z \in \mathcal{H}$,

$$\begin{aligned}
|\langle Tx\,;y\rangle| &\leq \tfrac{1}{4}\big(|\langle T(x+y)\,;(x+y)\rangle| + |\langle T(x-y)\,;(x-y)\rangle| \\
&\quad + |\langle T(x+iy)\,;(x+iy)\rangle| + |\langle T(x-iy)\,;(x-iy)\rangle|\big) \\
&\leq \tfrac{1}{4}w(T)\big(\|x+y\|^2 + \|x-y\|^2 + \|x+iy\|^2 + \|x-iy\|^2\big).
\end{aligned}$$

for every $x, y \in \mathcal{H}$. So, by the parallelogram law (cf. Proposition 1.A),

$$|\langle Tx\,;y\rangle| \leq w(T)\big(\|x\|^2 + \|y\|^2\big) \leq 2w(T)$$

whenever $\|x\| = \|y\| = 1$. Thus, since $\|T\| = \sup_{\|x\|=\|y\|=1} |\langle Tx\,;y\rangle|$ (see, e.g., [78, Corollary 5.71]), we get $\|T\| \leq 2w(T)$. $\qquad\square$

An operator $T \in \mathcal{B}[\mathcal{H}]$ is *spectraloid* if $r(T) = w(T)$. The next result is a straightforward application of Theorem 2.14.

Corollary 2.15. *Every normaloid operator is spectraloid.*

In a normed space, an operator T is normaloid if $r(T) = \|T\|$ or, equivalently, if $\|T^n\| = \|T\|^n$ for every $n \geq 1$ by Theorem 1.11. In a Hilbert space, $r(T) = \|T\|$ implies $r(T) = w(T)$ as in Corollary 2.15. Moreover, $r(T) = \|T\|$ also implies $w(T) = \|T\|$ by Theorem 2.14 again. Thus $w(T) = \|T\|$ is a property of every normaloid operator on a Hilbert space. In fact, this can be viewed as a third definition of a normaloid operator on a complex Hilbert space.

Theorem 2.16. $T \in \mathcal{B}[\mathcal{H}]$ *is normaloid if and only if* $w(T) = \|T\|$.

Proof. Half of the proof was given above. It remains to prove the other half:

$$w(T) = \|T\| \quad \text{implies} \quad r(T) = \|T\|.$$

Suppose $w(T) = \|T\|$ (and $T \neq O$ to avoid trivialities). The closure of the numerical range $W(T)^-$ is compact in $\mathbb{C}$ (since $W(T)$ is trivially bounded). Thus $\max_{\lambda \in W(T)^-} |\lambda| = \sup_{\lambda \in W(T)^-} |\lambda| = \sup_{\lambda \in W(T)} |\lambda| = w(T) = \|T\|$, and so there exists $\lambda \in W(T)^-$ such that $|\lambda| = \|T\|$. Since $W(T)$ is nonempty, λ is a point of adherence of $W(T)$, and hence there exists a sequence $\{\lambda_n\}$ with each λ_n in $W(T)$ for which $\lambda_n \to \lambda$. This means there exists a sequence $\{x_n\}$ of unit vectors in $\mathcal{H}$ (i.e., $\|x_n\| = 1$) such that $\lambda_n = \langle Tx_n\,;x_n\rangle \to \lambda$, where $|\lambda| = \|T\| \neq 0$. Thus if $S = \lambda^{-1}T \in \mathcal{B}[\mathcal{H}]$, then

$$\langle Sx_n\,;x_n\rangle \to 1.$$

Claim. $\|Sx_n\| \to 1$ and $\operatorname{Re}\langle Sx_n\,;x_n\rangle \to 1$.

Proof. $|\langle Sx_n\,;x_n\rangle| \leq \|Sx_n\| \leq \|S\| = 1$ for each n. But $\langle Sx_n\,;x_n\rangle \to 1$ implies $|\langle Sx_n\,;x_n\rangle| \to 1$ (and so $\|Sx_n\| \to 1$) and also $\operatorname{Re}\langle Sx_n\,;x_n\rangle \to 1$ (since $|\cdot|$ and $\operatorname{Re}(\cdot)$ are continuous functions), which concludes the proof of the claim.

Then $\|(I - S)x_n\|^2 = \|Sx_n - x_n\|^2 = \|Sx_n\|^2 - 2\operatorname{Re}\langle Sx_n\,;x_n\rangle + \|x_n\|^2 \to 0$, and so $1 \in \sigma_{AP}(S) \subseteq \sigma(S)$ (cf. Theorem 2.4). Hence $r(S) \geq 1$ which implies $r(T) = r(\lambda S) = |\lambda|\, r(S) \geq |\lambda| = \|T\|$, and therefore $r(T) = \|T\|$ (because $r(T) \leq \|T\|$ for every operator T). $\qquad\square$

Remark. Take $T \in \mathcal{B}[\mathcal{H}]$. If $w(T) = 0$, then $T = O$ (since the numerical radius is a norm — also by Theorem 2.14). In particular, if $w(T) = r(T) = 0$, then $T = O$. Thus, *if an operator is spectraloid and quasinilpotent, then it is the null operator*. Therefore, *the unique normal (or hyponormal, or normaloid, or spectraloid) quasinilpotent operator is the null operator*.

Corollary 2.17. *If there exists $\lambda \in W(T)$ such that $|\lambda| = \|T\|$, then T is normaloid and $\lambda \in \sigma_P(T)$. In other words, if there exists a unit vector x such that $\|T\| = |\langle Tx\,;x\rangle|$, then $r(T) = w(T) = \|T\|$ and $\langle Tx\,;x\rangle \in \sigma_P(T)$.*

Proof. If $\lambda \in W(T)$ is such that $|\lambda| = \|T\|$, then $w(T) = \|T\|$ (Theorem 2.14) and so T is normaloid (Theorem 2.16). Moreover, since $\lambda = \langle Tx\,;x\rangle$ for some unit vector x, we get $\|T\| = |\lambda| = |\langle Tx\,;x\rangle| \leq \|Tx\|\|x\| \leq \|T\|$, and hence $|\langle Tx\,;x\rangle| = \|Tx\|\|x\|$ (the Schwarz inequality becomes an identity), which implies $Tx = \alpha x$ for some $\alpha \in \mathbb{C}$ (see, e.g., [78, Problem 5.2]). Thus $\alpha \in \sigma_P(T)$. But $\alpha = \alpha\|x\|^2 = \langle \alpha x\,;x\rangle = \langle Tx\,;x\rangle = \lambda$. $\qquad\square$

2.6 Spectrum of a Compact Operator

The spectral theory of compact operators plays a central role in the Spectral Theorem for compact normal operators of the next chapter. Normal operators were defined on Hilbert spaces; thus we keep on working with compact operators on Hilbert spaces, as we did in Section 1.8, although the spectral theory of compact operators can be equally developed on nonzero complex Banach spaces. So we assume all operators in this section acting on a nonzero complex Hilbert space $\mathcal{H}$. The main result for characterizing the spectrum of a compact operator is the Fredholm Alternative of Corollary 1.20 which, in view of the classical partition of the spectrum in Section 2.2, can be restated as follows.

Theorem 2.18. (FREDHOLM ALTERNATIVE). *Take $T \in \mathcal{B}_\infty[\mathcal{H}]$. If $\lambda \in \mathbb{C}\backslash\{0\}$, then $\lambda \in \rho(T) \cup \sigma_P(T)$. Equivalently,*

$$\sigma(T)\backslash\{0\} = \sigma_P(T)\backslash\{0\}.$$

Moreover, if $\lambda \in \mathbb{C}\backslash\{0\}$, then $\dim\mathcal{N}(\lambda I - T) = \dim\mathcal{N}(\overline{\lambda} I - T^*) < \infty$ and so $\lambda \in \rho(T) \cup \sigma_{P_4}(T)$. Equivalently,

$$\sigma(T)\backslash\{0\} = \sigma_{P_4}(T)\backslash\{0\}.$$

Proof. Take a compact operator T in $\mathcal{B}[\mathcal{H}]$ and a nonzero scalar λ in $\mathbb{C}$. By Corollary 1.20 and the diagram of Section 2.2,

$$\lambda \in \rho(T) \cup \sigma_{P_1}(T) \cup \sigma_{R_1}(T) \cup \sigma_{P_4}(T).$$

Also by Corollary 1.20, $\mathcal{N}(\lambda I - T) = \{0\}$ if and only if $\mathcal{N}(\overline{\lambda} I - T^*) = \{0\}$, and so $\lambda \in \sigma_P(T)$ if and only if $\overline{\lambda} \in \sigma_P(T^*)$. Thus $\lambda \notin \sigma_{P_1}(T) \cup \sigma_{R_1}(T)$ by Theorem 2.6, and hence $\lambda \in \rho(T) \cup \sigma_{P_4}(T)$ or, equivalently, $\lambda \in \rho(T) \cup \sigma_P(T)$ (since $\lambda \notin \sigma_{P_1}(T) \cup \sigma_{P_2}(T) \cup \sigma_{P_3}(T)$). Therefore

$$\sigma(T)\backslash\{0\} = \sigma_P(T)\backslash\{0\} = \sigma_{P_4}(T)\backslash\{0\}. \qquad \square$$

The scalar 0 may be anywhere. In other words, if $T \in \mathcal{B}_\infty[\mathcal{H}]$, then $\lambda = 0$ may lie in $\sigma_P(T)$, $\sigma_R(T)$, $\sigma_C(T)$, or $\rho(T)$. However, if T is a compact operator on a nonzero space $\mathcal{H}$ and $0 \in \rho(T)$, then $\mathcal{H}$ must be finite-dimensional. Indeed, if $0 \in \rho(T)$, then $T^{-1} \in \mathcal{B}[\mathcal{H}]$ and so $I = T^{-1}T$ is compact (since $\mathcal{B}_\infty[\mathcal{H}]$ is an ideal of $\mathcal{B}[\mathcal{H}]$), which forces $\mathcal{H}$ to be finite-dimensional (cf. Proposition 1.Y in Section 1.9). The preceding theorem in fact is a rewriting of the Fredholm Alternative (and it is also referred to as the Fredholm Alternative). It will be applied often from now on. Here is a first application. Let $\mathcal{B}_0[\mathcal{H}]$ denote the class of all *finite-rank operators* on $\mathcal{H}$ (i.e., the class of all operators from $\mathcal{B}[\mathcal{H}]$ with a finite-dimensional range). Recall: $\mathcal{B}_0[\mathcal{H}] \subseteq \mathcal{B}_\infty[\mathcal{H}]$ (finite-rank operators are compact — cf. Proposition 1.X in Section 1.9). Let $\#A$ stand for the cardinality of a set A. Thus $\#A < \infty$ means "A is a finite set".

Corollary 2.19. *If $T \in \mathcal{B}_0[\mathcal{H}]$, then*

$$\sigma(T) = \sigma_P(T) = \sigma_{P_4}(T) \qquad and \qquad \#\sigma(T) < \infty.$$

Proof. In a finite-dimensional normed space an operator is injective if and only if it is surjective (see, e.g., [78, Problem 2.18]), and linear manifolds are closed (see, e.g., [78, Corollary 4.29]). So $\sigma(T) = \sigma_P(T) = \sigma_{P_4}(T)$ by the diagram of Section 2.2 (since $\sigma_{P_1}(T) = \sigma_{R_1}(T^*)^*$ according to Theorem 2.6). On the other hand, suppose $\dim \mathcal{H} = \infty$. Since $\mathcal{B}_0[\mathcal{H}] \subseteq \mathcal{B}_\infty[\mathcal{H}]$ we get, by Theorem 2.18,

$$\sigma(T)\backslash\{0\} = \sigma_P(T)\backslash\{0\} = \sigma_{P_4}(T)\backslash\{0\}.$$

Since $\dim \mathcal{R}(T) < \infty$ and $\dim \mathcal{H} = \infty$, then $\mathcal{R}(T)^- = \mathcal{R}(T) \neq \mathcal{H}$ and $\mathcal{N}(T) \neq \{0\}$ (by the *rank and nullity identity*: $\dim \mathcal{N}(T) + \dim \mathcal{R}(T) = \dim \mathcal{H}$ — see, e.g., [78, Problem 2.17]). Then $0 \in \sigma_{P_4}(T)$ (cf. diagram of Section 2.2) and so

$$\sigma(T) = \sigma_P(T) = \sigma_{P_4}(T).$$

If $\sigma_P(T)$ is an infinite set, then there is an infinite set of linearly independent eigenvectors of T (Theorem 2.3). Since every eigenvector of T lies in $\mathcal{R}(T)$, this implies $\dim \mathcal{R}(T) = \infty$ (because every linearly independent subset of a linear space is included in some Hamel basis — see, e.g., [78, Theorem 2.5]), which is a contradiction. Conclusion: $\sigma_P(T)$ must be a finite set. $\qquad \square$

In particular, the above result clearly holds if $\mathcal{H}$ is finite-dimensional since, as we saw above, $\dim \mathcal{H} < \infty$ implies $\mathcal{B}[\mathcal{H}] = \mathcal{B}_0[\mathcal{H}]$.

Corollary 2.20. *Take an arbitrary compact operator $T \in \mathcal{B}_\infty[\mathcal{H}]$.*

(a) *An infinite sequence of distinct points of $\sigma(T)$ converges to zero.*

(b) *0 is the only possible accumulation point of $\sigma(T)$.*

(c) *If $\lambda \in \sigma(T)\backslash\{0\}$, then λ is an isolated point of $\sigma(T)$.*

(d) *$\sigma(T)\backslash\{0\}$ is a discrete subset of $\mathbb{C}$.*

(e) *$\sigma(T)$ is countable.*

Proof. Let T be a compact operator on $\mathcal{H}$.

(a) Let $\{\lambda_n\}_{n=1}^\infty$ be an infinite sequence of distinct points in the spectrum $\sigma(T)$. Without loss of generality, suppose every λ_n is nonzero. Since T is compact and $0 \neq \lambda_n \in \sigma(T)$, we get $\lambda_n \in \sigma_P(T)$ by Theorem 2.18. Let $\{x_n\}_{n=1}^\infty$ be a sequence of eigenvectors associated with the eigenvalues $\{\lambda_n\}_{n=1}^\infty$ (i.e., $Tx_n = \lambda_n x_n$ with each $x_n \neq 0$). This is a sequence of linearly independent vectors by Theorem 2.3. For each $n \geq 1$, set

$$\mathcal{M}_n = \mathrm{span}\{x_i\}_{i=1}^n,$$

which is a subspace of $\mathcal{H}$ with $\dim \mathcal{M}_n = n$, and

$$\mathcal{M}_n \subset \mathcal{M}_{n+1}$$

for every $n \geq 1$ (because $\{x_i\}_{i=1}^{n+1}$ is linearly independent and so x_{n+1} lies in $\mathcal{M}_{n+1}\backslash\mathcal{M}_n$). From now on the argument is similar to the argument in the proof of Theorem 1.18. Since each $\mathcal{M}_n$ is a proper subspace of the Hilbert space $\mathcal{M}_{n+1}$, there exists a vector y_{n+1} in $\mathcal{M}_{n+1}$ with $\|y_{n+1}\| = 1$ for which $\frac{1}{2} < \inf_{u \in \mathcal{M}_n} \|y_{n+1} - u\|$. Writing $y_{n+1} = \sum_{i=1}^{n+1} \alpha_i x_i$ in $\mathcal{M}_{n+1}$ we get

$$(\lambda_{n+1}I - T)y_{n+1} = \sum_{i=1}^{n+1} \alpha_i(\lambda_{n+1} - \lambda_i)x_i = \sum_{i=1}^n \alpha_i(\lambda_{n+1} - \lambda_i)x_i \in \mathcal{M}_n.$$

Since $\lambda_n \neq 0$ for every n, take any pair of integers $1 \leq m < n$ and set

$$y = y_m - \lambda_m^{-1}(\lambda_m I - T)y_m + \lambda_n^{-1}(\lambda_n I - T)y_n,$$

so that $T(\lambda_m^{-1}y_m) - T(\lambda_n^{-1}y_n) = y - y_n$. Since y lies in $\mathcal{M}_{n-1}$,

$$\tfrac{1}{2} < \|y - y_n\| = \|T(\lambda_m^{-1}y_m) - T(\lambda_n^{-1}y_n)\|.$$

Thus the sequence $\{T(\lambda_n^{-1}y_n)\}$ has no convergent subsequence. Then, since T is compact, $\{\lambda_n^{-1}y_n\}$ has no bounded subsequence by Proposition 1.S. Hence $\sup_k |\lambda_k|^{-1} = \sup_k \|\lambda_k^{-1}y_k\| = \infty$, and so $\inf_k |\lambda_k| = 0$, for every subsequence $\{\lambda_k\}_{k=1}^\infty$ of $\{\lambda_n\}_{n=1}^\infty$. Therefore $\lambda_n \to 0$.

(b) Thus if $\lambda \neq 0$, then there is no sequence of distinct points in $\sigma(T)$ converging to λ. So $\lambda \neq 0$ is not an accumulation point of $\sigma(T)$.

(c) Therefore every λ in $\sigma(T)\backslash\{0\}$ is not an accumulation point of $\sigma(T)$. Equivalently, every λ in $\sigma(T)\backslash\{0\}$ is an isolated point of $\sigma(T)$.

(d) Hence $\sigma(T)\backslash\{0\}$ consists entirely of isolated points. In other words, $\sigma(T)\backslash\{0\}$ is a discrete subset of $\mathbb{C}$.

(e) Since *a discrete subset of a separable metric space is countable* (see, e.g., [78, Example 3.Q]), and since $\mathbb{C}$ is separable, $\sigma(T)\backslash\{0\}$ is countable. $\square$

Corollary 2.21. *If an operator* $T \in \mathcal{B}[\mathcal{H}]$ *is compact and normaloid, then* $\sigma_P(T) \neq \varnothing$ *and there exists* $\lambda \in \sigma_P(T)$ *such that* $|\lambda| = \|T\|$.

Proof. Suppose T is normaloid (i.e., $r(T) = \|T\|$). Thus $\sigma(T) = \{0\}$ only if $T = O$. If $T = O$ and $\mathcal{H} \neq \{0\}$, then $0 \in \sigma_P(T)$ and $\|T\| = 0$. If $T \neq O$, then $\sigma(T) \neq \{0\}$ and $\|T\| = r(T) = \max_{\lambda \in \sigma(T)} |\lambda|$. Thus there exists $\lambda \neq 0$ in $\sigma(T)$ such that $|\lambda| = \|T\|$. Moreover, if T is compact and $\sigma(T) \neq \{0\}$, then $\varnothing \neq \sigma(T)\backslash\{0\} \subseteq \sigma_P(T)$ by Theorem 2.18. Hence $r(T) = \max_{\lambda \in \sigma(T)} |\lambda| = \max_{\lambda \in \sigma_P(T)} |\lambda| = \|T\|$. So there exists $\lambda \in \sigma_P(T)$ with $|\lambda| = \|T\|$. $\square$

Corollary 2.22. *Every compact hyponormal operator is normal.*

Proof. Suppose $T \in \mathcal{B}[\mathcal{H}]$ is a compact hyponormal operator on a nonzero complex Hilbert space $\mathcal{H}$. By Corollary 2.21, $\sigma_P(T) \neq \varnothing$. Consider the subspace $\mathcal{M} = \left(\sum_{\lambda \in \sigma_P(T)} \mathcal{N}(\lambda I - T)\right)^-$ of Theorem 1.16 with $\{\lambda_\gamma\}_{\gamma \in \Gamma} = \sigma_P(T)$. Hence $\sigma_P(T|_{\mathcal{M}^\perp}) = \varnothing$. In fact, if there is a $\lambda \in \sigma_P(T|_{\mathcal{M}^\perp})$, then there exists $0 \neq x \in \mathcal{M}^\perp$ for which $\lambda x = T|_{\mathcal{M}^\perp} x = Tx$, and so $x \in \mathcal{N}(\lambda I - T) \subseteq \mathcal{M}$, which is a contradiction. Moreover, $T|_{\mathcal{M}^\perp}$ is compact and hyponormal (Propositions 1.O and 1.U). Thus if $\mathcal{M}^\perp \neq \{0\}$, then $\sigma_P(T|_{\mathcal{M}^\perp}) \neq \varnothing$ by Corollary 2.21, which is another contradiction. Therefore $\mathcal{M}^\perp = \{0\}$ and so $\mathcal{M} = \mathcal{H}$ (see Section 1.3). Then $T = T|_{\mathcal{H}} = T|_{\mathcal{M}}$ is normal by Theorem 1.16. $\square$

2.7 Supplementary Propositions

The residual spectrum of a normal operator on a Hilbert space is empty. This also happens for cohyponormal operators, as we will see in Proposition 2.A below. Such a result is a consequence of the inclusion $\mathcal{N}(\lambda I - T) \subseteq \mathcal{N}(\bar{\lambda} I - T^*)$ which holds for every hyponormal operator as in Lemma 1.13(a). So the identity $\mathcal{N}(\lambda I - T) = \mathcal{N}(\bar{\lambda} I - T^*)$ holds for every normal operator as in Lemma 1.13(b). This makes a difference as far as the Spectral Theorem for compact normal operators (next chapter) is concerned. Indeed, the inclusion in Lemma 1.13(a) is enough for proving Theorem 1.16, which in turn plays a crucial role in the Spectral Theorem for compact normal operators (Theorem 3.3, Chapter 3). However, every compact hyponormal operator is normal, as we saw above in Corollary 2.22 (and so the inclusion becomes an identity when applied to the Spectral Theorem for compact normal operators). An eigenvalue for any Hilbert-space operator satisfying the inclusion $\mathcal{N}(\lambda I - T) \subseteq \mathcal{N}(\bar{\lambda} I - T^*)$ is referred to as a *normal eigenvalue*.

Proposition 2.A. *Let $\mathcal{H} \neq \{0\}$ be a complex Hilbert space and let $\mathbb{T}$ denote the unit circle about the origin of the complex plane.*

(a) *If $H \in \mathcal{B}[\mathcal{H}]$ is hyponormal, then $\sigma_P(H)^* \subseteq \sigma_P(H^*)$ and $\sigma_R(H^*) = \varnothing$.*

(b) *If $N \in \mathcal{B}[\mathcal{H}]$ is normal, then $\sigma_P(N^*) = \sigma_P(N)^*$ and $\sigma_R(N) = \varnothing$.*

(c) *If $U \in \mathcal{B}[\mathcal{H}]$ is unitary, then $\sigma(U) \subseteq \mathbb{T}$.*

(d) *If $A \in \mathcal{B}[\mathcal{H}]$ is self-adjoint, then $\sigma(A) \subset \mathbb{R}$.*

(e) *If $Q \in \mathcal{B}[\mathcal{H}]$ is nonnegative, then $\sigma(Q) \subset [0, \infty)$.*

(f) *If $R \in \mathcal{B}[\mathcal{H}]$ is strictly positive, then $\sigma(R) \subset [\alpha, \infty)$ for some $\alpha > 0$.*

(g) *If $E \in \mathcal{B}[\mathcal{H}]$ is a nontrivial projection, then $\sigma(E) = \sigma_P(E) = \{0, 1\}$.*

Proposition 2.B. *Similarity preserves the spectrum and its parts, and so it preserves the spectral radius. That is, let $\mathcal{X}$ and $\mathcal{Y}$ be nonzero complex Banach spaces. For every $T \in \mathcal{B}[\mathcal{X}]$ and $W \in \mathcal{G}[\mathcal{X}, \mathcal{Y}]$,*

(a) $\sigma_P(T) = \sigma_P(WTW^{-1})$,

(b) $\sigma_R(T) = \sigma_R(WTW^{-1})$,

(c) $\sigma_C(T) = \sigma_C(WTW^{-1})$.

So $\sigma(T) = \sigma(WTW^{-1})$, $\rho(T) = \rho(WTW^{-1})$, and $r(T) = r(WTW^{-1})$. If W is an isometric isomorphism (in particular, a unitary operator), then the norm is also preserved: if $W \in \mathcal{G}[\mathcal{X}, \mathcal{Y}]$ is an isometry, then $\|T\| = \|WTW^{-1}\|$.

Proposition 2.C. $\sigma(ST) \backslash \{0\} = \sigma(TS) \backslash \{0\}$ *for every $S, T \in \mathcal{B}[\mathcal{X}]$.*

Proposition 2.D. *If $Q \in \mathcal{B}[\mathcal{H}]$ is nonnegative, then $\sigma(Q^{\frac{1}{2}}) = \sigma(Q)^{\frac{1}{2}}$.*

Proposition 2.E. *Take an arbitrary operator $T \in \mathcal{B}[\mathcal{X}]$ on a Banach space. Let d denote the usual distance in $\mathbb{C}$. If $\lambda \in \rho(T)$, then*

$$r((\lambda I - T)^{-1}) = [d(\lambda, \sigma(T))]^{-1}.$$

If $T \in \mathcal{B}[\mathcal{H}]$ on a Hilbert space is hyponormal and $\lambda \in \rho(T)$, then

$$\|(\lambda I - T)^{-1}\| = [d(\lambda, \sigma(T))]^{-1}.$$

Proposition 2.F. *Let $\{\mathcal{H}_k\}$ be a collection of Hilbert spaces, let $\{T_k\}$ be a (similarly indexed) bounded family of operators T_k in $\mathcal{B}[\mathcal{H}_k]$ (i.e., $\sup_k \|T_k\| < \infty$), and take the (orthogonal) direct sum $\bigoplus_k T_k$ in $\mathcal{B}[\bigoplus_k \mathcal{H}_k]$. Then*

(a) $\sigma_P(\bigoplus_k T_k) = \bigcup_k \sigma_P(T_k)$,

(b) $\sigma(\bigoplus_k T_k) = \bigcup_k \sigma(T_k)$ *if the collection $\{T_k\}$ is finite.*

In general (if the collection $\{T_k\}$ is not finite), then

(c) $(\bigcup_k \sigma(T_k))^- \subseteq \sigma(\bigoplus_k T_k)$, *and the inclusion may be proper.*

However, if $\|(\lambda I - T_k)^{-1}\| = [d(\lambda, \sigma(T_k))]^{-1}$ *for each* k *and every* $\lambda \in \rho(T_k)$,

(d) $(\bigcup_k \sigma(T_k))^- = \sigma(\bigoplus_k T_k)$, *which happens whenever each* T_k *is hyponormal.*

Proposition 2.G. *Let* $T \in \mathcal{B}[\mathcal{X}]$ *be an operator on a complex Banach space.*

(a) T *is normaloid if and only if there is a* $\lambda \in \sigma(T)$ *such that* $|\lambda| = \|T\|$.

(b) *If* T *is compact and normaloid, then* $\sigma_P(T) \neq \varnothing$ *and there is a* $\lambda \in \sigma_P(T)$ *such that* $|\lambda| = \|T\|$.

Proposition 2.H. *Let* $\mathcal{H}$ *be a complex Hilbert space.*

(a) $\sigma_R(T) \subseteq \{\lambda \in \mathbb{C} \colon |\lambda| < \|T\|\}$ *for every operator* $T \in \mathcal{B}[\mathcal{H}]$,

which is particularly relevant if T *is normaloid.*

(b) $\sigma_R(T) \subseteq \{\lambda \in \mathbb{C} \colon |\lambda| < 1\}$ *if* $T \in \mathcal{B}[\mathcal{H}]$ *is power bounded.*

Proposition 2.I. *If* $\mathcal{H}$ *and* $\mathcal{K}$ *are complex Hilbert spaces and* $T \in \mathcal{B}[\mathcal{H}]$, *then*

$$r(T) = \inf_{W \in \mathcal{G}[\mathcal{H},\mathcal{K}]} \|W T W^{-1}\| = \inf_{M \in \mathcal{G}[\mathcal{H}]} \|M T M^{-1}\| = \inf_{Q \in \mathcal{G}[\mathcal{H}]^+} \|Q T Q^{-1}\|,$$

where $\mathcal{G}[\mathcal{H}]^+$ *is the class of invertible nonnegative operators (cf. Section 1.5).*

　　An operator is uniformly stable if and only if it is similar to a strict contraction (by the spectral radius expression in the preceding proposition).

Proposition 2.J. *If* $T \in \mathcal{B}[\mathcal{X}]$ *is a nilpotent operator on a complex Banach space, then* $\sigma(T) = \sigma_P(T) = \{0\}$.

Proposition 2.K. *An operator* $T \in \mathcal{B}[\mathcal{H}]$ *on a complex Hilbert space is spectraloid if and only if* $w(T^n) = w(T)^n$ *for every* $n \geq 0$.

　　An operator $T \in \mathcal{B}[\mathcal{H}]$ on a separable infinite-dimensional Hilbert space $\mathcal{H}$ is *diagonalizable* if $Tx = \sum_{k=0}^{\infty} \alpha_k \langle x \,; e_k \rangle e_k$ for every $x \in \mathcal{H}$, for some orthonormal basis $\{e_k\}_{k=0}^{\infty}$ for $\mathcal{H}$ and some bounded sequence $\{\alpha_k\}_{k=0}^{\infty}$ of scalars.

Proposition 2.L. *If* $T \in \mathcal{B}[\mathcal{H}]$ *is diagonalizable and* $\mathcal{H}$ *is complex, then*

$$\sigma_P(T) = \{\lambda \in \mathbb{C} \colon \lambda = \alpha_k \text{ for some } k \geq 0\}, \qquad \sigma_R(T) = \varnothing, \qquad and$$

$$\sigma_C(T) = \{\lambda \in \mathbb{C} \colon \inf_k |\lambda - \alpha_k| = 0 \text{ and } \lambda \neq \alpha_k \text{ for all } k \geq 0\}.$$

　　An operator $S_+ \in \mathcal{B}[\mathcal{K}_+]$ on a Hilbert space $\mathcal{K}_+$ is a *unilateral shift*, and an operator $S \in \mathcal{B}[\mathcal{K}]$ on a Hilbert space $\mathcal{K}$ is a *bilateral shift*, if there exists an infinite sequence $\{\mathcal{H}_k\}_{k=0}^{\infty}$ and an infinite family $\{\mathcal{H}_k\}_{k=-\infty}^{\infty}$ of nonzero pairwise orthogonal subspaces of $\mathcal{K}_+$ and $\mathcal{K}$ such that $\mathcal{K}_+ = \bigoplus_{k=0}^{\infty} \mathcal{H}_k$ and $\mathcal{K} = \bigoplus_{k=-\infty}^{\infty} \mathcal{H}_k$ (cf. Section 1.3), and both S_+ and S map each $\mathcal{H}_k$ isometrically onto $\mathcal{H}_{k+1}$, so that each transformation $U_{+(k+1)} = S_+|_{\mathcal{H}_k} \colon \mathcal{H}_k \to \mathcal{H}_{k+1}$, and each transformation $U_{k+1} = S|_{\mathcal{H}_k} \colon \mathcal{H}_k \to \mathcal{H}_{k+1}$, is unitary, and therefore $\dim \mathcal{H}_{k+1} = \dim \mathcal{H}_k$. Such a common dimension is the *multiplicity of S_+ and S*. If $\mathcal{H}_k = \mathcal{H}$ for all k, then $\mathcal{K}_+ = \ell_+^2(\mathcal{H}) = \bigoplus_{k=0}^{\infty} \mathcal{H}$ and $\mathcal{K} = \ell^2(\mathcal{H}) = \bigoplus_{k=-\infty}^{\infty} \mathcal{H}$ are the direct orthogonal sums of countably infinite copies of a single nonzero Hilbert space $\mathcal{H}$, indexed either by the nonnegative integers or by all integers, which are precisely the Hilbert spaces consisting of all square-summable $\mathcal{H}$-valued sequences $\{x_k\}_{k=0}^{\infty}$ and of all square-summable $\mathcal{H}$-valued families $\{x_k\}_{k=-\infty}^{\infty}$. In this case (if $\mathcal{H}_k = \mathcal{H}$ for all k), $U_{+(k+1)} = S_+|_{\mathcal{H}} = U_+$ and $U_{k+1} = S|_{\mathcal{H}} = U$ for all k, where U_+ and U are any unitary operators on $\mathcal{H}$. In particular, if $U_+ = U = I$, the identity on $\mathcal{H}$, then S_+ and S are referred to as the *canonical* unilateral and bilateral shifts on $\ell_+^2(\mathcal{H})$ and on $\ell^2(\mathcal{H})$. The adjoint $S_+^* \in \mathcal{B}[\mathcal{K}_+]$ of $S_+ \in \mathcal{B}[\mathcal{K}_+]$ and the adjoint of $S^* \in \mathcal{B}[\mathcal{K}]$ of $S \in \mathcal{B}[\mathcal{K}]$ are referred to as a *backward unilateral shift* and as a *backward bilateral shift*. Writing $\bigoplus_{k=0}^{\infty} x_k$ for $\{x_k\}_{k=0}^{\infty}$ in $\bigoplus_{k=0}^{\infty} \mathcal{H}_k$, and $\bigoplus_{k=-\infty}^{\infty} x_k$ for $\{x_k\}_{k=-\infty}^{\infty}$ in $\bigoplus_{k=-\infty}^{\infty} \mathcal{H}_k$, it follows that $S_+ \colon \mathcal{K}_+ \to \mathcal{K}_+$ and $S_+^* \colon \mathcal{K}_+ \to \mathcal{K}_+$, and $S \colon \mathcal{K} \to \mathcal{K}$ and $S^* \colon \mathcal{K} \to \mathcal{K}$, are given by the formulas

$$S_+ x = 0 \oplus \bigoplus_{k=1}^{\infty} U_{+(k)} x_{k-1} \quad \text{and} \quad S_+^* x = \bigoplus_{k=0}^{\infty} U_{+(k+1)}^* x_{k+1}$$

for all $x = \bigoplus_{k=0}^{\infty} x_k$ in $\mathcal{K}_+ = \bigoplus_{k=0}^{\infty} \mathcal{H}_k$, with 0 being the origin of $\mathcal{H}_0$, where $U_{+(k+1)}$ is any unitary transformation of $\mathcal{H}_k$ onto $\mathcal{H}_{k+1}$ for each $k \geq 0$, and

$$Sx = \bigoplus_{k=-\infty}^{\infty} U_k x_{k-1} \quad \text{and} \quad S^* x = \bigoplus_{k=\infty}^{\infty} U_{k+1}^* x_{k+1}$$

for all $x = \bigoplus_{k=-\infty}^{\infty} x_k$ in $\mathcal{K} = \bigoplus_{k=-\infty}^{\infty} \mathcal{H}_k$, where, for each integer k, U_{k+1} is any unitary transformation of $\mathcal{H}_k$ onto $\mathcal{H}_{k+1}$. The spectrum of a bilateral shift is simpler than that of a unilateral shift, for bilateral shifts are unitary operators (i.e., besides being isometries as unilateral shifts are, bilateral shifts are normal too). For a full treatment of shifts on Hilbert spaces, see [56].

Proposition 2.M. *Let $\mathbb{D}$ and $\mathbb{T} = \partial \mathbb{D}$ denote the open unit disk and the unit circle about the origin of the complex plane, respectively. If $S_+ \in \mathcal{B}[\mathcal{K}_+]$ is a unilateral shift and $S \in \mathcal{B}[\mathcal{K}]$ is a bilateral shift on complex spaces, then*

(a) $\sigma_P(S_+) = \sigma_R(S_+^*) = \varnothing, \quad \sigma_R(S_+) = \sigma_P(S_+^*) = \mathbb{D}, \quad \sigma_C(S_+) = \sigma_C(S_+^*) = \mathbb{T}.$

(b) $\sigma(S) = \sigma(S^*) = \sigma_C(S^*) = \sigma_C(S) = \mathbb{T}.$

A *unilateral weighted shift* $T_+ = S_+ D_+$ in $\mathcal{B}[\ell_+^2(\mathcal{H})]$ is the product of a canonical unilateral shift S_+ and a diagonal operator $D_+ = \bigoplus_{k=0}^{\infty} \alpha_k I$, both in $\mathcal{B}[\ell_+^2(\mathcal{H})]$, where $\{\alpha_k\}_{k=0}^{\infty}$ is a bounded sequence of scalars. A *bilateral weighted shift* $T = SD$ in $\mathcal{B}[\ell^2(\mathcal{H})]$ is the product of a canonical bilateral shift S and a diagonal operator $D = \bigoplus_{k=-\infty}^{\infty} \alpha_k I$, both in $\mathcal{B}[\ell^2(\mathcal{H})]$, where $\{\alpha_k\}_{k=-\infty}^{\infty}$ is a bounded family of scalars.

Proposition 2.N. *Let $T_+ \in \mathcal{B}[\ell_+^2(\mathcal{H})]$ be a unilateral weighted shift and let $T \in \mathcal{B}[\ell^2(\mathcal{H})]$ be a bilateral weighted shift where $\mathcal{H}$ is a complex Hilbert space.*

(a) *If* $\alpha_k \to 0$ *as* $|k| \to \infty$, *then* T_+ *and* T *are compact and quasinilpotent.*
If, in addition, $\alpha_k \neq 0$ *for all* k, *then*

(b) $\sigma(T_+) = \sigma_R(T_+) = \sigma_{R_2}(T_+) = \{0\}$ *and* $\sigma(T_+^*) = \sigma_P(T_+^*) = \sigma_{P_2}(T_+^*) = \{0\}$,

(c) $\sigma(T) = \sigma_C(T) = \sigma_C(T^*) = \sigma(T^*) = \{0\}$.

Proposition 2.O. *Let* $T \in \mathcal{B}[\mathcal{H}]$ *be an operator on a complex Hilbert space and let* $\mathbb{D}$ *be the open unit disk about the origin of the complex plane.*

(a) *If* $T^n \xrightarrow{w} O$, *then* $\sigma_P(T) \subseteq \mathbb{D}$.

(b) *If* T *is compact and* $\sigma_P(T) \subseteq \mathbb{D}$, *then* $T^n \xrightarrow{u} O$.

(c) *The concepts of weak, strong, and uniform stability coincide for a compact operator on a complex Hilbert space.*

Proposition 2.P. *Take an operator* $T \in \mathcal{B}[\mathcal{H}]$ *on a complex Hilbert space and let* $\mathbb{T} = \partial\mathbb{D}$ *be the unit circle about the origin of the complex plane.*

(a) $T^n \xrightarrow{u} O$ *if and only if* $T^n \xrightarrow{w} O$ *and* $\sigma_C(T) \cap \mathbb{T} = \varnothing$.

(b) *If the continuous spectrum does not intersect the unit circle, then the concepts of weak, strong, and uniform stability coincide.*

The concepts of resolvent set $\rho(T)$ and spectrum $\sigma(T)$ of an operator T in the unital complex Banach algebra $\mathcal{B}[\mathcal{X}]$ introduced in Section 2.1, viz., $\rho(T) = \{\lambda \in \mathbb{C} \colon \lambda I - T$ has an inverse in $\mathcal{B}[\mathcal{X}]\}$ and $\sigma(T) = \mathbb{C}\backslash\rho(T)$, of course, hold in any unital complex Banach algebra $\mathcal{A}$, as does the concept of spectral radius $r(T) = \sup_{\lambda \in \sigma(T)} |\lambda|$, in relation to which the Gelfand–Beurling formula of Theorem 2.10, namely, $r(T) = \lim_n \|T^n\|^{\frac{1}{n}}$, holds in any unital complex Banach algebra (the proof being essentially the same as that of Theorem 2.10). A *component* of a set in a topological space is any maximal (in the inclusion ordering) connected subset of it. A *hole* of a compact set is any bounded component of its complement. Thus the holes of the spectrum $\sigma(T)$ are the bounded components of the resolvent set $\rho(T) = \mathbb{C}\backslash\sigma(T)$.

If $\mathcal{A}'$ is a closed unital subalgebra of a unital complex Banach algebra $\mathcal{A}$ (for instance, $\mathcal{A} = \mathcal{B}[\mathcal{X}]$ where $\mathcal{X}$ is a Banach space), and if $T \in \mathcal{A}'$, then let $\rho'(T)$ be the resolvent set of T with respect to $\mathcal{A}'$, let $\sigma'(T) = \mathbb{C}\backslash\rho'(T)$ be the spectrum of T with respect to $\mathcal{A}'$, and set $r'(T) = \sup_{\lambda \in \sigma'(T)} |\lambda|$. A *homomorphism* (or an *algebra homomorphism*) between two algebras is a linear transformation between them which also preserves the product operation. Let $\mathcal{A}'$ be a maximal (in the inclusion ordering) commutative subalgebra of a unital complex Banach algebra $\mathcal{A}$ (i.e., a commutative subalgebra of $\mathcal{A}$ not included in any other commutative subalgebra of $\mathcal{A}$). In fact, $\mathcal{A}'$ is trivially unital, and closed in $\mathcal{A}$ because the closure of a commutative subalgebra of a Banach algebra is again commutative since multiplication is continuous in $\mathcal{A}$. Consider the (unital complex commutative) Banach algebra $\mathbb{C}$ (of all complex numbers). Let $\widehat{\mathcal{A}'} = \{\Phi\colon \mathcal{A}' \to \mathbb{C} : \Phi \text{ is a homomorphism}\}$ stand for the collection of all algebra homomorphisms of $\mathcal{A}'$ onto $\mathbb{C}$.

Proposition 2.Q. *Let $\mathcal{A}$ be any unital complex Banach algebra (for instance, $\mathcal{A} = \mathcal{B}[\mathcal{X}]$). If $T \in \mathcal{A}'$, where $\mathcal{A}'$ is any closed unital subalgebra of $\mathcal{A}$, then*

(a) $\rho'(T) \subseteq \rho(T)$ *and* $r'(T) = r(T)$ *(invariance of the spectral radius). Hence $\partial\sigma'(T) \subseteq \partial\sigma(T)$ and $\sigma(T) \subseteq \sigma'(T)$.*

Thus $\sigma'(T)$ is obtained by adding to $\sigma(T)$ some holes of $\sigma(T)$.

(b) *If the unital subalgebra $\mathcal{A}'$ of $\mathcal{A}$ is a commutative, then*

$$\sigma'(T) = \big\{\Phi(T) \in \mathbb{C} : \Phi \in \widehat{\mathcal{A}'}\big\} \quad \text{for every} \quad T \in \mathcal{A}'.$$

Moreover, if $\mathcal{A}'$ is a maximal commutative subalgebra of $\mathcal{A}$, then

$$\sigma'(T) = \sigma(T).$$

Proposition 2.R. *If $\mathcal{A}$ is a unital complex Banach algebra, and if S, T in $\mathcal{A}$ commute (i.e., if $S, T \in \mathcal{A}$ and $ST = TS$), then*

$$\sigma(S + T) \subseteq \sigma(S) + \sigma(T) \quad \text{and} \quad \sigma(ST) \subseteq \sigma(S) \cdot \sigma(T).$$

If $\mathcal{M}$ is an invariant subspace for T, then it may happen that $\sigma(T|_{\mathcal{M}}) \not\subseteq \sigma(T)$. Sample: every unilateral shift is the restriction of a bilateral shift to an invariant subspace (see, e.g., [74, Lemma 2.14]). However, if $\mathcal{M}$ reduces T, then $\sigma(T|_{\mathcal{M}}) \subseteq \sigma(T)$ by Proposition 2.F(b). The *full spectrum* of $T \in \mathcal{B}[\mathcal{H}]$ (notation: $\sigma(T)^{\#}$) is the union of $\sigma(T)$ and all bounded components of $\rho(T)$ (i.e., $\sigma(T)^{\#}$ is the union of $\sigma(T)$ and all holes of $\sigma(T)$).

Proposition 2.S. *If $\mathcal{M}$ is T-invariant, then $\sigma(T|_{\mathcal{M}}) \subseteq \sigma(T)^{\#}$.*

Proposition 2.T. *Let $T \in \mathcal{B}[\mathcal{H}]$ and $S \in \mathcal{B}[\mathcal{K}]$ be operators on Hilbert spaces $\mathcal{H}$ and $\mathcal{K}$. If $\sigma(T) \cap \sigma(S) = \varnothing$, then for every bounded linear transformation $Y \in \mathcal{B}[\mathcal{H}, \mathcal{K}]$ there exists a unique bounded linear transformation $X \in \mathcal{B}[\mathcal{H}, \mathcal{K}]$ such that $XT - SX = Y$. In particular,*

$$\sigma(T) \cap \sigma(S) = \varnothing \text{ and } XT = SX \quad \Longrightarrow \quad X = O.$$

This is the Rosenblum Corollary, which will be used to prove the Fuglede Theorems in Chapter 3.

Notes: Again, as in Chapter 1, these are basic results needed throughout the text. We will not point out original sources here, but will instead refer to well-known secondary sources where proofs for these propositions can be found, as well as deeper discussions on them. Proposition 2.A holds independently of the forthcoming Spectral Theorem (Theorem 3.11) — see, for instance,

[78, Corollary 6.18]. A partial converse, however, needs the Spectral Theorem (see Proposition 3.D in Chapter 3). Proposition 2.B is a standard result (see, e.g., [58, Problem 75] and [78, Problem 6.10]), as is Proposition 2.C (see, e.g., [58, Problem 76]). Proposition 2.D also bypasses the Spectral Theorem of the next chapter: if Q is a nonnegative operator, then it has a unique nonnegative square root $Q^{\frac{1}{2}}$ by the Square Root Theorem (Proposition 1.M), $\sigma(Q^{\frac{1}{2}})^2 = \sigma((Q^{\frac{1}{2}})^2) = \sigma(Q)$ by the Spectral Mapping Theorem (Theorem 2.7), and $\sigma(Q) \subseteq [0, \infty)$ by Proposition 2.A(e). Proposition 2.E is a rather useful technical result (see, e.g., [78, Problem 6.14]). Proposition 2.F is a synthesis of some scattered results (cf. [31, Proposition I.5.1], [58, Solution 98], [64, Theorem 5.42], [78, Problem 6.37], and Proposition 2.E). Proposition 2.G(a) is straightforward: $r(T) \leq \|T\|$, $\sigma(T)$ is closed, and T is normaloid if and only if $r(T) = \|T\|$. Proposition 2.G(b) is its compact version (see, e.g., [78, Corollary 6.32]). For Proposition 2.H(a) see, for instance, [74, Section 0.2]. Proposition 2.H(b) has been raised in [74, Section 8.2]; for a proof see [94]. The spectral radius formula in Proposition 2.I (see, e.g., [48, p. 22] or [74, Corollary 6.4]) ensures that *an operator is uniformly stable if and only if it is similar to a strict contraction* (hint: use the equivalence between (a) and (b) in Corollary 2.11). For Propositions 2.J and 2.K see [78, Sections 6.3 and 6.4]. The examples of spectra in Propositions 2.L, 2.M, and 2.N are widely known (see, e.g., [78, Examples 6.C, 6.D, 6.E, 6.F, and 6.G]). Proposition 2.O deals with the equivalence between uniform and weak stabilities for compact operators, and it is extended to a wider class of operators in Proposition 2.P (see [75, Problems 8.8. and 8.9]). Proposition 2.Q(a) is readily verified, where the invariance of the spectral radius follows by the Gelfand–Beurling formula. Proposition 2.Q(b) is a key result for proving both Theorem 2.8 (the Spectral Mapping Theorem for normal operators) and also Lemma 5.43 of Chapter 5 (for the characterization of the Browder spectrum in Theorem 5.44) — see, e.g., [98, Theorems 0.3 and 0.4]. For Propositions 2.R, 2.S, and 2.T see [104, Theorem 11.22], [98, Theorem 0.8], and [98, Corollary 0.13], respectively. Proposition 2.T will play an important role in the proof of the Fuglede Theorems of the next chapter (specifically in Corollary 3.5 and Theorem 3.17).

Suggested Readings

Arveson [7]	Halmos [58]
Bachman and Narici [9]	Herrero [64]
Berberian [18]	Istrățescu [69]
Conway [30, 31, 33]	Kato [72]
Douglas [38]	Kubrusly [74, 75, 78]
Dowson [39]	Radjavi and Rosenthal [98]
Dunford and Schwartz [45]	Reed and Simon [99]
Fillmore [48]	Rudin [104]
Gustafson and Rao [52]	Taylor and Lay [111]

3

The Spectral Theorem

The Spectral Theorem is a milestone in the theory of Hilbert-space operators, providing a full statement about the nature and structure of normal operators.

"Most students of mathematics learn quite early and most mathematicians remember till quite late that every Hermitian matrix (and, in particular, every real symmetric matrix) may be put into diagonal form. A more precise statement of the result is that every Hermitian matrix is unitarily equivalent to a diagonal one. The spectral theorem is widely and correctly regarded as the generalization of this assertion to operators on Hilbert space." Paul Halmos [57]

For compact normal operators the Spectral Theorem can be investigated without requiring any knowledge of measure theory, and this leads to the concept of diagonalization. However, the Spectral Theorem for plain normal operators (the general case) requires some (elementary) measure theory.

3.1 Spectral Theorem for Compact Operators

Throughout this chapter $\mathcal{H}$ will denote a *nonzero complex* Hilbert space. A bounded weighted sum of projections was defined in the last paragraph of Section 1.4 as an operator T on $\mathcal{H}$ such that

$$Tx = \sum_{\gamma \in \Gamma} \lambda_\gamma E_\gamma x$$

for every $x \in \mathcal{H}$, where $\{E_\gamma\}_{\gamma \in \Gamma}$ is a resolution of the identity on $\mathcal{H}$ made up of nonzero projections (i.e., $\{E_\gamma\}_{\gamma \in \Gamma}$ is an orthogonal family of nonzero orthogonal projections such that $\sum_{\gamma \in \Gamma} E_\gamma x = x$ for every $x \in \mathcal{H}$), and $\{\lambda_\gamma\}_{\gamma \in \Gamma}$ is a bounded family of scalars. *Every bounded weighted sum of projections is normal* (Corollary 1.9). The spectrum of a bounded weighted sum of projections is characterized next as the closure of the set $\{\lambda_\gamma\}_{\gamma \in \Gamma}$,

$$\sigma(T) = \{\lambda \in \mathbb{C} \colon \lambda = \lambda_\gamma \text{ for some } \lambda \in \Gamma\}^{-}.$$

So, since T is normaloid, $\|T\| = r(T) = \sup_{\gamma \in \Gamma} |\lambda_\gamma|$.

© Springer Nature Switzerland AG 2020
C. S. Kubrusly, *Spectral Theory of Bounded Linear Operators*,
https://doi.org/10.1007/978-3-030-33149-8_3

Lemma 3.1. *If $T \in \mathcal{B}[\mathcal{H}]$ is a weighted sum of projections, then*

$$\sigma_P(T) = \{\lambda \in \mathbb{C} \colon \lambda = \lambda_\gamma \text{ for some } \gamma \in \Gamma\}, \qquad \sigma_R(T) = \varnothing, \qquad and$$

$$\sigma_C(T) = \{\lambda \in \mathbb{C} \colon \inf_{\gamma \in \Gamma} |\lambda - \lambda_\gamma| = 0 \text{ and } \lambda \neq \lambda_\gamma \text{ for all } \gamma \in \Gamma\}.$$

Proof. Let $T = \sum_{\gamma \in \Gamma} \lambda_\gamma E_\gamma$ be a bounded weighted sum of projections where $\{E_\gamma\}_{\gamma \in \Gamma}$ is a resolution of the identity on $\mathcal{H}$ and $\{\lambda_\gamma\}_{\gamma \in \Gamma}$ is a bounded sequence of scalars (see Section 1.4). Take an arbitrary x in $\mathcal{H}$. Since $\{E_\gamma\}_{\gamma \in \Gamma}$ is a resolution of the identity on $\mathcal{H}$, we get $x = \sum_{\gamma \in \Gamma} E_\gamma x$, and the general version of the Pythagorean Theorem (see the Orthogonal Structure Theorem of Section 1.3) ensures that $\|x\|^2 = \sum_{\gamma \in \Gamma} \|E_\gamma x\|^2$. Take any scalar $\lambda \in \mathbb{C}$. Thus $(\lambda I - T)x = \sum_{\gamma \in \Gamma} (\lambda - \lambda_\gamma) E_\gamma x$ and so $\|(\lambda I - T)x\|^2 = \sum_{\gamma \in \Gamma} |\lambda - \lambda_\gamma|^2 \|E_\gamma x\|^2$ by the same argument. If $\mathcal{N}(\lambda I - T) \neq \{0\}$, then there exists $x \neq 0$ in $\mathcal{H}$ such that $(\lambda I - T)x = 0$. Hence $\sum_{\gamma \in \Gamma} \|E_\gamma x\|^2 \neq 0$ and $\sum_{\gamma \in \Gamma} |\lambda - \lambda_\gamma|^2 \|E_\gamma x\|^2 = 0$, which implies $\|E_\alpha x\| \neq 0$ for some $\alpha \in \Gamma$ and $|\lambda - \lambda_\alpha| \|E_\alpha x\| = 0$, and therefore $\lambda = \lambda_\alpha$. Conversely, take an arbitrary $\alpha \in \Gamma$ and an arbitrary nonzero vector x in $\mathcal{R}(E_\alpha)$ (recall: $\mathcal{R}(E_\gamma) \neq \{0\}$ because $E_\gamma \neq O$ for every $\gamma \in \Gamma$). Since $\mathcal{R}(E_\alpha) \perp \mathcal{R}(E_\gamma)$ whenever $\alpha \neq \gamma$, we get $\mathcal{R}(E_\alpha) \perp \bigcup_{\alpha \neq \gamma \in \Gamma} \mathcal{R}(E_\gamma)$, and consequently $\mathcal{R}(E_\alpha) \subseteq (\bigcup_{\alpha \neq \gamma \in \Gamma} \mathcal{R}(E_\gamma))^\perp = \bigcap_{\alpha \neq \gamma \in \Gamma} \mathcal{R}(E_\gamma)^\perp = \bigcap_{\alpha \neq \gamma \in \Gamma} \mathcal{N}(E_\gamma)$ (see Lemma 1.4 and Theorem 1.8). Thus $x \in \mathcal{N}(E_\gamma)$ for every $\alpha \neq \gamma \in \Gamma$, which implies $\|(\lambda_\alpha I - T)x\|^2 = \sum_{\gamma \in \Gamma} |\lambda_\alpha - \lambda_\gamma|^2 \|E_\gamma x\|^2 = 0$, and therefore $\mathcal{N}(\lambda_\alpha I - T) \neq \{0\}$. Summing up: $\mathcal{N}(\lambda I - T) \neq \{0\}$ if and only if $\lambda = \lambda_\alpha$ for some $\alpha \in \Gamma$. Equivalently,

$$\sigma_P(T) = \{\lambda \in \mathbb{C} \colon \lambda = \lambda_\gamma \text{ for some } \gamma \in \Gamma\}.$$

Take an arbitrary $\lambda \in \mathbb{C}$ such that $\lambda \neq \lambda_\gamma$ for all $\gamma \in \Gamma$. The bounded weighted sum of projections $\lambda I - T = \sum_{\gamma \in \Gamma} (\lambda - \lambda_\gamma) E_\gamma$ has an inverse, and this inverse is the weighted sum of projections $(\lambda I - T)^{-1} = \sum_{\gamma \in \Gamma} (\lambda - \lambda_\gamma)^{-1} E_\gamma$ since

$$\sum_{\alpha \in \Gamma} (\lambda - \lambda_\alpha)^{-1} E_\alpha \sum_{\beta \in \Gamma} (\lambda - \lambda_\beta) E_\beta x$$
$$= \sum_{\alpha \in \Gamma} \sum_{\beta \in \Gamma} (\lambda - \lambda_\alpha)^{-1} (\lambda - \lambda_\beta) E_\alpha E_\beta x = \sum_{\gamma \in \Gamma} E_\gamma x = x$$

for every x in $\mathcal{H}$ (because the resolution of the identity $\{E_\gamma\}_{\gamma \in \Gamma}$ is an orthogonal family of continuous projections, and the inverse is unique). Moreover, the weighted sum of projections of $(\lambda I - T)^{-1}$ is bounded if and only if $\sup_{\gamma \in \Gamma} |\lambda - \lambda_\gamma|^{-1} < \infty$; that is, if and only if $\inf_{\gamma \in \Gamma} |\lambda - \lambda_\gamma| > 0$. Thus

$$\rho(T) = \{\lambda \in \mathbb{C} \colon \inf_{\gamma \in \Gamma} |\lambda - \lambda_\gamma| > 0\}.$$

Furthermore, since $\sigma_R(T) = \varnothing$ (a weighted sum of projections is normal — cf. Corollary 1.9 and Proposition 2.A(b)), we get $\sigma_C(T) = (\mathbb{C} \backslash \rho(T)) \backslash \sigma_P(T)$. $\square$

Compare with Proposition 2.L. Actually, Proposition 2.L is a particular case of Lemma 3.1, since $E_k x = \langle x ; e_k \rangle e_k$ for every $x \in \mathcal{H}$ defines the orthogonal projection E_k onto the unidimensional space spanned by e_k.

Lemma 3.2. *A weighted sum of projections $T \in \mathcal{B}[\mathcal{H}]$ is compact if and only if the following triple condition holds.*

(i) $\sigma(T)$ *is countable,*

(ii) 0 *is the only possible accumulation point of $\sigma(T)$, and*

(iii) $\dim \mathcal{R}(E_\gamma) < \infty$ *for every γ such that $\lambda_\gamma \neq 0$.*

Proof. Let $T = \sum_{\gamma \in \Gamma} \lambda_\gamma E_\gamma \in \mathcal{B}[\mathcal{H}]$ be a weighted sum of projections.

Claim. $\mathcal{R}(E_\gamma) \subseteq \mathcal{N}(\lambda_\gamma I - T)$ for every $\gamma \in \Gamma$.

Proof. Take any $\alpha \in \Gamma$. If $x \in \mathcal{R}(E_\alpha)$, then $x = E_\alpha x$ and so $Tx = TE_\alpha x = \sum_{\gamma \in \Gamma} \lambda_\gamma E_\gamma E_\alpha x = \lambda_\alpha E_\alpha x = \lambda_\alpha x$ (because $E_\gamma \perp E_\alpha$ whenever $\alpha \neq \gamma$). Hence $x \in \mathcal{N}(\lambda_\alpha I - T)$, which concludes the proof of the claimed inclusion.

If T is compact, then $\sigma(T)$ is countable and 0 is the only possible accumulation point of $\sigma(T)$ (Corollary 2.20), and $\dim \mathcal{N}(\lambda I - T) < \infty$ whenever $\lambda \neq 0$ (Theorem 1.19). Then $\dim \mathcal{R}(E_\gamma) < \infty$ for every γ for which $\lambda_\gamma \neq 0$, by the above claim. Conversely, if $T = O$, then T is trivially compact. Thus suppose $T \neq O$. Since T is normal (Corollary 1.9), $r(T) > 0$ (because normal operators are normaloid, i.e., $r(T) = \|T\|$) and so there exists $\lambda \neq 0$ in $\sigma_P(T)$ by Theorem 2.18. If $\sigma(T)$ is countable, then let $\{\lambda_k\}$ be any enumeration of the countable set $\sigma_P(T) \backslash \{0\} = \sigma(T) \backslash \{0\}$. Hence

$$Tx = \sum_k \lambda_k E_k x \quad \text{for every } x \in \mathcal{H}$$

(cf. Lemma 3.1), where $\{E_k\}$ is included in a resolution of the identity on $\mathcal{H}$ (and is itself a resolution of the identity on $\mathcal{H}$ if $0 \notin \sigma_P(T)$). If $\{\lambda_k\}$ is finite, say $\{\lambda_k\} = \{\lambda_k\}_{k=1}^n$, then $\mathcal{R}(T) = \sum_{k=1}^n \mathcal{R}(E_k)$. If $\dim \mathcal{R}(E_k) < \infty$ for each k, then $\dim \left(\sum_{k=1}^n \mathcal{R}(E_k) \right)^- < \infty$. Thus T is a finite-rank operator, and so compact (cf. Proposition 1.X). Now suppose $\{\lambda_k\}$ is countably infinite. Since $\sigma(T)$ is a compact set (Theorem 2.1), the infinite set $\{\lambda_k\}$ has an accumulation point in $\sigma(T)$ — this in fact is the Bolzano–Weierstrass Property which characterizes compact sets in metric spaces. If 0 is the only possible accumulation point of $\sigma(T)$, then 0 is the unique accumulation point of $\{\lambda_k\}$. Therefore, for each integer $n \geq 1$ consider the partition $\{\lambda_k\} = \{\lambda_{k'}\} \cup \{\lambda_{k''}\}$, where $\frac{1}{n} \leq |\lambda_{k'}|$ and $|\lambda_{k''}| < \frac{1}{n}$. Since $\{\lambda_{k'}\}$ is a finite subset of $\sigma(T)$ (it has no accumulation point), it follows that $\{\lambda_{k''}\}$ is an infinite subset of $\sigma(T)$. Set

$$T_n = \sum_{k'} \lambda_{k'} E_{k'} \in \mathcal{B}[\mathcal{H}] \quad \text{for each } n \geq 1.$$

As we have seen, $\dim \mathcal{R}(T_n) < \infty$; equivalently, T_n is a finite-rank operator. However, since $E_j \perp E_k$ whenever $j \neq k$ we get

$$\|(T - T_n)x\|^2 = \left\| \sum_{k''} \lambda_{k''} E_{k''} x \right\|^2 \leq \sup_{k''} |\lambda_{k''}|^2 \sum_{k''} \|E_{k''} x\|^2 \leq \tfrac{1}{n^2} \|x\|^2$$

for all $x \in \mathcal{H}$, and so $T_n \xrightarrow{u} T$. Then T is compact by Proposition 1.Z. $\square$

Thus every bounded weighted sum of projections is normal and, if it is compact, it has a countable set of distinct eigenvalues. The Spectral Theorem for compact operators ensures the converse: *every compact and normal operator T is a (countable) weighted sum of projections, whose weights are precisely the eigenvalues of T.*

Theorem 3.3. (SPECTRAL THEOREM FOR COMPACT OPERATORS). *If an operator $T \in \mathcal{B}[\mathcal{H}]$ on a (nonzero complex) Hilbert space $\mathcal{H}$ is compact and normal, then there exists a unique countable resolution of the identity $\{E_k\}$ on $\mathcal{H}$ and a bounded set of scalars $\{\lambda_k\}$ for which*

$$T = \sum_{k} \lambda_k E_k,$$

where $\{\lambda_k\} = \sigma_P(T)$ is the (nonempty) set of all (distinct) eigenvalues of T and each E_k is the orthogonal projection onto the eigenspace $\mathcal{N}(\lambda_k I - T)$ (i.e., $\mathcal{R}(E_k) = \mathcal{N}(\lambda_k I - T)$). If the above countable weighted sum of projections is infinite, then it converges in the (uniform) topology of $\mathcal{B}[\mathcal{H}]$.

Proof. If $T \in \mathcal{B}[\mathcal{H}]$ is compact and normal, then it has a nonempty point spectrum (Corollary 2.21). In fact, the heart of the matter is: T has enough eigenvalues so that its eigenspaces span the Hilbert space $\mathcal{H}$.

Claim. $\left(\sum_{\lambda \in \sigma_P(T)} \mathcal{N}(\lambda I - T) \right)^{-} = \mathcal{H}$.

Proof. Set $\mathcal{M} = \left(\sum_{\lambda \in \sigma_P(T)} \mathcal{N}(\lambda I - T) \right)^{-}$, which is a subspace of $\mathcal{H}$. Suppose $\mathcal{M} \neq \mathcal{H}$ and hence $\mathcal{M}^{\perp} \neq \{0\}$. Consider the restriction $T|_{\mathcal{M}^{\perp}}$ of T to $\mathcal{M}^{\perp}$. Since T is normal, $\mathcal{M}$ reduces T (Theorem 1.16). Thus $\mathcal{M}^{\perp}$ is T-invariant, and $T|_{\mathcal{M}^{\perp}} \in \mathcal{B}[\mathcal{M}^{\perp}]$ is normal (Proposition 1.Q). Since T is compact, $T|_{\mathcal{M}^{\perp}}$ is compact (Proposition 1.V). Then $T|_{\mathcal{M}^{\perp}}$ is a compact normal operator on the Hilbert space $\mathcal{M}^{\perp} \neq \{0\}$, and so $\sigma_P(T|_{\mathcal{M}^{\perp}}) \neq \varnothing$ by Corollary 2.21. Thus there exist $\lambda \in \mathbb{C}$ and $0 \neq x \in \mathcal{M}^{\perp}$ for which $T|_{\mathcal{M}^{\perp}} x = \lambda x$ and hence $Tx = \lambda x$. Then $\lambda \in \sigma_P(T)$ and $x \in \mathcal{N}(\lambda I - T) \subseteq \mathcal{M}$. This leads to a contradiction: $0 \neq x \in \mathcal{M} \cap \mathcal{M}^{\perp} = \{0\}$. Outcome: $\mathcal{M} = \mathcal{H}$.

Since T is compact, the nonempty set $\sigma_P(T)$ is countable (Corollaries 2.20 and 2.21) and bounded (because $T \in \mathcal{B}[\mathcal{H}]$ — cf. Theorem 2.1). Thus write

$$\sigma_P(T) = \{\lambda_k\},$$

where $\{\lambda_k\}$ is a finite or a countably infinite bounded subset of $\mathbb{C}$ consisting of all eigenvalues of T. Each $\mathcal{N}(\lambda_k I - T)$ is a subspace of $\mathcal{H}$ (since each $\lambda_k I - T$ is bounded). Moreover, since T is normal, Lemma 1.15(a) ensures that $\mathcal{N}(\lambda_k I - T) \perp \mathcal{N}(\lambda_j I - T)$ if $k \neq j$. Then $\{\mathcal{N}(\lambda_k I - T)\}$ is a sequence of pairwise orthogonal subspaces of $\mathcal{H}$ spanning $\mathcal{H}$; that is, $\mathcal{H} = \left(\sum_{k} \mathcal{N}(\lambda_k I - T) \right)^{-}$ by the above claim. So the sequence $\{E_k\}$ of the orthogonal projections onto each $\mathcal{N}(\lambda_k I - T)$ is a resolution of the identity on $\mathcal{H}$ (see Section 1.4 — recall: E_k onto $\mathcal{N}(\lambda_k I - T)$ is unique). Thus $x = \sum_{k} E_k x$. Since T is linear and continuous, $Tx = \sum_{k} T E_k x$ for every $x \in \mathcal{H}$. But $E_k x \in \mathcal{R}(E_k) = \mathcal{N}(\lambda_k I - T)$, and so $T E_k x = \lambda_k E_k x$, for each k and every $x \in \mathcal{H}$. Therefore

$$Tx = \sum_k \lambda_k E_k x \quad \text{for every } x \in \mathcal{H},$$

and hence T is a countable weighted sum of projections. If it is a finite weighted sum of projections, then we are done. On the other hand, suppose it is an infinite weighted sum of projections. In this case the above identity says

$$\sum_{k=1}^n \lambda_k E_k \xrightarrow{s} T \quad \text{as } n \to \infty.$$

So the sequence $\left\{ \sum_{k=1}^n \lambda_k E_k \right\}$ converges strongly to T. However, since T is compact, this convergence actually takes place in the uniform topology (i.e., the sequence $\left\{ \sum_{k=1}^n \lambda_k E_k \right\}$ converges uniformly to T). Indeed,

$$\left\| \left(T - \sum_{k=1}^n \lambda_k E_k \right) x \right\|^2 = \left\| \sum_{k=n+1}^\infty \lambda_k E_k x \right\|^2 = \sum_{k=n+1}^\infty |\lambda_k|^2 \|E_k x\|^2$$

$$\leq \sup_{k \geq n+1} |\lambda_k|^2 \sum_{k=n+1}^\infty \|E_k x\|^2 \leq \sup_{k \geq n+1} |\lambda_k|^2 \|x\|^2$$

for every integer $n \geq 1$ (reason: $\mathcal{R}(E_j) \perp \mathcal{R}(E_k)$ whenever $j \neq k$, and $x = \sum_{k=1}^\infty E_k x$ so that $\|x\|^2 = \sum_{k=1}^\infty \|E_k x\|^2$ — see Sections 1.3 and 1.4). Hence

$$0 \leq \left\| T - \sum_{k=1}^n \lambda_k E_k \right\| = \sup_{\|x\|=1} \left\| \left(T - \sum_{k=1}^n \lambda_k E_k \right) x \right\| \leq \sup_{k \geq n+1} |\lambda_k|$$

for all $n \in \mathbb{N}$. As T is compact, the infinite sequence $\{\lambda_n\}$ of distinct elements in $\sigma(T)$ must converge to zero (Corollary 2.20(a)). Since $\lim_n \lambda_n = 0$ we get $\lim_n \sup_{k \geq n+1} |\lambda_k| = \lim \sup_n |\lambda_n| = 0$ and so

$$\sum_{k=1}^n \lambda_k E_k \xrightarrow{u} T \quad \text{as } n \to \infty. \qquad \square$$

3.2 Diagonalizable Operators

If T in $\mathcal{B}[\mathcal{H}]$ is compact and normal, then the (orthogonal) family of orthogonal projections $\{E_\lambda\}_{\lambda \in \sigma_P(T)}$ onto each eigenspace $\mathcal{N}(\lambda I - T)$ is a countable resolution of the identity on $\mathcal{H}$, and T is a weighted sum of projections. This is what the Spectral Theorem for compact (normal) operators says. Thus we write

$$T = \sum_{\lambda \in \sigma_P(T)} \lambda E_\lambda,$$

which is a uniform limit, and so it implies pointwise convergence (i.e., $Tx = \sum_{\lambda \in \sigma_P(T)} \lambda E_\lambda x$ for every $x \in \mathcal{H}$). This is naturally identified with (i.e., it is unitarily equivalent to) an orthogonal direct sum of scalar operators,

$$T \cong \bigoplus_{\lambda \in \sigma_P(T)} \lambda I_\lambda,$$

with each I_λ denoting the identity on each eigenspace $\mathcal{N}(\lambda I - T)$. In fact, $I_\lambda = E_\lambda|_{\mathcal{R}(E_\lambda)}$, where $\mathcal{R}(E_\lambda) = \mathcal{N}(\lambda I - T)$. Thus under such an identification it is usual to write (a slight abuse of notation)

$$T = \bigoplus_{\lambda \in \sigma_P(T)} \lambda E_\lambda.$$

Each of these equivalent representations is referred to as the *spectral decomposition* of a compact normal operator T. The Spectral Theorem for compact (normal) operators can be restated in terms of an orthonormal basis for $\mathcal{N}(T)^\perp$ consisting of eigenvectors of T, as follows.

Corollary 3.4. *Let $T \in \mathcal{B}[\mathcal{H}]$ be compact and normal.*

(a) *For each $\lambda \in \sigma_P(T)\backslash\{0\}$ there is a finite orthonormal basis $\{e_j(\lambda)\}_{j=1}^{n_\lambda}$ for $\mathcal{N}(\lambda I - T)$ consisting entirely of eigenvectors of T.*

(b) *The set $\{e_k\} = \bigcup_{\lambda \in \sigma_P(T)\backslash\{0\}}\{e_j(\lambda)\}_{j=1}^{n_\lambda}$ is a countable orthonormal basis for $\mathcal{N}(T)^\perp$ made up of eigenvectors of T.*

(c) $Tx = \sum_{\lambda \in \sigma_P(T)\backslash\{0\}} \lambda \sum_{j=1}^{n_\lambda} \langle x\,; e_j(\lambda)\rangle e_j(\lambda)$ *for every $x \in \mathcal{H}$.*

(d) $Tx = \sum_k \lambda_k \langle x\,; e_k\rangle e_k$ *for every $x \in \mathcal{H}$, where $\{\lambda_k\}$ is a sequence containing all nonzero eigenvalues of T including multiplicity (i.e., finitely repeated according to the dimension of the respective eigenspace).*

Proof. The point spectrum $\sigma_P(T)$ is nonempty and countable by Theorem 3.3, and $\sigma_P(T) = \{0\}$ if and only if $T = O$ by Corollary 2.21 or, equivalently, if and only if $\mathcal{N}(T)^\perp = \{0\}$ (i.e., $\mathcal{N}(T) = \mathcal{H}$). If $T = O$, then the above assertions hold trivially ($\sigma_P(T)\backslash\{0\} = \varnothing$, $\{e_k\} = \varnothing$, $\mathcal{N}(T)^\perp = \{0\}$ and $Tx = 0x = 0$ for every $x \in \mathcal{H}$ because the empty sum is null). Thus suppose $T \neq O$ (so that $\mathcal{N}(T)^\perp \neq \{0\}$) and take an arbitrary $\lambda \neq 0$ in $\sigma_P(T)$.

(a) By Theorem 1.19 $\dim \mathcal{N}(\lambda I - T)$ is finite, say $\dim \mathcal{N}(\lambda I - T) = n_\lambda$ for a positive integer n_λ. This ensures the existence of a *finite* orthonormal basis $\{e_j(\lambda)\}_{j=1}^{n_\lambda}$ for the Hilbert space $\mathcal{N}(\lambda I - T) \neq \{0\}$, and so each $e_j(\lambda)$ is an eigenvector of T (since $0 \neq e_j(\lambda) \in \mathcal{N}(\lambda I - T)$), concluding the proof of (a).

Claim. $\bigcup_{\lambda \in \sigma_P(T)\backslash\{0\}}\{e_j(\lambda)\}_{j=1}^{n_\lambda}$ is an orthonormal basis for $\mathcal{N}(T)^\perp$.

Proof. By Theorem 3.3,

$$\mathcal{H} = \left(\sum_{\lambda \in \sigma_P(T)} \mathcal{N}(\lambda I - T)\right)^{-}.$$

Since $\{\mathcal{N}(\lambda I - T)\}_{\lambda \in \sigma_P(T)}$ is a nonempty family of orthogonal subspaces of $\mathcal{H}$ (by Lemma 1.15(a)) we get

$$\mathcal{N}(T) = \bigcap_{\lambda \in \sigma_P(T)\backslash\{0\}} \mathcal{N}(\lambda I - T)^\perp = \left(\sum_{\lambda \in \sigma_P(T)\backslash\{0\}} \mathcal{N}(\lambda I - T)\right)^\perp$$

(see, e.g., [78, Problem 5.8(b,e)]). Hence (see Section 1.3)

$$\mathcal{N}(T)^\perp = \left(\sum_{\lambda \in \sigma_P(T)\backslash\{0\}} \mathcal{N}(\lambda I - T)\right)^{-}.$$

Therefore, since $\{\mathcal{N}(\lambda I - T)\}_{\lambda \in \sigma_P(T)}$ is a family of orthogonal subspaces, the claimed result follows from (a).

(b) Since $\{e_k\} = \bigcup_{\lambda \in \sigma_P(T)\backslash\{0\}}\{e_j(\lambda)\}_{j=1}^{n_\lambda}$ is a countable set (a countable union of countable sets), the result in (b) follows from (a) and the above claim.

(c,d) Consider the decomposition $\mathcal{H} = \mathcal{N}(T) + \mathcal{N}(T)^{\perp}$ (see Section 1.3) and take an arbitrary $x \in \mathcal{H}$. Thus $x = u + v$ with $u \in \mathcal{N}(T)$ and $v \in \mathcal{N}(T)^{\perp}$. Consider the Fourier series expansion of v in terms of the orthonormal basis $\{e_k\} = \bigcup_{\lambda \in \sigma_P(T) \backslash \{0\}} \{e_j(\lambda)\}_{j=1}^{n_\lambda}$ for the Hilbert space $\mathcal{N}(T)^{\perp} \neq \{0\}$, namely, $v = \sum_k \langle v ; e_k \rangle e_k = \sum_{\lambda \in \sigma_P(T) \backslash \{0\}} \sum_{j=1}^{n_\lambda} \langle v ; e_j(\lambda) \rangle e_j(\lambda)$. Since the operator T is linear and continuous, and since $T e_j(\lambda) = \lambda e_j(\lambda)$ for every $j = 1, \ldots, n_\lambda$ and every $\lambda \in \sigma_P(T) \backslash \{0\}$, it follows that

$$Tx = Tu + Tv = Tv = \sum_{\lambda \in \sigma_P(T) \backslash \{0\}} \sum_{j=1}^{n_\lambda} \langle v ; e_j(\lambda) \rangle \, T e_j(\lambda)$$
$$= \sum_{\lambda \in \sigma_P(T) \backslash \{0\}} \lambda \sum_{j=1}^{n_\lambda} \langle v ; e_j(\lambda) \rangle \, e_j(\lambda).$$

However, $\langle x ; e_j(\lambda) \rangle = \langle u ; e_j(\lambda) \rangle + \langle v ; e_j(\lambda) \rangle = \langle v ; e_j(\lambda) \rangle$ because $u \in \mathcal{N}(T)$ and $e_j(\lambda) \in \mathcal{N}(T)^{\perp}$, concluding the proof of (c), which implies the result in (d). $\square$

Remark. If $T \in \mathcal{B}[\mathcal{H}]$ *is compact and normal, and if $\mathcal{H}$ is nonseparable, then* $\mathcal{N}(T)$ *is nonseparable and* $0 \in \sigma_P(T)$. Indeed, suppose $T \neq O$ (otherwise the above statement is trivial). Thus $\mathcal{N}(T)^{\perp} \neq \{0\}$ is separable (i.e., it has a countable orthonormal basis) by Corollary 3.4, and hence $\mathcal{N}(T)$ is nonseparable whenever $\mathcal{H} = \mathcal{N}(T) + \mathcal{N}(T)^{\perp}$ is nonseparable. Moreover, if $\mathcal{N}(T)$ is nonseparable, then $\mathcal{N}(T) \neq \{0\}$ tautologically, which means $0 \in \sigma_P(T)$.

The expression for a compact normal operator T in Corollary 3.4(c) (and, equivalently, the spectral decomposition of T) says T is a diagonalizable operator in the following sense: $T = T|_{\mathcal{N}(T)^{\perp}} \oplus O$ (since $\mathcal{N}(T)$ reduces T), where $T|_{\mathcal{N}(T)^{\perp}}$ in $\mathcal{B}[\mathcal{N}(T)^{\perp}]$ is a diagonal operator with respect to a countable orthonormal basis $\{e_k\}$ for the separable Hilbert space $\mathcal{N}(T)^{\perp}$.

Generalizing: an operator T in $\mathcal{B}[\mathcal{H}]$ (not necessarily compact) on a Hilbert space $\mathcal{H}$ (not necessarily separable) is *diagonalizable* if there is a resolution of the identity $\{E_\gamma\}_{\gamma \in \Gamma}$ on $\mathcal{H}$ and a bounded family of scalars $\{\lambda_\gamma\}_{\gamma \in \Gamma}$ for which

$$Tu = \lambda_\gamma u \quad \text{whenever} \quad u \in \mathcal{R}(E_\gamma).$$

Take an arbitrary $x = \sum_{\gamma \in \Gamma} E_\gamma x$ in $\mathcal{H}$. Since T is linear and continuous, $Tx = \sum_{\gamma \in \Gamma} T E_\gamma x = \sum_{\gamma \in \Gamma} \lambda_\gamma E_\gamma x$. Hence T is a weighted sum of projections (and consequently it is normal). Thus we write

$$T = \sum_{\gamma \in \Gamma} \lambda_\gamma E_\gamma \quad \text{or} \quad T = \bigoplus_{\gamma \in \Gamma} \lambda_\gamma E_\gamma.$$

Conversely, if T is a weighted sum of projections (i.e., $Tx = \sum_{\gamma \in \Gamma} \lambda_\gamma E_\gamma x$ for every $x \in \mathcal{H}$), then $Tu = \sum_{\gamma \in \Gamma} \lambda_\gamma E_\gamma u = \sum_{\gamma \in \Gamma} \lambda_\gamma E_\gamma E_\alpha u = \lambda_\alpha u$ for every $u \in \mathcal{R}(E_\alpha)$ (since $E_\gamma E_\alpha = O$ for $\gamma \neq \alpha$ and $u = E_\alpha u$ for $u \in \mathcal{R}(E_\alpha)$), and so T is diagonalizable. Conclusion:

An operator T on $\mathcal{H}$ is diagonalizable if and only if it is a weighted sum of projections for some bounded sequence of scalars $\{\lambda_\gamma\}_{\gamma \in \Gamma}$ and some resolution of the identity $\{E_\gamma\}_{\gamma \in \Gamma}$ on $\mathcal{H}$.

In this case $\{E_\gamma\}_{\gamma \in \Gamma}$ is said to *diagonalize* T. (See Proposition 3.A and recall: the identity on any Hilbert space $\mathcal{H}$ is trivially diagonalizable — it is actually a diagonal if $\mathcal{H}$ is the collection ℓ_Γ^2 of all square-summable families $\{\zeta_\gamma\}_{\gamma \in \Gamma}$ of scalars $\zeta_\gamma \in \mathbb{C}$ — and thus it is normal but not compact; cf. Proposition 1.Y.)

The implication (b) $\Rightarrow$ (d) in Corollary 3.5 below is the compact version of the Fuglede Theorem (Theorem 3.17) which will be developed in Section 3.5.

Corollary 3.5. *Suppose T is a compact operator on a Hilbert space $\mathcal{H}$.*

(a) *T is normal if and only if it is diagonalizable.*

Let $\{E_k\}$ be a resolution of the identity on $\mathcal{H}$ that diagonalizes a compact and normal operator $T \in \mathcal{B}[\mathcal{H}]$ into its spectral decomposition, and take any operator $S \in \mathcal{B}[\mathcal{H}]$. The following assertions are pairwise equivalent.

(b) *S commutes with T.*

(c) *S commutes with T^*.*

(d) *S commutes with every E_k.*

(e) *$\mathcal{R}(E_k)$ reduces S for every k.*

Proof. Take a compact operator T on $\mathcal{H}$. If T is normal, then it is diagonalizable by the Spectral Theorem. The converse is trivial because a diagonalizable operator is normal. This proves (a). From now on suppose T is compact and normal, and consider its spectral decomposition

$$T = \sum_k \lambda_k E_k$$

as in Theorem 3.3, where $\{E_k\}$ is a resolution of the identity on $\mathcal{H}$ and $\{\lambda_k\} = \sigma_P(T)$ is the set of all (distinct) eigenvalues of T. Observe that

$$T^* = \sum_k \overline{\lambda}_k E_k.$$

The central result is the implication (b) $\Rightarrow$ (e). The main idea behind the proof relies on Proposition 2.T (see also [98, Theorem 1.16]). Take any operator S on $\mathcal{H}$. Suppose $ST = TS$. Since

$$E_k T = \lambda_k E_k = T E_k$$

(recall: $E_k T = \sum_j \lambda_j E_k E_j = \lambda_k E_k = \sum_j \lambda_j E_j E_k = T E_k$), each $\mathcal{R}(E_k) = \mathcal{N}(\lambda_k I - T)$ reduces T (Proposition 1.I). Then

$$E_k T|_{\mathcal{R}(E_k)} = T E_k|_{\mathcal{R}(E_k)} = T|_{\mathcal{R}(E_k)} = \lambda_k I : \mathcal{R}(E_k) \to \mathcal{R}(E_k).$$

So on $\mathcal{R}(E_k)$,

$$\begin{aligned}(E_j S E_k) T|_{\mathcal{R}(E_k)} &= (E_j S E_k)(T E_k) = E_j S T E_k = E_j T S E_k \\ &= (T E_j)(E_j S E_k) = T|_{\mathcal{R}(E_j)}(E_j S E_k)\end{aligned}$$

whenever $j \neq k$. Moreover, since $\sigma(T|_{\mathcal{R}(E_k)}) = \{\lambda_k\}$ and $\lambda_k \neq \lambda_j$, we get

$$\sigma(TE_k) \cap \sigma(TE_j) = \varnothing,$$

and hence by Proposition 2.T for $j \neq k$

$$E_j SE_k = O.$$

Thus (as each E_k is self-adjoint and $\{E_k\}$ is a resolution of the identity)

$$\langle SE_k x\,; y \rangle = \left\langle SE_k x\,; \sum_j E_j y \right\rangle = \sum_j \langle E_j SE_k x\,; y \rangle = \langle E_k SE_k x\,; y \rangle$$

for every $x, y \in \mathcal{H}$. Then

$$SE_k = E_k SE_k.$$

Equivalently, $\mathcal{R}(E_k)$ is S-invariant (Proposition 1.I) for every k. Furthermore, $\mathcal{R}(E_k)^\perp = \left(\sum_{j \neq k} \mathcal{R}(E_j) \right)^- = \bigvee_{j \neq k} \mathcal{R}(E_j)$ (because $\{E_k\}$ is a resolution of the identity on $\mathcal{H}$). Since $\mathcal{R}(E_j)$ is S-invariant for every j, then $\mathcal{R}(E_k)^\perp$ is S-invariant as well. Therefore

$$\mathcal{R}(E_k) \text{ reduces } S \text{ for every } k.$$

This proves that (b) implies (e). But (e) is equivalent to (d) by Proposition 1.I. Also (d) implies (c). Indeed, if $SE_k = E_k S$ for every k, then $ST^* = \sum_k \overline{\lambda}_k SE_k = \sum_k \overline{\lambda}_k E_k S = T^* S$ (as S is linear and continuous). Thus (b) implies (c), and therefore (c) implies (b) because $T^{**} = T$. $\square$

3.3 Spectral Measure

If T is a compact normal operator, then its point spectrum is nonempty and countable, which may not hold for noncompact normal operators. But this is not the main role played by compact operators in the Spectral Theorem — we can deal with an uncountable weighted sum of projections $Tx = \sum_{\gamma \in \Gamma} \lambda_\gamma E_\gamma x$. What is actually special with a compact operator is that a compact normal operator not only has a nonempty point spectrum but it has enough eigenspaces to span $\mathcal{H}$; that is, $\left(\sum_{\lambda \in \sigma_P(T)} \mathcal{N}(\lambda I - T) \right)^- = \mathcal{H}$ (see proof of Theorem 3.3). This makes the difference, since a normal (noncompact) operator may have an empty point spectrum or it may have eigenspaces but not enough to span the whole space $\mathcal{H}$. The Spectral Theorem, however, survives the lack of compactness if the point spectrum is replaced with the whole spectrum (which is never empty). Such an approach for the general case of the Spectral Theorem (i.e., for normal, not necessarily compact operators) requires measure theory.

Let $\overline{\mathbb{R}} = \mathbb{R} \cup \{-\infty\} \cup \{+\infty\}$ stand for the *extended real line*. Take a measure space $(\Omega, \mathcal{A}_\Omega, \mu)$ where $\mathcal{A}_\Omega$ is a σ-algebra of subsets of a nonempty set Ω and $\mu \colon \mathcal{A}_\Omega \to \overline{\mathbb{R}}$ is a nonnegative measure on $\mathcal{A}_\Omega$, which means $\mu(\Lambda) \geq 0$ for every $\Lambda \in \mathcal{A}_\Omega$. Sets in $\mathcal{A}_\Omega$ are called *measurable* (or $\mathcal{A}_\Omega$-*measurable*). A nonnegative measure μ is *positive* if $\mu(\Omega) > 0$. It is *finite* if $\mu(\Omega) < \infty$ (i.e., $\mu \colon \mathcal{A}_\Omega \to \mathbb{R}$), and σ-*finite* if Ω is a countable union of measurable sets of finite measure.

Let $\mathbb{F}$ denote either the real line $\mathbb{R}$ or the complex plane $\mathbb{C}$. A *Borel σ-algebra* (or a σ-algebra of *Borel sets*) is the σ-algebra $\mathcal{A}_{\mathbb{F}}$ of subsets of $\mathbb{F}$ generated by the usual (metric) topology of $\mathbb{F}$ (i.e., the smallest σ-algebra of subsets of $\mathbb{F}$ including all open subsets of $\mathbb{F}$). *Borel sets* are precisely the sets in $\mathcal{A}_{\mathbb{F}}$ (i.e., the $\mathcal{A}_{\mathbb{F}}$-measurable sets). Thus all open, and consequently all closed, subsets of $\mathbb{F}$ are Borel sets (i.e., lie in $\mathcal{A}_{\mathbb{F}}$). If $\Omega \subseteq \mathbb{F}$ is a Borel subset of $\mathbb{F}$ (i.e., if $\Omega \in \mathcal{A}_{\mathbb{F}}$), then set $\mathcal{A}_{\Omega} = \wp(\Omega) \cap \mathcal{A}_{\mathbb{F}}$. This is the σ-algebra *of Borel subsets of* Ω, where $\wp(\Omega)$ is the *power set* of Ω (the collection of all subsets of Ω). A measure μ on $\mathcal{A}_{\Omega} = \wp(\Omega) \cap \mathcal{A}_{\mathbb{F}}$ is said to be a *Borel measure* if $\mu(K) < \infty$ for every compact set $K \subseteq \Omega$ (recall: compact subsets of Ω lie in $\mathcal{A}_{\Omega}$). Every finite measure is a Borel measure, and every Borel measure is σ-finite. The *support of a measure* μ on $\mathcal{A}_{\Omega}$ is the set $\mathrm{support}(\mu) = \Omega \backslash \bigcup \{ \Lambda \in \mathcal{A}_{\Omega} : \Lambda \text{ is open and } \mu(\Lambda) = 0 \}$, a closed set whose complement is the largest open set of measure zero. Equivalently, $\mathrm{support}(\mu) = \bigcap \{ \Lambda \in \mathcal{A}_{\Omega} : \Lambda \text{ is closed and } \mu(\Omega \backslash \Lambda) = 0 \}$, the smallest closed set whose complement has measure zero. There is a vast literature on measure theory. See, for instance, [12], [23], [53], [81], [102], or [103] (which simply reflects the author's preference but by no means exhausts any list).

Remark. Take an arbitrary subset Ω of $\mathbb{F}$. A set $K \subseteq \Omega \subseteq \mathbb{F}$ is compact if it is complete and totally bounded. (In $\Omega \subseteq \mathbb{F}$ total boundedness coincides with boundedness and so compactness in $\Omega \subseteq \mathbb{F}$ means complete and bounded.) Thus $K \subseteq \Omega$ is compact in the relative topology of Ω if and only if $K \subseteq \mathbb{F}$ is compact in the topology of $\mathbb{F}$ (where compact means closed and bounded). But *open* and *closed* in the preceding paragraph refer to the topology of Ω. Note: if Ω is closed (open) in $\mathbb{F}$, then *closed* (*open*) in Ω and in $\mathbb{F}$ coincide.

The *disjoint union* $\biguplus_{\gamma \in \Gamma} \Omega_{\gamma}$ of a collection $\{\Omega_{\gamma}\}_{\gamma \in \Gamma}$ of subsets of $\mathbb{C}$ is obtained by reindexing the elements of each set Ω_{γ} as follows. For every index $\gamma \in \Gamma$ write $\Omega_{\gamma} = \{\lambda_{\delta}(\gamma)\}_{\delta \in \Gamma_{\gamma}}$ so that if there is a point λ in $\Omega_{\alpha} \cap \Omega_{\beta}$ for $\alpha \neq \beta$ in Γ, then the same point λ is represented as $\lambda_{\omega}(\alpha) \in \Omega_{\alpha}$ for some $\omega \in \Gamma_{\alpha}$ and as $\lambda_{\tau}(\beta) \in \Omega_{\beta}$ for some $\tau \in \Gamma_{\beta}$, and these representations are interpreted as distinct elements. Thus under this definition the sets Ω_{α} and Ω_{β} become disjoint. (Regard the sets in $\{\Omega_{\gamma}\}_{\gamma \in \Gamma}$ as subsets of distinct copies of $\mathbb{C}$ so that Ω_{α} and Ω_{β} are regarded as disjoint for distinct indices $\alpha, \beta \in \Gamma$.) Unlike the ordinary union, the disjoint union can be viewed as the union of all Ω_{γ} considering *multiplicity of points* in it disregarding possible overlappings.

From now on suppose either (i) Ω is a nonempty Borel subset of the complex plane $\mathbb{C}$ and $\mathcal{A}_{\Omega}$ is the σ-algebra of all Borel subsets of Ω, or (ii) Ω is a nonempty set consisting of a disjoint union of Borel subsets Ω_{γ} of $\mathbb{C}$ and $\mathcal{A}_{\Omega}$ is the σ-algebra of all disjoint unions included in Ω made up of Borel subsets of $\mathbb{C}$, again referred to as the σ-algebra of Borel subsets of Ω.

Definition 3.6. Let Ω be a nonempty set and let $\mathcal{A}_{\Omega}$ be a σ-algebra of Borel subsets of Ω as above. A (complex) *spectral measure* in a (complex) Hilbert space $\mathcal{H}$ is a mapping $E : \mathcal{A}_{\Omega} \to \mathcal{B}[\mathcal{H}]$ satisfying the following axioms.

(a) $E(\Lambda)$ is an orthogonal projection for every $\Lambda \in \mathcal{A}_\Omega$,

(b) $E(\varnothing) = O$ and $E(\Omega) = I$,

(c) $E(\Lambda_1 \cap \Lambda_2) = E(\Lambda_1)E(\Lambda_2)$ for every $\Lambda_1, \Lambda_2 \in \mathcal{A}_\Omega$,

(d) $E\left(\bigcup_k \Lambda_k\right) = \sum_k E(\Lambda_k)$ (ordinary union) whenever $\{\Lambda_k\}$ is a countable collection of pairwise disjoint sets in $\mathcal{A}_\Omega$ (i.e., E is countably additive).

If $\{\Lambda_k\}$ is a countably infinite collection of pairwise disjoint sets in $\mathcal{A}_\Omega$, then the identity in (d) means convergence in the strong topology:

$$\sum_{k=1}^{n} E(\Lambda_k) \overset{s}{\longrightarrow} E\left(\bigcup_k \Lambda_k\right).$$

Indeed, since $\Lambda_j \cap \Lambda_k = \varnothing$ if $j \neq k$, it follows by properties (b) and (c) that $E(\Lambda_j)E(\Lambda_k) = E(\Lambda_j \cap \Lambda_k) = E(\varnothing) = O$ for $j \neq k$, and so $\{E(\Lambda_k)\}$ is an orthogonal sequence of orthogonal projections in $\mathcal{B}[\mathcal{H}]$. Then, according to Proposition 1.K, $\{\sum_{k=1}^{n} E(\Lambda_k)\}$ converges strongly to the orthogonal projection in $\mathcal{B}[\mathcal{H}]$ onto $\left(\sum_k \mathcal{R}(E(\Lambda_k))\right)^- = \bigvee \left(\bigcup_k \mathcal{R}(E(\Lambda_k))\right)$. Thus what property (d) says is: $E\left(\bigcup_k \Lambda_k\right)$ coincides with the orthogonal projection in $\mathcal{B}[\mathcal{H}]$ onto $\bigvee \left(\bigcup_k \mathcal{R}(E(\Lambda_k))\right)$. This generalizes the concept of resolution of the identity on $\mathcal{H}$. In fact, if $\{\Lambda_k\}$ is a partition of Ω, then the orthogonal sequence of orthogonal projections $\{E(\Lambda_k)\}$ is such that

$$\sum_{k=1}^{n} E(\Lambda_k) \overset{s}{\longrightarrow} E\left(\bigcup_k \Lambda_k\right) = E(\Omega) = I.$$

Let $E \colon \mathcal{A}_\Omega \to \mathcal{B}[\mathcal{H}]$ be a spectral measure into $\mathcal{B}[\mathcal{H}]$. For each pair of vectors $x, y \in \mathcal{H}$ consider the map $\pi_{x,y} \colon \mathcal{A}_\Omega \to \mathbb{C}$ defined for every $\Lambda \in \mathcal{A}_\Omega$ by

$$\pi_{x,y}(\Lambda) = \langle E(\Lambda)x \,;\, y \rangle = \langle x \,;\, E(\Lambda)y \rangle.$$

(Recall: $E(\Lambda)$ is self-adjoint.) The function $\pi_{x,y}$ is an ordinary complex-valued (countably additive) measure on $\mathcal{A}_\Omega$. For each $x, y \in \mathcal{H}$ consider the measure space $(\Omega, \mathcal{A}_\Omega, \pi_{x,y})$. Let $B(\Omega)$ denote the algebra of all complex-valued *bounded $\mathcal{A}_\Omega$-measurable functions* $\phi \colon \Omega \to \mathbb{C}$ on Ω. The integral $\int \phi \, d\pi_{x,y}$ of ϕ in $B(\Omega)$ with respect to the measure $\pi_{x,y}$ will also be denoted by $\int \phi(\lambda) \, d\pi_{x,y}$, or $\int \phi \, d\langle E_\lambda x \,;\, y \rangle$, or $\int \phi \, d\langle x \,;\, E_\lambda y \rangle$, or $\int \phi(\lambda) \, d\langle E_\lambda x \,;\, y \rangle$, or $\int \phi(\lambda) \, d\langle x \,;\, E_\lambda y \rangle$.

Lemma 3.7. *Let $E \colon \mathcal{A}_\Omega \to \mathcal{B}[\mathcal{H}]$ be a spectral measure. Every function ϕ in $B(\Omega)$ is $\pi_{x,y}$-integrable, and there is a unique operator $F \in \mathcal{B}[\mathcal{H}]$ for which*

$$\langle Fx \,;\, y \rangle = \int \phi(\lambda) \, d\langle E_\lambda x \,;\, y \rangle \quad \text{for every } x, y \in \mathcal{H}.$$

Proof. Let $\phi \colon \Omega \to \mathbb{C}$ be a (bounded $\mathcal{A}_\Omega$-measurable) function in $B(\Omega)$. Set $\|\phi\|_\infty = \sup_{\lambda \in \Omega} |\phi(\lambda)|$. Since $\langle E(\cdot)x \,;\, x \rangle = \|E(\cdot)x\|^2$ is a positive measure,

$$\left| \int \phi(\lambda) \, d\langle E_\lambda x \,;\, x \rangle \right| \leq \|\phi\|_\infty \int d\langle E_\lambda x \,;\, x \rangle = \|\phi\|_\infty \langle E(\Omega)x \,;\, x \rangle = \|\phi\|_\infty \|x\|^2$$

for $x \in \mathcal{H}$. Apply the polarization identity and the parallelogram law (Proposition 1.A (a$_1$,b)) for the sesquilinear form $\langle E(\varLambda) \cdot \, ; \cdot \rangle \colon \mathcal{H} \times \mathcal{H} \to \mathbb{C}$ to get

$$\left| \int \phi(\lambda) \, d\langle E_\lambda x \, ; y \rangle \right| \le \|\phi\|_\infty \big(\|x\|^2 + \|y\|^2 \big) \quad \text{for every } x, y \in \mathcal{H}.$$

So ϕ is $\pi_{x,y}$-integrable. Take the sesquilinear form $f \colon \mathcal{H} \times \mathcal{H} \to \mathbb{C}$ given by

$$f(x,y) = \int \phi(\lambda) \, d\langle E_\lambda x \, ; y \rangle \quad \text{for every } x, y \in \mathcal{H},$$

which is bounded ($\sup_{\|x\| \le 1, \|y\| \le 1} |f(x,y)| \le 2\|\phi\|_\infty < \infty$), and so is the linear functional $f(\cdot, y) \colon \mathcal{H} \to \mathbb{C}$ for each $y \in \mathcal{H}$. Thus, by the RIESZ REPRESENTATION THEOREM IN HILBERT SPACE (see, e.g., [78, Theorem 5.62]), *for each $y \in \mathcal{H}$ there is a unique $z_y \in \mathcal{H}$ for which $f(x,y) = \langle x \, ; z_y \rangle$ for every $x \in \mathcal{H}$.* This establishes a mapping $F^\# \colon \mathcal{H} \to \mathcal{H}$ that assigns to each $y \in \mathcal{H}$ a unique $z_y \in \mathcal{H}$ such that $f(x,y) = \langle x \, ; F^\# y \rangle$ for every $x \in \mathcal{H}$. As it is easy to show, $F^\#$ is unique and lies in $\mathcal{B}[\mathcal{H}]$. So the mapping $F = (F^\#)^*$ is the unique operator in $\mathcal{B}[\mathcal{H}]$ such that $\langle Fx \, ; y \rangle = f(x,y)$ for every $x, y \in \mathcal{H}$. $\square$

Notation. The unique $F \in \mathcal{B}[\mathcal{H}]$ for which $\langle Fx \, ; y \rangle = \int \phi(\lambda) \, d\langle E_\lambda x \, ; y \rangle$ for every $x, y \in \mathcal{H}$ as in Lemma 3.7 is usually denoted by

$$F = \int \phi(\lambda) \, dE_\lambda.$$

In particular, if ϕ is the characteristic function $\chi_\varLambda$ of a set $\varLambda \in \mathcal{A}_\Omega$, then $\langle E(\varLambda)x \, ; y \rangle = \pi_{x,y}(\varLambda) = \int_\varLambda d\pi_{x,y} = \int \chi_\varLambda \, d\pi_{x,y} = \int \chi_\varLambda(\lambda) \, d\langle E_\lambda x \, ; y \rangle = \int_\varLambda d\langle E_\lambda x \, ; y \rangle$ for every $x, y \in \mathcal{H}$. So the orthogonal projection $E(\varLambda) \in \mathcal{B}[\mathcal{H}]$ is denoted by

$$E(\varLambda) = \int \chi_\varLambda(\lambda) \, dE_\lambda = \int_\varLambda dE_\lambda.$$

Lemma 3.8. *Let $E \colon \mathcal{A}_\Omega \to \mathcal{B}[\mathcal{H}]$ be a spectral measure. If ϕ, ψ are functions in $B(\Omega)$ and $F = \int \phi(\lambda) \, dE_\lambda$, $G = \int \psi(\lambda) \, dE_\lambda$ are operators in $\mathcal{B}[\mathcal{H}]$, then*

(a) $\displaystyle F^* = \int \overline{\phi(\lambda)} \, dE_\lambda$ *and*   (b) $\displaystyle FG = \int \phi(\lambda)\psi(\lambda) \, dE_\lambda.$

Proof. Given ϕ and ψ in $B(\Omega)$, consider the operators $F = \int \phi(\lambda) \, dE_\lambda$ and $G = \int \psi(\lambda) \, dE_\lambda$ in $\mathcal{B}[\mathcal{H}]$ (cf. Lemma 3.7). Take an arbitrary pair $x, y \in \mathcal{H}$.

(a) Recall: $E(\varLambda)$ is self-adjoint for every $\varLambda \in \mathcal{A}_\Omega$. Thus we get

$$\langle F^*x \, ; y \rangle = \overline{\langle y \, ; F^*x \rangle} = \overline{\langle Fy \, ; x \rangle} = \int \overline{\phi(\lambda) \, d\langle E_\lambda y \, ; x \rangle} = \int \overline{\phi(\lambda)} \, d\langle E_\lambda x \, ; y \rangle.$$

(b) Set $\pi = \pi_{x,y} \colon \mathcal{A}_\Omega \to \mathbb{C}$. Since $\langle Fx \, ; y \rangle = \int \phi \, d\pi = \int \phi(\lambda) \, d\langle E_\lambda x \, ; y \rangle$,

$$\langle FGx \, ; y \rangle = \int \phi(\lambda) \, d\langle E_\lambda Gx \, ; y \rangle.$$

Consider the measure $\nu = \nu_{x,y} \colon \mathcal{A}_\Omega \to \mathbb{C}$ defined for every $\varLambda \in \mathcal{A}_\Omega$ by

$$\nu(\Lambda) = \langle E(\Lambda)Gx\,;y\rangle.$$

Thus

$$\langle FGx\,;y\rangle = \int \phi\,d\nu.$$

Since $\langle Gx\,;y\rangle = \int \psi\,d\pi = \int \psi(\lambda)\,d\langle E_\lambda x\,;y\rangle$,

$$\nu(\Lambda) = \langle Gx\,;E(\Lambda)y\rangle = \int \psi(\lambda)\,d\langle E_\lambda x\,;E(\Lambda)y\rangle.$$

Therefore, since $\langle E(\Delta)x\,;E(\Lambda)y\rangle = \langle E(\Lambda)E(\Delta)x\,;y\rangle = \langle E(\Delta\cap\Lambda)x\,;y\rangle = \pi(\Delta\cap\Lambda) = \pi|_\Lambda(\Delta\cap\Lambda)$ for every $\Lambda, \Delta \in \mathcal{A}_\Omega$, where the measure $\pi|_\Lambda\colon \mathcal{A}_\Lambda \to \mathbb{C}$ is the restriction of $\pi\colon \mathcal{A}_\Omega \to \mathbb{C}$ to $\mathcal{A}_\Lambda = \wp(\Lambda)\cap\mathcal{A}_\Omega$, we get

$$\int_\Lambda d\nu = \nu(\Lambda) = \int \psi(\lambda)\,d\langle E_\lambda x\,;E(\Lambda)y\rangle = \int \psi\,d\pi|_\Lambda = \int \chi_\Lambda \psi\,d\pi = \int_\Lambda \psi\,d\pi$$

for every $\Lambda \in \mathcal{A}_\Omega$. This is usually written as $d\nu = \psi\,d\pi$, meaning that $\psi = \frac{d\nu}{d\pi}$ is the Radon–Nikodým derivative of ν with respect to π. In fact, since $\int_\Lambda d\nu = \int_\Lambda \psi\,d\pi$ for every $\Lambda \in \mathcal{A}_\Omega$, then $\int \phi\,d\nu = \int \phi\psi\,d\pi$ for every $\phi \in B(\Omega)$. Hence

$$\langle FGx\,;y\rangle = \int \phi\,d\nu = \int \phi\psi\,d\pi = \int \phi(\lambda)\psi(\lambda)\,d\langle E_\lambda x\,;y\rangle. \qquad \square$$

Lemma 3.9. *If* $E\colon \mathcal{A}_\Omega \to \mathcal{B}[\mathcal{H}]$ *a spectral measure, and if* ϕ *is a function in* $B(\Omega)$, *then the operator* $F = \int\phi(\lambda)\,dE_\lambda$ *is normal.*

Proof. By Lemma 3.8,

$$F^*F = \int |\phi(\lambda)|^2\,dE_\lambda = FF^*. \qquad \square$$

Lemma 3.10. *Let* $E\colon \mathcal{A}_\Omega \to \mathcal{B}[\mathcal{H}]$ *be a spectral measure. If* $\varnothing \neq \Omega \subset \mathbb{C}$ *is compact, then the operator* $F = \int \lambda\,dE_\lambda$ *is well defined and normal in* $\mathcal{B}[\mathcal{H}]$. *Also*

$$p(F, F^*) = \int p(\lambda, \overline{\lambda})\,dE_\lambda$$

is again normal in $\mathcal{B}[\mathcal{H}]$ *for every polynomial* $p(\cdot\,,\cdot)\colon \Omega\times\Omega \to \mathbb{C}$ *in* λ *and* $\overline{\lambda}$.

Proof. Consider Lemmas 3.7, 3.8, 3.9. If $\Omega \subset \mathbb{C}$ is compact, then the identity map $\varphi\colon \Omega \to \Omega$ lies in $B(\Omega)$ and F defined for $\varphi(\lambda) = \lambda$ is normal. For any function $p(\cdot\,,\bar\cdot) = \sum_{i,j=0}^{n,m} \alpha_{i,j}(\cdot)^i\,\overline{(\cdot)}^j\colon \Omega \to \mathbb{C}$, which also lies in $B(\Omega)$, set $p(F, F^*) = \sum_{i,j=0}^{n,m} \alpha_{i,j} F^i F^{*j}$ in $\mathcal{B}[\mathcal{X}]$, so that $p(F, F^*)p(F, F^*)^* = p(F, F^*)^*p(F, F^*)$. $\square$

Remark. Lemma 3.10 still holds if $\Omega = \bigcup_{\gamma \in \Gamma} \Omega_\gamma$ is a disjoint union where each $\Omega_\gamma \subset \mathbb{C}$ is compact. Actually if $\lambda \in \Omega$, then $\lambda = \lambda \in \Omega_\beta$ for a unique $\beta \in \Gamma$. Thus let $\varphi\colon \Omega \to \mathbb{C}$ be given by $\varphi(\lambda) = \lambda \in \mathbb{C}$ for each $\lambda \in \Omega$. In this case $F = \int \varphi(\lambda)\,dE_\lambda$ and $p(\cdot\,,\cdot)$ is replaced by $p(\varphi(\cdot), \overline{\varphi(\cdot)})\colon \Omega \to \mathbb{C}$ in $B(\Omega)$.

The standard form of the Spectral Theorem (Theorem 3.15) states the converse of Lemma 3.9. A proof of it relies on Theorem 3.11. In fact, Theorems 3.11 and 3.15 can be thought of as equivalent versions of the Spectral Theorem.

3.4 Spectral Theorem: The General Case

Assume all Hilbert spaces are nonzero and complex. Consider the notation in the paragraph preceding Proposition 3.B. Take the *multiplication operator* M_ϕ in $\mathcal{B}[L^2(\Omega, \mu)]$, which is normal, where ϕ lies in $L^\infty(\Omega, \mu)$, Ω is a nonempty set, and μ is a positive measure on $\mathcal{A}_\Omega$ (cf. Proposition 3.B in Section 3.6):

$$M_\phi f = \phi f \quad \text{for every} \quad f \in L^2(\Omega, \mu).$$

Theorem 3.11. (SPECTRAL THEOREM – FIRST VERSION). *If T is a normal operator on a Hilbert space, then there is a positive measure μ on a σ-algebra $\mathcal{A}_\Omega$ of subsets of a set $\Omega \neq \varnothing$ and a function φ in $L^\infty(\Omega, \mu)$ such that T is unitarily equivalent to the multiplication operator M_φ on $L^2(\Omega, \mu)$, and $\|\varphi\|_\infty = \|T\|$.*

Proof. Let T be a normal operator on a Hilbert space $\mathcal{H}$. We split the proof into two parts. In part (a) we prove the theorem for the case where there exists a *star-cyclic vector* x for T, which means there exists a vector $x \in \mathcal{H}$ for which $\bigvee_{m,n}\{T^n T^{*m} x\} = \mathcal{H}$ (each index m and n runs over all nonnegative integers; that is, $(m, n) \in \mathbb{N}_0 \times \mathbb{N}_0$). In part (b) we consider the complementary case.

(a) Fix a unit vector $x \in \mathcal{H}$. Take a compact $\varnothing \neq \Omega \subset \mathbb{C}$. Let $P(\Omega)$ be the set of all polynomials $p(\cdot, \cdot) \colon \Omega \times \Omega \to \mathbb{C}$ in λ and $\overline{\lambda}$ (equivalently, $p(\cdot, \overline{\cdot}) \colon \Omega \to \mathbb{C}$ such that $\lambda \mapsto p(\lambda, \overline{\lambda})$). Let $C(\Omega)$ be the set of all complex-valued continuous functions on Ω. Thus $P(\Omega) \subset C(\Omega)$. By the STONE–WEIERSTRASS THEOREM, if Ω is compact, then $P(\Omega)$ is dense in the Banach space $(C(\Omega), \|\cdot\|_\infty)$:

$$P(\Omega)^- = C(\Omega) \quad \text{in the sup-norm topology.}$$

(For real and complex versions of the Stone–Weierstrass Theorem see, e.g., [45, Theorems IV.6.16,17], [35, Theorems 7.3.1,2], [86, Theorems III.1.1,4], or [30, Theorem V.8.1].) Consider the functional $\Phi \colon P(\Omega) \to \mathbb{C}$ given by

$$\Phi(p) = \langle p(T, T^*) x \, ; x \rangle \quad \text{for every} \quad p \in P(\Omega),$$

which is linear on the linear space $P(\Omega)$. Moreover,

$$|\Phi(p)| \leq \|p(T, T^*)\| \|x\|^2 = \|p(T, T^*)\| = r(p(T, T^*))$$

by the Schwarz inequality since $\|x\| = 1$ and $p(T, T^*)$ is normal (thus normaloid). From now on set $\Omega = \sigma(T)$. According to Theorem 2.8, and with $P(\sigma(T))$ viewed as a linear manifold of the Banach space $(C(\Omega), \|\cdot\|_\infty)$,

$$|\Phi(p)| \leq r(p(T, T^*)) = \sup_{\lambda \in \sigma(p(T, T^*))} |\lambda| = \sup_{\lambda \in \sigma(T)} |p(\lambda, \overline{\lambda})| = \|p\|_\infty$$

Then $\Phi \colon (P(\sigma(T)), \|\cdot\|_\infty) \to (\mathbb{C}, |\cdot|)$ is a linear contraction. Since $P(\sigma(T))$ is dense in $C(\sigma(T))$ and Φ is linear continuous, Φ has a unique linear continuous extension over $C(\sigma(T))$, denoted again by $\Phi \colon (C(\sigma(T)), \|\cdot\|_\infty) \to (\mathbb{C}, |\cdot|)$.

Claim. $\Phi \colon C(\sigma(T)) \to \mathbb{C}$ is a *positive functional* in the following sense:

$$0 \leq \Phi(\psi) \quad \text{for every} \quad \psi \in C(\sigma(T)) \text{ such that } 0 \leq \psi(\lambda) \text{ for every } \lambda \in \sigma(T).$$

Proof. Let $\psi \in C(\sigma(T))$ be such that $0 \le \psi(\lambda)$ for every $\lambda \in \sigma(T)$. Take an arbitrary $\varepsilon > 0$. Since Φ is a linear contraction and $P(\sigma(T))^- = C(\sigma(T))$, there is a polynomial $p_\varepsilon \in P(\sigma(T))$ such that $0 \le p_\varepsilon(\lambda, \overline{\lambda})$ for every $\lambda \in \sigma(T)$ and

$$|\Phi(p_\varepsilon) - \Phi(\psi)| \le \|p_\varepsilon - \psi\|_\infty < \varepsilon.$$

The Stone–Weierstrass Theorem for real functions ensures the existence of a polynomial q_ε in the real variables α, β with real coefficients for which

$$|q_\varepsilon(\alpha, \beta)^2 - p_\varepsilon(\lambda, \overline{\lambda})| < \varepsilon \quad \text{for every} \quad \lambda = \alpha + i\beta \in \sigma(T).$$

(Note: $0 \le p_\varepsilon(\lambda, \overline{\lambda})$ and the above may be replaced by $|q_\varepsilon(\alpha, \beta) - p_\varepsilon(\lambda, \overline{\lambda})^{\frac{1}{2}}|$). Take the Cartesian decomposition $T = A + iB$ of T, where A and B are commuting self-adjoint operators (Proposition 1.O). Then $q_\varepsilon(A, B)$ is self-adjoint, and so $q_\varepsilon(A, B)^2$ is a nonnegative operator commuting with T and T^*. Also,

$$
\begin{aligned}
|\langle q_\varepsilon(A, B)^2 x\,;x\rangle - \Phi(p_\varepsilon)| &= |\langle q_\varepsilon(A, B)^2 x\,;x\rangle - \langle p_\varepsilon(T, T^*)x\,;x\rangle| \\
&= |\langle (q_\varepsilon(A, B)^2 - p_\varepsilon(T, T^*))x\,;x\rangle| \\
&\le \|q_\varepsilon(A, B)^2 - p_\varepsilon(T, T^*)\| = r(q_\varepsilon(A, B)^2 - p_\varepsilon(T, T^*))
\end{aligned}
$$

since $q_\varepsilon(A, B)^2 - p_\varepsilon(T, T^*)$ is normal, thus normaloid (reason: $q_\varepsilon(A, B)^2$ is a self-adjoint operator commuting with the normal operator $p_\varepsilon(T, T^*)$). Hence

$$|\langle q_\varepsilon(A, B)^2 x\,;x\rangle - \Phi(p_\varepsilon)| \le \sup_{\lambda = \alpha + i\beta \in \sigma(T)} |q_\varepsilon(\alpha, \beta)^2 - p_\varepsilon(\lambda, \overline{\lambda})| \le \varepsilon$$

by Theorem 2.8 again. So

$$|\langle q_\varepsilon(A, B)^2 x\,;x\rangle - \Phi(\psi)| \le |\langle q_\varepsilon(A, B)^2 x\,;x\rangle - \Phi(p_\varepsilon)| + |\Phi(p_\varepsilon) - \Phi(\psi)| \le 2\varepsilon.$$

But $0 \le \langle p_\varepsilon(A, B)^2 x\,;x\rangle$ (since $O \le q_\varepsilon(A, B)^2$). Then for every $\varepsilon > 0$,

$$0 \le \langle p_\varepsilon(A, B)^2 x\,;x\rangle \le 2\varepsilon + \Phi(\psi).$$

Therefore $0 \le \Phi(\psi)$. This completes the proof of the claimed results.

Take the *bounded, linear, positive* functional Φ on $C(\sigma(T))$. The RIESZ REPRESENTATION THEOREM IN $C(\Omega)$ (see, e.g., [53, Theorem 56.D], [103, Theorem 2.14], [102, Theorem 13.23], or [81, Theorem 12.5]) ensures the existence of a *finite positive* measure μ on the σ-algebra $\mathcal{A}_\Omega = \mathcal{A}_{\sigma(T)}$ of Borel subsets of the *compact* set $\Omega = \sigma(T)$ (where continuous functions are measurable) for which

$$\Phi(\psi) = \int \psi \, d\mu \quad \text{for every} \quad \psi \in C(\Omega).$$

Let $U : P(\Omega) \to \mathcal{H}$ be a mapping defined by

$$Up = p(T, T^*)x \quad \text{for every} \quad p \in P(\Omega).$$

Regard $P(\Omega)$ as a linear manifold of the Banach space $L^2(\Omega, \mu)$ equipped with the L^2-norm $\|\cdot\|_2$. So U is a linear transformation, and

$$
\begin{aligned}
\|p\|_2^2 = \int |p|^2 \, d\mu = \int p\overline{p} \, d\mu &= \Phi(p\overline{p}) = \langle p(T, T^*)\overline{p(T, T^*)}x\,;x\rangle \\
&= \langle p(T, T^*)p(T, T^*)^* x\,;x\rangle = \|p(T, T^*)^* x\|^2 = \|p(T, T^*)x\|^2 = \|Up\|^2
\end{aligned}
$$

for every $p \in P(\Omega)$ (by the definition of $p(T, T^*)$ in the proof of Theorem 2.8, which is normal as in Lemma 3.10). Then U is an isometry, thus injective, and hence it has an inverse on its range

$$\mathcal{R}(U) = \{p(T, T^*)x \colon p \in P(\Omega)\} \subseteq \mathcal{H},$$

say $U^{-1} \colon \mathcal{R}(U) \to P(\Omega)$. For each polynomial p in $P(\Omega)$ let q be the polynomial in $P(\Omega)$ defined by $q(\lambda, \overline{\lambda}) = \lambda p(\lambda, \overline{\lambda})$ for every $\lambda \in \Omega$. Let $\varphi \colon \Omega \to \Omega$ be the identity map (i.e., $\varphi(\lambda) = \lambda \subset \Omega$ for every $\lambda \in \Omega = \sigma(T) \subset \mathbb{C}$), which is bounded (i.e., $\varphi \in L^\infty(\Omega, \mu)$ since Ω is bounded). In this case we get

$$q = \varphi p \quad \text{and} \quad q(T, T^*)x = T p(T, T^*)x.$$

Thus

$$U^{-1}TUp = U^{-1}Tp(T, T^*)x = U^{-1}q(T, T^*)x = U^{-1}Uq = q = \varphi p = M_\varphi p$$

for every $p \in P(\Omega)$. Since $P(\Omega)$ is dense in the Banach space $(C(\Omega), \|\cdot\|_\infty)$, and since Ω is a compact set in $\mathbb{C}$ and μ is a finite positive measure on $\mathcal{A}_\Omega$, then $P(\Omega)$ is dense in the normed space $(C(\Omega), \|\cdot\|_2)$ as well, which in turn can be regarded as a dense linear manifold of the Banach space $(L^2(\Omega, \mu), \|\cdot\|_2)$, and so $P(\Omega)$ is dense in $(L^2(\Omega, \mu), \|\cdot\|_2)$. That is,

$$P(\Omega)^- = L^2(\Omega, \mu) \quad \text{in the } L^2\text{-norm topology.}$$

Moreover, since U is a linear isometry of $P(\Omega)$ onto $\mathcal{R}(U)$, which are linear manifolds of the Hilbert spaces $L^2(\Omega, \mu)$ and $\mathcal{H}$, then U extends by continuity to a unique linear isometry, also denoted by U, of the Hilbert space $P(\Omega)^- = L^2(\Omega, \mu)$ onto the range of U, viz., $\mathcal{R}(U) \subseteq \mathcal{H}$. (In fact, it extends onto the closure of $\mathcal{R}(U)$, but when U acts on $P(\Omega)^- = L^2(\Omega, \mu)$ its range is closed in $\mathcal{H}$ because every linear isometry of a Banach space into a normed space has a closed range; see, e.g., [78, Problem 4.41(d)].) So U is a unitary transformation (i.e., a surjective isometry) between the Hilbert spaces $L^2(\Omega, \mu)$ and $\mathcal{R}(U)$. If, however, x is a star-cyclic vector for T (i.e., if $\bigvee_{m,n}\{T^n T^{*m} x\} = \mathcal{H}$), then

$$\mathcal{R}(U) = \{p(T, T^*)x \colon p \in P(\Omega)\}^- = \bigvee_{m,n}\{T^n T^{*m} x\} = \mathcal{H}.$$

Hence U is a unitary transformation in $\mathcal{B}[L^2(\Omega, \mu), \mathcal{H}]$ for which

$$U^{-1}TU = M_\varphi.$$

Therefore the operator T on $\mathcal{H}$ is unitarily equivalent to the multiplication operator M_φ on $L^2(\Omega, \mu)$ for the identity map $\varphi \colon \Omega \to \Omega$ on $\Omega = \sigma(T)$ (i.e., $\varphi(\lambda) = \lambda$ for every $\lambda \in \sigma(T)$) which lies in $L^\infty(\Omega, \mu)$. Since T is normal (thus normaloid) and unitarily equivalent to M_φ (thus with the same norm as M_φ) we get $\|\varphi\|_\infty = \|T\|$. Indeed, according to Proposition 3.B,

$$\|\varphi\|_\infty = \text{ess sup}\,|\varphi| \leq \sup_{\lambda \in \sigma(T)} |\lambda| = r(T) = \|T\| = \|M_\varphi\| = \|\varphi\|_\infty.$$

(b) On the other hand, suppose there is no star-cyclic vector for T. Assume T and $\mathcal{H}$ are nonzero to avoid trivialities. Since T is normal, the nontrivial subspace $\mathcal{M}_x = \bigvee_{m,n}\{T^n T^{*m} x\}$ of $\mathcal{H}$ reduces T for an arbitrary unit vector x in $\mathcal{H}$. Consider the set $\mathfrak{S} = \{\bigoplus \mathcal{M}_x \colon \text{unit } x \in \mathcal{H}\}$ of all *orthogonal* direct sums of these subspaces, which is partially ordered (in the inclusion ordering). $\mathfrak{S}$ is not empty (if a unit $y \in \mathcal{M}_x{}^\perp$, then $\mathcal{M}_y = \bigvee_{m,n}\{T^n T^{*m} y\} \subseteq \mathcal{M}_x{}^\perp$ because $\mathcal{M}_x{}^\perp$ reduces T and so $\mathcal{M}_x \perp \mathcal{M}_y$). Moreover, every chain in $\mathfrak{S}$ has an upper bound in $\mathfrak{S}$ (the union of all orthogonal direct sums in a chain of orthogonal direct sums in $\mathfrak{S}$ is again an orthogonal direct sum in $\mathfrak{S}$). Thus Zorn's Lemma ensures that $\mathfrak{S}$ has a maximal element, say $\mathcal{M} = \bigoplus \mathcal{M}_x$, which coincides with $\mathcal{H}$ (otherwise it would not be maximal since $\mathcal{M} \oplus \mathcal{M}^\perp = \mathcal{H}$). Summing up:

> There exists an indexed family $\{x_\gamma\}_{\gamma \in \Gamma}$ of unit vectors in $\mathcal{H}$ generating a (similarly indexed) collection $\{\mathcal{M}_\gamma\}_{\gamma \in \Gamma}$ of orthogonal nontrivial subspaces of $\mathcal{H}$ such that each $\mathcal{M}_\gamma = \bigvee_{m,n}\{T^n T^{*m} x_\gamma\}$ reduces T, each x_γ is star-cyclic for $T|_{\mathcal{M}_\gamma}$, and $\mathcal{H} = \bigoplus_{\gamma \in \Gamma} \mathcal{M}_\gamma$.

Each restriction $T|_{\mathcal{M}_\gamma}$ is a normal operator on the Hilbert space $\mathcal{M}_\gamma$ (Proposition 1.Q). By part (a) there is a positive finite measure μ_γ on the σ-algebra $\mathcal{A}_{\Omega_\gamma}$ of Borel subsets of $\Omega_\gamma = \sigma(T|_{\mathcal{M}_\gamma}) \subseteq \mathbb{C}$ such that $T|_{\mathcal{M}_\gamma}$ in $\mathcal{B}[\mathcal{M}_\gamma]$ is unitarily equivalent to the multiplication operator M_{φ_γ} in $\mathcal{B}[L^2(\Omega_\gamma, \mu_\gamma)]$, where each $\varphi_\gamma \colon \Omega_\gamma \to \Omega_\gamma$ is the identity map ($\varphi_\gamma(\lambda) = \lambda$ for every $\lambda \in \Omega_\gamma$), which is bounded on $\Omega_\gamma = \sigma(T|_{\mathcal{M}_\gamma})$ (i.e., $\varphi_\gamma \in L^\infty(\Omega_\gamma, \mu_\gamma)$). Thus there is a unitary U_γ in $\mathcal{B}[\mathcal{M}_\gamma, L^2(\Omega_\gamma, \mu_\gamma)]$ for which $U_\gamma T|_{\mathcal{M}_\gamma} = M_{\varphi_\gamma} U_\gamma$. Take the external orthogonal direct sum $\bigoplus_{\gamma \in \Gamma} L^2(\Omega_\gamma, \mu_\gamma)$, and take the unitary $U = \bigoplus_{\gamma \in \Gamma} U_\gamma$ in $\mathcal{B}[\bigoplus_{\gamma \in \Gamma} \mathcal{M}_\gamma, \bigoplus_{\gamma \in \Gamma} L^2(\Omega_\gamma, \mu_\gamma)]$ such that $(\bigoplus_{\gamma \in \Gamma} U_\gamma)(\bigoplus_{\gamma \in \Gamma} T|_{\mathcal{M}_\gamma}) = \bigoplus_{\gamma \in \Gamma} U_\gamma T|_{\mathcal{M}_\gamma} = \bigoplus_\gamma M_{\varphi_\gamma} U_\gamma = (\bigoplus_{\gamma \in \Gamma} M_{\varphi_\gamma})(\bigoplus_{\gamma \in \Gamma} U_\gamma)$. Therefore

$$T = \bigoplus_{\gamma \in \Gamma} T|_{\mathcal{M}_\gamma} \text{ in } \mathcal{B}[\mathcal{H}] = \mathcal{B}[\bigoplus_{\gamma \in \Gamma} \mathcal{M}_\gamma] \text{ is}$$

unitarily equivalent to $\bigoplus_{\gamma \in \Gamma} M_{\varphi_\gamma}$ in $\mathcal{B}[\bigoplus_{\gamma \in \Gamma} L^2(\Omega_\gamma, \mu_\gamma)]$.

Let $\Omega = \bigcup_{\gamma \in \Gamma} \Omega_\gamma$ be the disjoint union of $\{\Omega_\gamma\}_{\gamma \in \Gamma}$. Note: $\Omega_\gamma = \sigma(T|_{\mathcal{M}_\gamma}) \subseteq \sigma(T)$ since each $\mathcal{M}_\gamma$ reduces T (Proposition 2.F(b)), and so $\Omega \subseteq \bigcup_{\gamma \in \Gamma} \sigma(T)$. Take the collection $\mathcal{A}_\Omega$ of all disjoint unions of Borel subsets of each Ω_γ; that is, the collection of all disjoint unions of the form $\Lambda = \bigcup_{\gamma \in \Gamma'} \Lambda_\gamma$ for any subset Γ' of Γ where Λ_γ is a Borel subset of Ω_γ (i.e., $\Lambda_\gamma \in \mathcal{A}_{\Omega_\gamma}$). This is the σ-algebra of Borel subsets of Ω. Let $\mu_\gamma \colon \mathcal{A}_{\Omega_\gamma} \to \mathbb{R}$ be the finite measure on each $\mathcal{A}_{\Omega_\gamma}$ whose existence was ensured in part (a). Consider the Borel subsets $\Lambda = \bigcup_{j \in \mathbb{J}} \Lambda_j$ of Ω made up of countable disjoint unions of Borel subsets of each Ω_j (i.e., $\Lambda_j \in \mathcal{A}_{\Omega_j}$). Let μ be the measure on $\mathcal{A}_\Omega$ given by $\mu(\Lambda) = \sum_{j \in \mathbb{J}} \mu_j(\Lambda_j)$ for every $\Lambda = \bigcup_{j \in \mathbb{J}} \Lambda_j$ in $\mathcal{A}_\Omega$ with $\{\Lambda_j\}_{j \in \mathbb{J}}$ being an arbitrary countable collection of sets with each $\Lambda_j \in \mathcal{A}_{\Omega_j}$ for any countable subset $\mathbb{J}$ of Γ. (For a proper subset Γ' of Γ we may identify $\bigcup_{\gamma \in \Gamma'} \Lambda_\gamma$ with $\bigcup_{\gamma \in \Gamma} \Lambda_\gamma$ by setting $\Lambda_\gamma = \varnothing$ for every $\gamma \in \Gamma \backslash \Gamma'$.) The measure μ is positive since each μ_j is (if $\Lambda_j = \Omega_j$, then $\mu_j(\Lambda_j) > 0$ for at least one j in one $\mathbb{J}$).

Consider the measure space $(\Omega, \mathcal{A}_\Omega, \mu)$.

Definition. Let $\varphi\colon\Omega\to\mathbb{C}$ be a function obtained from the identity maps $\varphi_\gamma\colon\Omega_\gamma\to\Omega_\gamma$ as follows. If $\boldsymbol\lambda$ lies in the *disjoint* union $\Omega=\bigcup_{\gamma\in\Gamma}\Omega_\gamma$, then $\boldsymbol\lambda=\lambda\in\Omega_\beta\subset\mathbb{C}$ for a unique index $\beta\in\Gamma$ and a unique $\lambda\in\Omega_\beta$. Thus for each $\boldsymbol\lambda\in\Omega$ set $\varphi(\boldsymbol\lambda)=\varphi_\beta(\lambda)=\lambda\in\mathbb{C}$. This can be viewed as a sort of identity map including multiplicity, referred to as an *identity map with multiplicity.*

Since $\Omega_\gamma=\sigma(T|_{\mathcal{M}_\gamma})\subseteq\sigma(T)$ for all $\gamma\in\Gamma$, φ is bounded. So $\varphi\in L^\infty(\Omega,\mu)$ and

$$\|\varphi\|_\infty=\operatorname{ess\,sup}|\varphi|\leq\sup_{\boldsymbol\lambda\in\Omega}|\varphi(\boldsymbol\lambda)|\leq\sup_{\lambda\in\sigma(T)}|\lambda|=r(T)\leq\|T\|.$$

(Actually, $r(T)=\|T\|$ because T is normaloid.) Now we show:

$$\bigoplus_{\gamma\in\Gamma}M_{\varphi_\gamma}\text{ in }\mathcal{B}[\bigoplus_\gamma L^2(\Omega_\gamma,\mu_\gamma)]\text{ is}$$
$$\text{unitarily equivalent to }M_\varphi\text{ in }\mathcal{B}[L^2(\Omega,\mu)].$$

First observe that the Hilbert spaces $L^2(\Omega,\mu)$ and $\bigoplus_{\gamma\in\Gamma}L^2(\Omega_\gamma,\mu_\gamma)$ are unitarily equivalent. In fact, direct sums and topological sums of families of externally orthogonal Hilbert spaces are unitarily equivalent, and therefore $\bigoplus_{\gamma\in\Gamma}L^2(\Omega_\gamma,\mu_\gamma)\cong(\sum_{\gamma\in\Gamma}L^2(\Omega_\gamma,\mu_\gamma))^-$. So take a unitary transformation $V\colon\bigoplus_{\gamma\in\Gamma}L^2(\Omega_\gamma,\mu_\gamma)\to(\sum_{\gamma\in\Gamma}L^2(\Omega_\gamma,\mu_\gamma))^-$. Next consider a transformation $W\colon\sum_{\gamma\in\Gamma}L^2(\Omega_\gamma,\mu_\gamma)\to L^2(\bigcup_{\gamma\in\Gamma}\Omega_\gamma,\mu)$ taking each function $\psi=\sum_{\gamma\in\Gamma}\psi_\gamma$ in $\sum_{\gamma\in\Gamma}L^2(\Omega_\gamma,\mu_\gamma)$ with $\psi_\gamma\in L^2(\Omega_\gamma,\mu_\gamma)$ to a function $W\psi$ in $L^2(\bigcup_{\gamma\in\Gamma}\Omega_\gamma,\mu)$ defined as follows. If $\boldsymbol\lambda\in\bigcup_{\gamma\in\Gamma}\Omega_\gamma$, then $\boldsymbol\lambda=\lambda\in\Omega_\beta$ for only one $\beta\in\Gamma$. Take this unique $\lambda\in\Omega_\beta\subset\mathbb{C}$ and set $(W\psi)(\boldsymbol\lambda)=(W\psi)|_{\Omega_\beta}(\boldsymbol\lambda)=\psi_\beta(\lambda)$. Regarding $\{\Omega_\gamma\}_{\gamma\in\Gamma}$ as a partition of the disjoint union Ω, this defines a linear and surjective map for which $\|W\psi\|^2=\int_\Omega|W\psi|^2\,d\mu=\sum_{\beta\in\Gamma}\int_{\Omega_\beta}|\psi_\beta|^2\,d\mu_\beta=\sum_{\gamma\in\Gamma}\|\psi_\gamma\|^2=\|\psi\|^2$, and so it is an isometry as well. Thus W is a unitary transformation of $(\sum_{\gamma\in\Gamma}L^2(\Omega_\gamma,\mu_\gamma))$ onto $L^2(\bigcup_{\gamma\in\Gamma}\Omega_\gamma,\mu)$. Then it extends by continuity to a unitary transformation, also denoted by W, over the whole Hilbert space $(\sum_{\gamma\in\Gamma}L^2(\Omega_\gamma,\mu_\gamma))^-$ onto the Hilbert space $L^2(\bigcup_{\gamma\in\Gamma}\Omega_\gamma,\mu)$. That is, $W\colon(\sum_{\gamma\in\Gamma}L^2(\Omega_\gamma,\mu_\gamma))^-\to L^2(\bigcup_{\gamma\in\Gamma}\Omega_\gamma,\mu)=L^2(\Omega,\mu)$ is unitary. So $(\sum_{\gamma\in\Gamma}L^2(\Omega_\gamma,\mu_\gamma))^-\cong L^2(\bigcup_{\gamma\in\Gamma}\Omega_\gamma,\mu)=L^2(\Omega,\mu)$ and hence by transitivity

$$\bigoplus_{\gamma\in\Gamma}L^2(\Omega_\gamma,\mu_\gamma)\cong L^2(\textstyle\bigcup_{\gamma\in\Gamma}\Omega_\gamma,\mu)=L^2(\Omega,\mu).$$

Now take any $\psi=\sum_{\gamma\in\Gamma}\psi_\gamma$ in $\sum_{\gamma\in\Gamma}L^2(\Omega_\gamma,\mu_\gamma)$. Also take any $\boldsymbol\lambda\in\Omega$ so that $\boldsymbol\lambda\in\Omega_\beta$ for one $\beta\in\Gamma$. Recall: $\varphi|_{\Omega_\beta}(\boldsymbol\lambda)=\varphi_\beta(\lambda)$ and $(W\psi)(\boldsymbol\lambda)=(W\psi)|_{\Omega_\beta}(\boldsymbol\lambda)=\psi_\beta(\lambda)$. Thus $(M_\varphi W\psi)(\boldsymbol\lambda)=(\varphi W\psi)(\boldsymbol\lambda)=\varphi_\gamma(\lambda)\psi_\beta(\lambda)=(\varphi_\beta\psi_\beta)(\lambda)$. On the other hand, since $V^{-1}\psi=\bigoplus_{\gamma\in\Gamma}\psi_\gamma$, it follows that $(\bigoplus_{\gamma\in\Gamma}M_{\varphi_\gamma})V^{-1}\psi=\bigoplus_{\gamma\in\Gamma}M_{\varphi_\gamma}\psi_\gamma=\bigoplus_{\gamma\in\Gamma}\varphi_\gamma\psi_\gamma$ and hence $V(\bigoplus_{\gamma\in\Gamma}M_{\varphi_\gamma})V^{-1}\psi=\sum_{\gamma\in\Gamma}\varphi_\gamma\psi_\gamma$. Thus $(WV(\bigoplus_{\gamma\in\Gamma}M_{\varphi_\gamma})V^{-1}\psi)(\boldsymbol\lambda)=(W\sum_{\gamma\in\Gamma}\varphi_\gamma\psi_\gamma)(\boldsymbol\lambda)=(\varphi_\beta\psi_\beta)(\lambda)$. Then $M_\varphi W=WV(\bigoplus_{\gamma\in\Gamma}M_{\varphi_\gamma})V^{-1}$ on $\sum_{\gamma\in\Gamma}L^2(\Omega_\gamma,\mu_\gamma)$, which extends by continuity over all $(\sum_{\gamma\in\Gamma}L^2(\Omega_\gamma,\mu_\gamma))^-$. Thus for the unitary transformation WV,

$$WV(\textstyle\bigoplus_{\gamma\in\Gamma}M_{\varphi_\gamma})=M_\varphi WV,$$

and so $\bigoplus_{\gamma \in \Gamma} M_{\varphi_\gamma}$ is unitarily equivalent to M_φ (i.e., $\bigoplus_\gamma M_{\varphi_\gamma} \cong M_\varphi$) through the unitary WV. Then $T \cong M_\varphi$ by transitivity since $T \cong \bigoplus_{\gamma \in \Gamma} M_{\varphi_\gamma}$. That is,

$$T \text{ in } \mathcal{B}[\mathcal{H}] \text{ is unitarily equivalent to } M_\varphi \text{ in } \mathcal{B}[L^2(\Omega, \mu)].$$

Finally, since $\|\varphi\|_\infty \leq \|T\|$ and $T \cong M_\varphi$ we get (Proposition 3.B)

$$\|\varphi\|_\infty \leq \|T\| = \|M_\varphi\| \leq \|\varphi\|_\infty. \qquad \square$$

Remark. The identity map with multiplicity $\varphi \colon \Omega \to \sigma(T) \subset \mathbb{C}$ may be neither injective nor surjective, but it is bounded. It takes disjoint unions in $\mathcal{A}_\Omega$ into ordinary unions in $\mathcal{A}_{\sigma(T)}$ so that $\varphi(\Omega)^- = \sigma(T)$ (see Proposition 2.F(d) since $T|_{\mathcal{M}_\gamma}$ is normal). Thus for each $\tilde\Lambda \subseteq \varphi(\Omega)$ there is one $\Lambda = \varphi^{-1}(\tilde\Lambda) \subseteq \Omega$ (e.g., a maximal) for which $\tilde\Lambda^- = \varphi(\Lambda)^-$ (since $\varphi(\varphi^{-1}(\tilde\Lambda)) = \tilde\Lambda$ for every $\tilde\Lambda \subseteq \varphi(\Omega)$), and so each set $\tilde\Lambda$ in $\mathcal{A}_{\sigma(T)}$ is such that $\tilde\Lambda^- = \varphi(\Lambda)^-$ for some set $\Lambda \in \mathcal{A}_\Omega$.

The following commutative diagram illustrates the proof of part (b).

$$
\begin{array}{ccc}
\mathcal{H} & \xrightarrow{\;\;T\;\;} & \mathcal{H} \\
{\scriptstyle U}\downarrow & & \downarrow{\scriptstyle U} \\
\bigoplus_{\gamma \in \Gamma} L^2(\Omega_\gamma, \mu_\gamma) & & \bigoplus_{\gamma \in \Gamma} L^2(\Omega_\gamma, \mu_\gamma) \\
{\scriptstyle V}\downarrow & & \downarrow{\scriptstyle V} \\
\big(\sum_{\gamma \in \Gamma} L^2(\Omega_\gamma, \mu_\gamma)\big)^- & & \big(\sum_{\gamma \in \Gamma} L^2(\Omega_\gamma, \mu_\gamma)\big)^- \\
{\scriptstyle W}\downarrow & & \downarrow{\scriptstyle W} \\
L^2(\Omega, \mu) & \xrightarrow{\;\;M_\varphi\;\;} & L^2(\Omega, \mu).
\end{array}
$$

Definition 3.12. A vector x in $\mathcal{H}$ is a *cyclic vector* for an operator T in $\mathcal{B}[\mathcal{H}]$ if $\mathcal{H}$ is the smallest invariant subspace for T that contains x. If T has a cyclic vector, then it is called a *cyclic operator*. A vector x in $\mathcal{H}$ is a *star-cyclic vector* for an operator T in $\mathcal{B}[\mathcal{H}]$ if $\mathcal{H}$ is the smallest reducing subspace for T that contains x. If T has a star-cyclic vector, then it is called a *star-cyclic operator*.

Therefore a vector x in $\mathcal{H}$ is *cyclic* for an operator T in $\mathcal{B}[\mathcal{H}]$ if and only if $\bigvee_n \{T^n x\} = \mathcal{H}$, where $\bigvee_n \{T^n x\} = \bigvee \big(\bigcup_n \{T^n x\}\big) = \{p(T)x \in \mathcal{H} \colon p$ is a polynomial$\}^-$ (i.e., $x \in \mathcal{H}$ is *cyclic* for $T \in \mathcal{B}[\mathcal{H}]$ if and only if $\{p(T)x \in \mathcal{H} \colon p$ is a polynomial$\}^- = \mathcal{H}$, which means $\{Sx \in \mathcal{H} \colon S \in \mathcal{P}(T)\}^- = \mathcal{H}$, where $\mathcal{P}(T)$ is the algebra of all polynomials in T with complex coefficients). Similarly, $x \in \mathcal{H}$ is *star-cyclic* for $T \in \mathcal{B}[\mathcal{H}]$ if and only if $\{Sx \in \mathcal{H} \colon S \in C^*(T)\}^- = \mathcal{H}$, where $C^*(T)$ is the C*-algebra generated by T (i.e., the smallest C*-algebra of operators from $\mathcal{B}[\mathcal{H}]$ containing T and the identity I — see, e.g., [30, Proposition IX.3.2]). Recall: $C^*(T) = \mathcal{P}(T, T^*)^-$, the closure in $\mathcal{B}[\mathcal{H}]$ of the set $\mathcal{P}(T, T^*)$ of all polynomials in two (noncommuting) variables T and T^* (in any order) with complex coefficients (see, e.g., [6, p. 1] — C*-algebra will be reviewed in Section 4.1), and $\{Sx \in \mathcal{H} \colon S \in \mathcal{P}(T, T^*)\}^- = \bigvee \big(\bigcup_{m,n} \{T^m T^{*n} x\} \cup \{T^{*n} T^m x\}\big)$.

Remark. Reducing subspaces are invariant. So *a cyclic vector is star-cyclic,* and *a cyclic operator is star-cyclic.* A normed space is separable if and only if it is spanned by a countable set (see, e.g., [78, Proposition 4.9]). Hence the subspace $\{Sx \in \mathcal{H}: S \in \mathcal{P}(T, T^*)\}^- = \bigvee(\bigcup_{m,n}\{T^m T^{*n}x\} \cup \{T^{*n}T^m x\})$ is separable, and so is $\{Sx \in \mathcal{H}: S \in \mathcal{C}^*(T)\}^- = \{Sx \in \mathcal{H}: S \in \mathcal{P}(T, T^*)^-\}^-$. Thus *if* T *is star-cyclic, then* $\mathcal{H}$ *is separable* (see also [30, Proposition IX.3.3]). Therefore

$$T \in \mathcal{B}[\mathcal{H}] \text{ is cyclic} \implies T \in \mathcal{B}[\mathcal{H}] \text{ is star-cyclic} \implies \mathcal{H} \text{ is separable.}$$

If T is *normal,* then $\bigvee(\bigcup_{m,n}\{T^m T^{*n}x\} \cup \{T^{*n}T^m x\}) = \bigvee(\bigcup_{m,n}\{T^n T^{*m}x\})$ is the smallest subspace of $\mathcal{H}$ that contains x and reduces T. Thus $x \in \mathcal{H}$ is a *star-cyclic vector* for a *normal operator* T if and only if $\bigvee_{m,n}\{T^n T^{*m}x\} = \mathcal{H}$ (cf. proof of Theorem 3.11). An important result from [20] reads as follows. *If* T *is a normal operator, then* T *is cyclic whenever it is star-cyclic.* Hence

$$T \in \mathcal{B}[\mathcal{H}] \text{ is normal and cyclic} \iff T \in \mathcal{B}[\mathcal{H}] \text{ is normal and star-cyclic.}$$

It can also be shown along this line that x *is a cyclic vector for a normal operator* T *if and only if it is a cyclic vector for* T^* [58, Problem 164].

Consider the proof of Theorem 3.11. Part (a) deals with the case where a normal operator T is star-cyclic, while part (b) deals with the case where T is not star-cyclic. In the former case $\mathcal{H}$ must be separable by the above remark. As we show next if $\mathcal{H}$ is separable, then the measure in Theorem 3.11 can be taken to be finite. Indeed, the positive measure $\mu: \mathcal{A}_\Omega \to \mathbb{R}$ constructed in part (a) is finite. However, the positive measure $\mu: \mathcal{A}_\Omega \to \overline{\mathbb{R}}$ constructed in part (b) may fail to be even σ-finite (see, e.g., [58, p. 66]). But finiteness can be restored by separability.

Corollary 3.13. *If* T *is a normal operator on a* separable *Hilbert space* $\mathcal{H}$, *then there is a* finite *positive measure* μ *on a* σ-algebra $\mathcal{A}_\Omega$ *of subsets of a nonempty set* Ω *and a function* φ *in* $L^\infty(\Omega, \mu)$ *such that* T *is unitarily equivalent to the multiplication operator* M_φ *on* $L^2(\Omega, \mu)$.

Proof. Consider the argument in part (b) of the proof of Theorem 3.11. Since the collection $\{\mathcal{M}_\gamma\}$ is made up of orthogonal subspaces, we may take one unit vector from each subspace by the Axiom of Choice to get an orthonormal set $\{e_\gamma\}$ (of unit vectors), which is in one-to-one correspondence with the set $\{\mathcal{M}_\gamma\}$. Hence $\{\mathcal{M}_\gamma\}$ and $\{e_\gamma\}$ have the same cardinality. Since $\{e_\gamma\}$ is an orthonormal set, its cardinality is not greater than the cardinality of any orthonormal basis for $\mathcal{H}$, and so the cardinality of $\{\mathcal{M}_\gamma\}$ does not surpass the (Hilbert) dimension of $\mathcal{H}$. If $\mathcal{H}$ is separable, then $\{\mathcal{M}_\gamma\}$ is countable. In this case, for notational convenience replace the index γ in the index set Γ with an integer index n in a countable index set, say $\mathbb{N}$, throughout the proof of part (b). Since the orthogonal collection $\{\mathcal{M}_n\}$ of subspaces is now countable, it follows that the family of sets $\{\Omega_n\}$, where each $\Omega_n = \sigma(T|_{\mathcal{M}_n})$, is countable. Recall: each positive measure $\mu_n: \mathcal{A}_{\Omega_n} \to \mathbb{R}$ inherited from part (a) is finite. Take the positive measure μ on $\mathcal{A}_\Omega$ defined in part (b). Thus $\mu(\Lambda) = \sum_{j\in\mathbb{J}} \mu_j(\Lambda_j)$ for every set $\Lambda = \biguplus_{j\in\mathbb{J}} \Lambda_j$ in $\mathcal{A}_\Omega$ (each Λ_j in $\mathcal{A}_{\Omega_j}$ and so $\Lambda_j \subseteq \Omega_j$) where $\mathbb{J}$ is any subset of $\mathbb{N}$. But now $\Omega = \biguplus_{n\in\mathbb{N}} \Omega_n$ and hence $\mu(\Omega) = \sum_{n\in\mathbb{N}} \mu_n(\Omega_n) = \sum_{n\in\mathbb{N}} \mu(\Omega_n)$, where $\mu(\Omega_n) = \mu_n(\Omega_n) < \infty$ for each $n \in \mathbb{N}$. (Again, we are identifying each Ω_n with $\biguplus_{m\in\mathbb{N}} \Omega_m$ when $\Omega_m = \varnothing$ for every $m \neq n$.) Conclusion: if $\mathcal{H}$ is separable, then μ is σ-finite. However, we can actually get a finite measure. Indeed, construct the measure μ by multiplying each original finite measure μ_n by a positive scalar so that $\sum_{n\in\mathbb{N}} \mu_n(\Omega_n) < \infty$ (e.g., such that $0 < \mu_n(\Omega_n) \leq 2^{-n}$ for each $n \in \mathbb{N}$). In this case $\mu(\Omega) = \sum_{n\in\mathbb{N}} \mu_n(\Omega_n) < \infty$, and so $\mu\colon \mathcal{A}_\Omega \to \mathbb{R}$ is now finite (rather than just σ-finite). Thus all measures involved in the proof of Theorem 3.11 are finite whenever $\mathcal{H}$ is separable. $\square$

Remark. Corollary 3.13 forces the Hilbert space $L^2(\Omega,\mu)$ with that particular finite measure μ to be separable (since it is unitarily equivalent to a separable Hilbert space $\mathcal{H}$). Recall that there exist nonseparable L^2 spaces with finite measures, and also non-σ-finite measures for which L^2 is separable (see, e.g., [23, pp. 174, 192, 376] and [59, pp. 2, 3]).

Corollary 3.14. *Let T be a normal operator on a Hilbert space $\mathcal{H}$ so that $T \cong M_\varphi$ on $L^2(\Omega,\mu)$, where φ lies in $L^\infty(\Omega,\mu)$ with μ being a positive measure on a σ-algebra $\mathcal{A}_\Omega$ of subsets of a nonempty set Ω as in Theorem 3.11. The assertions*

(a) *$\mathcal{H}$ is separable,*

(b) *μ is finite,*

(c) *T is star-cyclic,*

(d) *T is cyclic,*

are related as follows: (c) $\Longleftrightarrow$ (d) $\Longrightarrow$ (a) $\Longrightarrow$ (b). *If* (c) *holds true, then the constant function $\psi = 1$ ($\psi(\lambda) = 1$ for all $\lambda \in \Omega = \sigma(T) \subset \mathbb{C}$) lies in $L^2(\Omega,\mu)$ and is star-cyclic for M_φ, and so M_φ has a star-cyclic vector in $L^\infty(\Omega,\mu)$.*

Proof. By Theorem 3.11, T is unitarily equivalent to a multiplication operator M_φ on $L^2(\Omega,\mu)$ for some φ in $L^\infty(\Omega,\mu)$. Let U be a unitary transformation in $\mathcal{B}[L^2(\Omega,\mu),\mathcal{H}]$ for which $U^*TU = M_\varphi$. Thus ψ in $L^2(\Omega,\mu)$ is a cyclic (star-cyclic) vector for the normal operator M_φ if and only if $x = U\psi$ in $\mathcal{H}$ is a cyclic (star-cyclic) vector for the normal operator T (in fact, $\bigvee_n \{T^n x\} = U \bigvee_n \{M_\varphi^n \psi\}$ and $\bigvee_{m,n}\{T^n T^{*m} x\} = U \bigvee_{m,n}\{M_\varphi^n M_\varphi^{*m}\psi\}$). Corollary 3.13 says that (a) $\Longrightarrow$ (b), and according to the remark preceding Corollary 3.13, (c) $\Longleftrightarrow$ (d) $\Longrightarrow$ (a). Now if (c) holds, then according to part (a) in the proof of Theorem 3.11, μ is finite and $\varphi\colon \Omega \to \Omega$ is the identity map ($\varphi(\lambda) = \lambda$) on the compact set $\Omega = \sigma(T)$. A trivial induction shows that $M_\varphi^n\psi = \varphi^n\psi$ and $M_\varphi^{*m}\psi = \overline{\varphi}^m\psi$, and so $\bigvee_{m,n}\{M_\varphi^n M_\varphi^{*m}\psi\} = \bigvee_{m,n}\{\varphi^n\overline{\varphi}^m\psi\} = \bigvee_{m,n}\{\lambda^n\overline{\lambda}^m\psi\}$ for every $\psi \in L^2(\Omega,\mu)$ and λ in Ω. Since μ is finite, the constant function 1 ($1(\lambda) = 1$ for all $\lambda \in \Omega$) lies in $L^2(\Omega,\mu) \cap L^\infty(\Omega,\mu)$. Set $\psi = 1$ and let $P(\Omega)$

denote the set of all polynomials $p(\cdot,\bar{\cdot})\colon \Omega \to \mathbb{C}$ such that $\lambda \mapsto p(\lambda,\bar{\lambda})$. Thus $L^2(\Omega,\mu) = P(\Omega)^- = \bigvee_{m,n}\{\lambda^n\bar{\lambda}^m\} = \bigvee_{m,n}\{\varphi^n\bar{\varphi}^m 1\} = \bigvee_{m,n}\{M_\varphi^n M_\varphi^{*m}\psi\}$ (cf. proof of Theorem 3.11(a)) and so $\psi = 1$ is a star-cyclic vector for M_φ. $\qquad\square$

Remark. Let K be a compact subset of $\mathbb{C}$. By the Stone–Weierstrass Theorem (cf. proof of Theorem 3.11), $P(K)^- = C(K)$ in the sup-norm, where $C(K)$ is the set of all complex-valued continuous functions on K and $P(K)$ is the set of all polynomials $p(\cdot,\cdot)\colon K \times K \to \mathbb{C}$ in λ and $\bar{\lambda}$ (or, $p(\cdot,\bar{\cdot})\colon K \to \mathbb{C}$ such that $\lambda \mapsto p(\lambda,\bar{\lambda})$). Let $P'(K)$ denote the set of all polynomials $p(\cdot)\colon K \to \mathbb{C}$ in one variable λ. The LAVRENTIEV THEOREM says that *if K is compact, then $P'(K)^- = C(K)$ in the sup-norm if and only if $\mathbb{C}\backslash K$ is connected and $K^\circ = \varnothing$* (i.e., *if and only if the compact set K has empty interior and no holes*), which extends the classical WEIERSTRASS APPROXIMATION THEOREM for (complex-valued) polynomials on compact intervals of the real line, viz., $P'([\alpha,\beta])^- = C([\alpha,\beta])$ *in the sup-norm* (see [23, p. 18], or [30, Proposition VII.5.3] and [31, Theorem V.14.20]; for real-valued functions the Weierstrass Theorem holds for every compact set in the real line — see [35, p. 139]). This underlines the difference between cyclic and star-cyclic vectors. Note that *if T is normal and $P'(\sigma(T))^- = C(\sigma(T))$, then T is reductive* (Proposition 3.J). Bram's result [20] not only shows that normal operators are cyclic if and only if they are star-cyclic, but also shows that M_φ has a cyclic vector in $L^\infty(\Omega,\mu)$, as we saw in Corollary 3.14, where $\Omega = \sigma(T)$ and μ is finite (as in part (a) of the proof of Theorem 3.11) — see also [31, Theorem V.14.21].

Theorem 3.15. (SPECTRAL THEOREM – SECOND VERSION). *If $T \in \mathcal{B}[\mathcal{H}]$ is normal, then there is a unique spectral measure $E\colon \mathcal{A}_{\sigma(T)} \to \mathcal{B}[\mathcal{H}]$ such that*

$$T = \int \lambda \, dE_\lambda.$$

If Λ is a nonempty relatively open subset of $\sigma(T)$, then $E(\Lambda) \neq O$.

Proof. Consider the proof of Theorem 3.11. We split this proof into four parts.

(a) Let μ be any positive measure on a σ-algebra $\mathcal{A}_\Omega$ of Borel subsets of a set $\Omega \neq \varnothing$ included in the disjoint union $\biguplus_{\gamma \in \Gamma} \sigma(T)$, take the multiplication operator $M_\phi \in \mathcal{B}[L^2(\Omega,\mu)]$ for any $\phi \in L^\infty(\Omega,\mu)$, and let $\chi_\Lambda\colon \Omega \to \{0,1\}$ be the characteristic function of $\Lambda \in \mathcal{A}_\Omega$, which is bounded (i.e., $\chi_\Lambda \in L^\infty(\Omega,\mu)$). Set

$$E'(\Lambda) = M_{\chi_\Lambda} \quad \text{for every} \quad \Lambda \in \mathcal{A}_\Omega.$$

As is readily verified, $E'\colon \mathcal{A}_\Omega \to \mathcal{B}[L^2(\Omega,\mu)]$ is a spectral measure in $L^2(\Omega,\mu)$ (cf. Definition 3.6). Take an arbitrary $f \in L^2(\Omega,\mu)$. Since a characteristic function of a measurable set is a measurable function,

$$\int \chi_\Lambda \, d\langle E'_\lambda f\,; f \rangle = \int \chi_\Lambda \, d\pi'_{f,f} = \int_\Lambda d\pi'_{f,f} = \pi'_{f,f}(\Lambda) = \langle E'(\Lambda)f\,; f \rangle$$

$$= \langle M_{\chi_\Lambda} f\,; f \rangle = \int \chi_\Lambda f \bar{f} \, d\mu = \int \chi_\Lambda |f|^2 \, d\mu$$

for every $\Lambda \in \mathcal{A}_\Omega$. Let $\Psi(\Omega)$ be the collection of all *simple functions* $\psi \colon \Omega \to \mathbb{C}$ of the form $\psi = \sum_i \alpha_i \chi_{\Lambda_i}$ for all *finite* measurable partitions $\{\Lambda_i\}$ of Ω (so Ω is the *ordinary* union of sets Λ_i in $\mathcal{A}_\Omega$) and any *finite* (similarly indexed) set of complex numbers $\{\alpha_i\}$. Since each $\psi \in \Psi(\Omega)$ is $\mathcal{A}_\Omega$-measurable and bounded,

$$\int \psi\, d\langle E'_{\boldsymbol{\lambda}} f \,;\, f\rangle = \sum_i \alpha_i \int \chi_{\Lambda_i}\, d\langle E'_{\boldsymbol{\lambda}} f \,;\, f\rangle = \sum_i \alpha_i \int \chi_{\Lambda_i} |f|^2 d\mu$$
$$= \int \sum_i \alpha_i \chi_{\Lambda_i} |f|^2\, d\mu = \int \psi |f|^2 d\mu$$

for every $\psi \in \Psi(\Omega)$ (by the previously displayed expression). This also holds for the identity map with multiplicity $\varphi \colon \Omega \to \mathbb{C}$ (since $\inf_{\psi \in \Psi(\Omega)} \|\varphi - \psi\|_\infty = 0$, then $\varphi \in \Psi(\Omega)^- \subseteq L^\infty(\Omega, \mu)$, and so apply extension by continuity.) Hence

$$\int \varphi\, d\langle E'_{\boldsymbol{\lambda}} f \,;\, f\rangle = \int \varphi |f|^2 d\mu = \int \varphi f \overline{f}\, d\mu = \langle M_\varphi f \,;\, f\rangle$$

for every $f \in L^2(\Omega, \mu)$. By the polarization identity (Proposition 1.A) we get $\langle Sf \,;\, g\rangle = \frac{1}{4}[\langle S(f+g) \,;\, (f+g)\rangle - \langle S(f-g) \,;\, (f-g)\rangle + i\langle S(f+ig) \,;\, (f+ig)\rangle - i\langle S(f-ig) \,;\, (f-ig)\rangle]$ for every $f, g \in L^2(\Omega, \mu)$ and every $S \in \mathcal{B}[L^2(\Omega, \mu)]$. So replacing S with $E'(\Lambda)$ on the one hand, and with M_φ on the other hand,

$$\int \varphi\, d\langle E'_{\boldsymbol{\lambda}} f \,;\, g\rangle = \langle M_\varphi f \,;\, g\rangle$$

for every $f, g \in L^2(\Omega, \mu)$, which means

$$M_\varphi = \int \varphi\, dE'_{\boldsymbol{\lambda}} = \int \varphi(\boldsymbol{\lambda})\, dE'_{\boldsymbol{\lambda}} = \int \boldsymbol{\lambda}\, dE'_{\boldsymbol{\lambda}}.$$

Uniqueness of the spectral measure $E' \colon \mathcal{A}_\Omega \to \mathcal{B}[L^2(\Omega, \mu)]$ is proved as follows. Let $P(\Omega)$ be the set of polynomials $p(\varphi(\cdot), \overline{\varphi(\cdot)})$ on Ω (see remark following Lemma 3.10). If $E'' \colon \mathcal{A}_\Omega \to \mathcal{B}[L^2(\Omega, \mu)]$ is a spectral measure in $L^2(\Omega, \mu)$ such that $M_\varphi = \int \boldsymbol{\lambda}\, dE''_{\boldsymbol{\lambda}} = \int \boldsymbol{\lambda}\, dE'_{\boldsymbol{\lambda}}$, then $p(T, T^*) = \int p(\lambda, \overline{\lambda})\, dF''_{\boldsymbol{\lambda}} = \int p(\lambda, \overline{\lambda})\, dF'_{\boldsymbol{\lambda}}$ for every $p \in P(\Omega)$ by Lemma 3.10. So $\int p\, d\langle E''_{\boldsymbol{\lambda}} x; y\rangle = \int p\, d\langle E'_{\boldsymbol{\lambda}} x; y\rangle$ for every $x, y \in \mathcal{H}$, for every $p \in P(\Omega)$. Fix an arbitrary $x \in \mathcal{H}$. In particular,

$$\int p\, d\langle E''_{\boldsymbol{\lambda}} x; x\rangle = \int p\, d\langle E'_{\boldsymbol{\lambda}} x; x\rangle$$

for each $p \in P(\Omega)$. Let K be the closure of the image under φ of the union of the supports of the measures E' and E'', which is compact since $K \subseteq \sigma(T) = \varphi(\Omega)^-$. Let $P(K)$ be the set of all complex-valued polynomials on K, which is dense in $(L^2(K, \nu), \|\cdot\|_2)$ for every finite positive measure ν on $\mathcal{A}_\Omega$ with support in $\varphi^{-1}(K)$ (by the Stone–Weierstrass Theorem for complex functions as applied in the proof of Theorem 3.11 part (a)). Since the positive measures $\langle E'(\cdot)x \,;\, x\rangle = \|E'(\cdot)x\|^2$ and $\langle E''(\cdot)x \,;\, x\rangle = \|E''(\cdot)x\|^2$ on $\mathcal{A}_\Omega$ are finite, we get

$$\int \chi_\Lambda\, d\langle E''_{\boldsymbol{\lambda}} x; x\rangle = \int \chi_\Lambda\, d\langle E'_{\boldsymbol{\lambda}} x; x\rangle$$

for the characteristic function χ_Λ of an arbitrary set $\Lambda \in \mathcal{A}_\Omega$. Thus

$$\langle E''(\Lambda)x\,;x\rangle = \int_\Lambda d\langle E''_\lambda x\,;x\rangle = \int_\Lambda d\langle E'_\lambda x\,;x\rangle = \langle E'(\Lambda)x\,;x\rangle,$$

and so $\langle (E''(\Lambda) - E'(\Lambda))x\,;x\rangle = 0$, for every $\Lambda \in \mathcal{A}_\Omega$ and every $x \in \mathcal{H}$. Since $\mathcal{H}$ is complex (or since $E''(\Lambda) - E'(\Lambda)$ is self-adjoint), then $E''(\Lambda) - E'(\Lambda) = O$ for every $\Lambda \in \mathcal{A}_\Omega$ (see, e.g., [78, Problem 5.4 or Corollary 5.80]). Thus E' is unique. This completes the proof for the normal operator $M_\varphi \in \mathcal{B}[L^2(\Omega,\mu)]$: *there is a unique spectral measure* $E' \colon \mathcal{A}_\Omega \to \mathcal{B}[L^2(\Omega,\mu)]$ *such that* $M_\varphi = \int \varphi(\lambda)\,dE'_\lambda$.

(b) If $T \in \mathcal{B}[\mathcal{H}]$ is normal, then $T \cong M_\varphi \in \mathcal{B}[L^2(\Omega,\mu)]$ for some positive measure μ on a σ-algebra $\mathcal{A}_\Omega$ of subsets of a nonempty disjoint union $\Omega = \bigcup_{\gamma \in \Gamma}\Omega_\gamma$ by Theorem 3.11. So there is a unitary $U \in \mathcal{B}[\mathcal{H}, L^2(\Omega,\mu)]$ for which $T = U^*M_\varphi U$. Since E' is a spectral measure in $L^2(\Omega,\mu)$, the map $\hat{E}\colon \mathcal{A}_\Omega \to \mathcal{B}[\mathcal{H}]$ defined by

$$\hat{E}(\Lambda) = U^*E'(\Lambda)U \quad \text{for every } \Lambda \in \mathcal{A}_\Omega$$

is a spectral measure in $\mathcal{H}$ (cf. Definition 3.6). Then we get from (a)

$$\langle Tx\,;y\rangle = \langle M_\varphi Ux\,;Uy\rangle = \int \varphi\,d\langle E'_\lambda Ux\,;Uy\rangle = \int \varphi\,d\langle \hat{E}_\lambda x\,;y\rangle$$

for every $x,y \in \mathcal{H}$, which means

$$T = \int \varphi\,d\hat{E}_\lambda = \int \varphi(\lambda)\,d\hat{E}_\lambda = \int \lambda\,d\hat{E}_\lambda.$$

Since $\hat{E}(\Lambda) \cong E'(\Lambda)$, uniqueness of E' implies uniqueness of $\hat{E}$. Thus *there exists a unique spectral measure* $\hat{E}\colon \mathcal{A}_\Omega \to \mathcal{B}[\mathcal{H}]$ *such that* $T = \int \varphi(\lambda)\,d\hat{E}_\lambda$.

(c) Let $\Omega = \bigcup_{\gamma \in \Gamma}\Omega_\gamma$ be the disjoint union ($\Omega_\gamma = \sigma(T|_{\mathcal{M}_\gamma})$), take the identity map with multiplicity $\varphi\colon \Omega \to \sigma(T)$, and consider the inverse image $\varphi^{-1}(\tilde{\Lambda}) = \{\lambda \in \Omega \colon \varphi(\lambda) \in \tilde{\Lambda}\} = \bigcup_{\gamma \in \Gamma}\{\tilde{\Lambda} \cap \Omega_\gamma\} \in \mathcal{A}_\Omega$ of $\tilde{\Lambda}$ in $\mathcal{A}_{\sigma(T)}$. For any singleton $\{\lambda\} \in \mathcal{A}_{\sigma(T)}$ set $\lambda = \varphi^{-1}(\{\lambda\}) = \{\lambda \in \Omega \colon \varphi(\lambda) = \lambda\} = \bigcup_{\gamma \in \Gamma}\{\{\lambda\} \cap \Omega_\gamma\} \in \mathcal{A}_\Omega$ (this accounts for the multiplicity of λ). With $\mathcal{A}_{\varphi(\Omega)} = \wp(\varphi(\Omega)) \cap \mathcal{A}_{\sigma(T)}$ set

$$E(\tilde{\Lambda}) = \int \chi_{\varphi^{-1}(\tilde{\Lambda})}\,d\hat{E}_\lambda = \int_{\varphi^{-1}(\tilde{\Lambda})} d\hat{E}_\lambda = \hat{E}(\varphi^{-1}(\tilde{\Lambda})) \quad \text{for every } \tilde{\Lambda} \in \mathcal{A}_{\varphi(\Omega)}.$$

Since $\varphi(\Omega)^- = \sigma(T)$, extend the map $E\colon \mathcal{A}_{\varphi(\Omega)} \to \mathcal{B}[\mathcal{H}]$ over $\mathcal{A}_{\sigma(T)}$, denoted again by E. This also defines a spectral measure $E\colon \mathcal{A}_{\sigma(T)} \to \mathcal{B}[\mathcal{H}]$ in $\mathcal{H}$ for

$$T = \int \varphi(\lambda)\,d\hat{E}_\lambda = \int \lambda\,dE_\lambda,$$

which is unique because $\hat{E}$ is unique. Then, according to (b), *there exists a unique spectral measure* $E\colon \mathcal{A}_{\sigma(T)} \to \mathcal{B}[\mathcal{H}]$ *such that* $T = \int \lambda\,dE_\lambda$.

Note: As $E(\{\lambda\}) = \hat{E}(\lambda)$ for $\lambda \in \sigma(T)$, $\dim \mathcal{R}(E(\{\lambda\})) = \dim \mathcal{R}(\hat{E}(\lambda))$, and so $\dim \mathcal{R}(E(\{\lambda\}))$ may be any cardinal number, but $\dim \mathcal{R}(\hat{E}(\{\lambda\})) \in \{0,1\}$ since $\dim \mathcal{R}(\hat{E}(\{\lambda\})) = \dim \mathcal{R}(E'(\{\lambda\})) = \dim \mathcal{R}(M_{\chi_{\{\lambda\}}}) = \dim L^2(\{\lambda\},\mu|_{\{\lambda\}})$. Then the multiplicity of λ is transferred from the disjoint union Ω to $\dim \mathcal{R}(E(\{\lambda\}))$.

(d) Next we show: *if* $\varnothing \neq \tilde{\Lambda} \in \mathcal{A}_{\sigma(T)}$ *is relatively open in* $\sigma(T)$, *then* $E(\tilde{\Lambda}) \neq O$. Take $\tilde{\Lambda}$ in $\mathcal{A}_{\sigma(T)}$ and consider the set $\Lambda = \varphi^{-1}(\tilde{\Lambda}) = \bigcup_{\gamma \in \Gamma}\{\tilde{\Lambda} \cap \Omega_\gamma\}$ in $\mathcal{A}_\Omega$. If

$\varnothing \neq \tilde{\Lambda} \subseteq \sigma(T)$ is relatively open in the closed set $\sigma(T)$, then $\tilde{\Lambda} \cap \Omega_\gamma$ in $\mathcal{A}_{\Omega_\gamma}$ are relatively open in the closed subsets $\Omega_\gamma = \sigma(T|_{\mathcal{M}_\gamma})$ and nonempty for some γ in Γ, and so $\mu_\gamma(\tilde{\Lambda} \cap \Omega_\gamma) > 0$ (since Ω_γ is the support of each positive measure μ_γ on $\mathcal{A}_{\Omega_\gamma}$). Thus $\mu(\Lambda) > 0$ (cf. definitions of μ_γ and μ — proof of Theorem 3.11 part (b)). However, since $\chi_\Lambda \in L^2(\Omega, \mu)$ and since $\langle E'(\Delta)f ; f \rangle = \int \chi_\Delta |f|^2 d\mu$ for each $\Delta \in \mathcal{A}_\Omega$ and every $f \in L^2(\Omega, \mu)$ (cf. definition of E' in part (b)),

$$\|E'(\Lambda)\chi_\Lambda\|^2 = \langle E'(\Lambda)\chi_\Lambda ; \chi_\Lambda \rangle = \int \chi_\Lambda |\chi_\Lambda|^2 d\mu = \int_\Lambda d\mu = \mu(\Lambda) > 0.$$

Thus $\|E(\tilde{\Lambda})\| = \|\hat{E}(\varphi^{-1}(\tilde{\Lambda}))\| = \|\hat{E}(\Lambda)\| = \|E'(\Lambda)\| \neq 0$ and so $E(\tilde{\Lambda}) \neq O$. $\square$

The representations $T = \int \varphi(\lambda) \, d\hat{E}_\lambda = \int \lambda \, dE_\lambda$ in the above proof are also called *spectral decompositions* of the normal operator T (see Section 3.2).

3.5 Fuglede Theorems and Reducing Subspaces

Let $\mathcal{A}_{\sigma(T)}$ be the σ-algebra of Borel subsets of the spectrum $\sigma(T)$ of T. The following lemma sets up the essential features for proving the next theorem.

Lemma 3.16. *Let $T = \int \lambda \, dE_\lambda$ be the spectral decomposition of a normal operator $T \in \mathcal{B}[\mathcal{H}]$ as in Theorem 3.15. If $\Lambda \in \mathcal{A}_{\sigma(T)}$, then*

(a) *$E(\Lambda)T = TE(\Lambda)$. Equivalently, $\mathcal{R}(E(\Lambda))$ reduces T.*

(b) *Moreover, if $\Lambda \neq \varnothing$, then $\sigma(T|_{\mathcal{R}(E(\Lambda))}) \subseteq \Lambda^-$.*

Proof. Let $T = \int \lambda \, dE_\lambda$ be the spectral decomposition of T. Take $\tilde{\Lambda} \in \mathcal{A}_{\sigma(T)}$.

(a) Since $E(\tilde{\Lambda}) = \int \chi_{\tilde{\Lambda}} \, dE_\lambda$ and $T = \int \lambda \, dE_\lambda$, we get by Lemma 3.8

$$E(\tilde{\Lambda})T = \int \lambda \chi_{\tilde{\Lambda}}(\lambda) \, dE_\lambda = \int \chi_{\tilde{\Lambda}}(\lambda) \, \lambda \, dE_\lambda = TE(\tilde{\Lambda}),$$

which is equivalent to saying that $\mathcal{R}(E(\tilde{\Lambda}))$ reduces T (by Proposition 1.I).

(b) Let $\mathcal{A}_\Omega$ be the σ-algebra of Borel subsets of Ω as in Theorem 3.11. Take an arbitrary nonempty $\Lambda \in \mathcal{A}_\Omega$. Recall (cf. proof of Theorem 3.15):

$$\int \lambda \, dE_\lambda = \int \varphi(\lambda) \, d\hat{E}_\lambda = T \cong M_\varphi = \int \varphi(\lambda) \, dE'_\lambda,$$

where $E' \colon \mathcal{A}_\Omega \to \mathcal{B}[L^2(\Omega, \mu)]$ and $\hat{E} \colon \mathcal{A}_\Omega \to \mathcal{B}[\mathcal{H}]$ are given by $E'(\Lambda) = M_{\chi_\Lambda}$ and $\hat{E}(\Lambda) = U^* E'(\Lambda) U$ for the same unitary transformation $U \colon \mathcal{H} \to L^2(\Omega, \mu)$ such that $T = U^* M_\varphi U$. So, since $\hat{E}(\Lambda)$ and $E'(\Lambda)$ reduce T and M_φ as in (a),

$$T|_{\mathcal{R}(\hat{E}(\Lambda))} \cong M_\varphi|_{\mathcal{R}(M_{\chi_\Lambda})}.$$

Indeed, since $\mathcal{R}(\hat{E}(\Lambda)) = U^* \mathcal{R}(E'(\Lambda)U)$,

$$T|_{\mathcal{R}(\hat{E}(\Lambda))} = (U^* M_\varphi U)|_{U^* \mathcal{R}(E'(\Lambda)U)} = U^* M_\varphi|_{\mathcal{R}(E'(\Lambda))} U = U^* M_\varphi|_{\mathcal{R}(M_{\chi_\Lambda})} U.$$

Let $\mu|_\Lambda\colon \mathcal{A}_\Lambda \to \mathbb{C}$ be the restriction of $\mu\colon \mathcal{A}_\Omega \to \mathbb{C}$ to the sub-σ-algebra $\mathcal{A}_\Lambda = \mathcal{A}_\Omega \cap \wp(\Lambda)$ so that $\mathcal{R}(M_{\chi_\Lambda})$ is unitarily equivalent to $L^2(\Lambda, \mu|_\Lambda)$. Explicitly,

$$\mathcal{R}(M_{\chi_\Lambda}) = \{f \in L^2(\Omega, \mu)\colon f = \chi_\Lambda g \text{ for some } g \in L^2(\Omega, \mu)\} \cong L^2(\Lambda, \mu|_\Lambda).$$

Consider the following notation: set $M_{\varphi\chi_\Lambda} = M_\varphi|_{L^2(\Lambda, \mu|_\Lambda)}$ in $\mathcal{B}[L^2(\Lambda, \mu|_\Lambda)]$ (not in $\mathcal{B}[L^2(\Omega, \mu)]$). Since $\mathcal{R}(E'(\Lambda)) = \mathcal{R}(M_{\chi_\Lambda})$ reduces M_φ,

$$M_\varphi|_{\mathcal{R}(M_{\chi_\Lambda})} \cong M_{\varphi\chi_\Lambda}.$$

Hence $T|_{\mathcal{R}(\hat{E}(\Lambda))} \cong M_{\varphi\chi_\Lambda}$ by transitivity, and therefore (cf. Proposition 2.B)

$$\sigma(T|_{\mathcal{R}(\hat{E}(\Lambda))}) = \sigma(M_{\varphi\chi_\Lambda}).$$

Since $\Lambda \neq \varnothing$, we get by Proposition 3.B (for $\varphi\colon \Omega \to \mathbb{C}$ such that $\varphi(\lambda) = \lambda$)

$$\sigma(M_{\varphi\chi_\Lambda}) \subseteq \bigcap\{\varphi(\Delta)^- \in \mathbb{C}\colon \Delta \in \mathcal{A}_\Lambda \text{ and } \mu|_\Lambda(\Lambda \backslash \Delta) = 0\}$$
$$= \bigcap\{\varphi(\Delta \cap \Lambda)^- \in \mathbb{C}\colon \Delta \in \mathcal{A}_\Omega \text{ and } \mu(\Lambda \backslash \Delta) = 0\} \subseteq \varphi(\Lambda)^-.$$

Take $\varnothing \neq \tilde{\Lambda} \in \mathcal{A}_{\sigma(T)}$. Let $E\colon \mathcal{A}_{\sigma(T)} \to \mathcal{B}[\mathcal{H}]$ be the spectral measure for which $E(\tilde{\Lambda}) = \hat{E}(\varphi^{-1}(\tilde{\Lambda}))$ if $\tilde{\Lambda} \in \mathcal{A}_{\varphi(\Omega)}$ (cf. proof of Theorem 3.15 part (c) — also see the remark after Theorem 3.11). Since $\tilde{\Lambda}^- = \varphi(\Lambda)^-$ for *some* $\Lambda \in \mathcal{A}_\Omega$, we get

$$\sigma(T|_{\mathcal{R}(E(\tilde{\Lambda}))}) = \sigma(T|_{\mathcal{R}(E(\varphi(\Lambda)))}) = \sigma(T|_{\mathcal{R}(\hat{E}(\Lambda))}) = \sigma(M_{\varphi\chi_\Lambda}) \subseteq \varphi(\Lambda)^- = \tilde{\Lambda}^-. \qquad \square$$

The next proof uses the same argument used in the proof of Corollary 3.5.

Theorem 3.17. (FUGLEDE THEOREM). *Let $T = \int \lambda \, dE_\lambda$ be the spectral decomposition of a normal operator in $\mathcal{B}[\mathcal{H}]$. If $S \in \mathcal{B}[\mathcal{H}]$ commutes with T, then S commutes with $E(\Lambda)$ for every $\Lambda \in \mathcal{A}_{\sigma(T)}$.*

Proof. Take any Λ in $\mathcal{A}_{\sigma(T)}$. Suppose $\varnothing \neq \Lambda \neq \sigma(T)$; otherwise the result is trivially verified with $E(\Lambda) = O$ or $E(\Lambda) = I$. Set $\Lambda' = \sigma(T) \backslash \Lambda$ in $\mathcal{A}_{\sigma(T)}$. Let Δ and Δ' be arbitrary nonempty measurable closed subsets of Λ and Λ'. That is, $\varnothing \neq \Delta \subseteq \Lambda$ and $\varnothing \neq \Delta' \subseteq \Lambda'$ lie in $\mathcal{A}_{\sigma(T)}$ and are closed in $\mathbb{C}$, and so they are closed relative to the compact set $\sigma(T)$. By Lemma 3.16(a) we get

$$E(\Lambda)T|_{\mathcal{R}(E(\Lambda))} = TE(\Lambda)|_{\mathcal{R}(E(\Lambda))} = T|_{\mathcal{R}(E(\Lambda))}\colon \mathcal{R}(E(\Lambda)) \to \mathcal{R}(E(\Lambda))$$

for every $\Lambda \in \mathcal{A}_{\sigma(T)}$. This holds in particular for Δ and Δ'. Take any operator S on $\mathcal{H}$ such that $ST = TS$. Thus, on $\mathcal{R}(E(\Delta))$,

(*) $(E(\Delta')SE(\Delta))T|_{\mathcal{R}(E(\Delta))} = (E(\Delta')SE(\Delta))(TE(\Delta)) = E(\Delta')STE(\Delta) = $
$E(\Delta')TSE(\Delta) = (TE(\Delta'))(E(\Delta')SE(\Delta)) = T|_{\mathcal{R}(E(\Delta'))}(E(\Delta')SE(\Delta)).$

Since the sets Δ and Δ' are closed and nonempty, Lemma 3.16(b) says

$$\sigma(T|_{\mathcal{R}(E(\Delta))}) \subseteq \Delta \quad \text{and} \quad \sigma(T|_{\mathcal{R}(E(\Delta'))}) \subseteq \Delta',$$

and hence, as $\Delta \subseteq \Lambda$ and $\Delta' \subseteq \Lambda' = \sigma(T) \backslash \Lambda$,

(**) $$\sigma(T|_{\mathcal{R}(E(\Delta))}) \cap \sigma(T|_{\mathcal{R}(E(\Delta'))}) = \varnothing.$$

Therefore, according to Proposition 2.T, $(*)$ and $(**)$ imply

$$E(\Delta')SE(\Delta) = O.$$

This identity holds for all $\Delta \in \mathcal{C}_\Lambda$ and all $\Delta' \in \mathcal{C}_{\Lambda'}$, where

$$\mathcal{C}_\Lambda = \big\{\Delta \in \mathcal{A}_{\sigma(T)}\colon \Delta \text{ is a closed subset of } \Lambda\big\}.$$

Now, according to Proposition 3.C (in the inclusion ordering),

$$\mathcal{R}(E(\Lambda)) = \sup_{\Delta \in \mathcal{C}_\Lambda} \mathcal{R}(E(\Delta)) \quad \text{and} \quad \mathcal{R}(E(\Lambda')) = \sup_{\Delta' \in \mathcal{C}_{\Lambda'}} \mathcal{R}(E(\Delta'))$$

and so (in the extension ordering — see, e.g., [78, Problem 1.17]),

$$E(\Lambda) = \sup_{\Delta \in \mathcal{C}_\Lambda} E(\Delta) \quad \text{and} \quad E(\Lambda') = \sup_{\Delta' \in \mathcal{C}_{\Lambda'}} E(\Delta').$$

Then

$$E(\Lambda')SE(\Lambda) = O.$$

Since $E(\Lambda) + E(\Lambda') = E(\Lambda \cup \Lambda') = I$, this leads to

$$SE(\Lambda) = E(\Lambda \cup \Lambda')SE(\Lambda) = (E(\Lambda) + E(\Lambda'))SE(\Lambda) = E(\Lambda)SE(\Lambda).$$

Thus $\mathcal{R}(E(\Lambda))$ is S-invariant (cf. Proposition 1.I). Since this holds for every Λ in $\mathcal{A}_{\sigma(T)}$, it holds for $\Lambda' = \sigma(T)\backslash\Lambda$ so that $\mathcal{R}(E(\Lambda'))$ is S-invariant. But $\mathcal{R}(E(\Lambda)) \perp \mathcal{R}(E(\Lambda'))$ (i.e., $E(\Lambda)E(\Lambda') = O$) and $\mathcal{H} = \mathcal{R}(E(\Lambda)) + \mathcal{R}(E(\Lambda'))$ (since $\mathcal{H} = \mathcal{R}(E(\Lambda \cup \Lambda')) = \mathcal{R}(E(\Lambda) + E(\Lambda')) \subseteq \mathcal{R}(E(\Lambda)) + \mathcal{R}(E(\Lambda')) \subseteq \mathcal{H}$). Hence $\mathcal{R}(E(\Lambda')) = \mathcal{R}(E(\Lambda))^\perp$, and so $\mathcal{R}(E(\Lambda))^\perp$ is S-invariant as well. Then

$$\mathcal{R}(E(\Lambda)) \text{ reduces } S,$$

which, by Proposition 1.I, is equivalent to

$$SE(\Lambda) = E(\Lambda)S. \qquad \square$$

A *subspace* is a closed linear manifold of a normed space. *Invariant subspaces* were discussed in Section 1.1 ($\mathcal{M}$ is T-invariant if $T(\mathcal{M}) \subseteq \mathcal{M}$). A subspace $\mathcal{M}$ of $\mathcal{X}$ is *nontrivial* if $\{0\} \neq \mathcal{M} \neq \mathcal{X}$. *Reducing subspaces* (i.e., subspaces $\mathcal{M}$ for which both $\mathcal{M}$ and $\mathcal{M}^\perp$ are T-invariant or, equivalently, subspaces $\mathcal{M}$ that are invariant for both T and T^*) and *reducible operators* (i.e., operators that have a nontrivial reducing subspace) were discussed in Section 1.5. *Hyperinvariant subspaces* were defined in Section 1.9 (subspaces that are invariant for every operator commuting with a given operator). An operator is *nonscalar* if it is not a (complex) multiple of the identity; otherwise it is a *scalar operator* (Section 1.7). Scalar operators are normal. A projection E is *nontrivial* if $O \neq E \neq I$. Equivalently, *a projection is nontrivial if and only if it is nonscalar*. (Indeed, if $E^2 = E$, then it follows at once that $O \neq E \neq I$ if and only if $E \neq \alpha I$ for every $\alpha \in \mathbb{C}$.)

Corollary 3.18. *Consider operators acting on a complex Hilbert space of dimension greater than 1.*

(a) *Every normal operator has a nontrivial reducing subspace.*

(b) *Every nonscalar normal operator has a nontrivial hyperinvariant subspace which reduces every operator that commutes with it.*

(c) *An operator is reducible if and only if it commutes with a nonscalar normal operator.*

(d) *An operator is reducible if and only if it commutes with a nontrivial orthogonal projection.*

(e) *An operator is reducible if and only if it commutes with a nonscalar operator that also commutes with its adjoint.*

Proof. Take $\Lambda \in \mathcal{A}_{\sigma(T)}$. Consider the spectral decomposition $T = \int \lambda\, dE_\lambda$ of a normal operator T as in Theorem 3.15. By Theorem 3.17, if $ST = TS$, then $SE(\Lambda) = E(\Lambda)S$ or, equivalently, each subspace $\mathcal{R}(E(\Lambda))$ reduces S, which means $\{\mathcal{R}(E(\Lambda))\}$ is a family of reducing subspaces for every operator that commutes with T. If $\sigma(T)$ has just one point, say $\sigma(T) = \{\lambda\}$, then $T = \lambda I$ (by uniqueness of the spectral measure). Thus T is a scalar operator and so every subspace of $\mathcal{H}$ reduces T. If $\dim \mathcal{H} > 1$, then in this case the scalar T has many *nontrivial* reducing subspaces. So if T is nonscalar, then $\sigma(T)$ has more than one point (and $\dim \mathcal{H} > 1$). If λ, λ_0 lie in $\sigma(T)$ and $\lambda \neq \lambda_0$, then let

$$\mathbb{D}_\lambda = B_{\frac{1}{2}|\lambda - \lambda_0|}(\lambda) = \left\{\zeta \in \mathbb{C} : |\zeta - \lambda| < \tfrac{1}{2}|\lambda - \lambda_0|\right\}$$

be the open disk of radius $\frac{1}{2}|\lambda - \lambda_0|$ centered at λ. Set

$$D_\lambda = \sigma(T) \cap \mathbb{D}_\lambda \quad \text{and} \quad D'_\lambda = \sigma(T) \backslash \mathbb{D}_\lambda.$$

The pair of sets $\{D_\lambda, D'_\lambda\}$ is a measurable partition of $\sigma(T)$ — both D_λ and D'_λ lie in $\mathcal{A}_{\sigma(T)}$. Since D_λ and $\sigma(T) \backslash \mathbb{D}_\lambda^-$ are nonempty relatively open subsets of $\sigma(T)$ and $\sigma(T) \backslash \mathbb{D}_\lambda^- \subseteq D'_\lambda$, it follows by Theorem 3.15 that $E(D_\lambda) \neq O$ and $E(D'_\lambda) \neq O$. Then $I = E(\sigma(T)) = E(D_\lambda \cup D'_\lambda) = E(D_\lambda) + E(D'_\lambda)$, and so $E(D_\lambda) = I - E(D'_\lambda) \neq I$. Therefore

$$O \neq E(D_\lambda) \neq I.$$

Equivalently, $\mathcal{R}(E(D_\lambda))$ is nontrivial:

$$\{0\} \neq \mathcal{R}(E(D_\lambda)) \neq \mathcal{H}.$$

This proves (a) (for the particular case of $S = T$), and also proves (b) — the nontrivial subspace $\mathcal{R}(E(\mathcal{D}_\lambda))$ reduces every S that commutes with T. Thus according to (b), if T commutes with a nonscalar normal operator, then it is reducible. The converse follows by Proposition 1.I, since every nontrivial orthogonal projection is a nonscalar normal operator. This proves (c). Since a projection P is nontrivial if and only if $\{0\} \neq \mathcal{R}(P) \neq \mathcal{H}$, an operator T is reducible if and only if it commutes with a nontrivial orthogonal projection (by

Proposition 1.I), which proves (d). Since every nontrivial orthogonal projection is self-adjoint and nonscalar, it follows by (d) that if T is reducible, then it commutes with a nonscalar operator that also commutes with its adjoint. Conversely, suppose an operator T is such that $LT = TL$ and $LT^* = T^*L$ (and so $L^*T = TL^*$) for some nonscalar operator L. Then $(L + L^*)T = T(L + L^*)$ where $L + L^*$ is self-adjoint (thus normal). If this self-adjoint $L + L^*$ is not scalar, then T commutes with the nonscalar normal $L + L^*$. On the other hand, if this self-adjoint $L + L^*$ is scalar, then the nonscalar L must be normal. Indeed, if $L + L^* = \alpha I$, then $L^*L = LL^* = \alpha L - L^2$ and therefore L is normal. Again in this case, T commutes with the nonscalar normal L. Thus in any case, T commutes with a nonscalar normal operator and so T is reducible according to (c). This completes the proof of (e). $\qquad\square$

The equivalent assertions in Corollary 3.5 which hold for compact normal operators remain equivalent for the general case (for plain normal — not necessarily compact — operators). Again the implication (a) $\Rightarrow$ (c) in Corollary 3.19 below is the Fuglede Theorem (Theorem 3.17).

Corollary 3.19. *Let* $T = \int \lambda\, dE_\lambda$ *be a normal operator in* $\mathcal{B}[\mathcal{H}]$ *and take any operator* S *in* $\mathcal{B}[\mathcal{H}]$. *The following assertions are pairwise equivalent.*

(a) *S commutes with T.*

(b) *S commutes with T^*.*

(c) *S commutes with every $E(\Lambda)$.*

(d) *$\mathcal{R}(E(\Lambda))$ reduces S for every Λ.*

Proof. Let $T = \int \lambda\, dE_\lambda$ be the spectral decomposition of a normal operator in $\mathcal{B}[\mathcal{H}]$. Take any $\Lambda \in \mathcal{A}_{\sigma(T)}$ and an arbitrary $S \subset \mathcal{B}[\mathcal{H}]$. First we show:
$$ST = TS \quad\Longleftrightarrow\quad SE(\Lambda) = E(\Lambda)S \quad\Longleftrightarrow\quad ST^* = T^*S.$$
If $ST = TS$, then $SE(\Lambda) = E(\Lambda)S$ by Theorem 3.17. So, for every $x, y \in \mathcal{H}$, $\pi_{x,S^*y}(\Lambda) = \langle E(\Lambda)x\,;S^*y \rangle = \langle SE(\Lambda)x\,;y \rangle = \langle E(\Lambda)Sx\,;y \rangle = \pi_{Sx,y}(\Lambda)$. Hence
$$\langle ST^*x\,;y \rangle = \langle T^*x\,;S^*y \rangle = \int \overline{\lambda}\, d\langle E_\lambda x\,;S^*y \rangle$$
$$= \int \overline{\lambda}\, d\langle SE_\lambda x\,;y \rangle = \int \overline{\lambda}\, d\langle E_\lambda Sx\,;y \rangle = \langle T^*Sx\,;y \rangle$$
for every $x, y \in \mathcal{H}$ (cf. Lemmas 3.7 and 3.8(a)). Thus
$$ST = TS \quad\Longrightarrow\quad SE(\Lambda) = E(\Lambda)S \quad\Longrightarrow\quad ST^* = T^*S \quad\Longleftrightarrow\quad S^*T = TS^*.$$
Since these hold for every $S \in \mathcal{B}[\mathcal{H}]$, and since $S^{**} = S$, we get
$$S^*T = TS^* \quad\Longrightarrow\quad ST = TS.$$
Therefore (a), (b), and (c) are equivalent, and (d) is equivalent to (c) by Proposition 1.I (as in the final part of the proof of Theorem 3.17). $\qquad\square$

Corollary 3.20. (FUGLEDE–PUTNAM THEOREM). *Suppose T_1 in $\mathcal{B}[\mathcal{H}]$ and T_2 in $\mathcal{B}[\mathcal{K}]$ are normal operators. If $X \in \mathcal{B}[\mathcal{H},\mathcal{K}]$ intertwines T_1 to T_2, then X intertwines T_1^* to T_2^* (i.e., if $XT_1 = T_2 X$, then $XT_1^* = T_2^* X$).*

Proof. Take $T_1 \in \mathcal{B}[\mathcal{H}]$, $T_2 \in \mathcal{B}[\mathcal{K}]$, and $X \in \mathcal{B}[\mathcal{H},\mathcal{K}]$. Consider the operators

$$T = T_1 \oplus T_2 = \begin{pmatrix} T_1 & O \\ O & T_2 \end{pmatrix} \quad \text{and} \quad S = \begin{pmatrix} O & O \\ X & O \end{pmatrix}$$

in $\mathcal{B}[\mathcal{H} \oplus \mathcal{K}]$. If T_1 and T_2 are normal, then T is normal. If $XT_1 = T_2 X$, then $ST = TS$ and so $ST^* = T^*S$ by Corollary 3.19. Hence $XT_1^* = T_2^* X$. □

3.6 Supplementary Propositions

Let Γ be an arbitrary (not necessarily countable) nonempty index set. We have defined a diagonalizable operator in Section 3.2 as follows.

(1) An operator T on a Hilbert space $\mathcal{H}$ is *diagonalizable* if there is a resolution of the identity $\{E_\gamma\}_{\gamma \in \Gamma}$ on $\mathcal{H}$ and a bounded family of scalars $\{\lambda_\gamma\}_{\gamma \in \Gamma}$ such that $Tu = \lambda_\gamma u$ whenever $u \in \mathcal{R}(E_\gamma)$.

As we saw in Section 3.2, this can be equivalently stated as follows.

(1′) An operator T on a Hilbert space $\mathcal{H}$ is *diagonalizable* if it is a weighted sum of projections for some bounded family of scalars $\{\lambda_\gamma\}_{\gamma \in \Gamma}$ and some resolution of the identity $\{E_\gamma\}_{\gamma \in \Gamma}$ on $\mathcal{H}$.

Thus *every diagonalizable operator is normal* by Corollary 1.9, and what the Spectral Theorem for compact normal operators (Theorem 3.3) says is precisely the converse for compact normal operators: *a compact operator is normal if and only if it is diagonalizable* — cf. Corollary 3.5(a) — and so (being normal) a diagonalizable operator is such that $r(T) = \|T\| = \sup_{\gamma \in \Gamma} |\lambda_\gamma|$ (cf. Lemma 3.1). However, a diagonalizable operator on a separable Hilbert space was defined in Section 2.7 (Proposition 2.L), whose natural extension to arbitrary Hilbert space (not necessarily separable) in terms of summable families (see Section 1.3), rather than summable sequences, reads as follows.

(2) An operator T on a Hilbert space $\mathcal{H}$ is *diagonalizable* if there is an orthonormal basis $\{e_\gamma\}_{\gamma \in \Gamma}$ for $\mathcal{H}$ and a bounded family $\{\lambda_\gamma\}_{\gamma \in \Gamma}$ of scalars such that $Tx = \sum_{\gamma \in \Gamma} \lambda_\gamma \langle x \,; e_\gamma \rangle e_\gamma$ for every $x \in \mathcal{H}$.

There is no ambiguity here. Both definitions coincide: the resolution of the identity $\{E_\gamma\}_{\gamma \in \Gamma}$ behind the statement in (2) is given by $E_\gamma x = \langle x \,; e_\gamma \rangle e_\gamma$ for every $x \in \mathcal{H}$ and every $\gamma \in \Gamma$ according to the Fourier series expansion $x = \sum_{\gamma \in \Gamma} \langle x \,; e_\gamma \rangle e_\gamma$ of each $x \in \mathcal{H}$ in terms of the orthonormal basis $\{e_\gamma\}_{\gamma \in \Gamma}$. Now consider the Banach space ℓ_Γ^∞ made up of all bounded families $a = \{\alpha_\gamma\}_{\gamma \in \Gamma}$ of complex numbers and also the Hilbert space ℓ_Γ^2 consisting of all square-summable families $z = \{\zeta_\gamma\}_{\gamma \in \Gamma}$ of complex numbers indexed by Γ. The definition of a diagonal operator in $\mathcal{B}[\ell_\Gamma^2]$ goes as follows.

(3) A *diagonal operator* in $\mathcal{B}[\ell_\Gamma^2]$ is a mapping $D \colon \ell_\Gamma^2 \to \ell_\Gamma^2$ such that $Dz = \{\alpha_\gamma \zeta_\gamma\}_{\gamma \in \Gamma}$ for every $z = \{\zeta_\gamma\}_{\gamma \in \Gamma} \in \ell_\Gamma^2$, where $a = \{\alpha_\gamma\}_{\gamma \in \Gamma} \in \ell_\Gamma^\infty$ is a (bounded) family of scalars. (Notation: $D = \operatorname{diag}\{\alpha_\gamma\}_{\gamma \in \Gamma}$.)

A diagonal operator D is clearly diagonalizable by the resolution of the identity $E_\gamma z = \langle z \, ; f_\gamma \rangle f_\gamma = \zeta_\gamma f_\gamma$ given by the *canonical orthonormal basis* $\{f_\gamma\}_{\gamma \in \Gamma}$ for ℓ_Γ^2, viz., $f_\gamma = \{\delta_{\gamma,\beta}\}_{\beta \in \Gamma}$ with $\delta_{\gamma,\beta} = 1$ if $\beta = \gamma$ and $\delta_{\gamma,\beta} = 0$ if $\beta \neq \gamma$. So D is normal with $r(D) = \|D\| = \|a\|_\infty = \sup_{\gamma \in \Gamma} |\alpha_\gamma|$. Consider the Fourier series expansion of an arbitrary x in $\mathcal{H}$ in terms of any orthonormal basis $\{e_\gamma\}_{\gamma \in \Gamma}$ for $\mathcal{H}$, namely, $x = \sum_{\gamma \in \Gamma} \langle x \, ; e_\gamma \rangle e_\gamma$. Thus we can identify x in $\mathcal{H}$ with the square-summable family $\{\langle x \, ; e_\gamma \rangle\}_{\gamma \in \Gamma}$ in ℓ_Γ^2 (i.e., x is the image of $\{\langle x \, ; e_\gamma \rangle\}_{\gamma \in \Gamma}$ under a unitary transformation between the unitarily equivalent spaces $\mathcal{H}$ and ℓ_Γ^2). Therefore if $Tx = \sum_{\gamma \in \Gamma} \lambda_\gamma \langle x \, ; e_\gamma \rangle e_\gamma$, then we can identify Tx with the square-summable family $\{\lambda_\gamma \langle x \, ; e_\gamma \rangle\}_{\gamma \in \Gamma}$ in ℓ_Γ^2, and so we can identify T with a diagonal operator D that takes $z = \{\langle x \, ; e_\gamma \rangle\}_{\gamma \in \Gamma}$ into $Dz = \{\lambda_\gamma \langle x \, ; e_\gamma \rangle\}_{\gamma \in \Gamma}$. This means: T and D are *unitarily equivalent* — there is a unitary transformation $U \in \mathcal{B}[\mathcal{H}, \ell_\Gamma^2]$ for which $UT = DU$. It is in this sense that T is said to be a *diagonal operator with respect to the orthonormal basis* $\{e_\gamma\}_{\gamma \in \Gamma}$. Thus definition (2) can be rephrased as follows.

(2′) An operator T on a Hilbert space $\mathcal{H}$ is *diagonalizable* if it is a diagonal operator with respect to some orthonormal basis for $\mathcal{H}$.

Proposition 3.A. *Let T be an operator on a nonzero complex Hilbert space. The following assertions are equivalent.*

(a) *T is diagonalizable in the sense of definition (1) above.*

(b) *$\mathcal{H}$ has an orthogonal basis made up of eigenvectors of T.*

(c) *T is diagonalizable in the sense of definition (2) above.*

(d) *T is unitarily equivalent to a diagonal operator.*

Take a measure space $(\Omega, \mathcal{A}_\Omega, \mu)$ where $\mathcal{A}_\Omega$ is a σ-algebra of subsets of a nonempty set Ω and μ is a positive measure on $\mathcal{A}_\Omega$. Let $L^\infty(\Omega, \mathbb{C}, \mu)$ — short notation: $L^\infty(\Omega, \mu)$, $L^\infty(\mu)$, or just L^∞ — denote the Banach space of all *essentially bounded $\mathcal{A}_\Omega$-measurable complex-valued functions* $f \colon \Omega \to \mathbb{C}$ on Ω equipped with the usual sup-norm (i.e., $\|f\|_\infty = \operatorname{ess\,sup} |f|$ for $f \in L^\infty$). Let $L^2(\Omega, \mathbb{C}, \mu)$ — short notation: $L^2(\Omega, \mu)$, $L^2(\mu)$, or just L^2 — denote the Hilbert space of all *square-integrable $\mathcal{A}_\Omega$-measurable complex-valued functions* $f \colon \Omega \to \mathbb{C}$ on Ω equipped with the usual L^2-norm (i.e., $\|f\|_2 = \left(\int |f|^2 d\mu \right)^{\frac{1}{2}}$ for $f \in L^2$). *Multiplication operators* on L^2 will be considered next (see, e.g., [58, Chapter 7]). Multiplication operators are prototypes of normal operators according to the first version of the Spectral Theorem (Theorem 3.11).

Proposition 3.B. *Let $\phi \colon \Omega \to \mathbb{C}$ and $f \colon \Omega \to \mathbb{C}$ be complex-valued functions on a nonempty set Ω. Let μ be a positive measure on a σ-algebra $\mathcal{A}_\Omega$ of* *subsets of Ω. Suppose $\phi \in L^\infty$. If $f \in L^2$, then $\phi f \in L^2$. Thus consider the mapping $M_\phi \colon L^2 \to L^2$ defined by*

$$M_\phi f = \phi f \quad \text{for every} \quad f \in L^2.$$

That is, $(M_\phi f)(\lambda) = \phi(\lambda) f(\lambda)$ μ-a.e. for $\lambda \in \Omega$. This is called the multiplication operator *on L^2, which is linear and bounded (i.e., $M_\phi \in \mathcal{B}[L^2]$). Moreover:*

(a) $M_\phi^* = M_{\overline{\phi}}$ *(i.e., $(M_\phi^* f)(\lambda) = \overline{\phi(\lambda)} f(\lambda)$ μ-a.e. for $\lambda \in \Omega$).*

(b) M_ϕ *is a normal operator.*

(c) $\|M_\phi\| \leq \|\phi\|_\infty = \operatorname{ess\,sup} |\phi| = \inf_{\Lambda \in \mathcal{N}_\Omega} \sup_{\lambda \in \Omega \setminus \Lambda} |\phi(\lambda)|$,

   where $\mathcal{N}_\Omega$ denotes the collection of all $\Lambda \in \mathcal{A}_\Omega$ such that $\mu(\Lambda) = 0$.

(d) $\sigma(M_\phi) \subseteq \operatorname{ess} \mathcal{R}(\phi) =$ *essential range of ϕ*

$$= \left\{ \alpha \in \mathbb{C} \colon \mu(\{\lambda \in \Omega \colon |\phi(\lambda) - \alpha| < \varepsilon\}) > 0 \text{ for every } \varepsilon > 0 \right\}$$

$$= \left\{ \alpha \in \mathbb{C} \colon \mu(\phi^{-1}(V_\alpha)) > 0 \text{ for every Borel neighborhood } V_\alpha \text{ of } \alpha \right\}$$

$$= \bigcap \left\{ \phi(\Lambda)^- \in \mathbb{C} \colon \Lambda \in \mathcal{A}_\Omega \text{ and } \mu(\Omega \setminus \Lambda) = 0 \right\} \subseteq \phi(\Omega)^- = \mathcal{R}(\phi)^-.$$

If in addition μ is σ-finite, then

(e) $\|M_\phi\| = \|\phi\|_\infty$,

(f) $\sigma(M_\phi) = \operatorname{ess} \mathcal{R}(\phi)$.

   Particular case: *Suppose* (i) *Ω is a nonempty bounded subset of $\mathbb{C}$ and* (ii) *$\varphi \colon \Omega \to \Omega = \mathcal{R}(\varphi) \subseteq \mathbb{C}$ is the identity map on Ω (i.e., $\varphi(\lambda) = \lambda$ for every $\lambda \in \Omega$). So $\operatorname{ess} \mathcal{R}(\varphi) = \operatorname{ess} \Omega = \bigcap \{\Lambda^- \in \mathbb{C} \colon \Lambda \in \mathcal{A}_\Omega$ and $\mu(\Omega \setminus \Lambda) = 0\} \subseteq \Omega^-$. Let support $(\mu) = \bigcap \{\Lambda \in \mathcal{A}_\Omega \colon \Lambda$ is closed in Ω and $\mu(\Omega \setminus \Lambda) = 0\}$ be the sup-port of μ. As $\Lambda^- \cap \Omega$ is closed in Ω, support $(\mu) = \operatorname{ess} \mathcal{R}(\varphi) \cap \Omega = \operatorname{ess} \Omega \cap \Omega$. This is the smallest relatively closed subset of Ω with complement of measure zero (i.e., $\Omega \setminus$support (μ) is the largest relatively open subset of Ω of measure zero). If $\varnothing \neq \Lambda \in \mathcal{A}_\Omega$ is relatively open and if $\Lambda \subseteq$ support (μ), then $\mu(\Lambda) > 0$. (Indeed, if $\varnothing \neq \Lambda \subseteq$ support (μ) is open in Ω and if $\mu(\Lambda) = 0$, then the set $[\Omega \setminus$support $(\mu)] \cup \Lambda$ is a relatively open subset of Ω larger than $\Omega \setminus$support (μ) and of measure zero, which is a contradiction.) Moreover, if in addition Ω is closed in $\mathbb{C}$ (i.e., if $\Omega \neq \varnothing$ is compact), then $\operatorname{ess} \mathcal{R}(\varphi) = \operatorname{ess} \Omega \subseteq \Omega$ and hence support $(\mu) = \operatorname{ess} \mathcal{R}(\varphi) = \operatorname{ess} \Omega$ is compact. Therefore if Ω is nonempty and compact and if μ is σ-finite, then $\sigma(M_\varphi) = \operatorname{ess} \mathcal{R}(\varphi) = \operatorname{ess} \Omega =$ support (μ).*

   If $\Omega = \mathbb{N}$, the set of all positive integers, if $\mathcal{A}_\Omega = \wp(\mathbb{N})$, the power set of $\mathbb{N}$, and if μ is the counting measure (assigning to each finite set the number of elements in it, which is σ-finite), then a multiplication operator is reduced to a diagonal operator on $L^2(\mathbb{N}, \mathbb{C}, \mu) = \ell_{\mathbb{N}}^2 = \ell_+^2$, where the bounded $\wp(\mathbb{N})$-meas-urable function $\phi \colon \mathbb{N} \to \mathbb{C}$ in $L^\infty(\mathbb{N}, \mathbb{C}, \mu) = \ell_{\mathbb{N}}^\infty = \ell_+^\infty$ is the bounded sequence $\{\alpha_k\}$ with $\alpha_k = \phi(k)$ for each $k \in \mathbb{N}$, so that $\|M_\phi\| = \|\phi\|_\infty = \sup_k |\alpha_k|$ and

$\sigma(\mathcal{M}_\phi) = \mathcal{R}(\phi)^- = \{\alpha_k\}^-$. What the Spectral Theorem says is: a normal operator is unitarily equivalent to a multiplication operator (Theorem 3.11), which turns out to be a diagonal operator in the compact case (Theorem 3.3).

Proposition 3.C. *Let Ω be a nonempty compact subset of the complex plane $\mathbb{C}$ and let $E \colon \mathcal{A}_\Omega \to \mathcal{B}[\mathcal{H}]$ be a spectral measure (cf. Definition 3.6). For each $\Lambda \in \mathcal{A}_\Omega$ set $\mathcal{C}_\Lambda = \{\Delta \in \mathcal{A}_\Omega \colon \Delta$ is a closed subset of $\Lambda\}$. Then $\mathcal{R}(E(\Lambda))$ is the smallest subspace of $\mathcal{H}$ such that $\bigcup_{\Delta \in \mathcal{C}_\Lambda} \mathcal{R}(E(\Delta)) \subseteq \mathcal{R}(E(\Lambda))$.*

Proposition 3.D. *Let $\mathcal{H}$ be a complex Hilbert space and let $\mathbb{T}$ be the unit circle about the origin of the complex plane. If $T \in \mathcal{B}[\mathcal{H}]$ is normal, then*

(a) *T is unitary if and only if $\sigma(T) \subseteq \mathbb{T}$,*

(b) *T is self-adjoint if and only if $\sigma(T) \subseteq \mathbb{R}$,*

(c) *T is nonnegative if and only if $\sigma(T) \subseteq [0, \infty)$,*

(d) *T is strictly positive if and only if $\sigma(T) \subseteq [\alpha, \infty)$ for some $\alpha > 0$,*

(e) *T is an orthogonal projection if and only if $\sigma(T) \subseteq \{0, 1\}$.*

If $T = \left(\begin{smallmatrix} 1 & 0 \\ 0 & 0 \end{smallmatrix}\right)$ and $S = \left(\begin{smallmatrix} 2 & 1 \\ 1 & 1 \end{smallmatrix}\right)$ in $\mathcal{B}[\mathbb{C}^2]$, then $S^2 - T^2 = \left(\begin{smallmatrix} 4 & 3 \\ 3 & 2 \end{smallmatrix}\right)$. Therefore $O \leq T \leq S$ does not imply $T^2 \leq S^2$. However, the converse holds.

Proposition 3.E. *If $T, S \in \mathcal{B}[\mathcal{H}]$ are nonnegative operators, then*

$$T \leq S \quad \text{implies} \quad T^{\frac{1}{2}} \leq S^{\frac{1}{2}}.$$

Proposition 3.F. *If $T = \int \lambda \, dE_\lambda$ is the spectral decomposition of a nonnegative operator T on a complex Hilbert space, then $T^{\frac{1}{2}} = \int \lambda^{\frac{1}{2}} \, dE_\lambda$.*

Proposition 3.G. *Let $T = \int \lambda \, dE_\lambda$ be the spectral decomposition of a normal operator $T \in \mathcal{B}[\mathcal{H}]$. Take an arbitrary $\lambda \in \sigma(T)$.*

(a) *$\mathcal{N}(\lambda I - T) = \mathcal{R}(E(\{\lambda\}))$.*

(b) *$\lambda \in \sigma_P(T)$ if and only if $E(\{\lambda\}) \neq O$.*

(c) *If λ is an isolated point of $\sigma(T)$, then $\lambda \in \sigma_P(T)$.*

(Every isolated point of the spectrum of a normal operator is an eigenvalue.)

(d) *If $\mathcal{H}$ is separable, then $\sigma_P(T)$ is countable.*

Recall: $X \in \mathcal{B}[\mathcal{H}, \mathcal{K}]$ *intertwines* $T \in \mathcal{B}[\mathcal{H}]$ to $S \in \mathcal{B}[\mathcal{K}]$ if $XT = SX$. If an invertible X intertwines T to S, then T and S are *similar*. If, in addition, X is unitary, then T and S are *unitarily equivalent* (cf. Section 1.9).

Proposition 3.H. *Let $T_1 \in \mathcal{B}[\mathcal{H}]$ and $T_2 \in \mathcal{B}[\mathcal{K}]$ be normal operators. If $X \in \mathcal{B}[\mathcal{H}, \mathcal{K}]$ intertwines T_1 to T_2 (i.e., $XT_1 = T_2X$), then*

(a) $\mathcal{N}(X)$ *reduces* T_1 *and* $\mathcal{R}(X)^-$ *reduces* T_2

so that $T_1|_{\mathcal{N}(X)^\perp} \in \mathcal{B}[\mathcal{N}(X)^\perp]$ *and* $T_2|_{\mathcal{R}(X)^-} \in \mathcal{B}[\mathcal{R}(X)^-]$. *Moreover*,

(b) $T_1|_{\mathcal{N}(X)^\perp}$ *and* $T_2|_{\mathcal{R}(X)^-}$ *are unitarily equivalent.*

A special case of Proposition 3.H: *if a quasiinvertible* (i.e., injective with dense range) *bounded linear transformation intertwines two normal operators, then they are unitarily equivalent.* This happens in particular if X is invertible.

Proposition 3.I. *Two similar normal operators are unitarily equivalent.*

A subset of a metric space is *nowhere dense* if its closure has empty interior. Thus the spectrum $\sigma(T)$ of $T \in \mathcal{B}[\mathcal{H}]$ is nowhere dense if and only if its interior $\sigma(T)^\circ$ is empty. The full spectrum $\sigma(T)^\#$ was defined in Section 2.7: the union of the spectrum and its holes. So $\sigma(T) = \sigma(T)^\#$ if and only if $\rho(T)$ is connected. An operator $T \in \mathcal{B}[\mathcal{H}]$ is reductive if all its invariant subspaces are reducing (Section 1.5). In particular, a normal operator is reductive if and only if its restriction to every invariant subspace is normal (Proposition 1.Q). Normal reductive operators are also called *completely normal.*

Proposition 3.J. *Suppose* $T \in \mathcal{B}[\mathcal{H}]$ *is a normal operator. If* $\sigma(T) = \sigma(T)^\#$ *and* $\sigma(T)^\circ = \varnothing$, *then* T *is reductive. If* T *is reductive, then* $\sigma(T)^\circ = \varnothing$.

There are normal reductive operators for which the spectrum is different from the full spectrum. Example: the unitary diagonal U in Proposition 3.L(j) below is reductive (so $\sigma(U)^\circ = \varnothing$) with $\sigma(U) = \mathbb{T}$. On the other hand, there are normal nonreductive operators with the same property. Example: a bilateral shift S is a nonreductive unitary operator with $\sigma(S) = \mathbb{T}$ (so $\sigma(S)^\circ = \varnothing$). In fact, S has an invariant subspace $\mathcal{M}$ for which the restriction $S|_\mathcal{M}$ is a unilateral shift (see, e.g., [74, Proposition 2.13]), which is not normal, and so $\mathcal{M}$ does not reduce S by Proposition 1.Q (see Proposition 3.L(b,c) below).

Proposition 3.K. *Let* T *be a diagonalizable operator on a nonzero complex Hilbert space. The following assertions are equivalent.*

(a) *Every nontrivial invariant subspace for* T *contains an eigenvector.*

(b) *Every nontrivial invariant subspace* $\mathcal{M}$ *for* T *is spanned by the eigenvectors of* $T|_\mathcal{M}$ *(i.e., if* $\mathcal{M}$ *is* T*-invariant, then* $\mathcal{M} = \bigvee_{\lambda \in \sigma_P(T|_\mathcal{M})} \mathcal{N}(\lambda I - T|_\mathcal{M})$).

(c) T *is reductive.*

There are diagonalizable operators which are not reductive. In fact, since a subset of $\mathbb{C}$ is a separable metric space, it includes a countable dense subset. Let Λ be any countable dense subset of an arbitrary nonempty compact subset Ω of $\mathbb{C}$. Let $\{\lambda_k\}$ be an enumeration of Λ so that $\sup_k |\lambda_k| < \infty$ (since Ω is bounded). Consider the diagonal operator $D = \mathrm{diag}\{\lambda_k\}$ in $\mathcal{B}[\ell_+^2]$. Then

$$\sigma(D) = \Lambda^- = \Omega$$

according to Proposition 2.L. Thus *every closed and bounded subset of the complex plane is the spectrum of a diagonal operator on ℓ_+^2.* Hence if $\Omega^\circ \neq \varnothing$, then the diagonal D is not reductive by Proposition 3.J.

Let μ and η be measures on a σ-algebra $\mathcal{A}_\Omega$. If $\{\eta(\Lambda) = 0 \Rightarrow \mu(\Lambda) = 0\}$ for $\Lambda \in \mathcal{A}_\Omega$, then μ is *absolutely continuous* with respect to η (notation: $\mu \ll \eta$). If $\{\eta(\{\lambda\}) = 0 \Rightarrow \mu(\{\lambda\}) = 0\}$ for every singleton $\{\lambda\} \in \mathcal{A}_\Omega$, then μ is *continuous* with respect to η. If there is a measurable partition $\{\Lambda, \Delta\}$ of Ω with $\eta(\Lambda) = \mu(\Delta) = 0$, then η and μ are *singular* (notation: $\eta \perp \mu$ or $\mu \perp \eta$). If $\mu \perp \eta$ and if the set $\Lambda = \{\lambda_k\} \in \mathcal{A}_\Omega$ in the above partition is countable (union of measurable singletons), then μ is *discrete* with respect to η. If μ and η are σ-finite, then

$$\mu = \mu_a + \mu_s = \mu_a + \mu_{sc} + \mu_{sd},$$

where μ_a, μ_s, μ_{sc}, and μ_{sd} are absolutely continuous, singular, singular-continuous, and singular-discrete (i.e., discrete) with respect to a reference measure η. This is the LEBESGUE DECOMPOSITION THEOREM (see, e.g., [81, Chapter 7]). Take the σ-algebra $\mathcal{A}_\mathbb{T}$ of Borel sets in the unit circle $\mathbb{T}$ and the normalized Lebesgue measure η on $\mathcal{A}_\mathbb{T}$ ($\eta(\mathbb{T}) = 1$). A unitary operator is *absolutely continuous, continuous, singular, singular-continuous,* or *singular-discrete* (i.e., *discrete*) if its scalar spectral measure is absolutely continuous, continuous, singular, singular-continuous, or singular-discrete with respect to η. By the Lebesgue Decomposition Theorem and by the Spectral Theorem, every unitary operator U on a Hilbert space $\mathcal{H}$ is uniquely decomposed as the direct sum

$$U = U_a \oplus U_s = U_a \oplus U_{sc} \oplus U_{sd}$$

of an absolutely continuous unitary U_a on $\mathcal{H}_a$, a singular unitary U_s on $\mathcal{H}_s$, a singular-continuous unitary U_{sc} on $\mathcal{H}_{sd}$, and a discrete unitary U_{sd} on $\mathcal{H}_{sd}$, with $\mathcal{H}$ decomposed as $\mathcal{H} = \mathcal{H}_a \oplus \mathcal{H}_s = \mathcal{H}_a \oplus \mathcal{H}_{sc} \oplus \mathcal{H}_{sd}$ (orthogonal direct sums).

Proposition 3.L. *Let $U \in \mathcal{B}[\mathcal{H}]$ be a unitary operator on a Hilbert space $\mathcal{H}$ (thus $\sigma(U) \subseteq \mathbb{T}$ and so $\sigma(U)^\circ = \varnothing$). Let φ be the identity map on $\mathbb{T}$.*

(a) *Direct sums of bilateral shifts are bilateral shifts.*

(b) *Every bilateral shift $S \in \mathcal{B}[\mathcal{H}]$ has an invariant subspace $\mathcal{M}$ for which the restriction $S|_\mathcal{M} \in \mathcal{B}[\mathcal{M}]$ is a unilateral shift.*

(c) *A bilateral shift is a unitary operator which is not reductive.*

(d) *$U = S \oplus W$ where S is a bilateral shift and W is a reductive unitary operator.*

(e) *U is reductive if and only if it has no bilateral shift S as a direct summand.*

(f) *If $\sigma(U) \neq \mathbb{T}$, then U is reductive.*

(g) *A unitary operator is absolutely continuous if and only if it is a direct summand of a bilateral shift (or a bilateral shift itself).*

(h) *A singular unitary operator is reductive.*

(i) U *is continuous if and only if* $\sigma_P(U) = \varnothing$. *If* U *is discrete, then* $\sigma_P(U) \neq \varnothing$.

(j) *If* $\{\lambda_k\}_{k \in \mathbb{N}}$ *is an enumeration of* $\mathbb{Q} \cap [0, 1)$, *then* $D = \mathrm{diag}(\{e^{2\pi i \lambda_k}\})$ *on* ℓ_+^2 *is a reductive singular-discrete unitary diagonal operator with* $\sigma(D) = \mathbb{T}$.

(k) *If* M_φ *is the bilateral shift on* $L^2(\mathbb{T}, \eta)$ *shifting the orthonormal basis* $\{\lambda^k\}_{k \in \mathbb{Z}}$ *and* $\Upsilon \in \mathcal{A}_{\mathbb{T}}$ *with* $0 < \eta(\Upsilon) < 1$, *then* $L^2(\Upsilon, \eta)$ *reduces* M_φ *and* $U = M_\varphi|_{L^2(\Upsilon, \eta)}$ *is an absolutely continuous unitary with* $\sigma(U) = \Upsilon \neq \mathbb{T}$, *thus reductive.*

(ℓ) *The multiplication operator* M_φ *on* $L^2(\mathbb{T}, \mu)$ *is a singular-continuous unitary if* μ *is the Borel–Stieltjes measure on* $\mathcal{A}_{\mathbb{T}}$ *generated by the Cantor function on the Cantor set* $\Gamma \in \mathcal{A}_{\mathbb{T}}$ *with* $\sigma(M_\varphi) = \Gamma \neq \mathbb{T}$, *thus reductive.*

Notes: Diagonalizable operators are considered in Proposition 3.A (see, e.g., [78, Problem 6.25]). Multiplication operators are considered in Proposition 3.B (see, e.g., [30, Example IX.2.6], [32, Proposition I.2.6], [58, Chapter 7], and [98, p. 13]). The technical result in Proposition 3.C is needed for proving the Fuglede Theorem — see [98, Proposition 1.4]. The next propositions depend on the Spectral Theorem. Proposition 3.D considers the converse of Proposition 2.A (see, e.g., [78, Problem 6.26]). For Proposition 3.E see, for instance, [75, Problem 8.12]. Proposition 3.F is a consequence of Lemma 3.10 and Proposition 1.M (see, e.g., [78, Problem 6.45]). Proposition 3.G is a classical result — see [104, Theorem 12.29 and Exercise 18(b) p. 325)]; also [98, Proposition 1.2], [78, Problem 6.28]. Propositions 3.H and 3.I follow from Corollary 3.20 (see, e.g., [30, Propositions IX.6.10 and IX.6.11]). Reductive normal operators are considered in Propositions 3.J and 3.K (see, e.g., [39, Theorems 13.2 and 13.33] and [98, Theorems 1.23 and 1.25]). Proposition 3.L is a collection of results on decomposition of unitary operators. For items (a) and (b) see [74, Propositions 2.11 and 2.13]; item (b) and Proposition 1.Q imply (c); for (d) see [48, 1.VI p. 18]; item (e) follows from (d) — see also [98, Proposition 1.11] and [39, Theorem 13.14]; Propositions 2.A and 3.J lead to item (f) — see also [98, Theorem 1.23] and [39, Theorem 13.2]; for item (g) see [48, Exercise 6.8 p. 56]; items (e) and (g) imply (h) — see also [39, Theorem 13.15]; item (i) is a consequence of Proposition 3.G; for item (j) see [39, Example 13.5]; item (k) follows from [98, Theorem 3.6] and (g); for item (ℓ) see Proposition 3.B, [81, Problem 7.15(c)], and remark after Lemma 4.7 (next section).

Suggested Readings

Arveson [6, 7]

Bachman and Narici [9]

Berberian [15]

Conway [30, 32]

Douglas [38]

Dowson [39]

Dunford and Schwartz [46]

Halmos [54, 58]

Helmberg [63]

Kubrusly [78]

Radjavi and Rosenthal [98]

Reed and Simon [99]

Rudin [104]

Weidmann [113]

4

Functional Calculi

Take a fixed operator T in the operator algebra $\mathcal{B}[\mathcal{X}]$. Suppose another operator $\psi(T)$ in $\mathcal{B}[\mathcal{X}]$ can be associated to each function $\psi\colon \sigma(T) \to \mathbb{C}$ in a suitable function algebra $F(\sigma(T))$. This chapter explores the map $\Phi_T\colon F(\sigma(T)) \to \mathcal{B}[\mathcal{X}]$ defined by $\Phi_T(\psi) = \psi(T)$ for every ψ in $F(\sigma(T))$ — a map $\psi \mapsto \psi(T)$ taking each function ψ in a function algebra $F(\sigma(T))$ to the operator $\psi(T)$ in the operator algebra $\mathcal{B}[\mathcal{X}]$. If Φ_T is linear and preserves the product operation (i.e., if it is an *algebra homomorphism*), then it is referred to as a *functional calculus* for T (also called *operational calculus*). Sections 4.1 and 4.2 deal with the case where $\mathcal{X}$ is a Hilbert space and T is a normal operator. Sections 4.3 and 4.4 deal with the case where $\mathcal{X}$ is a Banach space ψ is an analytic function.

4.1 Rudiments of Banach and C*-Algebras

Let $\mathcal{A}$ and $\mathcal{B}$ be algebras over the same scalar field $\mathbb{F}$. If $\mathbb{F} = \mathbb{C}$, then these are referred to as *complex algebras*. A linear transformation $\Phi\colon \mathcal{A} \to \mathcal{B}$ (of the linear space $\mathcal{A}$ into the linear space $\mathcal{B}$) that preserves products (i.e., such that $\Phi(xy) = \Phi(x)\Phi(y)$ for every x, y in $\mathcal{A}$) is a *homomorphism* (or an *algebra homomorphism*) of $\mathcal{A}$ into $\mathcal{B}$. A *unital algebra* (or an *algebra with identity*) is an algebra with an identity element (i.e., with a neutral element under multiplication). An element x in a unital algebra $\mathcal{A}$ is *invertible* if there is an $x^{-1} \in \mathcal{A}$ such that $x^{-1}x = x\,x^{-1} = 1$, where $1 \in \mathcal{A}$ denotes the identity in $\mathcal{A}$. A *unital homomorphism* between unital algebras is one that takes the identity of $\mathcal{A}$ to the identity of $\mathcal{B}$. If Φ is an isomorphism (of the linear space $\mathcal{A}$ *onto* the linear space $\mathcal{B}$) and also a homomorphism (of the algebra $\mathcal{A}$ *onto* the algebra $\mathcal{B}$), then it is an *algebra isomorphism of $\mathcal{A}$ onto $\mathcal{B}$*. In this case $\mathcal{A}$ and $\mathcal{B}$ are said to be *isomorphic algebras*. A *normed algebra* is an algebra $\mathcal{A}$ which is also a normed space whose norm satisfies the operator norm property, viz., $\|xy\| \leq \|x\|\,\|y\|$ for every x, y in $\mathcal{A}$. The identity element of a unital normed algebra has norm $1 \in \mathbb{F}$. A *Banach algebra* is a complete normed algebra. The *spectrum* $\sigma(x)$ of an element $x \in \mathcal{A}$ in a *unital complex Banach algebra* $\mathcal{A}$ is the complement of the set $\rho(x) = \{\lambda \in \mathbb{C}\colon \lambda 1 - x \text{ has an inverse in } \mathcal{A}\}$, and its *spectral radius* is the number $r(x) = \sup_{\lambda \in \sigma(x)} |\lambda|$. An *involution* $*\colon \mathcal{A} \to \mathcal{A}$ on an algebra $\mathcal{A}$ is a map $x \mapsto x^*$ such that $(x^*)^* = x$, $(xy)^* = y^*x^*$, and

© Springer Nature Switzerland AG 2020
C. S. Kubrusly, *Spectral Theory of Bounded Linear Operators*,
https://doi.org/10.1007/978-3-030-33149-8_4

$(\alpha x + \beta y)^* = \overline{\alpha}x^* + \overline{\beta}y^*$ for all scalars α, β and every x, y in $\mathcal{A}$. A *-*algebra* (or an *involutive algebra*) is an algebra equipped with an involution. If $\mathcal{A}$ and $\mathcal{B}$ are *-algebras, then a *-*homomorphism* (a *-*isomorphism*) between them is an algebra homomorphism (an algebra isomorphism) $\varPhi\colon \mathcal{A} \to \mathcal{B}$ that preserves involution (i.e., $\varPhi(x^*) = \varPhi(x)^*$ in $\mathcal{B}$ for every x in $\mathcal{A}$ — we use the same notation for involutions on $\mathcal{A}$ and $\mathcal{B}$). An element x in a *-algebra is *Hermitian* if $x^* = x$ and *normal* if $x^*x = xx^*$. An element x in a unital *-algebra is *unitary* if $x^*x = xx^* = 1 \in \mathcal{A}$. A C*-*algebra* is a Banach *-algebra $\mathcal{A}$ for which

$$\|x^*x\| = \|x\|^2 \quad \text{for every } x \in \mathcal{A}.$$

In a *-algebra $(x^*)^*$ is usually denoted by x^{**}. The origin 0 in a *-algebra and the identity 1 in a unital *-algebra are Hermitian. Indeed, since $(\alpha x)^* = \overline{\alpha}x^*$ for every x and every scalar α, we get $0^* = 0$ by setting $\alpha = 0$. Since $1^*x = (1^*x)^{**} = (x^*1)^* = (1x^*)^* = x1^*$ and $(x^*1)^* = x^{**} = x$, we also get $1^*x = x1^* = x$, and so (by uniqueness of the identity) $1^* = 1$. In a unital C*-algebra the expression $\|1\| = 1$ is a theorem (rather than an axiom) since $\|1\|^2 = \|1^*1\| = \|1^2\| = \|1\|$, hence $\|1\| = 1$ (because $\|1\| \neq 0$ as $1 \neq 0$ in $\mathcal{A}$).

Elementary properties of Banach algebras, C*-algebras, and algebra homomorphisms needed in the sequel are brought together in the next lemma.

Lemma 4.1. *Let $\mathcal{A}$ and $\mathcal{B}$ be algebras over the same scalar field and let $\varPhi\colon \mathcal{A} \to \mathcal{B}$ be a homomorphism. Take an arbitrary $x \in \mathcal{A}$.*

(a) *If $\mathcal{A}$ is a complex *-algebra, then $x = a + ib$ with $a^* = a$ and $b^* = b$ in $\mathcal{A}$.*

(b) *If $\mathcal{A}$, $\mathcal{B}$, and $\varPhi$ are unital and x is invertible, then $\varPhi(x)^{-1} = \varPhi(x^{-1})$.*

(c) *If $\mathcal{A}$ is unital and normed, and x is invertible, then $\|x\|^{-1} \leq \|x^{-1}\|$.*

(d) *If $\mathcal{A}$ is a unital *-algebra and x is invertible, then $(x^{-1})^* = (x^*)^{-1}$.*

(e) *If $\mathcal{A}$ is a C*-algebra, then $\|x^*\| = \|x\|$ (so $\|x^*x\| = \|x\|^2 = \|x^*\|^2 = \|xx^*\|$).*

(f) *If $\mathcal{A}$ is a unital C*-algebra and $x^*x = 1$, then $\|x\| = 1$.*

Suppose $\mathcal{A}$ and $\mathcal{B}$ are unital complex Banach algebras.

(g) *If $\varPhi$ is a unital homomorphism, then $\sigma(\varPhi(x)) \subseteq \sigma(x)$.*

(h) *If $\varPhi$ is an injective unital homomorphism, then $\sigma(\varPhi(x)) = \sigma(x)$.*

(i) $r(x) = \lim_n \|x^n\|^{\frac{1}{n}}$.

Suppose $\mathcal{A}$ and $\mathcal{B}$ are unital complex C-algebras.*

(j) *If $x^* = x$, then $r(x) = \|x\|$.*

(k) *If $\varPhi$ is a unital *-homomorphism, then $\|\varPhi(x)\| \leq \|x\|$.*

(ℓ) *If $\varPhi$ is an injective unital *-homomorphism, then $\|\varPhi(x)\| = \|x\|$.*

Proof. Let $\varPhi\colon \mathcal{A} \to \mathcal{B}$ be a homomorphism and take an arbitrary $x \in \mathcal{A}$.

(a) This is the *Cartesian decomposition* (as in Proposition 1.O). If $\mathcal{A}$ is a $*$-algebra, then set $a = \frac{1}{2}(x^* + x) \in \mathcal{A}$ and $b = \frac{i}{2}(x^* - x) \in \mathcal{A}$. Thus $a^* = \frac{1}{2}(x + x^*) = a$, $b^* = \frac{-i}{2}(x - x^*) = b$, and $a + ib = \frac{1}{2}(x^* + x - x^* + x) = x$.

(b) If $\mathcal{A}$ and $\mathcal{B}$ are unital algebras, Φ is a unital homomorphism, and x is invertible, then $1 = \Phi(1) = \Phi(x^{-1}x) = \Phi(xx^{-1}) = \Phi(x^{-1})\Phi(x) = \Phi(x)\Phi(x^{-1})$.

(c) If $\mathcal{A}$ is a unital normed algebra, then $1 = \|1\| = \|x^{-1}x\| \le \|x^{-1}\|\,\|x\|$ whenever x is invertible.

(d) If $\mathcal{A}$ is a unital $*$-algebra and if $x^{-1}x = xx^{-1} = 1$, then $x^*(x^{-1})^* = (x^{-1}x)^* = (xx^{-1})^* = (x^{-1})^*x^* = 1^* = 1$, and so $(x^*)^{-1} = (x^{-1})^*$.

(e) Let $\mathcal{A}$ be a C*-algebra. The result is trivial for $x = 0$ because $0^* = 0$. Since $\|x\|^2 = \|x^*x\| \le \|x^*\|\,\|x\|$, it follows that $\|x\| \le \|x^*\|$ if $x \ne 0$. Replacing x with x^* and since $x^{**} = x$ we get $\|x^*\| \le \|x\|$.

(f) Particular case of (e) since $\|1\| = 1$.

(g) Let $\mathcal{A}$ and $\mathcal{B}$ be unital complex Banach algebras, and let Φ be a unital homomorphism. If $\lambda \in \rho(x)$, then $\lambda 1 - x$ is invertible in $\mathcal{A}$. Since Φ is unital, $\Phi(\lambda 1 - x) = \lambda 1 - \Phi(x)$ is invertible in $\mathcal{B}$ according to (b). Hence $\lambda \in \rho(\Phi(x))$. Thus $\rho(x) \subseteq \rho(\Phi(x))$. Therefore $\mathbb{C} \setminus \rho(\Phi(x)) \subseteq \mathbb{C} \setminus \rho(x)$.

(h) If $\Phi \colon \mathcal{A} \to \mathcal{B}$ is injective, then it has an inverse $\Phi^{-1} \colon \mathcal{R}(\Phi) \to \mathcal{B}$ on its range $\mathcal{R}(\Phi) = \Phi(\mathcal{A}) \subseteq \mathcal{B}$ which, being the image of a unital algebra under a unital homomorphism, is again a unital algebra. Moreover, Φ^{-1} itself is a unital homomorphism. Indeed, if $y = \Phi(u)$ and $z = \Phi(v)$ are arbitrary elements in $\mathcal{R}(\Phi)$, then $\Phi^{-1}(yz) = \Phi^{-1}(\Phi(u)\Phi(v)) = \Phi^{-1}(\Phi(uv)) = uv = \Phi^{-1}(y)\Phi^{-1}(z)$ and so Φ^{-1} is a homomorphism (since inverses of linear transformations are linear) which is trivially unital (since Φ is unital). Now apply (b) so that $\Phi^{-1}(y)$ is invertible in $\mathcal{A}$ whenever y is invertible in $\mathcal{R}(\Phi) \subseteq \mathcal{B}$. If $\lambda \in \rho(\Phi(x))$, then $y = \lambda 1 - \Phi(x) = \Phi(\lambda 1 - x)$ is invertible in $\mathcal{R}(\Phi)$ and hence $\Phi^{-1}(y) = \Phi^{-1}(\Phi(\lambda 1 - x)) = \lambda 1 - x$ is invertible in $\mathcal{A}$. Thus $\lambda \in \rho(x)$. Therefore $\rho(\Phi(x)) \subseteq \rho(x)$. This implies $\rho(x) = \rho(\Phi(x))$ according to (g).

(i) The proof of the *Gelfand–Beurling formula* follows similarly to the proof of the particular case in the complex Banach algebra $\mathcal{B}[\mathcal{X}]$ as in Theorem 2.10.

(j) If $\mathcal{A}$ is a C*-algebra and if $x^* = x$, then $\|x\|^2 = \|x^*x\| = \|x^2\|$ and so, by induction, $\|x\|^{2n} = \|x^{2n}\|$ for every integer $n \ge 1$. If $\mathcal{A}$ is unital and complex, then $r(x) = \lim_n \|x^n\|^{\frac{1}{n}} = \lim_n \|x^{2n}\|^{\frac{1}{2n}} = \lim_n (\|x\|^{2n})^{\frac{1}{2n}} = \|x\|$ by (i).

(k) Let $\mathcal{A}$ and $\mathcal{B}$ be complex unital C*-algebras and Φ a $*$-homomorphism. Then $\|\Phi(x)\|^2 = \|\Phi(x)^*\Phi(x)\| = \|\Phi(x^*)\Phi(x)\| = \|\Phi(x^*x)\|$ for every $x \in \mathcal{A}$. Since x^*x is Hermitian, $\Phi(x^*x)$ is Hermitian because Φ is a $*$-homomorphism. Hence $\|\Phi(x^*x)\| = r(\Phi(x^*x)) \le r(x^*x) = \|x^*x\| = \|x\|^2$ according to (j) and (g).

(ℓ) Consider the setup and the argument of the preceding proof and suppose in addition that the unital $*$-homomorphism $\Phi \colon \mathcal{A} \to \mathcal{B}$ is injective. Thus apply (h) instead of (g) to get (ℓ) instead of (k). $\qquad\square$

Remark. A unital ∗-homomorphism Φ between unital complex C*-algebras is a contraction by Lemma 4.1(k), and so it is a contractive unital ∗-homomorphism. If it is injective, then Φ is an isometry by Lemma 4.1(ℓ), and so an isometric unital ∗-homomorphism. Since an isometry of a Banach space into a normed space has a closed range (see, e.g., [78, Problem 4.41(d)]), if a unital ∗-homomorphism Φ between unital complex C*-algebras $\mathcal{A}$ and $\mathcal{B}$ is injective, then the unital ∗-algebra $\mathcal{R}(\Phi) = \Phi(\mathcal{A}) \subseteq \mathcal{B}$, being closed in the Banach space $\mathcal{B}$, is itself a Banach space, and so $\mathcal{R}(\Phi)$ is a unital complex C*-algebra. Thus if $\Phi\colon \mathcal{A} \to \mathcal{B}$ between unital complex C*-algebras is an injective unital ∗-homomorphism, then $\Phi\colon \mathcal{A} \to \mathcal{R}(\Phi)$ is an *isometric isomorphism* (i.e., an injective and surjective linear isometry), and therefore $\mathcal{A}$ and $\mathcal{R}(\Phi)$ are *isometrically isomorphic* unital complex C*-algebras.

These are the elementary results on Banach and C*-algebras required in this chapter. For a thorough treatment of Banach algebra see, e.g., [11], [26], [38], [104], and [106], and for C*-algebra see, e.g., [6], [30], [34], [49], and [96].

Throughout this and the next section $\mathcal{H}$ will stand for a *nonzero complex* Hilbert space and so $\mathcal{B}[\mathcal{H}]$ is a (unital complex) C*-algebra, where involution is the adjoint operation. Let $\mathbb{C}^\Omega$ denote the collection of all functions from a nonempty set Ω to the complex plane $\mathbb{C}$, and let $F(\Omega) \subseteq \mathbb{C}^\Omega$ be an algebra of complex-valued functions on Ω where addition, scalar multiplication, and product are pointwise defined as usual. Take the map $*\colon F(\Omega) \to F(\Omega)$ assigning to each function ψ in $F(\Omega)$ the function ψ^* in $F(\Omega)$ given by $\psi^*(\lambda) = \overline{\psi(\lambda)}$ for $\lambda \in \Omega$. This (i.e., complex conjugation) defines an involution on $F(\Omega)$, and so $F(\Omega)$ is a commutative ∗-algebra. If in addition $F(\Omega)$ is endowed with a norm that makes it complete (thus a Banach algebra) and if $\|\psi^*\psi\| = \|\psi\|^2$, then $F(\Omega)$ is a (complex) C*-algebra. (For instance, if $F(\Omega)$ is the algebra of all complex-valued bounded functions on Ω equipped with the sup-norm $\|\cdot\|_\infty$, since $(\psi^*\psi)(\lambda) = \psi^*(\lambda)\psi(\lambda) = \overline{\psi(\lambda)}\psi(\lambda) = |\psi(\lambda)|^2$ for every $\lambda \in \Omega$).

Take any operator T in $\mathcal{B}[\mathcal{H}]$. A first attempt towards a *functional calculus* for T alludes to polynomials in T. To each polynomial $p\colon \sigma(T) \to \mathbb{C}$ with complex coefficients, $p(\lambda) = \sum_{i=0}^{n} \alpha_i \lambda^i$ for $\lambda \in \sigma(T)$, associate the operator

$$p(T) = \sum_{i=0}^{n} \alpha_i T^i.$$

Let $P'(\sigma(T))$ be the unital algebra of all polynomials (in one variable with complex coefficients) on $\sigma(T)$. The map $\Phi_T\colon P'(\sigma(T)) \to \mathcal{B}[\mathcal{H}]$ given by

$$\Phi_T(p) = p(T) \quad \text{for each} \quad p \in P'(\sigma(T))$$

is a unital homomorphism from $P'(\sigma(T))$ to $\mathcal{B}[\mathcal{H}]$. In fact, if $q(\lambda) = \sum_{j=0}^{m} \beta_j \lambda^j$, then $(pq)(\lambda) = \sum_{i=0}^{n} \sum_{j=0}^{m} \alpha_i \beta_j \lambda^{i+j}$ so that $\Phi_T(pq) = \sum_{i=0}^{n} \sum_{j=0}^{m} \alpha_i \beta_j T^{i+j} = \Phi_T(p)\Phi_T(q)$, and also $\Phi_T(1) = 1(T) = T^0 = I$. This can be extended from finite power series (i.e., from polynomials) to infinite power series: if a sequence $\{p_n\}$ with $p_n(\lambda) = \sum_{k=0}^{n} \alpha_k \lambda^k$ for each n converges in some sense on $\sigma(T)$ to a limit ψ, denoted by $\psi(\lambda) = \sum_{k=0}^{\infty} \alpha_k \lambda^k$, then we may set $\psi(T) = \sum_{k=0}^{\infty} \alpha_k T^k$ if the series converges in some topology of $\mathcal{B}[\mathcal{H}]$. For instance, if $\alpha_k = \frac{1}{k!}$ for each $k \geq 0$, then $\{p_n\}$ converges pointwise to the exponential function ψ such that $\psi(\lambda) = e^\lambda$ for every $\lambda \in \sigma(T)$ (on any $\sigma(T) \subseteq \mathbb{C}$), and so we define

$$e^T = \psi(T) = \sum_{k=0}^{\infty} \tfrac{1}{k!} T^k$$

(where the series converges uniformly in $\mathcal{B}[\mathcal{H}]$). However, unlike in the preceding paragraph, complex conjugation does not define an involution on $P'(\sigma(T))$ (and so $P'(\sigma(T))$ is not a $*$-algebra under it). Indeed, $p^*(\lambda) = \overline{p(\lambda)}$ is not a polynomial in λ — the continuous function $\overline{p(\cdot)} \colon \sigma(T) \to \mathbb{C}$ given by $\overline{p(\lambda)} = \sum_{i=0}^{n} \overline{\alpha}_i \overline{\lambda}^i$ is not a polynomial in λ (for $\sigma(T) \not\subseteq \mathbb{R}$, even if $\sigma(T)^* = \sigma(T)$).

The extension of the mapping Φ_T to some larger algebras of functions is the focus of this chapter.

4.2 Functional Calculus for Normal Operators

Let $\mathcal{A}_{\sigma(T)}$ be the σ-algebra of Borel subsets of the spectrum $\sigma(T)$ of an operator T on a Hilbert space $\mathcal{H}$. Let $B(\sigma(T))$ be the Banach space of all bounded $\mathcal{A}_{\sigma(T)}$-measurable complex-valued functions $\psi \colon \sigma(T) \to \mathbb{C}$ on $\sigma(T)$ equipped with the sup-norm. Let $L^\infty(\sigma(T), \mu)$ be the Banach space of all essentially bounded (i.e., μ-essentially bounded) $\mathcal{A}_{\sigma(T)}$-measurable complex-valued functions $\psi \colon \sigma(T) \to \mathbb{C}$ on $\sigma(T)$ for any positive measure μ on $\mathcal{A}_{\sigma(T)}$ also equipped with the sup-norm $\| \cdot \|_\infty$ (i.e., μ-essential sup-norm). In both cases we have unital commutative C*-algebras, where involution is defined as complex conjugation, and the identity element is the characteristic function $1 = \chi_{\sigma(T)}$ of $\sigma(T)$. (Note: we will also deal with the identity map on $\sigma(T)$ given by $\varphi(\lambda) = \lambda \chi_{\sigma(T)}$.) If T is normal, then let $E \colon \mathcal{A}_{\sigma(T)} \to \mathcal{B}[\mathcal{H}]$ be the unique spectral measure of its spectral decomposition $T = \int \lambda \, dE_\lambda$ as in Theorem 3.15.

Theorem 4.2. *Let $T \in \mathcal{B}[\mathcal{H}]$ be a normal operator and consider its spectral decomposition $T = \int \lambda \, dE_\lambda$. For every function $\psi \in B(\sigma(T))$ there is a unique operator $\psi(T) \in \mathcal{B}[\mathcal{H}]$ given by*

$$\psi(T) = \int \psi(\lambda) \, dE_\lambda,$$

which is normal. Moreover, the mapping $\Phi_T \colon B(\sigma(T)) \to \mathcal{B}[\mathcal{H}]$ such that $\Phi_T(\psi) = \psi(T)$ is a unital $*$-homomorphism.

Proof. The operator $\psi(T)$ is unique in $\mathcal{B}[\mathcal{H}]$ and normal for every $\psi \in B(\sigma(T))$ by Lemmas 3.7 and 3.9. The mapping $\Phi_T \colon B(\sigma(T)) \to \mathcal{B}[\mathcal{H}]$ is a unital $*$-homomorphism by Lemma 3.8 and by the linearity of the integral. Indeed,

(a) $\Phi_T(\alpha \psi + \beta \phi) = (\alpha \psi + \beta \phi)(T) = \int (\alpha \psi(\lambda) + \beta \phi(\lambda)) \, dE_\lambda$
$\qquad = \alpha \psi(T) + \beta \phi(T) = \alpha \Phi_T(\psi) + \beta \Phi_T(\phi),$

(b) $\Phi_T(\psi \phi) = (\psi \phi)(T) = \int \psi(\lambda) \phi(\lambda) \, dE_\lambda = \psi(T) \phi(T) = \Phi_T(\psi) \Phi_T(\phi),$

(c) $\Phi_T(1) = 1(T) = \int dE_\lambda = I$,

(d) $\Phi_T(\psi^*) = \psi^*(T) = \int \overline{\psi(\lambda)}\, dE_\lambda = \psi(T)^* = \Phi_T(\psi)^*$,

for every $\psi, \phi \in B(\sigma(T))$ and every $\alpha, \beta \in \mathbb{C}$, where the constant function 1 (i.e., $1(\lambda) = 1$ for every $\lambda \in \sigma(T)$) is the identity in the algebra $B(\sigma(T))$. $\quad\square$

Let $P(\sigma(T))$ be the set of all polynomials $p(\cdot\,,\cdot)\colon \sigma(T) \times \sigma(T) \to \mathbb{C}$ in λ and $\overline{\lambda}$ (or $p(\cdot\,,\bar{\cdot})\colon \sigma(T) \to \mathbb{C}$ such that $\lambda \mapsto p(\lambda, \overline{\lambda})$). Let $C(\sigma(T))$ be the set of all complex-valued continuous (thus $\mathcal{A}_{\sigma(T)}$-measurable) functions on $\sigma(T)$. So

$$P'(\sigma(T)) \subset P(\sigma(T)) \subset C(\sigma(T)) \subset B(\sigma(T)),$$

where the last inclusion follows from the Weierstrass Theorem since $\sigma(T)$ is compact — cf. proof of Theorem 2.2. Except for $P'(\sigma(T))$, these are all unital $*$-subalgebras of $B(\sigma(T))$ and the restriction of the mapping Φ_T to any of them remains a unital $*$-homomorphism into $\mathcal{B}[\mathcal{H}]$. Thus Theorem 4.2 holds for all these unital $*$-subalgebras of the unital C*-algebra $B(\sigma(T))$. The algebra $P'(\sigma(T))$ is a unital subalgebra of $P(\sigma(T))$, and the restriction of Φ_T to it is a unital homomorphism into $\mathcal{B}[\mathcal{H}]$, and so Theorem 4.2 also holds for $P'(\sigma(T))$ by replacing $*$-homomorphism with plain homomorphism. So if T is normal and p lies in $P'(\sigma(T))$, say $p(\lambda) = \sum_{i=0}^{n} \alpha_i \lambda^i$, then by Theorem 4.2

$$p(T) = \int p(\lambda)\, dE_\lambda = \sum_{i=0}^{n} \alpha_i \int \lambda^i\, dE_\lambda = \sum_{i=0}^{n} \alpha_i T^i,$$

and for p in $P(\sigma(T))$, say $p(\lambda, \overline{\lambda}) = \sum_{i,j=0}^{n,m} \alpha_{i,j} \lambda^i \overline{\lambda}^j$ (cf. Lemma 3.10), we get

$$p(T, T^*) = \int p(\lambda, \overline{\lambda})\, dE_\lambda = \sum_{i,j=0}^{n,m} \alpha_{i,j} \int \lambda^i \overline{\lambda}^j\, dE_\lambda = \sum_{i,j=0}^{n,m} \alpha_{i,j} T^i T^{*j}.$$

The *quotient space of a set modulo an equivalence relation is a partition of the set.* So we may regard $B(\sigma(T))$ as consisting of equivalence classes from $L^\infty(\sigma(T), \mu)$ — bounded functions are essentially bounded with respect to any measure μ. On the other hand, shifting from restriction to extension, it is sensible to argue that the preceding theorem might be extended from $B(\sigma(T))$ to some $L^\infty(\sigma(T), \mu)$, where bounded would be replaced by essentially bounded. But essentially bounded with respect to what measure μ?

Recall from Section 3.6: a measure μ on a σ-algebra is *absolutely continuous* with respect to a measure ν on the same σ-algebra if $\nu(\Lambda) = 0$ implies $\mu(\Lambda) = 0$ for measurable sets Λ. Two measures on the same σ-algebra are *equivalent* if they are absolutely continuous with respect to each other, which means they have the same sets of measure zero. That is, μ and ν are equivalent if $\nu(\Lambda) = 0$ if and only if $\mu(\Lambda) = 0$ for every measurable set Λ. (Note: two positive equivalent measures are not necessarily both finite or both not finite.)

Remark. Let $E\colon \mathcal{A}_{\sigma(T)} \to \mathcal{B}[\mathcal{H}]$ be the spectral measure of the spectral decomposition of a normal operator T and consider the statements of Lemmas 3.7, 3.8, and 3.9 for $\Omega = \sigma(T)$. *Lemma 3.7 remains true if $B(\sigma(T))$ is replaced by*

$L^\infty(\sigma(T), \mu)$ *for every positive measure* $\mu\colon \mathcal{A}_{\sigma(T)} \to \overline{\mathbb{R}}$ *equivalent to the spectral measure* E. The proof is essentially the same as that of Lemma 3.7. In fact, the spectral integrals of functions in $L^\infty(\sigma(T), \mu)$ coincide whenever the functions are equal μ-almost everywhere, and so a function in $B(\sigma(T))$ has the same integral as any representative from its equivalence class in $L^\infty(\sigma(T), \mu)$. Moreover, since μ and E have exactly the same sets of measure zero, we get

$$\|\phi\|_\infty = \operatorname{ess\,sup} |\phi| = \inf_{\Lambda \in \mathcal{N}_{\sigma(T)}} \sup_{\lambda \in \sigma(T) \backslash \Lambda} |\phi(\lambda)|$$

for every $\phi \in L^\infty(\sigma(T), \mu)$, where $\mathcal{N}_{\sigma(T)}$ denotes the collection of all $\Lambda \in \mathcal{A}_{\sigma(T)}$ for which $\mu(\Lambda) = 0$ or, equivalently, for which $E(\Lambda) = O$. Thus the sup-norm $\|\phi\|_\infty$ of $\phi\colon \sigma(T) \to \mathbb{C}$ with respect to μ dominates the sup-norm with respect to $\pi_{x,x}$ for every $x \in \mathcal{H}$. Then the proof of Lemma 3.7 remains unchanged if $B(\sigma(T))$ is replaced by $L^\infty(\sigma(T), \mu)$ for any positive measure μ equivalent to E. (Reason: the inequality $|f(x,x)| \leq \|\phi\|_\infty \int d\pi_{x,x} = \|\phi\|_\infty \|x\|^2$ for $x \in \mathcal{H}$ still holds for the sesquilinear form $f\colon \mathcal{H} \times \mathcal{H} \to \mathbb{C}$ in the proof of Lemma 3.7 if $\phi \in L^\infty(\sigma(T), \mu)$.) Hence for each $\phi \in L^\infty(\sigma(T), \mu)$ there is a unique operator $F \in \mathcal{B}[\mathcal{H}]$ such that $F = \int \phi(\lambda)\, dE_\lambda$ as in Lemma 3.7. This ensures that *Lemmas 3.8 and 3.9 also hold if $B(\sigma(T))$ is replaced by $L^\infty(\sigma(T), \mu)$.*

Theorem 4.3. *Let* $T \in \mathcal{B}[\mathcal{H}]$ *be a normal operator and consider its spectral decomposition* $T = \int \lambda\, dE_\lambda$. *If* μ *is a positive measure on* $\mathcal{A}_{\sigma(T)}$ *equivalent to the spectral measure* $E\colon \mathcal{A}_{\sigma(T)} \to \mathcal{B}[\mathcal{H}]$, *then for every* $\psi \in L^\infty(\sigma(T), \mu)$ *there is a unique operator* $\psi(T) \in \mathcal{B}[\mathcal{H}]$ *given by*

$$\psi(T) = \int \psi(\lambda)\, dE_\lambda,$$

which is normal. Moreover, the mapping $\Phi_T\colon L^\infty(\sigma(T), \mu) \to \mathcal{B}[\mathcal{H}]$ *such that* $\Phi_T(\psi) = \psi(T)$ *is a unital $*$-homomorphism.*

Proof. If $\mu\colon \mathcal{A}_{\sigma(T)} \to \overline{\mathbb{R}}$ and $E\colon \mathcal{A}_{\sigma(T)} \to \mathcal{B}[\mathcal{H}]$ are equivalent measures, then the proof of Theorem 4.2 still holds according to the above remark. $\qquad \square$

Sometimes it is convenient that a positive measure μ on $\mathcal{A}_{\sigma(T)}$ equivalent to the spectral measure E be *finite*. For example, if ψ lies in $L^\infty(\sigma(T), \mu)$ and μ is finite, then ψ also lies in $L^2(\sigma(T), \mu)$ since $\|\psi\|_2^2 = \int |\psi|^2 d\mu \leq \|\psi\|_\infty^2\, \mu(\sigma(T)) < \infty$. This is important if an essentially constant function is expected to be in $L^2(\sigma(T), \mu)$ as will be the case in the proof of Lemma 4.7 below.

Definition 4.4. A *scalar spectral measure* for a normal operator is a positive *finite* measure equivalent to the spectral measure of its spectral decomposition.

Let T be a normal operator. If $\hat{E}$ is its unique spectral measure on $\mathcal{A}_\Omega$ as in the proof of Theorem 3.15 part (b), then a positive *finite* measure μ on $\mathcal{A}_\Omega$ is a scalar spectral measure for T if, for any $\Lambda \in \mathcal{A}_\Omega$, $\{\mu(\Lambda) = 0 \iff \hat{E}(\Lambda) = O\}$. This also applies to the unique spectral measure E on $\mathcal{A}_{\sigma(T)}$ as in the proof of Theorem 3.15 part (c). Take the collections $\{\hat{E}(\Lambda)\}_{\Lambda \in \mathcal{A}_\Omega}$ and $\{E(\tilde{\Lambda})\}_{\tilde{\Lambda} \in \mathcal{A}_{\sigma(T)}}$ of all images of $\hat{E}$ and E. For each $x, y \in \mathcal{H}$ consider the scalar-valued (complex) measure of Section 3.3 acting on $\mathcal{A}_\Omega$ or on $\mathcal{A}_{\sigma(T)}$, namely,

$$\hat{\pi}_{x,y}(\Lambda) = \langle \hat{E}(\Lambda)x \, ; y \rangle \text{ for } \Lambda \in \mathcal{A}_\Omega \quad \text{and} \quad \pi_{x,y}(\tilde{\Lambda}) = \langle E(\tilde{\Lambda})x \, ; y \rangle \text{ for } \tilde{\Lambda} \in \mathcal{A}_{\sigma(T)}.$$

Definition 4.5. A *separating vector* for a collection $\{C_\gamma\}$ of operators in $\mathcal{B}[\mathcal{H}]$ is a vector $e \in \mathcal{H}$ such that $e \notin \mathcal{N}(C)$ for every $O \neq C \in \{C_\gamma\}$.

Lemma 4.6. *If* $T \in \mathcal{B}[\mathcal{H}]$ *is normal, then* $e \in \mathcal{H}$ *is a separating vector for* $\{\hat{E}(\Lambda)\}_{\Lambda \in \mathcal{A}_\Omega}$ *if and only if* $\hat{\pi}_{e,e} \colon \mathcal{A}_\Omega \to \mathbb{R}$ *is a scalar spectral measure for* T.

Proof. Let $\hat{E} \colon \mathcal{A}_\Omega \to \mathcal{B}[\mathcal{H}]$ be the spectral measure for a normal operator T in $\mathcal{B}[\mathcal{H}]$. Since $\hat{\pi}_{x,x}(\Lambda) = \|\hat{E}(\Lambda)x\|^2$ for every $\Lambda \in \mathcal{A}_\Omega$ and every $x \in \mathcal{H}$, the measure $\hat{\pi}_{x,x} \colon \mathcal{A}_\Omega \to \mathbb{R}$ is positive and finite for every $x \in \mathcal{H}$. Recall: $e \in \mathcal{H}$ is a separating vector for the collection $\{\hat{E}(\Lambda)\}_{\Lambda \in \mathcal{A}_\Omega}$ if and only if $\hat{E}(\Lambda)e \neq 0$ whenever $\hat{E}(\Lambda) \neq O$ or, equivalently, if and only if $\{\hat{E}(\Lambda)e \neq 0 \iff \hat{E}(\Lambda) \neq O\}$. Conclusion: $e \in \mathcal{H}$ is a separating vector for the collection $\{\hat{E}(\Lambda)\}_{\Lambda \in \mathcal{A}_\Omega}$ if and only if $\{\hat{\pi}_{e,e}(\Lambda) = 0 \iff \hat{E}(\Lambda) = O\}$, which is to say if and only if $\hat{E}$ and $\hat{\pi}_{e,e}$ are equivalent measures. Since $\hat{\pi}_{e,e}$ is positive and finite, then by Definition 4.4 this equivalence means $\hat{\pi}_{e,e}$ is a scalar spectral measure for T. $\qquad\square$

Lemma 4.7. *If* T *is a normal operator on a separable Hilbert space* $\mathcal{H}$, *then there is a separating vector* $e \in \mathcal{H}$ *for* $\{\hat{E}(\Lambda)\}_{\Lambda \in \mathcal{A}_\Omega}$. (*So there is a vector* e *in* $\mathcal{H}$ *such that* $\hat{\pi}_{e,e} \colon \mathcal{A}_\Omega \to \mathbb{R}$ *is a scalar spectral measure for* T *by Lemma 4.6*).

Proof. Take the positive measure μ on $\mathcal{A}_\Omega$ of Theorem 3.11. If $\mathcal{H}$ is separable, then μ is finite (Corollary 3.13). Let M_φ be the multiplication operator on $L^2(\Omega, \mu)$ with $\varphi \in L^\infty(\Omega, \mu)$ being the identity map with multiplicity. Let $\chi_\Lambda \in L^\infty(\Omega, \mu)$ be the characteristic function of each $\Lambda \in \mathcal{A}_\Omega$. Set $E'(\Lambda) = M_{\chi_\Lambda}$ for every $\Lambda \in \mathcal{A}_\Omega$ so that $E' \colon \mathcal{A}_\Omega \to \mathcal{B}[L^2(\Omega, \mu)]$ is the spectral measure in $L^2(\Omega, \mu)$ for the spectral decomposition (cf. proof of Theorem 3.15 part (a))

$$M_\varphi = \int \varphi(\lambda) \, dE'_\lambda$$

of the normal operator $M_\varphi \in \mathcal{B}[L^2(\Omega, \mu)]$. Hence, for each $f \in L^2(\Omega, \mu)$,

$$\pi'_{f,f}(\Lambda) = \langle E'(\Lambda)f \, ; f \rangle = \langle M_{\chi_\Lambda}f \, ; f \rangle = \int \chi_\Lambda f \overline{f} \, d\mu = \int_\Lambda |f|^2 d\mu$$

for every $\Lambda \in \mathcal{A}_\Omega$. In particular, the function $1 = \chi_\Omega$ (i.e., the constant function $1(\lambda) = 1$ for all $\lambda \in \Omega$ — the characteristic function χ_Ω of Ω) lies in $L^2(\Omega, \mu)$ because μ is finite. This is a separating vector for $\{E'(\Lambda)\}_{\Lambda \in \mathcal{A}_\Omega}$ since $E'(\Lambda)1 = M_{\chi_\Lambda}1 = \chi_\Lambda \neq 0$ for every $\Lambda \neq \varnothing$ in $\mathcal{A}_\Omega$. Thus $\pi'_{1,1}$ is a scalar spectral measure for M_φ by Lemma 4.6. If $T \in \mathcal{B}[\mathcal{H}]$ is normal, then $T \cong M_\varphi$ by Theorem 3.11. Let $\hat{E} \colon \mathcal{A}_\Omega \to \mathcal{B}[\mathcal{H}]$ be the spectral measure in $\mathcal{H}$ for T which is given by $\hat{E}(\Lambda) = U^*E'(\Lambda)U$ for each $\Lambda \in \mathcal{A}_\Omega$, where $U \in \mathcal{B}[\mathcal{H}, L^2(\Omega, \mu)]$ is unitary (cf. proof of Theorem 3.15 part (b)). Set $e = U^*1 \in \mathcal{H}$ (with $1 = \chi_\Omega$ in $L^2(\Omega, \mu)$).

This e is a separating vector for $\{\hat{E}(\Lambda)\}_{\Lambda \in \mathcal{A}_\Omega}$ since $\hat{E}(\Lambda)e = U^*E'(\Lambda)UU^*1 = U^*E'(\Lambda)1 \neq 0$ if $\hat{E}(\Lambda) \neq O$ (because $E'(\Lambda)1 \neq 0$ with $E'(\Lambda) = U\hat{E}(\Lambda)U^* \neq O$). Thus $\hat{\pi}_{e,e}$ is a scalar spectral measure for T (Lemma 4.6). $\qquad\qquad\square$

Remarks. (a) Claim: $\mu = \pi'_{1,1} = \hat{\pi}_{e,e}$ are scalar spectral measures on $\mathcal{A}_\Omega$ for both M_φ and T. Actually, by the above proof the scalar spectral measure $\pi'_{1,1}$ for M_φ is precisely the measure μ of Theorem 3.11. In fact, for every $\Lambda \in \mathcal{A}_\Omega$

$$\mu(\Lambda) = \int_\Lambda d\mu = \|\chi_\Lambda\|^2 = \|M_{\chi_\Lambda}1\|^2 = \|E'(\Lambda)1\|^2 = \pi'_{1,1}(\Lambda) = \langle E'(\Lambda)1\,;1\rangle$$

$$= \langle U\hat{E}(\Lambda)U^*Ue\,;Ue\rangle = \langle \hat{E}(\Lambda)e\,;e\rangle = \|\hat{E}(\Lambda)e\|^2 = \hat{\pi}_{e,e}(\Lambda).$$

(b) Take the spectral measure $E \colon \mathcal{A}_{\sigma(T)} \to \mathcal{B}[\mathcal{H}]$ with $E(\tilde{\Lambda}) = \hat{E}(\varphi^{-1}(\tilde{\Lambda}))$ for $\tilde{\Lambda}$ in $\mathcal{A}_{\varphi(\Omega)} \subseteq \mathcal{A}_{\sigma(T)}$, where $\varphi(\Omega)^- = \sigma(T)$ (cf. proof of Theorem 3.15 part (c) and the remark after Theorem 3.11). If $E(\tilde{\Lambda}) \neq O$, then $E(\tilde{\Lambda})e = \hat{E}(\varphi^{-1}(\tilde{\Lambda}))e \neq 0$: the same e is a separating vector for $\{E(\tilde{\Lambda})\}_{\tilde{\Lambda} \in \mathcal{A}_{\sigma(T)}}$. So $\pi_{e,e}(\tilde{\Lambda}) = \langle E(\tilde{\Lambda})e\,;e\rangle$ for every $\tilde{\Lambda} \in \mathcal{A}_{\sigma(T)}$ defines a scalar spectral measure $\pi_{e,e}$ on $\mathcal{A}_{\sigma(T)}$ for T.

Theorem 4.8. *Let $T \in \mathcal{B}[\mathcal{H}]$ be a normal operator on a separable Hilbert space $\mathcal{H}$ and consider its spectral decomposition $T = \int \lambda\, dE_\lambda$. Let μ be an arbitrary scalar spectral measure for T. For every $\psi \in L^\infty(\sigma(T), \mu)$ there is a unique operator $\psi(T) \in \mathcal{B}[\mathcal{H}]$ given by*

$$\psi(T) = \int \psi(\lambda)\, dE_\lambda,$$

which is normal. Moreover, the mapping $\Phi_T \colon L^\infty(\sigma(T), \mu) \to \mathcal{B}[\mathcal{H}]$ such that $\Phi_T(\psi) = \psi(T)$ is an isometric unital $*$-homomorphism.

Proof. All but the fact that Φ_T is an isometry is a particular case of Theorem 4.3. We show that Φ_T is an isometry. By Lemma 3.8, the uniqueness in Lemma 3.7, and the uniqueness of the expression for $\psi(T)$, we get for every $x \in \mathcal{H}$

$$\|\Phi_T(\psi)x\|^2 = \langle \psi(T)^*\psi(T)x\,;x\rangle = \int \overline{\psi(\lambda)}\,\psi(\lambda)\, d\langle E_\lambda x\,;x\rangle = \int |\psi(\lambda)|^2\, d\pi_{x,x}.$$

Suppose $\Phi_T(\psi) = O$. Then $\int |\psi(\lambda)|^2\, d\pi_{x,x} = 0$ for every $x \in \mathcal{H}$. By Lemma 4.7 (and the above remarks), there is a separating vector $e \in \mathcal{H}$ for $\{E(\Lambda)\}_{\Lambda \in \mathcal{A}_{\sigma(T)}}$. Take the scalar spectral measure $\pi_{e,e}$. Thus $\int |\psi(\lambda)|^2\, d\pi_{e,e} = 0$ so that $\psi = 0$ in $L^\infty(\sigma(T), \pi_{e,e})$, which means $\psi = 0$ in $L^\infty(\sigma(T), \mu)$ for every measure μ equivalent to $\pi_{e,e}$. In particular, this holds for every scalar spectral measure μ for T. Then $\{0\} = \mathcal{N}(\Phi_T) \subseteq L^\infty(\sigma(T), \mu)$, and hence the linear transformation $\Phi_T \colon L^\infty(\sigma(T), \mu) \to \mathcal{B}[\mathcal{H}]$ is injective, and is so isometric by Lemma 4.1(ℓ). $\qquad\square$

If T is normal on a *separable* Hilbert space $\mathcal{H}$, then $\Phi_T \colon L^\infty(\sigma(T), \mu) \to \mathcal{B}[\mathcal{H}]$ is an isometric unital $*$-homomorphism between C*-algebras (Theorem 4.8). Thus $\Phi_T \colon L^\infty(\sigma(T), \mu) \to \mathcal{R}(\Phi_T)$ is an isometric unital $*$-isomorphism between the C*-algebras $L^\infty(\sigma(T), \mu)$ and $\mathcal{R}(\Phi_T) \subseteq \mathcal{B}[\mathcal{H}]$ (see remark after Lemma 4.1), which is the von Neumann algebra generated by T (see Proposition 4.A).

Theorem 4.9. *Let* $T \in \mathcal{B}[\mathcal{H}]$ *be a normal operator on a separable Hilbert space* $\mathcal{H}$ *and consider its spectral decomposition* $T = \int \lambda \, dE_\lambda$. *For every* $\psi \in C(\sigma(T))$ *there is a unique operator* $\psi(T) \in \mathcal{B}[\mathcal{H}]$ *given by*

$$\psi(T) = \int \psi(\lambda) \, dE_\lambda,$$

which is normal. Moreover, the mapping $\Phi_T \colon C(\sigma(T)) \to \mathcal{R}(\Phi_T) \subseteq \mathcal{B}[\mathcal{H}]$ *such that* $\Phi_T(\psi) = \psi(T)$ *is an isometric unital* $*$-*isomorphism, where* $\mathcal{R}(\Phi_T)$ *is the C*-algebra generated by* T *and the identity* I *in* $\mathcal{B}[\mathcal{H}]$.

Proof. The set $C(\sigma(T))$ is a unital $*$-subalgebra of the C*-algebra $B(\sigma(T))$, which can be viewed as a subalgebra of the C*-algebra $L^\infty(\sigma(T), \mu)$ for every measure μ (in terms of equivalence classes in $L^\infty(\sigma(T), \mu)$ of each function in $B(\sigma(T))$). So $\Phi_T \colon C(\sigma(T)) \to \mathcal{B}[\mathcal{H}]$, the restriction of $\Phi_T \colon L^\infty(\sigma(T), \mu) \to \mathcal{B}[\mathcal{H}]$ to $C(\sigma(T))$, is an isometric unital $*$-homomorphism under the same assumptions as in Theorem 4.8. Hence $\Phi_T \colon C(\sigma(T)) \to \mathcal{R}(\Phi_T)$ is an isometric unital $*$-isomorphism (cf. paragraph preceding the theorem statement). Thus it remains to show that $\mathcal{R}(\Phi_T)$ is the C*-algebra generated by T and the identity I. Indeed, take $\mathcal{R}(\Phi_T) = \Phi_T(C(\sigma(T)))$. Since $\Phi_T(P(\sigma(T))) = \mathcal{P}(T, T^*)$, which is the algebra of all polynomials in T and T^*, since $P(\sigma(T))^- = C(\sigma(T))$ in the sup-norm topology by the Stone–Weierstrass Theorem (cf. proof of Theorem 3.11), and since $\Phi_T \colon C(\sigma(T)) \to \mathcal{B}[\mathcal{H}]$ is an isometry (thus with a closed range), we get $\Phi_T(C(\sigma(T))) = \Phi_T(P(\sigma(T))^-) = \Phi_T(P(\sigma(T)))^- = \mathcal{P}(T, T^*)^- = \mathcal{C}^*(T)$, where $\mathcal{C}^*(T)$ stands for the C*-algebra generated by T and I (see the paragraph following Definition 3.12, where T is now normal). $\qquad \square$

The mapping Φ_T of Theorems 4.2, 4.3, 4.8, and 4.9 such that $\Phi_T(\psi) = \psi(T)$ is referred to as the *functional calculus* for T, and the theorems themselves are referred to as the Functional Calculus for Normal Operators. Since $\sigma(T)$ is compact, for any positive measure μ on $\mathcal{A}_{\sigma(T)}$

$$C(\sigma(T)) \subset B(\sigma(T)) \subset L^\infty(\sigma(T), \mu).$$

(Again, the last inclusion is interpreted in terms of the equivalence classes in $L^\infty(\sigma(T), \mu)$ of each representative in $B(\sigma(T))$ as in the remark preceding Theorem 4.3.) Thus ψ lies in $L^\infty(\sigma(T), \mu)$ as in Theorem 4.3 or 4.8 whenever ψ satisfies the assumptions of Theorems 4.2 and 4.9, and so the next result applies to the settings of the preceding Functional Calculus Theorems. It says that $\psi(T)$ *commutes with every operator that commutes with* T.

Corollary 4.10. *Let* $T = \int \lambda \, dE_\lambda$ *be a normal operator in* $\mathcal{B}[\mathcal{H}]$. *Take any* ψ *in* $L^\infty(\sigma(T), \mu)$ *where* μ *is any positive measure on* $\mathcal{A}_{\sigma(T)}$ *equivalent to the spectral measure* E. *Consider the operator* $\psi(T)$ *in* $\mathcal{B}[\mathcal{H}]$ *of Theorem 4.3, viz.,*

$$\psi(T) = \int \psi(\lambda) \, dE_\lambda.$$

If S *in* $\mathcal{B}[\mathcal{H}]$ *commutes with* T, *then* S *commutes with* $\psi(T)$.

Proof. If $ST = TS$, then by Theorem 3.17 (cf. proof of Corollary 3.19),

$$\langle S\psi(T)x\,;y\rangle = \langle \psi(T)x\,;S^*y\rangle = \int \psi(\lambda)\,d\langle E_\lambda x\,;S^*y\rangle$$

$$= \int \psi(\lambda)\,d\langle SE_\lambda x\,;y\rangle = \int \psi(\lambda)\,d\langle E_\lambda Sx\,;y\rangle = \langle \psi(T)Sx\,;y\rangle$$

for every $x, y \in \mathcal{H}$ (see Lemma 3.7), which means $S\psi(T) = \psi(T)S$.   □

If T is a normal operator, then the Spectral Mapping Theorem for Normal Operators in Theorem 2.8 (which deals with polynomials) does not depend on the Spectral Theorem. Moreover, for a normal operator the Spectral Mapping Theorem for Polynomials in Theorem 2.7 is a particular case of Theorem 2.8. However, extensions from polynomials to bounded or continuous functions do depend on the Spectral Theorem via Theorems 4.8 or 4.9, as we will see next. The forthcoming Theorems 4.11 and 4.12 are also referred to as SPECTRAL MAPPING THEOREMS for bounded or continuous functions of normal operators.

Theorem 4.11. *If T is a normal operator on a* separable *Hilbert space, if μ is a scalar spectral measure for T, and if ψ lies in $L^\infty(\sigma(T), \mu)$, then*

$$\sigma(\psi(T)) = \operatorname{ess}\mathcal{R}(\psi) = \operatorname{ess}\{\psi(\lambda) \in \mathbb{C}\colon \lambda \in \sigma(T)\} = \operatorname{ess}\psi(\sigma(T)).$$

Proof. If the theorem holds for a scalar spectral measure, then it holds for any scalar spectral measure. Suppose T is a normal operator on a separable Hilbert space. Let $\mu\colon \mathcal{A}_\Omega \to \mathbb{R}$ be the positive measure of Theorem 3.11, which is a scalar spectral measure for $\mathcal{M}_\varphi \cong T$ (cf. remarks following the proof of Lemma 4.7). Also let $\pi_{e,e}\colon \mathcal{A}_{\sigma(T)} \to \mathbb{R}$ be a scalar spectral measure for T (cf. remarks following the proof of Theorem 4.7 again). Take any ψ in $L^\infty(\sigma(T), \pi_{e,e})$. By Theorem 4.8, $\Phi_T\colon L^\infty(\sigma(T), \pi_{e,e}) \to \mathcal{B}[\mathcal{H}]$ is an isometric (thus injective) unital $*$-isomorphism between C*-algebras and $\Phi_T(\psi) = \psi(T)$. Then

$$\sigma(\psi) = \sigma(\Phi_T(\psi)) = \sigma(\psi(T))$$

by Lemma 4.1(h). Consider the multiplication operator $\mathcal{M}_\varphi$ on $L^2(\Omega, \mu)$ as in Theorem 3.11. Since $T \cong \mathcal{M}_\varphi$, $L^2(\Omega, \mu)$ is separable. Since $\varphi(\Omega)^- = \sigma(T)$ (cf. remark after Theorem 3.11), set $\psi' = \psi \circ \varphi$ in $L^\infty(\Omega, \mu)$. Since $\psi'(\lambda) = \psi(\lambda)$ for $\lambda \in \Omega$, $\sigma(\psi) = \sigma(\psi')$ in both algebras and $\operatorname{ess}\mathcal{R}(\psi') = \operatorname{ess}\mathcal{R}(\psi)$. Since μ is a scalar spectral measure for $\mathcal{M}_\varphi$ (cf. remarks following the proof of Lemma 4.7 once again), we may apply Theorem 4.8 to $\mathcal{M}_\varphi$. So using the same argument,

$$\sigma(\psi') = \sigma(\psi'(\mathcal{M}_\varphi)).$$

Claim. $\psi'(\mathcal{M}_\varphi) = \mathcal{M}_{\psi'}$.

Proof. Let E' be the spectral measure for $\mathcal{M}_\varphi$ where $E'(\Lambda) = \mathcal{M}_{\chi_\Lambda}$ with χ_Λ being the characteristic function of Λ in $\mathcal{A}_\Omega$ as in the proof of Lemma 4.7. Take the scalar-valued measure $\pi'_{f,g}$ given by $\pi'_{f,g}(\Lambda) = \langle E'(\Lambda)f\,;g\rangle$ for every Λ in $\mathcal{A}_\Omega$ and each $f, g \in L^2(\Omega, \mu)$. Recalling the remarks following the proof of Lemma 4.7 for the fourth time, we get $\mu = \pi'_{1,1}$. Note: for every Λ in $\mathcal{A}_\Omega$,

$$\int_\Lambda d\pi'_{f,g} = \pi'_{f,g}(\Lambda) = \langle E'(\Lambda)f\,;g\rangle = \langle \chi_\Lambda f\,;g\rangle = \int_\Lambda f\,\overline{g}\,d\mu = \int_\Lambda f\,\overline{g}\,d\pi'_{1,1}.$$

So $d\pi'_{f,g} = f\overline{g}\,d\pi'_{1,1}$ ($f\overline{g}$ is the Radon–Nikodým derivative of $\pi'_{f,g}$ with respect to $\pi'_{1,1}$; see proof of Lemma 3.8). Since $\psi'(M_\varphi) = \int \psi'(\lambda)\,dE'_\lambda$ (Theorem 4.8),

$$\langle \psi'(M_\varphi)f\,;g\rangle = \int \psi'\,d\,\langle E'_\lambda f\,;g\rangle = \int \psi'\,d\pi'_{f,g}$$

$$= \int \psi' f\overline{g}\,d\pi'_{1,1} = \int \psi' f\overline{g}\,d\mu = \langle M_{\psi'}f\,;g\rangle$$

for every $f,g \in L^2(\Omega,\mu)$. Hence $\psi'(M_\varphi) = M_{\psi'}$, proving the claimed identity. Therefore (since μ is finite and so σ-finite — cf. Proposition 3.B(f)) we get

$$\sigma(\psi(T)) = \sigma(\psi) = \sigma(\psi') = \sigma(\psi'(M_\varphi)) = \sigma(M_{\psi'}) = \operatorname{ess}\mathcal{R}(\psi') = \operatorname{ess}\mathcal{R}(\psi). \ \square$$

Theorem 4.12. *If T is a normal operator on a separable Hilbert space, then*

$$\sigma(\psi(T)) = \psi(\sigma(T)) \quad \text{for every} \quad \psi \in C(\sigma(T)).$$

Proof. Consider the setup of the previous proof with $\pi_{e,e}$ on $\mathcal{A}_\sigma(T)$ and $\mu = \hat\pi_{e,e}$ on $\mathcal{A}_\Omega$ (as in Theorem 3.11) which are scalar spectral measures for T and for $M_\varphi \cong T$, respectively. Thus all that has been said so far about μ holds for every finite positive measure equivalent to it; in particular, for every scalar spectral measure for T. Recall: $C(\sigma(T)) \subset B(\sigma(T)) \subset L^\infty(\sigma(T),\nu)$ for every measure ν on $\mathcal{A}_\sigma(T)$. Take an arbitrary $\psi \in C(\sigma(T))$. Since $\sigma(T)$ is compact in $\mathbb{C}$ and ψ is continuous, $\mathcal{R}(\psi) = \psi(\sigma(T))$ is compact (see, e.g., [78, Theorem 3.64]) and so $\mathcal{R}(\psi)^- = \mathcal{R}(\psi)$. From Proposition 3.B we get

$$\operatorname{ess}\mathcal{R}(\psi) = \{\alpha \in \mathbb{C}\colon \tilde\mu(\psi^{-1}(B_\varepsilon(\alpha))) > 0 \text{ for all } \varepsilon > 0\} \subseteq \mathcal{R}(\psi),$$

where $B_\varepsilon(\alpha)$ denotes the open ball of radius ε centered at α. Also, $\pi_{e,e}(\Lambda) > 0$ for every nonempty $\Lambda \in \mathcal{A}_{\sigma(T)}$ open relative to $\sigma(T) = \sigma(M_\varphi)$ (cf. Theorem 3.15). If there is an α in $\mathcal{R}(\psi)\backslash\operatorname{ess}\mathcal{R}(\psi)$, then there is an $\varepsilon > 0$ such that $\pi_{e,e}(\psi^{-1}(B_\varepsilon(\alpha))) = 0$. Since ψ is continuous, the inverse image $\psi^{-1}(B_\varepsilon(\alpha))$ of the open set $B_\varepsilon(\alpha)$ in $\mathbb{C}$ must be open in $\sigma(T)$, and hence $\psi^{-1}(B_\varepsilon(\alpha)) = \varnothing$. But this is a contradiction: $\varnothing \neq \psi^{-1}(\{\alpha\}) \subseteq \psi^{-1}(B_\varepsilon(\alpha))$ because $\alpha \in \mathcal{R}(\psi)$. Thus $\mathcal{R}(\psi)\backslash\operatorname{ess}\mathcal{R}(\psi) = \varnothing$, and so $\mathcal{R}(\psi) = \operatorname{ess}\mathcal{R}(\psi)$. Then, by Theorem 4.11,

$$\sigma(\psi(T)) = \operatorname{ess}\mathcal{R}(\psi) = \mathcal{R}(\psi) = \psi(\sigma(T)). \qquad \square$$

4.3 Analytic Functional Calculus: Riesz Functional Calculus

The approach here is rather different from Section 4.2. However, Proposition 4.J will show that the Functional Calculus for Normal Operators of the previous section coincides with the Riesz Functional Calculus of this section when restricted to a normal operator on a Hilbert space (for analytic functions).

Measure theory was the means for Chapter 3 and Section 4.2 where operators necessarily act on Hilbert spaces. In contrast, the tool for the present section is complex analysis where the integrals are Riemann integrals (unlike the measure-theoretic integrals we have been dealing with so far) and operators may act on Banach spaces. For standard results of complex analysis used in this section the reader is referred, for instance, to [2], [14], [23], [29], and [103].

An *arc* is a *continuous* function from a nondegenerate closed interval of the real line into the complex plane, say $\alpha \colon [0,1] \to \mathbb{C}$. Let $\varUpsilon$ be the range of an arc, $\varUpsilon = \mathcal{R}(\alpha) = \{\alpha(t) \in \mathbb{C} \colon t \in [0,1]\}$, which is connected and compact (continuous image of connected sets is connected, and of compact sets is compact). Suppose $\varUpsilon \subset \mathbb{C}$ is nowhere dense to avoid space-filling curves. In this case $\varUpsilon$ is referred to as a *curve* generated by an arc α. Let $\mathbb{G} = \{\#X \colon \varnothing \neq X \in \wp([0,1])\}$ be the set of all cardinal numbers of nonempty subsets of the interval $[0,1]$. Given an arc $\alpha \colon [0,1] \to \varUpsilon$, set $m(t) = \#\{s \in [0,1) \colon \alpha(s) = \alpha(t)\} \in \mathbb{G}$ for each $t \in [0,1)$. Thus $m(t) \in \mathbb{G}$ is, for each $t \in [0,1)$, the cardinality of the subset of the interval $[0,1]$ consisting of all points $s \in [0,1)$ for which $\alpha(s) = \alpha(t)$. This $m(t)$ is the *multiplicity of the point* $\alpha(t) \in \varUpsilon$. Consider the function $m \colon [0,1) \to \mathbb{G}$ defined by $t \mapsto m(t)$, which is referred to as the *multiplicity of the arc* α. A pair $(\varUpsilon, m)$ consisting of the range $\varUpsilon$ and the multiplicity m of an arc is called a *directed pair* if the direction in which the arc traverses the curve $\varUpsilon$, according to the natural order of the interval $[0,1]$, is taken into account. An *oriented curve* generated by an arc $\alpha \colon [0,1] \to \varUpsilon$ is the directed pair $(\varUpsilon, m)$, and the arc α itself is referred to as a *parameterization* of the oriented curve $(\varUpsilon, m)$ (which is not unique for $(\varUpsilon, m)$). If an oriented curve $(\varUpsilon, m)$ is such that $\alpha(0) = \alpha(1)$ for some parameterization α (and so for all parameterizations), then it is called a *closed curve*. A parameterization α of an oriented curve $(\varUpsilon, m)$ is injective on $[0,1)$ if and only if $m = 1$ (i.e., $m(t) = 1$ for all $t \in [0,1)$). If one parameterization of $(\varUpsilon, 1)$ is injective, then so are all of them. An oriented curve $(\varUpsilon, 1)$ parameterized by an injective arc α on $[0,1)$ is a *simple curve* (or a *Jordan curve*).

Thus an oriented curve consists of a directed pair $(\varUpsilon, m)$ where $\varUpsilon \subset \mathbb{C}$ is the (nowhere dense) range of a continuous function $\alpha \colon [0,1] \to \mathbb{C}$ called a *parameterization* of $(\varUpsilon, m)$, and m is a $\mathbb{G}$-valued function that assigns to each point t of $[0,1)$ the multiplicity of the point $\alpha(t)$ in $\varUpsilon$ (i.e., each $m(t)$ says how often α traverses the point $\lambda = \alpha(t)$ of $\varUpsilon$). For notational simplicity we refer to $\varUpsilon$ itself as an oriented curve when the multiplicity m is either clear in the context or is immaterial. In this case we simply say $\varUpsilon$ is an oriented curve.

By a *partition P of the interval* $[0,1]$ we mean a finite sequence $\{t_j\}_{j=0}^n$ of points in $[0,1]$ such that $0 = t_0$, $t_{j-1} < t_j$ for $1 \leq j \leq n$, and $t_n = 1$. The *total variation* of a function $\alpha \colon [0,1] \to \mathbb{C}$ is the supremum of the set $\{v \in \mathbb{R} \colon v = \sum_{j=1}^n |\alpha(t_j) - \alpha(t_{j-1})|\}$ taken over all partitions P of $[0,1]$. A function $\alpha \colon [0,1] \to \mathbb{C}$ is of *bounded variation* if its total variation is finite. An oriented curve $\varUpsilon$ is *rectifiable* if some (and so any) parameterization α of it is of bounded variation, and its *length* $\ell(\varUpsilon)$ is the total variation of α. An arc $\alpha \colon [0,1] \to \mathbb{C}$

is *continuously differentiable* (or *smooth*) if it is differentiable on the open interval $(0, 1)$, it has one-sided derivatives at the end points 0 and 1, and its derivative $\alpha': [0, 1] \to \mathbb{C}$ is continuous on $[0, 1]$. In this case the oriented curve Υ parameterized by α is also referred to as *continuously differentiable* (or *smooth*), and as is known from advanced calculus, Υ is rectifiable with length

$$\ell(\Upsilon) = \int_\Upsilon |d\lambda| = \int_0^1 |d\alpha(t)| = \int_0^1 |\alpha'(t)| dt.$$

The linear space of all continuously differentiable functions from $[0, 1]$ to $\mathbb{C}$ is denoted by $C^1([0, 1])$. If there is a partition of $[0, 1]$ such that each restriction $\alpha|_{[t_{j-1}, t_j]}$ of the continuous α is continuously differentiable (i.e., $\alpha|_{[t_{j-1}, t_j]}$ is in $C^1([t_{j-1}, t_j])$), then $\alpha: [0, 1] \to \mathbb{C}$ is a *piecewise continuously differentiable arc* (or *piecewise smooth arc*), and the curve Υ parameterized by α is an oriented *piecewise continuously differentiable curve* (or an oriented *piecewise smooth curve*) which is rectifiable with length $\ell(\Upsilon) = \int_\Upsilon |d\lambda| = \sum_{j=1}^n \int_{t_{j-1}}^{t_j} |\alpha'_{[t_{j-1}, t_j]}(t)| dt$.

Every continuous function $\psi: \Upsilon \to \mathbb{C}$ has a Riemann–Stieltjes integral associated with a parameterization $\alpha: [0, 1] \to \Upsilon$ of a rectifiable curve Υ, viz., $\int_\Upsilon \psi(\lambda) \, d\lambda = \int_0^1 \psi(\alpha(t)) \, d\alpha(t)$. If in addition α is smooth, then

$$\int_\Upsilon \psi(\lambda) \, d\lambda = \int_0^1 \psi(\alpha(t)) \, d\alpha(t) = \int_0^1 \psi(\alpha(t)) \, \alpha'(t) \, dt.$$

Now consider a *bounded* function $f: \Upsilon \to \mathcal{Y}$ of an oriented curve Υ to a Banach space $\mathcal{Y}$. Let $\alpha: [0, 1] \to \Upsilon$ be a parameterization of Υ and let $P = \{t_j\}_{j=0}^n$ be a partition of the interval $[0, 1]$. The norm of a partition P is the number $\|P\| = \max_{1 \le j \le n}(t_j - t_{j-1})$. A Riemann–Stieltjes sum for the function f with respect to α based on a given partition P is a vector

$$\Sigma(f, \alpha, P) = \sum_{j=1}^n f(\alpha(\tau_j)) \, (\alpha(t_j) - \alpha(t_{j-1})) \ \in \ \mathcal{Y}$$

with $\tau_j \in [t_{j-1}, t_j]$. Suppose there is a unique vector $\Sigma \in \mathcal{Y}$ with the following property: for every $\varepsilon > 0$ there is a $\delta_\varepsilon > 0$ such that if P is an arbitrary partition of $[0, 1]$ with $\|P\| \le \delta_\varepsilon$, then $\|\Sigma(f, \alpha, P) - \Sigma\| \le \varepsilon$ for all Riemann–Stieltjes sums $\Sigma(f, \alpha, P)$ for f with respect to α based on P. If such a vector Σ exists, it is called the *Riemann–Stieltjes integral* of f with respect to α, denoted by

$$\Sigma = \int_\Upsilon f(\lambda) \, d\lambda = \int_0^1 f(\alpha(t)) \, d\alpha(t) \ \in \ \mathcal{Y}.$$

If the $\mathcal{Y}$-valued function $f: \Upsilon \to \mathcal{Y}$ is *continuous* and the curve Υ is *rectifiable*, then it can be verified (exactly as in the case of complex-valued functions; see, e.g., [23, Proposition 5.3 and p. 385]) the existence of the Riemann–Stieltjes integral of f with respect to α. If α_1 and α_2 are parameterizations of Υ, then for every $\delta > 0$ there are partitions P_1 and P_2 with norm less than δ for which $\Sigma(f, \alpha_1, P_1) = \Sigma(f, \alpha_2, P_2)$. Given a rectifiable oriented curve Υ and a continuous $\mathcal{Y}$-valued function f on Υ, then the integral $\int_\Upsilon f(\lambda) \, d\lambda$ does not depend on the parameterization of Υ. We refer to it as *the integral of f over Υ*.

Let $C[\Upsilon, \mathcal{Y}]$ and $B[\Upsilon, \mathcal{Y}]$ be the linear spaces of all $\mathcal{Y}$-valued continuous functions and of all bounded functions on Υ equipped with the sup-norm, which are Banach spaces because Υ is compact and $\mathcal{Y}$ is complete. Since Υ is compact, then $C[\Upsilon, \mathcal{Y}] \subseteq B[\Upsilon, \mathcal{Y}]$. The transformation $\int_\Upsilon \colon C[\Upsilon, \mathcal{Y}] \to \mathcal{Y}$ assigning to each function f in $C[\Upsilon, \mathcal{Y}]$ its integral $\int_\Upsilon f(\lambda)\, d\lambda$ in $\mathcal{Y}$ is linear and continuous (i.e., linear and bounded — $\int_\Upsilon \in \mathcal{B}[\,C[\Upsilon, \mathcal{Y}], \mathcal{Y}\,]$). Indeed,

$$\left\| \int_\Upsilon f(\lambda)\, d\lambda \right\| \leq \sup_{\lambda \in \Upsilon} \|f(\lambda)\| \int_\Upsilon |d\lambda(t)| = \|f\|_\infty\, \ell(\Upsilon).$$

If $\mathcal{Z}$ is a Banach space and $S \in \mathcal{B}[\mathcal{Y}, \mathcal{Z}]$, then (recall: Sf is continuous)

$$S \int_\Upsilon f(\lambda)\, d\lambda = \int_\Upsilon Sf(\lambda)\, d\lambda.$$

The *reverse* (or the *opposite*) of an oriented curve Υ is an oriented curve denoted by $-\Upsilon$ obtained from Υ by reversing the orientation of Υ. To reverse the orientation of Υ means to reverse the order of the domain $[0, 1]$ of its parameterization α or, equivalently, to replace $\alpha(t)$ with $\alpha(-t)$ for t running over $[-1, 0]$. Reversing the orientation of a curve Υ has the effect of replacing $\int_\Upsilon f(\lambda)\, d\lambda = \int_0^1 f(\alpha(t))\, d\alpha(t)$ with $\int_1^0 f(\alpha(t))\, d\alpha(t) = \int_{-\Upsilon} f(\lambda)\, d\lambda$, and so

$$\int_{-\Upsilon} f(\lambda)\, d\lambda = -\int_\Upsilon f(\lambda)\, d\lambda.$$

The preceding arguments and results all remain valid if the continuous function f is defined on a subset Λ of $\mathbb{C}$ that includes the curve Υ (by simply replacing it with its restriction to Υ, which remains continuous). In particular, if Λ is a nonempty open subset of $\mathbb{C}$ and Υ is a rectifiable curve included in Λ, and if $f \colon \Lambda \to \mathcal{Y}$ is a continuous function on the open set Λ, then f has an integral over $\Upsilon \subset \Lambda \subseteq \mathbb{C}$, viz., $\int_\Upsilon f(\lambda)\, d\lambda$.

A topological space is *disconnected* if it is the union of two disjoint non-empty subsets that are both open and closed. Otherwise the topological space is said to be *connected*. A subset of a topological space is called a *connected set* (or a *connected subset*) if, as a topological subspace, it is a connected topological space itself. Take any nonempty subset Λ of $\mathbb{C}$. A *component* of Λ is a maximal connected subset of Λ, which coincides with the union of all connected subsets of Λ containing a given point of Λ. Thus any two components of Λ are disjoint. Since the closure of a connected set is connected, any component of Λ is closed relative to Λ. By a *region* (or a *domain*) we mean a nonempty connected open subset of $\mathbb{C}$. Every open subset of $\mathbb{C}$ is uniquely expressed as a countable union of disjoint regions which are the components of it (see, e.g., [23, Proposition 3.9]). The closure of a region is sometimes called a *closed region*. Different regions may have the same closure (sample: a punctured open disk).

Let Υ be a closed rectifiable oriented curve in $\mathbb{C}$. A classical and important result in complex analysis establishes the value of the integral

$$w_\Upsilon(\zeta) = \tfrac{1}{2\pi i} \int_\Upsilon \frac{1}{\lambda - \zeta} \, d\lambda \quad \text{for every} \quad \zeta \in \mathbb{C} \backslash \Upsilon.$$

This integral has an integer value which is constant on each component of $\mathbb{C} \backslash \Upsilon$ and zero on the unbounded component of $\mathbb{C} \backslash \Upsilon$. The number $w_\Upsilon(\zeta)$ is referred to as the *winding number* of Υ about ζ. *If a closed rectifiable oriented curve Υ is a simple curve, then $\mathbb{C} \backslash \Upsilon$ has only two components, just one of them is bounded, and Υ is their common boundary* (this is the JORDAN CURVE THEOREM). In this case (i.e., for a *simple* closed rectifiable oriented curve Υ) the winding number $w_\Upsilon(\zeta)$ takes on only three values for each $\zeta \in \mathbb{C} \backslash \Upsilon$, either 0 or ± 1. If $w_\Upsilon(\zeta) = 1$ for every ζ in the bounded component of $\mathbb{C} \backslash \Upsilon$, then Υ is said to be *positively* (i.e., counterclockwise) *oriented*; otherwise if $w_\Upsilon(\zeta) = -1$, then Υ is said to be *negatively* (i.e., clockwise) *oriented*. For every ζ in the unbounded component of $\mathbb{C} \backslash \Upsilon$ we get $w_\Upsilon(\zeta) = 0$. If Υ is positively oriented, then the reverse curve $-\Upsilon$ is negatively oriented, and vice versa. These notions can be extended as follows. A finite union $\Upsilon = \bigcup_{j=1}^m \Upsilon_j$ of *disjoint* closed rectifiable oriented *simple* curves Υ_j is called a *path*, and its *winding number* $w_\Upsilon(\zeta)$ about $\zeta \in \mathbb{C} \backslash \Upsilon$ is defined by $w_\Upsilon(\zeta) = \sum_{j=1}^m w_{\Upsilon_j}(\zeta)$. A path Υ is *positively oriented* if for every $\zeta \in \mathbb{C} \backslash \Upsilon$ the winding number $w_\Upsilon(\zeta)$ is either 0 or 1, and *negatively oriented* if for every $\zeta \in \mathbb{C} \backslash \Upsilon$ the winding number $w_\Upsilon(\zeta)$ is either 0 or -1. If a path $\Upsilon = \bigcup_{j=1}^m \Upsilon_j$ is positively oriented, then the reverse path $-\Upsilon = \bigcup_{j=1}^m -\Upsilon_j$ is negatively oriented. The *inside* (notation: ins Υ) and the *outside* (notation: out Υ) of a positively oriented path Υ are the sets

$$\text{ins } \Upsilon = \big\{ \zeta \in \mathbb{C} \backslash \Upsilon : w_\Upsilon(\zeta) = 1 \big\} \quad \text{and} \quad \text{out } \Upsilon = \big\{ \zeta \in \mathbb{C} \backslash \Upsilon : w_\Upsilon(\zeta) = 0 \big\}.$$

From now on all paths are positively oriented. If $\Upsilon = \bigcup_{j=1}^m \Upsilon_j$ is a path and if there is a finite subset $\{\Upsilon_{j'}\}_{j'=1}^n$ of Υ such that $\Upsilon_{j'} \subset \text{ins } \Upsilon_{j'+1}$, then these nested (disjoint closed rectifiable oriented) simple curves $\{\Upsilon_{j'}\}$ are oppositely oriented, with Υ_n being positively oriented because $w_\Upsilon(\zeta)$ is either 0 or 1 for every $\zeta \in \mathbb{C} \backslash \Upsilon$. An open subset of $\mathbb{C}$ is a *Cauchy domain* (or a *Jordan domain*) if it has finitely many components whose closures are pairwise disjoint, and its boundary is a path. The closure of a Jordan domain is sometimes referred to as a *Jordan closed region*. If Υ is a path, then $\{\Upsilon, \text{ins } \Upsilon, \text{out } \Upsilon\}$ is a partition of $\mathbb{C}$. Since a path Υ is a closed set in $\mathbb{C}$ (it is a finite union of closed sets), and since it is the common boundary of ins Υ and out Υ, it follows that ins Υ *and* out Υ *are open sets in* $\mathbb{C}$, and their closures are given by the union with their common boundary,

$$(\text{ins } \Upsilon)^- = \Upsilon \cup \text{ins } \Upsilon \quad \text{and} \quad (\text{out } \Upsilon)^- = \Upsilon \cup \text{out } \Upsilon.$$

If Υ is a path, $\mathcal{Y}$ is a Banach space, and $f : \Upsilon \to \mathcal{Y}$ is a continuous function (and so f has an integral over each closed rectifiable oriented simple curve Υ_j), then the integral of the $\mathcal{Y}$-valued function f over the path $\Upsilon \subset \mathbb{C}$ is defined by

$$\int_\Upsilon f(\lambda) \, d\lambda = \sum_{j=1}^m \int_{\Upsilon_j} f(\lambda) \, d\lambda.$$

Again, if $f: \Lambda \to \mathcal{Y}$ is a continuous function on a nonempty open subset Λ of $\mathbb{C}$ that includes a path Υ, then we define the integral of f over the path Υ as the integral of the restriction of f to Υ over $\Upsilon \subset \Lambda \subseteq \mathbb{C}$; that is, as the integral of $f|_{\Upsilon}: \Upsilon \to \mathcal{Y}$ (which is continuous as well) over the path Υ:

$$\int_{\Upsilon} f(\lambda) \, d\lambda = \int_{\Upsilon} f|_{\Upsilon}(\lambda) \, d\lambda.$$

Therefore if we are given a nonempty open subset Λ of $\mathbb{C}$, a Banach space $\mathcal{Y}$, and a continuous function $f: \Lambda \to \mathcal{Y}$, then the above identity establishes how to define the integral of f over an arbitrary path Υ included in Λ.

The CAUCHY INTEGRAL FORMULA says: *if* $\psi: \Lambda \to \mathbb{C}$ *is an analytic function on a nonempty open subset* Λ *of* $\mathbb{C}$ *including a path* Υ *and its inside* ins Υ *(i.e.,* $\Upsilon \cup$ ins $\Upsilon \subset \Lambda \subseteq \mathbb{C}$*), then*

$$\psi(\zeta) = \tfrac{1}{2\pi i} \int_{\Upsilon} \frac{\psi(\lambda)}{\lambda - \zeta} \, d\lambda \quad \textit{for every} \quad \zeta \in \text{ins } \Upsilon.$$

In fact this is the most common application of the general case of the Cauchy Integral Formula (see, e.g., [23, Problem 5.O]). What the basic assumption $\Upsilon \cup$ ins $\Upsilon \subset \Lambda \subseteq \mathbb{C}$ (i.e., (ins $\Upsilon)^- \subset \Lambda \subseteq \mathbb{C}$) says is simply this: $\Upsilon \subset \Lambda \subseteq \mathbb{C}$ and $\mathbb{C} \backslash \Lambda \subset$ out Υ (i.e., $w_{\Upsilon}(\zeta) = 0$ for every $\zeta \in \mathbb{C} \backslash \Lambda$), and the assumption $\zeta \in$ ins $\Upsilon \subset \Lambda$ is equivalent to saying $\zeta \in \Lambda \backslash \Upsilon$ and $w_{\Upsilon}(\zeta) = 1$.

Since the function $\psi(\cdot)[(\cdot) - \zeta]^{-1}: \Upsilon \to \mathbb{C}$ for $\zeta \in$ ins Υ is continuous on the path Υ, the above integral may be generalized to a $\mathcal{Y}$-valued function by considering the following special case in a complex Banach space $\mathcal{Y}$. Moreover, throughout this chapter $\mathcal{X}$ will also denote a nonzero complex Banach space.

Set $\mathcal{Y} = \mathcal{B}[\mathcal{X}]$, the Banach algebra of all operators on $\mathcal{X}$. Take an operator T in $\mathcal{B}[\mathcal{X}]$ and let $R_T: \rho(T) \to \mathcal{G}[\mathcal{X}]$ be its resolvent function, defined by $R_T(\lambda) = (\lambda I - T)^{-1}$ for every λ in the resolvent set $\rho(T)$, where $\rho(T)$ is an open and nonempty subset of $\mathbb{C}$ (cf. Section 2.1). Let $\psi: \Lambda \to \mathbb{C}$ be an analytic function on a nonempty open subset Λ of $\mathbb{C}$ for which the intersection $\Lambda \cap \rho(T)$ is again nonempty. Recall from the proof of Theorem 2.2 that $R_T: \rho(T) \to \mathcal{G}[\mathcal{X}]$ is continuous, and so is the product function $\psi R_T: \Lambda \cap \rho(T) \to \mathcal{B}[\mathcal{X}]$ defined by $(\psi R_T)(\lambda) = \psi(\lambda) R_T(\lambda) \in \mathcal{B}[\mathcal{X}]$ for every λ in the open subset $\Lambda \cap \rho(T)$ of $\mathbb{C}$. Then we can define the integral of ψR_T over any path $\Upsilon \subset \Lambda \cap \rho(T)$,

$$\int_{\Upsilon} \psi(\lambda) \, R_T(\lambda) \, d\lambda = \int_{\Upsilon} \psi(\lambda) \, (\lambda I - T)^{-1} \, d\lambda \ \in \ \mathcal{B}[\mathcal{X}].$$

Thus, for any complex number $\zeta \in$ ins Υ, the integral of the $\mathbb{C}$-valued function $\psi(\cdot)[(\cdot) - \zeta]^{-1}: \Lambda \backslash \{\zeta\} \to \mathbb{C}$ over any path Υ such that (ins $\Upsilon)^- \subset \Lambda$ is generalized, for any operator $T \in \mathcal{B}[\mathcal{X}]$, to the integral of the $\mathcal{B}[\mathcal{X}]$-valued function $\psi(\cdot)[(\cdot)I - T]^{-1}: \Lambda \cap \rho(T) \to \mathcal{B}[\mathcal{X}]$ over any path Υ such that $\Upsilon \subset \Lambda \cap \rho(T)$.

As we will see next, the function $\psi R_T: \Lambda \cap \rho(T) \to \mathcal{B}[\mathcal{X}]$ is analytic, and the definition of a Banach-space-valued analytic function on a nonempty open subset of $\mathbb{C}$ is exactly the same as that of a scalar-valued analytic function.

If Ω is a compact set included in an open subset Λ of $\mathbb{C}$ (so $\Omega \subset \Lambda \subseteq \mathbb{C}$), then there exists a path $\Upsilon \subset \Lambda$ for which $\Omega \subset \text{ins}\, \Upsilon$ and $\mathbb{C}\backslash\Lambda \subset \text{out}\, \Upsilon$ (see, e.g., [30, p. 200]). As we have already seen, this condition is equivalent to $\Omega \subset \text{ins}\, \Upsilon \subset (\text{ins}\, \Upsilon)^- \subset \Lambda$. Take an operator $T \in \mathcal{B}[\mathcal{X}]$. Set $\Omega = \sigma(T)$, which is compact. Now take an open subset Λ of $\mathbb{C}$ such that $\sigma(T) \subset \Lambda$. Let Υ be any path for which $\Upsilon \subset \Lambda$ and $\sigma(T) \subset \text{ins}\, \Upsilon \subset \Lambda$. In other words, for a given open set Λ including $\sigma(T)$ let the path Υ satisfy the following assumption:

$$\sigma(T) \subset \text{ins}\, \Upsilon \subset (\text{ins}\, \Upsilon)^- \subset \Lambda.$$

Since $\rho(T) = \mathbb{C}\backslash\sigma(T)$, we get $\Upsilon \subset \Lambda \cap \rho(T) \neq \varnothing$ by the above inclusions.

Lemma 4.13. *If $\psi\colon \Lambda \to \mathbb{C}$ is an analytic function on a nonempty open subset Λ of $\mathbb{C}$, then the integral*

$$\int_\Upsilon \psi(\lambda)\, R_T(\lambda)\, d\lambda = \int_\Upsilon \psi(\lambda)\, (\lambda I - T)^{-1}\, d\lambda$$

does not depend on the choice of any path Υ that satisfies the assumption

$$\sigma(T) \subset \text{ins}\, \Upsilon \subset (\text{ins}\, \Upsilon)^- \subset \Lambda.$$

Proof. Let $\mathcal{X}$ be a Banach space and take an operator T in the Banach algebra $\mathcal{B}[\mathcal{X}]$. Let $R_T\colon \rho(T) \to \mathcal{G}[\mathcal{X}]$ be its resolvent function and let $\psi\colon \Lambda \to \mathbb{C}$ be an analytic function on an open set $\Lambda \subseteq \mathbb{C}$ properly including $\sigma(T)$. So there is a path Υ satisfying the above assumption. Suppose Υ' and Υ'' are distinct positively oriented paths satisfying the above assumption. We show that

$$\int_{\Upsilon'} \psi(\lambda)\, R_T(\lambda)\, d\lambda = \int_{\Upsilon''} \psi(\lambda)\, R_T(\lambda)\, d\lambda.$$

Claim 1. The product $\psi R_T\colon \Lambda \cap \rho(T) \to \mathcal{B}[\mathcal{X}]$ is analytic on $\Lambda \cap \rho(T)$.

Proof. First recall: $\Lambda \cap \rho(T) \neq \varnothing$. By the resolvent identity of Section 2.1, if λ and ζ are distinct points in $\rho(T)$, then

$$\frac{R_T(\lambda) - R_T(\zeta)}{\lambda - \zeta} + R_T(\zeta)^2 = \big(R_T(\zeta) - R_T(\lambda)\big) R_T(\zeta)$$

(cf. proof of Theorem 2.2, Claim 2). Since $R_T\colon \rho(T) \to \mathcal{G}[\mathcal{X}]$ is continuous,

$$\lim_{\lambda \to \zeta} \left\| \frac{R_T(\lambda) - R_T(\zeta)}{\lambda - \zeta} + R_T(\zeta)^2 \right\| \leq \lim_{\lambda \to \zeta} \left\| R_T(\zeta) - R_T(\lambda) \right\| \left\| R_T(\zeta) \right\| = 0$$

for every $\zeta \in \rho(T)$. Thus the resolvent function $R_T\colon \rho(T) \to \mathcal{G}[\mathcal{X}]$ is analytic on $\rho(T)$ and so it is analytic on the nonempty open set $\Lambda \cap \rho(T)$. Since the function $\psi\colon \Lambda \to \mathbb{C}$ also is analytic on $\Lambda \cap \rho(T)$, and since the product of analytic functions on the same domain is again analytic, the product function $\psi R_T\colon \Lambda \cap \rho(T) \to \mathcal{B}[\mathcal{X}]$ given by $(\psi R_T)(\lambda) = \psi(\lambda) R_T(\lambda) \in \mathcal{B}[\mathcal{X}]$ for every $\lambda \in \Lambda \cap \rho(T)$ is analytic on $\Lambda \cap \rho(T)$, concluding the proof of Claim 1.

The CAUCHY THEOREM says: *if* $\psi \colon \Lambda \to \mathbb{C}$ *is analytic on a nonempty open subset Λ of $\mathbb{C}$ that includes a path Υ and its inside* ins Υ, *then*

$$\int_{\Upsilon} \psi(\lambda)\, d\lambda = 0$$

(see, e.g., [23, Problem 5.O]). This can be extended from scalar-valued functions to Banach-space-valued functions. Let $\mathcal{Y}$ be a complex Banach space, let $f \colon \Lambda \to \mathcal{Y}$ be an analytic function on a nonempty open subset Λ of $\mathbb{C}$ including a path Υ and its inside ins Υ, and consider the integral $\int_{\Upsilon} f(\lambda)\, d\lambda$.

Claim 2. If $f \colon \Lambda \to \mathcal{Y}$ is analytic on a given nonempty open subset Λ of $\mathbb{C}$ and if Υ is any path for which $(\text{ins } \Upsilon)^- \subset \Lambda$, then

$$\int_{\Upsilon} f(\lambda)\, d\lambda = 0.$$

Proof. Take an arbitrary nonzero ξ in $\mathcal{Y}^*$ (i.e., take a nonzero bounded linear functional $\xi \colon \mathcal{Y} \to \mathbb{C}$), and consider the composition $\xi \circ f \colon \Lambda \to \mathbb{C}$ of ξ with an analytic function $f \colon \Lambda \to \mathcal{Y}$, which is again analytic on Λ. Indeed,

$$\left| \frac{\xi(f(\lambda)) - \xi(f(\zeta))}{\lambda - \zeta} - \xi(f'(\zeta)) \right| \le \|\xi\| \left| \frac{f(\lambda) - f(\zeta)}{\lambda - \zeta} - f'(\zeta) \right|$$

for every pair of distinct points λ and ζ in Λ. Since both ξ and the integral are linear and continuous,

$$\xi \left(\int_{\Upsilon} f(\lambda)\, d\lambda \right) = \xi \left(\sum_{j=1}^m \int_{\Upsilon_j} f(\lambda)\, d\lambda \right) = \sum_{j=1}^m \xi \left(\int_{\Upsilon_j} f(\lambda)\, d\lambda \right)$$

$$= \sum_{j=1}^m \int_{\Upsilon_j} \xi(f(\lambda))\, d\lambda = \int_{\Upsilon} \xi(f(\lambda))\, d\lambda.$$

By the Cauchy Theorem for scalar-valued functions we get $\int_{\Upsilon} \xi(f(\lambda))\, d\lambda = 0$ (because $\xi \circ f$ is analytic on Λ). Therefore $\xi(\int_{\Upsilon} f(\lambda)\, d\lambda) = 0$. Since this holds for every $\xi \in \mathcal{Y}^*$, the Hahn–Banach Theorem (see, e.g., [78, Corollary 4.64]) ensures the claimed result, viz., $\int_{\Upsilon} f(\lambda)\, d\lambda = 0$.

Consider the nonempty open subset ins $\Upsilon' \cap$ ins Υ'' of $\mathbb{C}$, which includes $\sigma(T)$. Let Δ be the (finite) union of open components of ins $\Upsilon' \cap$ ins Υ'' including $\sigma(T)$. Let $\Upsilon \subseteq \Upsilon' \cup \Upsilon''$ be the path consisting of the boundary of Δ positively oriented, and so ins $\Upsilon = \Delta$ and $\Delta^- = (\text{ins } \Upsilon)^- = \Delta \cup \Upsilon$. Observe that

$$\sigma(T) \subset \text{ins } \Upsilon \subset (\text{ins } \Upsilon)^- \subseteq (\text{ins } \Upsilon' \cap \text{ins } \Upsilon'')^- \subset \Lambda.$$

Thus $\Upsilon \subset \Delta^- \cap \rho(T) \ne \varnothing$. Set $\tilde{\Delta} = \text{ins } \Upsilon' \backslash \Delta$. If $\tilde{\Delta} \ne \varnothing$, then orient the boundary of $\tilde{\Delta}$ so as to make it a positively oriented path $\tilde{\Upsilon} \subseteq \Upsilon' \cup (-\Upsilon)$. (Note: If ins $\Upsilon' \subset$ ins Υ'', then $\Delta = \text{ins } \Upsilon'$ so that $\tilde{\Delta} = \varnothing$, which implies $\tilde{\Upsilon} = \varnothing$.) Since ins $\Upsilon' = \Delta \cup \tilde{\Delta}$ with $\Delta \cap \tilde{\Delta} = \varnothing$ and Υ' and Υ are positively oriented, and so is $\widetilde{\varUpsilon}$ if it is not empty, then $\int_{\varUpsilon'} f(\lambda)\, d\lambda = \int_{\varUpsilon} f(\lambda)\, d\lambda + \int_{\widetilde{\varUpsilon}} f(\lambda)\, d\lambda$ for every $\mathcal{B}[\mathcal{X}]$-valued continuous function f whose domain includes $(\text{ins}\,\varUpsilon')^-$. Hence

$$\int_{\varUpsilon'} \psi(\lambda)\, R_T(\lambda)\, d\lambda = \int_{\varUpsilon} \psi(\lambda)\, R_T(\lambda)\, d\lambda + \int_{\widetilde{\varUpsilon}} \psi(\lambda)\, R_T(\lambda)\, d\lambda.$$

But ψR_T is analytic on $\varLambda \cap \rho(T)$ by Claim 1 and so, since $\widetilde{\varDelta}^- = (\text{ins}\,\widetilde{\varUpsilon})^- \subset \varLambda \cap \rho(T)$, we get $\int_{\widetilde{\varUpsilon}} \psi(\lambda)\, R_T(\lambda)\, d\lambda = 0$ by Claim 2. Therefore

$$\int_{\varUpsilon'} \psi(\lambda)\, R_T(\lambda)\, d\lambda = \int_{\varUpsilon} \psi(\lambda)\, R_T(\lambda)\, d\lambda.$$

Similarly (by exactly the same argument, redefining $\widetilde{\varDelta}$ as $\text{ins}\,\varUpsilon'' \backslash \varDelta$),

$$\int_{\varUpsilon''} \psi(\lambda)\, R_T(\lambda)\, d\lambda = \int_{\varUpsilon} \psi(\lambda)\, R_T(\lambda)\, d\lambda. \qquad \square$$

Definition 4.14. Take an arbitrary $T \in \mathcal{B}[\mathcal{X}]$. If $\psi\colon \varLambda \to \mathbb{C}$ is analytic on an open set $\varLambda \subseteq \mathbb{C}$ including the spectrum $\sigma(T)$ of T, and if $\varUpsilon$ is any path such that $\sigma(T) \subset \text{ins}\,\varUpsilon \subset (\text{ins}\,\varUpsilon)^- \subset \varLambda$, then define the operator $\psi(T)$ in $\mathcal{B}[\mathcal{X}]$ by

$$\psi(T) = \tfrac{1}{2\pi i} \int_{\varUpsilon} \psi(\lambda)\, R_T(\lambda)\, d\lambda = \tfrac{1}{2\pi i} \int_{\varUpsilon} \psi(\lambda)\, (\lambda I - T)^{-1}\, d\lambda.$$

After defining the integral of ψR_T over any path $\varUpsilon$ such that $\varUpsilon \subset \varLambda \cap \rho(T)$, its invariance for any path $\varUpsilon$ such that $\sigma(T) \subset \text{ins}\,\varUpsilon \subset (\text{ins}\,\varUpsilon)^- \subset \varLambda$ was ensured in Lemma 4.13. Thus the above definition of the operator $\psi(T)$ is clearly motivated by the Cauchy Integral Formula.

Let T be an arbitrary operator in $\mathcal{B}[\mathcal{X}]$. A complex-valued function ψ is said to be *analytic on* $\sigma(T)$ (or *analytic on a neighborhood of* $\sigma(T)$) if it is analytic on an open set including $\sigma(T)$. By a *path about* $\sigma(T)$ we mean a path $\varUpsilon$ whose inside (properly) includes $\sigma(T)$ (i.e., a path for which $\sigma(T) \subset \text{ins}\,\varUpsilon$). So what is behind Definition 4.14 is: if ψ is analytic on $\sigma(T)$, then the integral in Definition 4.14 exists as an operator in $\mathcal{B}[\mathcal{X}]$, and the integral does not depend on the path about $\sigma(T)$ whose closure of its inside is included in the open set upon which ψ is defined (i.e., it is included in the domain of ψ).

Lemma 4.15. *If ϕ and ψ are analytic on $\sigma(T)$ and if $\varUpsilon$ is any path about $\sigma(T)$ such that $\sigma(T) \subset \text{ins}\,\varUpsilon \subset (\text{ins}\,\varUpsilon)^- \subset \varLambda$, where the nonempty open set $\varLambda \subseteq \mathbb{C}$ is the intersection of the domains of ϕ and ψ, then*

$$\phi(T)\psi(T) = \tfrac{1}{2\pi i} \int_{\varUpsilon} \phi(\lambda)\psi(\lambda)\, R_T(\lambda)\, d\lambda,$$

and so $\phi(T)\psi(T) = (\phi\psi)(T)$ since $\phi\psi$ is analytic on $\sigma(T)$.

Proof. Let ϕ and ψ be analytic on $\sigma(T)$. Thus $\phi\psi$ is analytic on the intersection $\varLambda$ of their domains. Let $\varUpsilon$ and $\varUpsilon'$ be arbitrary paths such that

$$\sigma(T) \subset \text{ins } \Upsilon \subset (\text{ins } \Upsilon)^- \subset \text{ins } \Upsilon' \subseteq \Lambda.$$

Thus by Definition 4.14 and by the resolvent identity (cf. Section 2.1),

$$\phi(T)\psi(T) = \left(\tfrac{1}{2\pi i} \int_{\Upsilon} \phi(\lambda)\, R_T(\lambda)\, d\lambda\right)\left(\tfrac{1}{2\pi i} \int_{\Upsilon'} \psi(\zeta)\, R_T(\zeta)\, d\zeta\right)$$

$$= -\tfrac{1}{4\pi^2} \int_{\Upsilon} \int_{\Upsilon'} \phi(\lambda)\psi(\zeta)\, R_T(\lambda)\, R_T(\zeta)\, d\zeta\, d\lambda$$

$$= -\tfrac{1}{4\pi^2} \int_{\Upsilon} \int_{\Upsilon'} \phi(\lambda)\psi(\zeta)\, (\zeta - \lambda)^{-1}\big(R_T(\lambda) - R_T(\zeta)\big)\, d\zeta\, d\lambda$$

$$= -\tfrac{1}{4\pi^2} \int_{\Upsilon} \int_{\Upsilon'} \phi(\lambda)\psi(\zeta)\, (\zeta - \lambda)^{-1} R_T(\lambda)\, d\zeta\, d\lambda$$

$$+ \tfrac{1}{4\pi^2} \int_{\Upsilon'} \int_{\Upsilon} \phi(\lambda)\psi(\zeta)\, (\zeta - \lambda)^{-1} R_T(\zeta)\, d\lambda\, d\zeta$$

$$= -\tfrac{1}{4\pi^2} \int_{\Upsilon} \phi(\lambda)\Big(\int_{\Upsilon'} \psi(\zeta)(\zeta - \lambda)^{-1}\, d\zeta\Big) R_T(\lambda)\, d\lambda$$

$$+ \tfrac{1}{4\pi^2} \int_{\Upsilon'} \psi(\zeta)\Big(\int_{\Upsilon} \phi(\lambda)(\zeta - \lambda)^{-1}\, d\lambda\Big) R_T(\zeta)\, d\zeta.$$

Since $\lambda \in \Upsilon$ we get $\lambda \in \text{ins } \Upsilon'$, and hence $\int_{\Upsilon'} \psi(\zeta)(\zeta - \lambda)^{-1}\, d\zeta = (2\pi i)\psi(\lambda)$ by the Cauchy Integral Formula. Moreover, since $\zeta \in \Upsilon'$ we get $\zeta \notin \text{ins } \Upsilon'$, and so the function $\phi(\cdot)[\zeta - (\cdot)]^{-1}$ is analytic on $\text{ins } \Upsilon'$ which includes Υ. Therefore $\int_{\Upsilon} \phi(\lambda)(\zeta - \lambda)^{-1}\, d\lambda = 0$ by the Cauchy Theorem. Thus

$$\phi(T)\psi(T) = \tfrac{1}{2\pi i} \int_{\Upsilon} \phi(\lambda)\psi(\lambda)\, R_T(\lambda)\, d\lambda. \qquad \square$$

Take an arbitrary operator $T \in \mathcal{B}[\mathcal{X}]$. Let $A(\sigma(T))$ denote the set of all *analytic functions on the spectrum* $\sigma(T)$ of T. That is, $\psi \in A(\sigma(T))$ if $\psi \colon \Lambda \to \mathbb{C}$ is analytic on an open set $\Lambda \subseteq \mathbb{C}$ including $\sigma(T)$. As is readily verified, $A(\sigma(T))$ is a unital algebra where the domain of the product of two functions in $A(\sigma(T))$ is the intersection of their domains, and the identity element 1 in $A(\sigma(T))$ is the constant function $1(\lambda) = 1$ for all $\lambda \in \Lambda$. The identity map also lies in $A(\sigma(T))$ (i.e., if $\varphi \colon \Lambda \to \Lambda \subseteq \mathbb{C}$ is such that $\varphi(\lambda) = \lambda$ for every $\lambda \in \Lambda$, then φ is in $A(\sigma(T))$). The next theorem is the main result of this section: the Analytic (or Holomorphic) Riesz (or Riesz–Dunford) Functional Calculus.

Theorem 4.16. RIESZ–DUNFORD FUNCTIONAL CALCULUS. *Let $\mathcal{X}$ be a nonzero complex Banach space. Take an operator T in $\mathcal{B}[\mathcal{X}]$. For every function ψ in $A(\sigma(T))$, let the operator $\psi(T)$ in $\mathcal{B}[\mathcal{X}]$ be defined as in Definition 4.14, viz.,*

$$\psi(T) = \tfrac{1}{2\pi i} \int_{\Upsilon} \psi(\lambda)\, R_T(\lambda)\, d\lambda.$$

The mapping $\Phi_T \colon A(\sigma(T)) \to \mathcal{B}[\mathcal{X}]$ such that $\Phi_T(\psi) = \psi(T)$ is a unital homomorphism, which is continuous when $A(\sigma(T))$ is equipped with the sup-norm. Moreover, (a) *since every polynomial p is such that $p \in A(\mathbb{C}) \supset A(\sigma(T))$,*

$$\Phi_T(p) = p(T) = \tfrac{1}{2\pi i} \int_\Upsilon \sum_{j=0}^m \alpha_j \lambda^j R_T(\lambda) \, d\lambda = \sum_{j=0}^m \alpha_j T^j,$$

with Υ being any circle about the origin such that $\sigma(T) \subset \mathrm{ins}\, \Upsilon \subset (\mathrm{ins}\, \Upsilon)^- \subset \Lambda$ where Λ is an open neighborhood of $\sigma(T)$, and with $p(\lambda) = \sum_{j=0}^m \alpha_j \lambda^j$ for $\lambda \in \Lambda$ where $\{\alpha_j\}_{j=0}^m$ is a finite set of complex coefficients. In particular,

$$\Phi_T(1) = 1(T) = \tfrac{1}{2\pi i} \int_\Upsilon R_T(\lambda) \, d\lambda = T^0 = I,$$

where $1 \in A(\sigma(T))$ is the constant function, $1(\lambda) = 1$ for every λ, and so Φ_T takes the identity element $1 \in A(\sigma(T))$ to the identity element $I \in \mathcal{B}[\mathcal{X}]$. Also

$$\Phi_T(\varphi) = \varphi(T) = \tfrac{1}{2\pi i} \int_\Upsilon \lambda\, R_T(\lambda) \, d\lambda = T,$$

where $\varphi \in A(\sigma(T))$ is the identity map (i.e., $\varphi(\lambda) = \lambda$ for every λ). (b) If $\psi(\lambda) = \sum_{k=0}^\infty \alpha_k \lambda^k$ is a power series expansion for $\varphi \in A(\sigma(T))$ on an open neighborhood Λ of $\sigma(T)$ with radius of convergence greater than $r(T)$, then

$$\Phi_T(\psi) = \psi(T) = \tfrac{1}{2\pi i} \int_\Upsilon \sum_{k=0}^\infty \alpha_k \lambda^k R_T(\lambda) \, d\lambda = \sum_{k=0}^\infty \alpha_k T^k,$$

where $\sum_{k=0}^\infty \alpha_k T^k$ converges in $\mathcal{B}[\mathcal{X}]$. (c) If $\{\psi_n\}$ is an arbitrary $A(\sigma(T))$-valued sequence converging uniformly on compact subsets of Λ to $\psi \in A(\sigma(T))$, then $\{\Phi_T(\psi_n)\}$ converges in $\mathcal{B}[\mathcal{X}]$ to $\Phi_T(\psi)$.

Proof. The integral $\int_\Upsilon (\cdot) R_T \, d\lambda \colon A(\sigma(T)) \to \mathcal{B}[\mathcal{X}]$ is a linear transformation between the linear spaces $A(\sigma(T))$ and $\mathcal{B}[\mathcal{X}]$, and so is $\Phi_T \colon A(\sigma(T)) \to \mathcal{B}[\mathcal{X}]$:

$$\Phi_T(\alpha\, \phi + \beta\, \psi) = (\alpha\, \phi + \beta\, \psi)(T) = \tfrac{1}{2\pi i} \int_\Upsilon (\alpha\, \phi + \beta\, \psi)(\lambda)\, R_T(\lambda)\, d\lambda$$

$$= \alpha\, \tfrac{1}{2\pi i} \int_\Upsilon \phi(\lambda)\, R_T(\lambda)\, d\lambda + \beta\, \tfrac{1}{2\pi i} \int_\Upsilon \psi(\lambda)\, R_T(\lambda)\, d\lambda = \alpha\, \Phi_T(\phi) + \beta\, \Phi_T(\psi)$$

for every $\alpha, \beta \in \mathbb{C}$ and every $\phi, \psi \in A(\sigma(T))$. Moreover, by Lemma 4.15,

$$\Phi_T(\phi\, \psi) = \tfrac{1}{2\pi i} \int_\Upsilon (\phi\, \psi)(\lambda)\, R_T(\lambda)\, d\lambda = \Phi_T(\phi)\, \Phi_T(\psi)$$

for every $\phi, \psi \in A(\sigma(T))$. Thus Φ is a homomorphism. Those integrals do not depend on the path Υ such that $\sigma(T) \subset \mathrm{ins}\, \Upsilon \subset (\mathrm{ins}\, \Upsilon)^- \subset \Lambda$. (a) First suppose $\psi \in A(\sigma(T))$ is analytic on an open neighborhood Λ of $\sigma(T)$ including a circle Υ about the origin with radius greater than the spectral radius $r(T)$ — e.g., if ψ is a polynomial. Since $r(T) < |\lambda|$ for $\lambda \in \Upsilon$, we get by Corollary 2.12(a)

$$(2\pi i)\, \psi(T) = \int_\Upsilon \psi(\lambda)\, R_T(\lambda)\, d\lambda = \int_\Upsilon \psi(\lambda) \sum_{k=0}^\infty \tfrac{1}{\lambda^{k+1}}\, T^k \, d\lambda$$

$$= I \int_\Upsilon \psi(\lambda)\, \tfrac{1}{\lambda}\, d\lambda + \sum_{k=1}^\infty T^k \int_\Upsilon \psi(\lambda)\, \tfrac{1}{\lambda^{k+1}}\, d\lambda,$$

because the integral is linear and continuous, where the series $\sum_{k=0}^\infty \tfrac{1}{\lambda^{k+1}}\, T^k$ converges uniformly (i.e., converges in the operator norm topology of $\mathcal{B}[\mathcal{X}]$). Consider two special cases of elementary polynomials. First let $\psi = 1$. So

$$1(T) = I\tfrac{1}{2\pi i}\int_\Upsilon \tfrac{1}{\lambda}\,d\lambda + \sum_{k=1}^\infty T^k \tfrac{1}{2\pi i}\int_\Upsilon \tfrac{1}{\lambda^{k+1}}\,d\lambda = I + O = I$$

for $\psi(\lambda) = 1(\lambda) = 1$ for all λ, since $\tfrac{1}{2\pi i}\int_\Upsilon \tfrac{1}{\lambda}\,d\lambda = 1(0) = 1$ by the Cauchy Integral Formula and, for $k \geq 1$, $\int_\Upsilon \tfrac{1}{\lambda^{k+1}}\,d\lambda = 0$. Indeed, with $\alpha(\theta) = \rho e^{i\theta}$ for $\theta \in [0, 2\pi]$ and ρ being the radius of the circle Υ, we get $\int_\Upsilon \tfrac{1}{\lambda}\,d\lambda = \int_0^{2\pi}\alpha(\theta)^{-1}\alpha'(\theta)\,d\theta = \int_0^{2\pi}\rho^{-1}e^{-i\theta}\rho i e^{i\theta}\,d\theta = i\int_0^{2\pi}d\theta = 2\pi i$ and, for $k \geq 1$, we also get $\int_\Upsilon \tfrac{1}{\lambda^{k+1}}\,d\lambda = \int_0^{2\pi}\alpha(\theta)^{-(k+1)}\alpha'(\theta)\,d\theta = \int_0^{2\pi}\rho^{-(k+1)}e^{-i(k+1)\theta}i\rho e^{i\theta}\,d\theta = i\rho^{-k}\int_0^{2\pi}e^{-ik\theta}\,d\theta = 0$. Now set $\psi = \varphi$, the identity map $\varphi(\lambda) = \lambda$ for every λ. Thus

$$\varphi(T) = I\tfrac{1}{2\pi i}\int_\Upsilon d\lambda + T\tfrac{1}{2\pi i}\int_\Upsilon \tfrac{1}{\lambda}\,d\lambda + \sum_{k=2}^\infty T^k \tfrac{1}{2\pi i}\int_\Upsilon \tfrac{1}{\lambda^k}\,d\lambda = O + T + O = T,$$

since $\tfrac{1}{2\pi i}\int_\Upsilon d\lambda = 0$ by the Cauchy Theorem (i.e., $\tfrac{1}{2\pi i}\int_\Upsilon d\lambda = \int_0^{2\pi}\alpha'(\theta)\,d\theta = i\rho\int_0^{2\pi}e^{i\theta}\,d\theta = 0$) and the other two integrals were computed above. Then

$$T^2 = \varphi(T)^2 = \tfrac{1}{2\pi i}\int_\Upsilon \varphi(\lambda)^2 R_T(\lambda)\,d\lambda = \tfrac{1}{2\pi i}\int_\Upsilon \lambda^2 R_T(\lambda)\,d\lambda$$

by Lemma 4.15, and so a trivial induction ensures

$$T^j = \tfrac{1}{2\pi i}\int_\Upsilon \lambda^j R_T(\lambda)\,d\lambda$$

for every integer $j \geq 0$. Therefore, by linearity of the integral,

$$p(T) = \sum_{j=0}^m \alpha_j T^j = \tfrac{1}{2\pi i}\int_\Upsilon p(\lambda)\,R_T(\lambda)\,d\lambda$$

if $p(\lambda) = \sum_{j=0}^m \alpha_j \lambda^j$. That is, if $p \in A(\sigma(T))$ is any polynomial. (b) Next we extend from polynomials (i.e., from finite power series) to infinite power series. A function $\psi \in A(\sigma(T))$ has a power series expansion $\psi(\lambda) = \sum_{k=0}^\infty \alpha_k \lambda^k$ on a neighborhood of $\sigma(T)$ if the radius of convergence of the series $\sum_{k=0}^\infty \alpha_k \lambda^k$ is greater than $r(T)$. So the polynomials $p_n(\lambda) = \sum_{k=0}^n \alpha_k \lambda^k$ make a sequence $\{p_n\}_{n=0}^\infty$ converging uniformly to ψ on an open neighborhood Λ of $\sigma(T)$ including a circle Υ about the origin with radius greater than the spectral radius $r(T)$. Thus $\sup_{\lambda \in \Lambda}|\psi(\lambda) - p_n(\lambda)| \to 0$ and so (since $\tfrac{1}{2\pi i}\int_\Upsilon R_T(\lambda)\,d\lambda = I$)

$$\|\Phi_T(p_n) - \Phi_T(\psi)\| = \|p_n(T) - \psi(T)\| = \left\|\tfrac{1}{2\pi i}\int_\Upsilon (p_n(\lambda) - \psi(\lambda))R_T(\lambda)\,d\lambda\right\|$$

$$\leq \sup_{\lambda \in \Upsilon}|p_n(\lambda) - \psi(\lambda)|\left\|\tfrac{1}{2\pi i}\int_\Upsilon R_T(\lambda)\,d\lambda\right\| = \sup_{\lambda \in \Upsilon}|p_n(\lambda) - \psi(\lambda)| \to 0$$

as $n \to \infty$, where (since $\tfrac{1}{2\pi i}\int_\Upsilon \lambda^k R_T(\lambda)\,d\lambda = T^k$)

$$\Phi_T(\psi) = \psi(T) = \tfrac{1}{2\pi i}\int_\Upsilon \psi(\lambda)\,R_T(\lambda)\,d\lambda = \tfrac{1}{2\pi i}\int_\Upsilon \left(\sum_{k=0}^\infty \alpha_k \lambda^k\right)R_T(\lambda)\,d\lambda$$

$$= \sum_{k=0}^\infty \alpha_k \left(\tfrac{1}{2\pi i}\int_\Upsilon \lambda^k R_T(\lambda)\,d\lambda\right) = \sum_{k=0}^\infty \alpha_k T^k,$$

because the integral is linear and continuous (i.e., $\int_\Upsilon \in \mathcal{B}[\,C[\Upsilon, \mathcal{B}[\mathcal{X}]], \mathcal{B}[\mathcal{X}]\,]$).
This means $\Phi_T(p_n) \to \Phi_T(\psi)$ in $\mathcal{B}[\mathcal{X}]$, where $\psi(T) = \sum_{k=0}^\infty \alpha_k T^k \in \mathcal{B}[\mathcal{X}]$ is the
uniform limit of the $\mathcal{B}[\mathcal{X}]$-valued sequence $\sum_{k=0}^n \alpha_k T^k$ (i.e., $\{\sum_{k=0}^n \alpha_k T^k\}_{n=0}^\infty$
converges in the operator norm topology). (c) Finally consider the general case
where the path $\Upsilon = \bigcup_{j=1}^m \Upsilon_j \subset \rho(T)$ is a finite union of disjoint closed rectifi-
able oriented simple curves $\Upsilon_j \subset \rho(T)$ such that $\sigma(T) \subset \mathrm{ins}\,\Upsilon \subset (\mathrm{ins}\,\Upsilon)^- \subset \Lambda$
for some open neighborhood Λ of $\sigma(T)$. Take an arbitrary $A(\sigma(T))$-valued se-
quence $\{\psi_n\}_{n=0}^\infty$ converging to $\psi \in A(\sigma(T))$ uniformly on compact subsets of
Λ (i.e., $\sup_{\lambda \in K \subset \Lambda}|\psi_n(\lambda) - \psi(\lambda)| \to 0$ as $n \to \infty$ where $\sup_{\lambda \in K \subset \Lambda}$ means sup
over all λ in every compact subset of Λ — called *compact convergence*). Thus

$$\|\Phi_T(\psi_n) - \Phi_T(\psi)\| = \tfrac{1}{2\pi i}\left\|\sum_{j=1}^m \int_{\Upsilon_j} (\psi_n(\lambda) - \psi(\lambda))R_T(\lambda)\,d\lambda\right\|$$

$$\leq \tfrac{m}{2\pi i}\sup_{\lambda \in \Upsilon}\|R_T(\lambda)\|\,\ell(\Upsilon)\sup_{\lambda \in K \subset \Lambda}|\psi_n(\lambda) - \psi(\lambda)| \to 0 \quad \text{as} \quad n \to \infty,$$

where $\sup_{\lambda \in \Upsilon_j}\|R_T(\lambda)\| \leq \sup_{\lambda \in \Upsilon}\|R_T(\lambda)\| < \infty$ and $\int_{\Upsilon_j}|d\lambda| = \ell(\Upsilon_j) \leq \ell(\Upsilon) < \infty$.
That is, $\|\Phi_T(\psi_n) - \Phi_T(\psi)\| \to 0$ whenever $\sup_{\lambda \in K \subset \Lambda}|\psi_n(\lambda) - \psi(\lambda)| \to 0$. So
$\|\psi_n - \psi\|_\infty = \sup_{\lambda \in \Lambda}|\psi_n(\lambda) - \psi(\lambda)| \to 0$ implies $\|\Phi_T(\psi_n) - \Phi_T(\psi)\| \to 0$. There-
fore $\Phi_T\colon (A(\sigma(T)), \|\cdot\|_\infty) \to (\mathcal{B}[\mathcal{X}], \|\cdot\|)$ is continuous. $\qquad\square$

Corollary 4.17. *Take any operator in $T \in \mathcal{B}[\mathcal{X}]$. If $\psi \in A(\sigma(T))$, then $\psi(T)$
commutes with every operator that commutes with T.*

Proof. This is the counterpart of Corollary 4.10. Let S be an operator in $\mathcal{B}[\mathcal{X}]$.

Claim. If $ST = TS$, then $R_T(\lambda)S = SR_T(\lambda)$ for every $\lambda \in \rho(T)$.

Proof. $S = S(\lambda I - T)(\lambda I - T)^{-1} = (\lambda I - T)S(\lambda I - T)^{-1}$ if $ST = TS$, and
so $R_T(\lambda)S = (\lambda I - T)^{-1}(\lambda I - T)S(\lambda I - T)^{-1} = SR_T(\lambda)$ for $\lambda \in \rho(T)$.

Therefore according to Definition 4.14, since S lies in $\mathcal{B}[\mathcal{X}]$,

$$S\psi(T) = S\tfrac{1}{2\pi i}\int_\Upsilon \psi(\lambda)\,R_T(\lambda)\,d\lambda = \tfrac{1}{2\pi i}\int_\Upsilon \psi(\lambda)\,SR_T(\lambda)\,d\lambda$$

$$= \tfrac{1}{2\pi i}\int_\Upsilon \psi(\lambda)\,R_T(\lambda)\,S\,d\lambda = \psi(T)\,S. \qquad\square$$

Theorem 2.7 is the Spectral Mapping Theorem for polynomials, which
holds for every Banach-space operator. For normal operators on a Hilbert
space, the Spectral Mapping Theorem was extended to larger classes of functions in Theorems 2.8, 4.11, and 4.12. Returning now to the general case of
arbitrary Banach-space operators, Theorem 2.7 can be extended to analytic
functions, and this is again referred to as the SPECTRAL MAPPING THEOREM.

Theorem 4.18. *Take an operator $T \in \mathcal{B}[\mathcal{X}]$ on a complex Banach space $\mathcal{X}$.
If ψ is analytic on the spectrum of T (i.e., if $\psi \in A(\sigma(T))$), then*

$$\sigma(\psi(T)) = \psi(\sigma(T)).$$

Proof. Take $T \in \mathcal{B}[\mathcal{X}]$ and $\psi \colon \Lambda \to \mathbb{C}$ in $A(\sigma(T))$. For an arbitrary $\lambda \in \sigma(T) \subset \Lambda$ consider the function $\phi = \phi_\lambda \colon \Lambda \to \mathbb{C}$ defined by $\phi(\zeta) = 0$ if $\zeta = \lambda$ and $\phi(\zeta) = \frac{\psi(\lambda)-\psi(\zeta)}{\lambda-\zeta}$ if $\zeta \in \Lambda\backslash\{\lambda\}$. Since the function $\psi(\lambda) - \psi(\cdot) = (\lambda - \cdot)\phi(\cdot) \colon \Lambda \to \mathbb{C}$ lies in $A(\sigma(T))$, Lemma 4.15, Theorem 4.16, and Corollary 4.17 ensure

$$\psi(\lambda)I - \psi(T) = \tfrac{1}{2\pi i}\int_\Upsilon [\psi(\lambda) - \psi(\zeta)]\, R_T(\zeta)\, d\zeta$$

$$= \tfrac{1}{2\pi i}\int_\Upsilon (\lambda - \zeta)\, \phi(\zeta)\, R_T(\zeta)\, d\zeta$$

$$= (\lambda I - T)\, \phi(T) = \phi(T)\,(\lambda I - T).$$

Suppose $\psi(\lambda) \in \rho(\psi(T))$. Then $\psi(\lambda)I - \psi(T)$ has an inverse $[\psi(\lambda)I - \psi(T)]^{-1}$ in $\mathcal{B}[\mathcal{X}]$. Since $\psi(\lambda)I - \psi(T) = \phi(T)\,(\lambda I - T)$, we get

$$[\psi(\lambda)I - \psi(T)]^{-1}\phi(T)\,(\lambda I - T) = [\psi(\lambda)I - \psi(T)]^{-1}[\psi(\lambda)I - \psi(T)] = I,$$

$$(\lambda I - T)\,\phi(T)\,[\psi(\lambda)I - \psi(T)]^{-1} = [\psi(\lambda)I - \psi(T)]\,[\psi(\lambda)I - \psi(T)]^{-1} = I.$$

Thus $\lambda I - T$ has a left and a right bounded inverse, and so it has an inverse in $\mathcal{B}[\mathcal{X}]$, which means $\lambda \in \rho(T)$. This is a contradiction. Therefore if $\lambda \in \sigma(T)$, then $\psi(\lambda) \in \sigma(\psi(T))$, and hence

$$\psi(\sigma(T)) = \big\{\psi(\lambda) \in \mathbb{C} \colon \lambda \in \sigma(T)\big\} \subseteq \sigma(\psi(T)).$$

Conversely, take an arbitrary $\zeta \notin \psi(\sigma(T))$, which means $\zeta - \psi(\lambda) \neq 0$ for every $\lambda \in \sigma(T)$, and consider the function $\phi'(\cdot) = \phi'_\zeta(\cdot) = \frac{1}{\zeta-\psi(\cdot)} \colon \Lambda' \to \mathbb{C}$, which lies in $A(\sigma(T))$ with $\Lambda' \subseteq \Lambda$. Take Υ such that $\sigma(T) \subset \mathrm{ins}\,\Upsilon \subseteq (\mathrm{ins}\,\Upsilon)^- \subset \Lambda'$ as in Definition 4.14. Then by Lemma 4.15 and Theorem 4.16

$$\phi'(T)\,[\zeta I - \psi(T)] = [\zeta I - \psi(T)]\,\phi'(T) = \tfrac{1}{2\pi i}\int_\Upsilon R_T(\lambda)\, d\lambda = I.$$

Thus $\zeta I - \psi(T)$ has a bounded inverse, $\phi'(T) \in \mathcal{B}[\mathcal{X}]$, and so $\zeta \in \rho(\psi(T))$. Equivalently, if $\zeta \in \sigma(\psi(T))$, then $\zeta \in \psi(\sigma(T))$. That is,

$$\sigma(\psi(T)) \subseteq \psi(\sigma(T)). \qquad \square$$

4.4 Riesz Idempotents and the Riesz Decomposition Theorem

A *clopen set* in a topological space is a set that is both open and closed in it. If a topological space has a nontrivial (i.e., proper and nonempty) clopen set, then (by definition) it is *disconnected*. Consider the spectrum $\sigma(T)$ of an operator $T \in \mathcal{B}[\mathcal{X}]$ on a complex Banach space $\mathcal{X}$. There are some different definitions of *spectral set*. We stick to the classical one [45, Definition VII.3.17]: A *spectral set* for T is a clopen set in $\sigma(T)$ (i.e., a subset of $\sigma(T) \subset \mathbb{C}$ which is both open and closed relative to $\sigma(T)$ — also called a *spectral subset* of $\sigma(T)$). For each spectral set $\Delta \subseteq \sigma(T)$ there is a function $\psi_\Delta \in A(\sigma(T))$ (analytic on a neighborhood of $\sigma(T)$) such that $\psi_\varnothing = 0$, $\psi_{\sigma(T)} = 1$ and, if $\varnothing \neq \Delta \neq \sigma(T)$,

$$\psi_\Delta(\lambda) = \left\{ \begin{array}{ll} 1, & \lambda \in \Delta, \\ 0, & \lambda \in \sigma(T) \backslash \Delta. \end{array} \right.$$

It does not matter which value $\psi_\Delta(\lambda)$ takes for those λ in a neighborhood of $\sigma(T)$ but not in $\sigma(T)$. If Δ is nontrivial (i.e., if $\varnothing \neq \Delta \neq \sigma(T)$), then $\sigma(T)$ must be disconnected (since $\psi_\Delta \in A(\sigma(T))$). Set $E_\Delta = E_\Delta(T) = \psi_\Delta(T)$. Thus

$$E_\Delta = \psi_\Delta(T) = \tfrac{1}{2\pi i} \int_\Upsilon \psi_\Delta(\lambda) \, R_T(\lambda) \, d\lambda = \tfrac{1}{2\pi i} \int_{\Upsilon_\Delta} R_T(\lambda) \, d\lambda$$

for any path Υ such that $\sigma(T) \subset$ ins Υ and (ins $\Upsilon)^-$ is included in an open neighborhood of $\sigma(T)$ as in Definition 4.14, and Υ_Δ is any path for which $\Delta \subseteq$ ins Υ_Δ and $\sigma(T) \backslash \Delta \subseteq$ out Υ_Δ. The operator $E_\Delta \in \mathcal{B}[\mathcal{X}]$ is referred to as the *Riesz idempotent* associated with Δ. In particular, the Riesz idempotent associated with an isolated point λ_0 of the spectrum of T, namely $E_{\{\lambda_0\}} = E_{\{\lambda_0\}}(T) = \psi_{\{\lambda_0\}}(T)$, will be denoted by

$$E_{\lambda_0} = \tfrac{1}{2\pi i} \int_{\Upsilon_{\lambda_0}} R_T(\lambda) \, d\lambda = \tfrac{1}{2\pi i} \int_{\Upsilon_{\lambda_0}} (\lambda I - T)^{-1} \, d\lambda,$$

where Υ_{λ_0} is any simple closed rectifiable positively oriented curve (e.g., any positively oriented circle) enclosing λ_0 but no other point of $\sigma(T)$.

Take an arbitrary operator T in $\mathcal{B}[\mathcal{X}]$. The collection of all spectral subsets of $\sigma(T)$ forms a (Boolean) algebra of subsets of $\sigma(T)$. Indeed, the empty set and whole set $\sigma(T)$ are spectral sets, complements relative to $\sigma(T)$ of spectral sets remain spectral sets, and finite unions of spectral sets are spectral sets, as are differences and intersections of spectral sets.

Lemma 4.19. *Let T be any operator in $\mathcal{B}[\mathcal{X}]$ and let $E_\Delta \in \mathcal{B}[\mathcal{X}]$ be the Riesz idempotent associated with an arbitrary spectral subset Δ of $\sigma(T)$.*

(a) *E_Δ is a projection with the following properties:*

(a_1) *$E_\varnothing = O$, $E_{\sigma(T)} = I$ and $E_{\sigma(T) \backslash \Delta} = I - E_\Delta$.*

If Δ_1 and Δ_2 are spectral subsets of $\sigma(T)$, then

(a_2) *$E_{\Delta_1 \cap \Delta_2} = E_{\Delta_1} E_{\Delta_2}$ and $E_{\Delta_1 \cup \Delta_2} = E_{\Delta_1} + E_{\Delta_2} - E_{\Delta_1} E_{\Delta_2}$.*

(b) *E_Δ commutes with every operator that commutes with T.*

(c) *The range $\mathcal{R}(E_\Delta)$ of E_Δ is a subspace of $\mathcal{X}$ which is S-invariant for every operator $S \in \mathcal{B}[\mathcal{X}]$ that commutes with T.*

Proof. (a) According to Lemma 4.15 a Riesz idempotent deserves its name: $E_\Delta = E_\Delta^2$ and so $E_\Delta \in \mathcal{B}[\mathcal{X}]$ is a projection. By the definition of the function ψ_Δ for an arbitrary spectral subset Δ of $\sigma(T)$ we get

(i) $\psi_\varnothing = 0,$ $\psi_{\sigma(T)} = 1,$ $\psi_{\sigma(T)\setminus\Delta} = 1 - \psi_\Delta,$

(ii) $\psi_{\Delta_1 \cap \Delta_2} = \psi_{\Delta_1}\psi_{\Delta_2}$ and $\psi_{\Delta_1 \cup \Delta_2} = \psi_{\Delta_1} + \psi_{\Delta_2} - \psi_{\Delta_1}\psi_{\Delta_2}.$

Since $E_\Delta = \psi_\Delta(T) = \frac{1}{2\pi i}\int_\Upsilon \psi_\Delta(\lambda)\,R_T(\lambda)\,d\lambda$ for every spectral subset Δ of $\sigma(T)$, where Υ is any path as in Definition 4.14, the identities in (a$_1$) and (a$_2$) follow from the identities in (i) and (ii), respectively (using the linearity of the integral, Lemma 4.15, and Theorem 4.16).

(b) Take any $S \in \mathcal{B}[\mathcal{X}]$. According to Corollary 4.17,

$$ST = TS \quad \text{implies} \quad SE_\Delta = E_\Delta S.$$

(c) Thus $\mathcal{R}(E_\Delta)$ is S-invariant. Moreover, $\mathcal{R}(E_\Delta) = \mathcal{N}(I - E_\Delta)$ because E_Δ is a projection. Since E_Δ is bounded, so is $I - E_\Delta$. Thus $\mathcal{N}(I - E_\Delta)$ is closed. Then $\mathcal{R}(E_\Delta)$ is a subspace (i.e., $\mathcal{R}(E_\Delta)$ is a *closed* linear manifold) of $\mathcal{X}$. $\square$

Theorem 4.20. *Take $T \in \mathcal{B}[\mathcal{X}]$. If Δ is any spectral subset of $\sigma(T)$, then*

$$\sigma(T|_{\mathcal{R}(E_\Delta)}) = \Delta.$$

Moreover, the map $\Delta \mapsto E_\Delta$ that assigns to each spectral subset of $\sigma(T)$ the associated Riesz idempotent $E_\Delta \in \mathcal{B}[\mathcal{X}]$ is injective.

Proof. Consider the results in Lemma 4.19(a$_1$). If $\Delta = \sigma(T)$, then $E_\Delta = I$ so that $\mathcal{R}(E_\Delta) = \mathcal{X}$, and hence $\sigma(T|_{\mathcal{R}(E_\Delta)}) = \sigma(T)$. On the other hand, if $\Delta = \varnothing$, then $E_\Delta = O$ so that $\mathcal{R}(E_\Delta) = \{0\}$, and hence $T|_{\mathcal{R}(E_\Delta)} \colon \{0\} \to \{0\}$ is the null operator on the zero space, which implies $\sigma(T|_{\mathcal{R}(E_\Delta)}) = \varnothing$. In both cases the identity $\sigma(T|_{\mathcal{R}(E_\Delta)}) = \Delta$ holds trivially. (Recall: the spectrum of a *bounded* linear operator is nonempty on every *nonzero* complex Banach space.) Thus suppose $\varnothing \neq \Delta \neq \sigma(T)$. Let $\psi_\Delta \subset \Lambda(\sigma(T))$ be the function defining the Riesz idempotent E_Δ associated with Δ,

$$\psi_\Delta(\lambda) = \begin{cases} 1, & \lambda \in \Delta, \\ 0, & \lambda \in \sigma(T)\setminus\Delta, \end{cases}$$

and so

$$E_\Delta = \psi_\Delta(T) = \frac{1}{2\pi i}\int_\Upsilon \psi_\Delta(\lambda)\,R_T(\lambda)\,d\lambda$$

for any path Υ as in Definition 4.14. If $\varphi \in A(\sigma(T))$ is the identity map,

$$\varphi(\lambda) = \lambda \quad \text{for every } \lambda \in \sigma(T),$$

then (cf. Theorem 4.16)

$$T = \varphi(T) = \frac{1}{2\pi i}\int_\Upsilon \varphi(\lambda)\,R_T(\lambda)\,d\lambda.$$

Claim 1. $\sigma(TE_\Delta) = \Delta \cup \{0\}.$

Proof. Since $TE_\Delta = \varphi(T)\psi_\Delta(T)$ and

$$\varphi(T)\psi_\Delta(T) = \tfrac{1}{2\pi i} \int_\Upsilon \varphi(\lambda)\psi_\Delta(\lambda)\, R_T(\lambda)\, d\lambda = \tfrac{1}{2\pi i} \int_\Upsilon (\varphi\psi_\Delta)(\lambda)\, R_T(\lambda)\, d\lambda,$$

and since $\varphi\psi_\Delta \in A(\sigma(T))$, the Spectral Mapping Theorem leads to

$$\sigma(TE_\Delta) = (\varphi\psi_\Delta)(\sigma(T)) = \varphi(\sigma(T))\,\psi_\Delta(\sigma(T)) = \sigma(T)\,\psi_\Delta(\sigma(T)) = \Delta \cup \{0\}$$

(cf. Lemma 4.15 and Theorems 4.16 and 4.18).

Claim 2. $\sigma(TE_\Delta) = \sigma(T|_{\mathcal{R}(E_\Delta)}) \cup \{0\}$.

Proof. Since E_Δ is a projection, $\mathcal{R}(E_\Delta)$ and $\mathcal{N}(E_\Delta)$ are algebraic complements, which means $\mathcal{R}(E_\Delta) \cap \mathcal{N}(E_\Delta) = \varnothing$ and $\mathcal{X} = \mathcal{R}(E_\Delta) \oplus \mathcal{N}(E_\Delta)$ — plain direct sum (no orthogonality in a pure Banach space setup). Thus, with respect to the complementary decomposition $\mathcal{X} = \mathcal{R}(E_\Delta) \oplus \mathcal{N}(E_\Delta)$, we get $E_\Delta = \left(\begin{smallmatrix} I & O \\ O & O \end{smallmatrix}\right) = I \oplus O$, and so $TE_\Delta = \left(\begin{smallmatrix} T|_{\mathcal{R}(E_\Delta)} & O \\ O & O \end{smallmatrix}\right) = T|_{\mathcal{R}(E_\Delta)} \oplus O$ because $E_\Delta T = TE_\Delta$. Then $\lambda I - TE_\Delta = (\lambda I - T|_{\mathcal{R}(E_\Delta)}) \oplus \lambda I$ for every $\lambda \in \mathbb{C}$, and hence $\lambda \in \rho(E_\Delta T)$ if and only if $\lambda \in \rho(T|_{\mathcal{R}(E_\Delta)})$ and $\lambda \neq 0$. Equivalently, $\lambda \in \sigma(E_\Delta T)$ if and only if $\lambda \in \sigma(T|_{\mathcal{R}(E_\Delta)})$ or $\lambda = 0$.

Claim 3. $\sigma(T|_{\mathcal{R}(E_\Delta)}) \cup \{0\} = \Delta \cup \{0\}$.

Proof. Claims 1 and 2.

Claim 4. $0 \in \Delta \iff 0 \in \sigma(T|_{\mathcal{R}(E_\Delta)})$.

Proof. Suppose $0 \in \Delta$. Thus there exists a function $\phi_0 \in A(\sigma(T))$ for which

$$\phi_0(\lambda) = \begin{cases} 0, & \lambda \in \Delta, \\ \lambda^{-1}, & \lambda \in \sigma(T) \backslash \Delta. \end{cases}$$

Observe that $\varphi\phi_0$ lies in $A(\sigma(T))$ and

$$\varphi\phi_0 = 1 - \psi_\Delta \quad \text{on} \quad \sigma(T),$$

and hence, since $\varphi(T) = T$ and $\phi_0(T) \in \mathcal{B}[\mathcal{X}]$,

$$T\phi_0(T) = \phi_0(T)T = I - \psi_\Delta(T) = I - E_\Delta.$$

If $0 \notin \sigma(T|_{\mathcal{R}(E_\Delta)})$, then $T|_{\mathcal{R}(E_\Delta)}$ has a bounded inverse on the Banach space $\mathcal{R}(E_\Delta)$. Therefore there exists an operator $[T|_{\mathcal{R}(E_\Delta)}]^{-1}$ on $\mathcal{R}(E_\Delta)$ such that $T|_{\mathcal{R}(E_\Delta)}[T|_{\mathcal{R}(E_\Delta)}]^{-1} = [T|_{\mathcal{R}(E_\Delta)}]^{-1}T|_{\mathcal{R}(E_\Delta)} = I$. Now consider the operator $[T|_{\mathcal{R}(E_\Delta)}]^{-1}E_\Delta$ in $\mathcal{B}[\mathcal{X}]$. Since $TE_\Delta = \varphi(T)\psi_\Delta(T) = \psi_\Delta(T)\varphi(T) = E_\Delta T$,

$$T\Big([T|_{\mathcal{R}(E_\Delta)}]^{-1}E_\Delta + \phi_0(T)\Big) = \Big([T|_{\mathcal{R}(E_\Delta)}]^{-1}E_\Delta + \phi_0(T)\Big)T = I \in \mathcal{B}[\mathcal{X}].$$

Then T has a bounded inverse on $\mathcal{X}$, namely $[T|_{\mathcal{R}(E_\Delta)}]^{-1}E_\Delta + \phi_0(T)$, and so $0 \in \rho(T)$, which is a contradiction (since $0 \in \Delta \subseteq \sigma(T)$). Thus

$$0 \in \Delta \quad \text{implies} \quad 0 \in \sigma(T|_{\mathcal{R}(E_\Delta)}).$$

Conversely, suppose $0 \notin \Delta$. Then there is a function $\phi_1 \in A(\sigma(T))$ for which

$$\phi_1(\lambda) = \begin{cases} \lambda^{-1}, & \lambda \in \Delta, \\ 0, & \lambda \in \sigma(T)\backslash\Delta. \end{cases}$$

Again, observe that $\varphi\phi_1$ lies in $A(\sigma(T))$ and

$$\varphi\phi_1 = \psi_\Delta \quad \text{on} \quad \sigma(T),$$

and hence, since $\varphi(T) = T$ and $\phi_1(T) \in \mathcal{B}[\mathcal{X}]$,

$$T\phi_1(T) = \phi_1(T)T = \psi_\Delta(T) = E_\Delta.$$

Since $\phi_1(T)\psi_\Delta(T) = \psi_\Delta(T)\phi_1(T)$ by Lemma 4.15, the operator $\phi_1(T)$ is $\mathcal{R}(E_\Delta)$-invariant. Hence $\phi_1(T)|_{\mathcal{R}(E_\Delta)}$ lies in $\mathcal{B}[\mathcal{R}(E_\Delta)]$ and is such that

$$T|_{\mathcal{R}(E_\Delta)}\phi_1(T)|_{\mathcal{R}(E_\Delta)} = \phi_1(T)|_{\mathcal{R}(E_\Delta)}T|_{\mathcal{R}(E_\Delta)} = E_\Delta|_{\mathcal{R}(E_\Delta)} = I \in \mathcal{B}[\mathcal{R}(E_\Delta)].$$

Then $T|_{\mathcal{R}(E_\Delta)}$ has a bounded inverse on $\mathcal{R}(E_\Delta)$, namely $\phi_1(T)|_{\mathcal{R}(E_\Delta)}$, and so $0 \in \rho(T|_{\mathcal{R}(E_\Delta)})$. Thus $0 \notin \Delta$ implies $0 \notin \sigma(T|_{\mathcal{R}(E_\Delta)})$. Equivalently,

$$0 \in \sigma(T|_{\mathcal{R}(E_\Delta)}) \quad \text{implies} \quad 0 \in \Delta.$$

This concludes the proof of Claim 4.

Claims 3 and 4 ensure the identity

$$\sigma(T|_{\mathcal{R}(E_\Delta)}) = \Delta.$$

Finally, consider the map $\Delta \mapsto E_\Delta$ assigning to each spectral subset Δ of $\sigma(T)$ the associated Riesz idempotent $E_\Delta \in \mathcal{B}[\mathcal{X}]$. We have already seen at the beginning of this proof that if $E_\Delta = O$, then $\sigma(T|_{\mathcal{R}(E_\Delta)}) = \varnothing$. Hence

$$E_\Delta = O \implies \Delta = \varnothing$$

by the above displayed identity (which holds for all spectral subsets of $\sigma(T)$). Take arbitrary spectral subsets Δ_1 and Δ_2 of $\sigma(T)$. From Lemma 4.19(a),

$$E_{\Delta_1\backslash\Delta_2} = E_{\Delta_1\cap(\sigma(T)\backslash\Delta_2)} = E_{\Delta_1}(I - E_{\Delta_2}) = E_{\Delta_1} - E_{\Delta_1}E_{\Delta_2},$$

where E_{Δ_1} and E_{Δ_2} commute since $E_{\Delta_1}E_{\Delta_2} = E_{\Delta_1\cap\Delta_2} = E_{\Delta_2}E_{\Delta_1}$, and so

$$(E_{\Delta_1} - E_{\Delta_2})^2 = E_{\Delta_1} + E_{\Delta_2} - 2E_{\Delta_1}E_{\Delta_2} = E_{\Delta_1\backslash\Delta_2} + E_{\Delta_2\backslash\Delta_1} = E_{\Delta_1\triangledown\Delta_2},$$

where $\Delta_1\triangledown\Delta_2 = (\Delta_1\backslash\Delta_2) \cup (\Delta_2\backslash\Delta_1)$ is the *symmetric difference*. Thus

$$E_{\Delta_1} = E_{\Delta_2} \implies E_{\Delta_1\triangledown\Delta_2} = O \implies \Delta_1\triangledown\Delta_2 = \varnothing \implies \Delta_1 = \Delta_2,$$

since $\Delta_1\triangledown\Delta_2 = \varnothing$ if and only if $\Delta_1 = \Delta_2$, which proves injectivity. $\qquad\square$

Remark. Let T be an operator in $\mathcal{B}[\mathcal{X}]$ and suppose $\sigma(T)$ is disconnected, so there are nontrivial sets in the algebra of all spectral subsets of $\sigma(T)$. We show that to each partition of $\sigma(T)$ into spectral sets there is a corresponding decomposition of the resolvent function R_T. Indeed, consider a finite partition $\{\varDelta_j\}_{j=1}^n$ of nontrivial spectral subsets of $\sigma(T)$; that is,

$$\varnothing \neq \varDelta_j \neq \sigma(T) \text{ are clopen subsets of } \sigma(T) \text{ for each } j = 1, \ldots, n$$

such that

$$\bigcup_{j=1}^n \varDelta_j = \sigma(T) \quad \text{with} \quad \varDelta_j \cap \varDelta_k = \varnothing \text{ whenever } j \neq k.$$

Consider the Riesz idempotent associated with each $\varDelta_j$, namely

$$E_{\varDelta_j} = \tfrac{1}{2\pi i} \int_{\varUpsilon_{\varDelta_j}} R_T(\lambda) \, d\lambda.$$

The results of Lemma 4.19 are readily extended for any integer $n \geq 2$. Thus each $E_{\varDelta_j}$ is a nontrivial projection $(O \neq E_{\varDelta_j} = E_{\varDelta_j}^2 \neq I)$ and

$$\sum_{j=1}^n E_{\varDelta_j} = I \quad \text{with} \quad E_{\varDelta_j} E_{\varDelta_k} = O \text{ whenever } j \neq k.$$

Take an arbitrary $\lambda \in \rho(T)$. For each $j = 1, \ldots, n$ set

$$T_j = TE_{\varDelta_j} \quad \text{and} \quad R_j(\lambda) = R_T(\lambda)E_{\varDelta_j}$$

in $\mathcal{B}[\mathcal{X}]$, and so

$$\sum_{j=1}^n T_j = T \quad \text{and} \quad \sum_{j=1}^n R_j(\lambda) = R_T(\lambda).$$

These operators commute. Indeed, $TR_T(\lambda) = R_T(\lambda)T$ and $TE_{\varDelta_j} = E_{\varDelta_j}T$, and hence $R_T(\lambda)E_{\varDelta_j} = E_{\varDelta_j}R_T(\lambda)$ (cf. the claim in the proof of Corollary 4.17 and Lemma 4.19(b)). Therefore, if $j \neq k$, then

$$T_j T_k = R_j(\lambda)R_k(\lambda) = T_k R_j(\lambda) = R_j(\lambda)T_k = O.$$

On the other hand, for every $j = 1, \ldots, n$,

$$T_j R_j(\lambda) = R_j(\lambda)T_j = TR_T(\lambda)E_{\varDelta_j}$$

and, since $(\lambda I - T)R_T(\lambda) = R_T(\lambda)(\lambda I - T) = I$,

$$\big(\lambda E_{\varDelta_j} - T_j\big) R_j(\lambda) = R_j(\lambda)\big(\lambda E_{\varDelta_j} - T_j\big) = E_{\varDelta_j}.$$

Moreover, by Claim 1 in the proof of Theorem 4.20,

$$\sigma(T_j) = \varDelta_j \cup \{0\}.$$

Clearly, the spectral radii are such that $r(T_j) \leq r(T)$. Since $T_j^k = T^k E_{\Delta_j}$ for every nonnegative integer k, we get (cf. Corollary 2.12)

$$R_j(\lambda) = (\lambda I - T)^{-1} E_{\Delta_j} = \frac{1}{\lambda} \sum_{k=0}^{\infty} \left(\frac{T}{\lambda}\right)^k E_{\Delta_j} = \frac{1}{\lambda} \sum_{k=0}^{\infty} \left(\frac{T_j}{\lambda}\right)^k$$

if λ in $\rho(T)$ is such that $r(T) < |\lambda|$ and, again by Corollary 2.12,

$$R_{T_j}(\lambda) = (\lambda I - T E_{\Delta_j})^{-1} = \frac{1}{\lambda} \sum_{k=0}^{\infty} \left(\frac{T E_{\Delta_j}}{\lambda}\right)^k = \frac{1}{\lambda} \sum_{k=0}^{\infty} \left(\frac{T_j}{\lambda}\right)^k$$

if λ in $\rho(T_j)$ is such that $r(T_j) < |\lambda|$, where the above series converge uniformly in $\mathcal{B}[\mathcal{X}]$. Hence if $r(T) < |\lambda|$, then

$$R_j(\lambda) = R_T(\lambda) E_{\Delta_j} = R_{T E_{\Delta_j}}(\lambda) = R_{T_j}(\lambda)$$

where each R_j can actually be extended to be analytic in $\rho(T_j)$.

The properties of Riesz idempotents in Lemma 4.19 and Theorem 4.20 lead to a major result in operator theory which reads as follows. *Operators whose spectra are disconnected have nontrivial hyperinvariant subspaces.* (For the original statement of the Riesz Decomposition Theorem see [100, p. 421].)

Corollary 4.21. (RIESZ DECOMPOSITION THEOREM). *Let $T \in \mathcal{B}[\mathcal{X}]$ be an operator on a complex Banach space $\mathcal{X}$. If $\sigma(T) = \Delta_1 \cup \Delta_2$, where Δ_1 and Δ_2 are disjoint nonempty closed sets, then T has a complementary pair $\{\mathcal{M}_1, \mathcal{M}_2\}$ of nontrivial hyperinvariant subspaces, viz., $\mathcal{M}_1 = \mathcal{R}(E_{\Delta_1})$ and $\mathcal{M}_2 = \mathcal{R}(E_{\Delta_2})$, for which $\sigma(T|_{\mathcal{M}_1}) = \Delta_1$ and $\sigma(T|_{\mathcal{M}_2}) = \Delta_2$.*

Proof. If Δ_1 and Δ_2 are disjoint closed sets in $\mathbb{C}$ and $\sigma(T) = \Delta_1 \cup \Delta_2$, then they are both clopen subsets of $\sigma(T)$ (i.e., spectral subsets of $\sigma(T)$), and hence the Riesz idempotents E_{Δ_1} and E_{Δ_2} associated with them are such that their ranges $\mathcal{M}_1 = \mathcal{R}(E_{\Delta_1})$ and $\mathcal{M}_2 = \mathcal{R}(E_{\Delta_2})$ are subspaces of $\mathcal{X}$ which are hyperinvariant for T by Lemma 4.19(c). Since Δ_1 and Δ_2 are also nontrivial (i.e., $\varnothing \neq \Delta_1 \neq \sigma(T)$ and $\varnothing \neq \Delta_2 \neq \sigma(T)$), the spectrum $\sigma(T)$ is disconnected (because they are both clopen subsets of $\sigma(T)$) and the projections E_{Δ_1} and E_{Δ_2} are nontrivial (i.e., $O \neq E_{\Delta_1} \neq I$ and $O \neq E_{\Delta_2} \neq I$ by the injectivity of Theorem 4.20), and so the subspaces $\mathcal{M}_1$ and $\mathcal{M}_2$ are also nontrivial (i.e., $\{0\} \neq \mathcal{R}(E_{\Delta_1}) \neq \mathcal{X}$ and $\{0\} \neq \mathcal{R}(E_{\Delta_2}) \neq \mathcal{X}$). Thus by Lemma 4.19(a) we get $E_{\Delta_1} + E_{\Delta_2} = I$ and $E_{\Delta_1} E_{\Delta_2} = O$, which means the operators E_{Δ_1} and E_{Δ_2} are complementary projections (not necessarily orthogonal even if $\mathcal{X}$ were a Hilbert space), and so their ranges are complementary subspaces (i.e., $\mathcal{M}_1 + \mathcal{M}_2 = \mathcal{X}$ and $\mathcal{M}_1 \cap \mathcal{M}_2 = \{0\}$) as in Section 1.4. Finally, according to Theorem 4.20, $\sigma(T|_{\mathcal{M}_1}) = \Delta_1$ and $\sigma(T|_{\mathcal{M}_2}) = \Delta_2$. $\square$

Remark. Since $\mathcal{M}_1 = \mathcal{R}(E_{\Delta_1})$ and $\mathcal{M}_2 = \mathcal{R}(E_{\Delta_2})$ are complementary subspaces (i.e., $\mathcal{M}_1 + \mathcal{M}_2 = \mathcal{X}$ and $\mathcal{M}_1 \cap \mathcal{M}_2 = \{0\}$), $\mathcal{X}$ can be identified with the direct sum $\mathcal{M}_1 \oplus \mathcal{M}_2$ (not necessarily an orthogonal direct sum even if $\mathcal{X}$

were a Hilbert space), which means there exists an isomorphism (the natural one) $\Psi\colon \mathcal{X}\to \mathcal{M}_1\oplus \mathcal{M}_2$ between the normed spaces $\mathcal{X}$ and $\mathcal{M}_1\oplus \mathcal{M}_2$ (see, e.g., [78, Theorem 2.14]). The normed space $\mathcal{M}_1\oplus \mathcal{M}_2$ is not necessarily complete and the invertible linear transformation Ψ is not necessarily bounded. In other words, $\mathcal{X}\sim \mathcal{M}_1\oplus \mathcal{M}_2$, where $\sim$ stands for *algebraic similarity* (i.e., isomorphic equivalence) and $\oplus$ stands for direct (not necessarily orthogonal) sum. Thus, since $\mathcal{M}_1$ and $\mathcal{M}_2$ are both T-invariant (recall: both E_{Δ_1} and E_{Δ_2} commute with T by Lemma 4.19), it follows that $T\sim T|_{\mathcal{M}_1}\oplus T|_{\mathcal{M}_2}$, which means $\Psi T\Psi^{-1}=T|_{\mathcal{M}_1}\oplus T|_{\mathcal{M}_2}$. (If $\mathcal{X}$ were Hilbert and the subspaces orthogonal, then we might say they would reduce T.)

A pair Δ_1 and Δ_2 of subsets of $\sigma(T)$ are *complementary spectral sets* for T if, besides being spectral sets (i.e., subsets of $\sigma(T)$ which are both open and closed relative to $\sigma(T)$) they also form a nontrivial partition of $\sigma(T)$; that is, $\varnothing\neq \Delta_1\neq \sigma(T)$, $\varnothing\neq \Delta_2\neq \sigma(T)$, $\Delta_1\cup \Delta_2=\sigma(T)$, and $\Delta_1\cap \Delta_2=\varnothing$. Thus by the Riesz Decomposition Theorem, *for every pair of complementary spectral sets, the ranges of their Riesz idempotents are complementary nontrivial hyperinvariant subspaces for T, and the spectra of the restrictions of T to those ranges coincide with themselves.*

Corollary 4.22. *Let Δ_1 and Δ_2 be complementary spectral sets for an operator $T\in \mathcal{B}[\mathcal{X}]$, and let E_{Δ_1} and E_{Δ_2} be the Riesz idempotents associated with them. If $\lambda\in \Delta_1$, then*

$$E_{\Delta_2}(\mathcal{N}(\lambda I-T))=\{0\},$$

$$E_{\Delta_2}(\mathcal{R}(\lambda I-T))=\mathcal{R}(E_{\Delta_2}),$$

$$E_{\Delta_1}(\mathcal{N}(\lambda I-T))=\mathcal{N}(\lambda I-T)\subseteq \mathcal{R}(E_{\Delta_1}),$$

$$E_{\Delta_1}(\mathcal{R}(\lambda I-T))=\mathcal{R}((\lambda I-T)E_{\Delta_1})\subseteq \mathcal{R}(E_{\Delta_1}),$$

so that

$$\mathcal{R}(\lambda I-T)=E_{\Delta_1}(\mathcal{R}(\lambda I-T))+\mathcal{R}(E_{\Delta_2}).$$

If $\dim \mathcal{R}(E_{\Delta_1})<\infty$, then $\dim \mathcal{N}(\lambda I-T)<\infty$ and $\mathcal{R}(\lambda I-T)$ is closed.

Proof. Let T be an operator on a complex Banach space $\mathcal{X}$.

Claim 1. If Δ_1 and Δ_2 are complementary spectral sets and $\lambda\in \mathbb{C}$, then

(a) $\mathcal{N}(\lambda I-T)=E_{\Delta_1}(\mathcal{N}(\lambda I-T))+E_{\Delta_2}(\mathcal{N}(\lambda I-T)),$

(b) $\mathcal{R}(\lambda I-T)=E_{\Delta_1}(\mathcal{R}(\lambda I-T))+E_{\Delta_2}(\mathcal{R}(\lambda I-T)).$

Proof. Let E_{Δ_1} and E_{Δ_2} be the Riesz idempotents associated with Δ_1 and Δ_2. By Lemma 4.19(a) $\mathcal{R}(E_{\Delta_1})$ and $\mathcal{R}(E_{\Delta_2})$ are complementary subspaces (i.e., $\mathcal{X}=\mathcal{R}(E_{\Delta_1})+\mathcal{R}(E_{\Delta_2})$ and $\mathcal{R}(E_{\Delta_1})\cap \mathcal{R}(E_{\Delta_2})=\varnothing$, since Δ_1 and Δ_2 are complementary spectral sets, and so E_{Δ_1} and E_{Δ_2} are complementary projections). Thus (cf. Section 1.1) there is a unique decomposition

$$x = u + v = E_{\Delta_1}x + E_{\Delta_2}x$$

for every vector $x \in \mathcal{X}$, with $u \in \mathcal{R}(E_{\Delta_1})$ and $v \in \mathcal{R}(E_{\Delta_2})$, where $u = E_{\Delta_1}x$ and $v = E_{\Delta_2}x$ (since $E_{\Delta_1}(\mathcal{R}(E_{\Delta_2})) = \{0\}$ and $E_{\Delta_2}(\mathcal{R}(E_{\Delta_1})) = \{0\}$ because $E_{\Delta_1}E_{\Delta_2} = E_{\Delta_2}E_{\Delta_1} = O$). So we get the decompositions in (a) and (b).

Claim 2. If $\lambda \in \sigma(T)\backslash\Delta$ for some spectral set Δ of $\sigma(T)$, then

$$E_\Delta(\mathcal{N}(\lambda I - T)) = \{0\} \quad \text{and} \quad E_\Delta(\mathcal{R}(\lambda I - T)) = \mathcal{R}(E_\Delta).$$

Proof. Let Δ be any spectral set of $\sigma(T)$ and let E_Δ be the Riesz idempotent associated with it. Since $\mathcal{R}(E_\Delta)$ is an invariant subspace for T (Lemma 4.19), take the restriction $T|_{\mathcal{R}(E_\Delta)} \in \mathcal{B}[\mathcal{R}(E_\Delta)]$ of T to the Banach space $\mathcal{R}(E_\Delta)$. If $\lambda \in \sigma(T)\backslash\Delta$, then $\lambda \in \rho(T|_{\mathcal{R}(E_\Delta)})$. So $(\lambda I - T)|_{\mathcal{R}(E_\Delta)} = \lambda I|_{\mathcal{R}(E_\Delta)} - T|_{\mathcal{R}(E_\Delta)}$ has a bounded inverse, where $I|_{\mathcal{R}(E_\Delta)} = E_\Delta|_{\mathcal{R}(E_\Delta)}$ stands for the identity on $\mathcal{R}(E_\Delta)$. Thus, since $\mathcal{R}(E_\Delta)$ is T-invariant,

$$
\begin{aligned}
R_{T|_{\mathcal{R}(E_\Delta)}}(\lambda)\,&(\lambda I - T)E_\Delta \\
&= (\lambda I|_{\mathcal{R}(E_\Delta)} - T|_{\mathcal{R}(E_\Delta)})^{-1}(\lambda I|_{\mathcal{R}(E_\Delta)} - T|_{\mathcal{R}(E_\Delta)})E_\Delta \\
&= I|_{\mathcal{R}(E_\Delta)}E_\Delta = E_\Delta,
\end{aligned}
$$

where $R_{T|_{\mathcal{R}(E_\Delta)}} \colon \rho(T|_{\mathcal{R}(E_\Delta)}) \to \mathcal{G}[\mathcal{R}(E_\Delta)]$ is the resolvent function of $T|_{\mathcal{R}(E_\Delta)}$. Now take an arbitrary $x \in \mathcal{N}(\lambda I - T)$. Since $E_\Delta T = T E_\Delta$,

$$E_\Delta x = R_{T|_{\mathcal{R}(E_\Delta)}}(\lambda)\,(\lambda I - T)E_\Delta x = R_{T|_{\mathcal{R}(E_\Delta)}}(\lambda)\,E_\Delta(\lambda I - T)x = 0.$$

Therefore,

$$E_\Delta(\mathcal{N}(\lambda I - T)) = \{0\}.$$

Moreover, since

$$(\lambda I - T)E_\Delta = \lambda E_\Delta - T E_\Delta = \lambda I|_{\mathcal{R}(E_\Delta)} - T|_{\mathcal{R}(E_\Delta)},$$

it follows that if $\lambda \in \sigma(T)\backslash\Delta$, then $\lambda \in \rho(T|_{\mathcal{R}(E_\Delta)})$ by Theorem 4.20 (where $T|_{\mathcal{R}(E_\Delta)} \in \mathcal{B}[\mathcal{R}(E_\Delta)]$), and so (see diagram of Section 2.2)

$$\mathcal{R}((\lambda I - T)E_\Delta) = \mathcal{R}(\lambda I|_{\mathcal{R}(E_\Delta)} - T|_{\mathcal{R}(E_\Delta)}) = \mathcal{R}(E_\Delta).$$

Since $T E_\Delta = E_\Delta T$, and $A(\mathcal{R}(B)) = \mathcal{R}(AB)$ for all operators A and B,

$$E_\Delta(\mathcal{R}(\lambda I - T)) = \mathcal{R}(E_\Delta(\lambda I - T)) = \mathcal{R}((\lambda I - T)E_\Delta) = \mathcal{R}(E_\Delta).$$

This concludes the proof of Claim 2.

From now on suppose $\lambda \in \Delta_1$. In this case

$$E_{\Delta_2}(\mathcal{N}(\lambda I - T)) = \{0\}$$

according to Claim 2, and so we get by Claim 1(a)

$$\mathcal{N}(\lambda I - T) = E_{\Delta_1}(\mathcal{N}(\lambda I - T)) \subseteq \mathcal{R}(E_{\Delta_1}).$$

Hence

$$\dim \mathcal{R}(E_{\Delta_1}) < \infty \quad \Longrightarrow \quad \dim \mathcal{N}(\lambda I - T) < \infty.$$

Again, according to Claim 2 we get

$$E_{\Delta_2}(\mathcal{R}(\lambda I - T)) = \mathcal{R}(E_{\Delta_2}).$$

Thus, from Claim 1(b),

$$\mathcal{R}(\lambda I - T) = E_{\Delta_1}(\mathcal{R}(\lambda I - T)) + \mathcal{R}(E_{\Delta_2}),$$

where (since $TE_\Delta = E_\Delta T$ and $A(\mathcal{R}(B)) = \mathcal{R}(AB)$ for all A and B),

$$E_{\Delta_1}(\mathcal{R}(\lambda I - T)) = \mathcal{R}(E_{\Delta_1}(\lambda I - T)) = \mathcal{R}((\lambda I - T)E_{\Delta_1}) \subseteq \mathcal{R}(E_{\Delta_1}).$$

But $\mathcal{R}(E_{\Delta_2})$ is a subspace (i.e., a closed linear manifold) of $\mathcal{X}$ (Lemma 4.19). If $\mathcal{R}(E_{\Delta_1})$ is finite-dimensional, then $E_{\Delta_1}(\mathcal{R}(\lambda I - T))$ is finite-dimensional, and so $\mathcal{R}(\lambda I - T)$ is the sum of a finite-dimensional linear manifold and a closed linear manifold, which is closed (cf. Proposition 1.C). Hence,

$$\dim \mathcal{R}(E_{\Delta_1}) < \infty \quad \Longrightarrow \quad \mathcal{R}(\lambda I - T) \text{ is closed.} \qquad \square$$

Remark. Let λ_0 be an isolated point of the spectrum of an operator T in $\mathcal{B}[\mathcal{X}]$. A particular case of the preceding corollary for $\Delta_1 = \{\lambda_0\}$ says: if $\dim \mathcal{R}(E_{\lambda_0}) < \infty$, then $\dim \mathcal{N}(\lambda_0 I - T) < \infty$ and $\mathcal{R}(\lambda_0 I - T)$ is closed. We will prove the converse in Theorem 5.19 (next chapter):

$$\dim \mathcal{R}(E_{\lambda_0}) < \infty \quad \Longleftrightarrow \quad \dim \mathcal{N}(\lambda_0 I - T) < \infty \text{ and } \mathcal{R}(\lambda_0 I - T) \text{ is closed.}$$

Moreover, $\lambda_0 I - T|_{\mathcal{R}(E_{\lambda_0})}$ is a quasinilpotent operator on $\mathcal{R}(E_{\lambda_0})$. Indeed, $\sigma(T|_{\mathcal{R}(E_{\lambda_0})}) = \{\lambda_0\}$ according to Theorem 4.20 and so, by the Spectral Mapping Theorem (Theorem 2.7),

$$\sigma(\lambda_0 I - T|_{\mathcal{R}(E_{\lambda_0})}) = \{0\}.$$

Corollary 4.23. *Let T be a compact operator on a complex Banach space $\mathcal{X}$. If $\lambda \in \sigma(T) \backslash \{0\}$, then λ is an isolated point of $\sigma(T)$ such that*

$$\dim \mathcal{R}(E_\lambda) < \infty,$$

where $E_\lambda \in \mathcal{B}[\mathcal{X}]$ is the Riesz idempotent associated with it, and

$$\{0\} \neq \mathcal{N}(\lambda I - T) \subseteq \mathcal{R}(E_\lambda) \subseteq \mathcal{N}((\lambda I - T)^n)$$

for some positive integer n such that

$$\mathcal{R}(E_\lambda) \not\subseteq \mathcal{N}((\lambda I - T)^{n-1}).$$

Proof. Take a compact operator $T \in \mathcal{B}_\infty[\mathcal{X}]$ on a nonzero complex Banach space $\mathcal{X}$. Consider the results of Section 2.6, which were proved on a Hilbert space but, as we had mentioned there, still hold on a Banach space. Suppose $\lambda \in \sigma(T) \backslash \{0\}$, and so λ is an isolated point of $\sigma(T)$ by Corollary 2.20. Take the restriction $T|_{\mathcal{R}(E_\lambda)}$ in $\mathcal{B}[\mathcal{R}(E_\lambda)]$ of T to the Banach space $\mathcal{R}(E_\lambda)$ (cf. Lemma 4.19), which is again compact (Proposition 1.V). According to Theorem 4.20, $\sigma(T|_{\mathcal{R}(E_\lambda)}) = \{\lambda\} \neq \{0\}$. Thus $0 \in \rho(T|_{\mathcal{R}(E_\lambda)})$. Hence the compact operator $T|_{\mathcal{R}(E_\lambda)}$ in $\mathcal{B}_\infty[\mathcal{R}(E_\lambda)]$ is invertible, which implies (cf. Proposition 1.Y)

$$\dim \mathcal{R}(E_\lambda) < \infty.$$

Set $m = \dim \mathcal{R}(E_\lambda)$. Note that $m \neq 0$ because $E_\lambda \neq O$ by the injectivity of Theorem 4.20. Since $\sigma(T|_{\mathcal{R}(E_\lambda)}) = \{\lambda\}$ we get $\sigma(\lambda I - T|_{\mathcal{R}(E_\lambda)}) = \{0\}$ by the Spectral Mapping Theorem (cf. Theorem 2.7). So the operator $\lambda I - T|_{\mathcal{R}(E_\lambda)}$ is quasinilpotent on the m-dimensional space $\mathcal{R}(E_\lambda)$, and therefore $\lambda I - T|_{\mathcal{R}(E_\lambda)}$ in $\mathcal{B}[\mathcal{R}(E_\lambda)]$ is a nilpotent operator for which

$$(\lambda I - T|_{\mathcal{R}(E_\lambda)})^m = O.$$

Indeed, this is a purely finite-dimensional algebraic result which is obtained together with the well-known Cayley–Hamilton Theorem (see, e.g., [55, Theorem 58.2]). Thus since E_λ is a projection commuting with T whose range is T-invariant (cf. Lemma 4.19(c) again),

$$(\lambda I - T)^m E_\lambda = O.$$

Moreover, since $\mathcal{X} \neq \{0\}$ we get $O \neq (\lambda I - T)^0 = I$, and so (since $E_\lambda \neq O$)

$$(\lambda I - T)^0 E_\lambda \neq O.$$

Thus there exists an integer $n \in [1, m]$ such that

$$(\lambda I - T)^n E_\lambda = O \quad \text{and} \quad (\lambda I - T)^{n-1} E_\lambda \neq O.$$

In other words,

$$\mathcal{R}(E_\lambda) \subseteq \mathcal{N}((\lambda I - T)^n) \quad \text{and} \quad \mathcal{R}(E_\lambda) \nsubseteq \mathcal{N}(\lambda I - T)^{n-1}.$$

(Recall: $AB = O$ if and only if $\mathcal{R}(B) \subseteq \mathcal{N}(A)$, for all operators A and B.) The remaining assertions are readily verified. In fact, $\{0\} \neq \mathcal{N}(\lambda I - T)$ since $\lambda \in \sigma(T) \backslash \{0\}$ is an eigenvalue of the compact operator T (by the Fredholm Alternative of Theorem 2.18), and $\mathcal{N}(\lambda I - T) \subseteq \mathcal{R}(E_\lambda)$ by Corollary 4.22. $\square$

Remark. According to Corollary 4.23 and Proposition 4.F, if $T \in \mathcal{B}_\infty[\mathcal{X}]$ (i.e., if T is compact) and $\lambda \in \sigma(T) \backslash \{0\}$, then λ is a pole of the resolvent function $R_T \colon \rho(T) \to \mathcal{G}[\mathcal{X}]$, and the integer n in the statement of the preceding result is the order of the pole λ. Moreover, it follows from Proposition 4.G that

$$\mathcal{R}(E_\lambda) = \mathcal{N}((\lambda I - T)^n).$$

4.5 Supplementary Propositions

If $\mathcal{H}$ is a nonzero complex Hilbert space, then $\mathcal{B}[\mathcal{H}]$ is a unital C*-algebra where the involution $*$ is the adjoint operation (see Section 4.1). Every $T \in \mathcal{B}[\mathcal{H}]$ determines a C*-subalgebra of $\mathcal{B}[\mathcal{H}]$, denoted by $C^*(T)$ and referred to as the C*-algebra generated by T, which is the smallest C*-algebra of operators from $\mathcal{B}[\mathcal{H}]$ containing T and the identity I, and which coincides with the closure $P(T, T^*)^-$ in $\mathcal{B}[\mathcal{H}]$ of all polynomials in T and T^* with complex coefficients (see, e.g., [6, p. 1] — cf. paragraph following Definition 3.12; see also proof of Theorem 4.9). The GELFAND–NAIMARK THEOREM asserts the converse. *Every unital C*-algebra is isometrically $*$-isomorphic to a C*-subalgebra of $\mathcal{B}[\mathcal{H}]$* (see, e.g., [6, Theorem 1.7.3] — i.e., for every abstract unital C*-algebra $\mathcal{A}$ there exists an involution-preserving isometric algebra isomorphism of $\mathcal{A}$ onto a C*-subalgebra of $\mathcal{B}[\mathcal{H}]$).

Consider a set $\mathcal{S} \subseteq \mathcal{B}[\mathcal{H}]$ of operators on a Hilbert space $\mathcal{H}$. The *commutant* $\mathcal{S}'$ of $\mathcal{S}$ is the set $\mathcal{S}' = \{T \in \mathcal{B}[\mathcal{H}] : TS = ST \text{ for every } S \in \mathcal{S}\}$ of all operators that commute with every operator in $\mathcal{S}$, which is a unital subalgebra of $\mathcal{B}[\mathcal{H}]$. The *double commutant* $\mathcal{S}''$ of $\mathcal{S}$ is the unital algebra $\mathcal{S}'' = (\mathcal{S}')'$. The DOUBLE COMMUTANT THEOREM reads as follows. *If $\mathcal{A}$ is a unital C*-subalgebra of $\mathcal{B}[\mathcal{H}]$, then $\mathcal{A}^- = \mathcal{A}''$*, where $\mathcal{A}^-$ stands for the weak closure of $\mathcal{A}$ in $\mathcal{B}[\mathcal{H}]$ (which coincides with the strong closure — see [30, Theorem IX.6.4]). A *von Neumann algebra* $\mathcal{A}$ is a C*-subalgebra of $\mathcal{B}[\mathcal{H}]$ such that $\mathcal{A} = \mathcal{A}''$, which is unital and weakly (thus strongly) closed. Take a normal operator on a *separable* Hilbert space $\mathcal{H}$. Let $\mathcal{A}^*(T)$ be the *von Neumann algebra generated by* $T \in \mathcal{B}[\mathcal{H}]$, which is the intersection of all von Neumann algebras containing T, and coincides with the weak closure of $\mathcal{P}(T, T^*)$ (compare with the above paragraph).

Proposition 4.A. *Consider the setup of Theorem 4.8. The range of Φ_T coincides with the von Neumann algebra generated by T (i.e., $\mathcal{R}(\Phi_T) = \mathcal{A}^*(T)$).*

Proposition 4.B. *Take $T \in \mathcal{B}[\mathcal{X}]$. If $\psi \in A(\sigma(T))$ and $\phi \in A(\sigma(\psi(T)))$, then the composition $(\phi \circ \psi) \in A(\sigma(T))$ and $\phi(\psi(T)) = (\phi \circ \psi)(T)$. That is,*

$$\phi(\psi(T)) = \tfrac{1}{2\pi i} \int_{\Upsilon} \phi(\psi(\lambda)) \, R_T(\lambda) \, d\lambda,$$

where Υ is any path about $\sigma(T)$ for which $\sigma(T) \subset \text{ins } \Upsilon \subset (\text{ins } \Upsilon)^- \subset \Lambda$, and the open set $\Lambda \subseteq \mathbb{C}$ is the domain of $\phi \circ \psi$.

Proposition 4.C. *Let $E_\Delta \in \mathcal{B}[\mathcal{X}]$ be the Riesz idempotent associated with a spectral subset Δ of $\sigma(T)$. The point, residual, and continuous spectra of the restriction $T|_{\mathcal{R}(E_\Delta)} \in \mathcal{B}[\mathcal{R}(E_\Delta)]$ of $T \in \mathcal{B}[\mathcal{X}]$ to $\mathcal{R}(E_\Delta)$ are given by*

$$\sigma_P(T|_{\mathcal{R}(E_\Delta)}) = \Delta \cap \sigma_P(T), \quad \sigma_R(T|_{\mathcal{R}(E_\Delta)}) = \Delta \cap \sigma_R(T), \quad \sigma_C(T|_{\mathcal{R}(E_\Delta)}) = \Delta \cap \sigma_C(T).$$

Proposition 4.D. *If λ_0 is an isolated point of the spectrum $\sigma(T)$ of an operator $T \in \mathcal{B}[\mathcal{X}]$, then the range of the Riesz idempotent $E_{\lambda_0} \in \mathcal{B}[\mathcal{X}]$ associated* *with* λ_0 *is given by*

$$\mathcal{R}(E_{\lambda_0}) = \{x \in \mathcal{X} \colon \|(\lambda_0 I - T)^n x\|^{\frac{1}{n}} \to 0\}.$$

Proposition 4.E. *If* λ_0 *is an isolated point of the spectrum* $\sigma(T)$ *of an operator* $T \in \mathcal{B}[\mathcal{X}]$, *then* (*with convergence in* $\mathcal{B}[\mathcal{X}]$)

$$R_T(\lambda) = (\lambda I - T)^{-1} = \sum_{k=-\infty}^{\infty} (\lambda - \lambda_0)^k T_k$$
$$= \sum_{k=1}^{\infty} (\lambda - \lambda_0)^{-k} T_{-k} + T_0 + \sum_{k=1}^{\infty} (\lambda - \lambda_0)^k T_k$$

for every λ *in the punctured disk* $B_{\delta_0}(\lambda_0) \backslash \{\lambda_0\} = \{\lambda \in \mathbb{C} \colon 0 < |\lambda - \lambda_0| < \delta_0\} \subseteq$ $\rho(T)$, *where* $\delta_0 = d(\lambda_0, \sigma(T) \backslash \{\lambda_0\})$ *is the distance from* λ_0 *to the rest of the spectrum, and* $T_k \in \mathcal{B}[\mathcal{X}]$ *is such that*

$$T_k = \frac{1}{2\pi i} \int_{\Upsilon_{\lambda_0}} (\lambda - \lambda_0)^{-(k+1)} R_T(\lambda) \, d\lambda$$

for every $k \in \mathbb{Z}$, *where* Υ_{λ_0} *is any positively oriented circle centered at* λ_0 *with radius less than* δ_0.

An isolated point of $\sigma(T)$ is a *singularity* (or an *isolated singularity*) of the resolvent function $R_T \colon \rho(T) \to \mathcal{G}[\mathcal{X}]$. The expansion of R_T in Proposition 4.E is the *Laurent expansion* of R_T about an isolated point of the spectrum. Since $E_{\lambda_0} = \frac{1}{2\pi i} \int_{\Upsilon_{\lambda_0}} R_T(\lambda) \, d\lambda$ we get, for every positive integer m (i.e., for $m \in \mathbb{N}$),

$$T_{-m} = E_{\lambda_0}(T - \lambda_0 I)^{m-1} = (T - \lambda_0 I)^{m-1} E_{\lambda_0},$$

and so $T_{-1} = E_{\lambda_0}$, the Riesz idempotent associated with the isolated point λ_0. The notion of poles associated with isolated singularities of analytic functions is extended as follows. An isolated point of $\sigma(T)$ is a *pole of order* n of R_T, for some $n \in \mathbb{N}$, if $T_{-n} \neq O$ and $T_k = O$ for every $k < -n$ (i.e., the order of a pole is the largest positive integer $|k|$ such that $T_{-|k|} \neq O$). Otherwise, if the number of nonzero coefficients T_k with negative indices is infinite, then the isolated point λ_0 of $\sigma(T)$ is said to be an *essential singularity* of R_T.

Proposition 4.F. *Let* λ_0 *be an isolated point of the spectrum* $\sigma(T)$ *of* $T \in \mathcal{B}[\mathcal{X}]$, *and let* $E_{\lambda_0} \in \mathcal{B}[\mathcal{X}]$ *be the Riesz idempotent associated with* λ_0. *The isolated point* λ_0 *is a* pole of order n *of* R_T *if and only if*

(a) $\qquad (\lambda_0 I - T)^n E_{\lambda_0} = O \quad$ *and* $\quad (\lambda_0 I - T)^{n-1} E_{\lambda_0} \neq O.$

Therefore, if λ_0 *is a pole of order* n *of* R_T, *then*

(b) $\qquad \{0\} \neq \mathcal{R}((\lambda_0 I - T)^{n-1} E_{\lambda_0}) \subseteq \mathcal{N}(\lambda_0 I - T),$

and so λ_0 *is an eigenvalue of* T (*i.e.,* $\lambda_0 \in \sigma_P(T)$). *Actually* (*by item* (a) *and Corollary 4.22*), λ_0 *is a pole of order* 1 *of* R_T *if and only if*

(c) $$\{0\} \neq \mathcal{R}(E_{\lambda_0}) = \mathcal{N}(\lambda_0 I - T).$$

Proposition 4.G. *Take* $T \in \mathcal{B}[\mathcal{X}]$. *If* λ_0 *is a pole of order* n *of* R_T, *then*

$$\mathcal{R}(E_{\lambda_0}) = \mathcal{N}((\lambda_0 I - T)^n) \quad and \quad \mathcal{R}(E_\Delta) = \mathcal{R}((\lambda_0 I - T)^n),$$

where E_{λ_0} *denotes the Riesz idempotent associated with* λ_0 *and* E_Δ *denotes the Riesz idempotent associated with the spectral set* $\Delta = \sigma(T) \backslash \{\lambda_0\}$.

Proposition 4.H. *Let* E_Δ *be the Riesz idempotent associated with a spectral subset* Δ *of* $\sigma(T)$. *If* $\dim \mathcal{R}(E_\Delta) < \infty$, *then* Δ *is a finite set of poles.*

Proposition 4.I. *Take* $T \in \mathcal{B}[\mathcal{X}]$. *Let* Δ *be a spectral subset of* $\sigma(T)$. *Consider the Riesz idempotent* $E_\Delta \in \mathcal{B}[\mathcal{X}]$ *associated with* Δ. *If* $\psi \in A(\sigma(T))$, *then* $\psi \in A(\sigma(T|_{\mathcal{R}(E_\Delta)}))$ *and* $\psi(T)|_{\mathcal{R}(E_\Delta)} = \psi(T|_{\mathcal{R}(E_\Delta)})$. *Thus*

$$\frac{1}{2\pi i} \int_{\Upsilon_\Delta} \psi(\lambda) R_T(\lambda) \, d\lambda \bigg|_{\mathcal{R}(E_\Delta)} = \frac{1}{2\pi i} \int_{\Upsilon_\Delta} \psi(\lambda) R_{T|_{\mathcal{R}(E_\Delta)}}(\lambda) \, d\lambda.$$

For a normal operator on a complex Hilbert space, the functional calculi of Sections 4.2 and 4.3 (cf. Theorems 4.2 and 4.16) coincide for every analytic function on a neighborhood of the spectrum.

Proposition 4.J. *If* $T = \int \lambda \, dE_\lambda \in \mathcal{B}[\mathcal{H}]$ *is the spectral decomposition of a normal operator* T *on a Hilbert space* $\mathcal{H}$ *and if* $\psi \in A(\sigma(T))$, *then*

$$\psi(T) = \int \psi(\lambda) \, dE_\lambda = \frac{1}{2\pi i} \int_\Upsilon \psi(\lambda) \, R_T(\lambda) \, d\lambda,$$

with the first integral as in Lemma 3.7 and the second as in Definition 4.14.

Consider Definition 3.6. The notion of *spectral measure* $E \colon \mathcal{A}_\Omega \to \mathcal{B}[\mathcal{X}]$ (carrying the same properties set forth in Definition 3.6) can be extended to a (complex) Banach space $\mathcal{X}$ where the projections $E(\Lambda)$ for $\Lambda \in \mathcal{A}_\Omega$ are not orthogonal but are bounded. If an operator $T \in \mathcal{B}[\mathcal{X}]$ is such that $E(\Lambda)T = TE(\Lambda)$ and $\sigma(T|_{\mathcal{R}(E(\Lambda))}) \subseteq \Lambda^-$ (see Lemma 3.16), then T is called a *spectral operator* (cf. [47, Definition XV.2.5]). Take $\psi \in B(\Omega)$, a bounded $\mathcal{A}_\Omega$-measurable complex-valued function on Ω. An integral $\int \psi(\lambda) \, dE_\lambda$ can be defined in such a Banach-space setting as the uniform limit of integrals $\int \phi_n(\lambda) \, dE_\lambda = \sum_{i=1}^n \alpha_i E(\Lambda_i)$ of measurable *simple functions* $\phi_n = \sum_{i=1}^n \alpha_i \chi_{\Lambda_i}$ (i.e., of finite linear combinations of characteristic functions χ_{Λ_i} of measurable sets $\Lambda_i \in \mathcal{A}_\Omega$) — see, e.g., [46, pp. 891–893]. If a spectral operator $T \in \mathcal{B}[\mathcal{X}]$ is such that $T = \int \lambda \, dE_\lambda$, then it is said to be of *scalar type* (cf. [47, Definition XV.4.1]). Proposition 4.J holds for Banach-space *spectral operators of scalar type*. In this case, the preceding integral $\int \lambda \, dE_\lambda = \int \varphi(\lambda) \, dE_\lambda$ (of the function $\varphi(\lambda) = \lambda \in \mathbb{C}$ for every $\lambda \in \Omega$, defined as the limit of a sequence $\sum_{i=1}^n \beta_i E(\Lambda_i)$ of integrals of simple functions) coincides with the integral of Theorem 3.15 if $\mathcal{X}$

is a Hilbert space, and if the spectral measure $E\colon \mathcal{A}_\Omega \to \mathcal{B}[\mathcal{X}]$, with $\Omega = \sigma(T)$, takes on orthogonal projections only.

Proposition 4.K. *If* $T = \int \lambda \, dE_\lambda \in \mathcal{B}[\mathcal{H}]$ *is the spectral decomposition of a normal operator* T *on a Hilbert space* $\mathcal{H}$, *then the orthogonal projection* $E(\Lambda)$ *coincides with the Riesz idempotent* E_Λ *for every clopen* Λ *in* $\mathcal{A}_{\sigma(T)}$:

$$E(\Lambda) = \int_\Lambda dE_\lambda = \tfrac{1}{2\pi i} \int_{\Upsilon_\Lambda} R_T(\lambda) \, d\lambda = E_\Lambda,$$

where Υ_Δ *is any path for which* $\Delta \subseteq$ *ins* Υ_Δ *and* $\sigma(T) \backslash \Delta \subseteq$ *out* Υ_Δ.

In particular, let $T = \sum_k \lambda_k E_k$ be the spectral decomposition of a *compact normal* operator T on a Hilbert space, with $\{\lambda_k\} = \sigma_P(T)$ and $\{E_k\}$ being a countable resolution of the identity where each orthogonal projection E_k is such that $\mathcal{R}(E_k) = \mathcal{N}(\lambda_k I - T)$ (Theorem 3.3). Take any nonzero eigenvalue λ_k in $\sigma(T) \backslash \{0\} = \sigma_P(T) \backslash \{0\} = \{\lambda_k\} \backslash \{0\}$ so that $\dim \mathcal{N}(\lambda_k I - T) < \infty$ (cf. Theorems 1.19 and 2.18 and Corollary 2.20). Each orthogonal projection E_k coincides with the Riesz idempotent E_{λ_k} associated with each isolated point $0 \neq \lambda_k$ of $\sigma(T)$ according to Proposition 4.K. Now we come back to a Banach-space setting: according to Corollary 4.23 and Proposition 4.G, $\mathcal{N}(\lambda_k I - T) = \mathcal{N}((\lambda_k I - T)^{n_k}) = \mathcal{R}(E_{\lambda_k})$ where the integer $n_k > 0$ is the order of the pole λ_k and the integer $0 < \dim \mathcal{R}(E_{\lambda_k}) = \dim \mathcal{N}(\lambda_k I - T) < \infty$ is the finite multiplicity of the eigenvalue λ_k. In general, $\dim \mathcal{R}(E_{\lambda_k})$ is sometimes referred to as the *algebraic multiplicity* of the isolated point λ_k of the spectrum, while $\dim \mathcal{N}(\lambda_k I - T)$ — *the multiplicity* of the eigenvalue λ_k — is sometimes called the *geometric multiplicity* of λ_k. In the preceding case (i.e., if T is compact and normal on a Hilbert space) these multiplicities are finite and coincide.

Proposition 4.L. *Every isolated point of the spectrum of a hyponormal operator is an eigenvalue. In fact, if* λ_0 *is an isolated point of* $\sigma(T)$, *then*

$$\{0\} \neq \mathcal{R}(E_{\lambda_0}) \subseteq \mathcal{N}(\lambda_0 I - T)$$

for every hyponormal operator $T \in \mathcal{B}[\mathcal{H}]$, *where* E_{λ_0} *is the Riesz idempotent associated with the isolated point* λ_0 *of the spectrum of* T.

Notes: For Proposition 4.A see, for instance, [30, Theorem IX.8.10]. Proposition 4.B is the composition counterpart of the product-preserving result of Lemma 4.15 (see, e.g., [66, Theorem 5.3.2] or [45, Theorem VII.3.12]). For Proposition 4.C see [45, Exercise VII.5.18] or [111, 1st edn. Theorem 5.7-B]. Proposition 4.D follows from Theorem 4.20 since, according to Theorem 2.7, $\lambda_0 I - T|_{\mathcal{R}(E_{\lambda_0})}$ is quasinilpotent and so $\lim_n \|(\lambda I - T)^n x\|^{\frac{1}{n}} = 0$ whenever x lies in $\mathcal{R}(E_\lambda)$. For the converse see [111, 1st edn. Lemma 5.8-C]. The Laurent expansion in Proposition 4.E and also Proposition 4.F are standard results (see [30, Lemma VII.6.11, Proposition 6.12, and Corollary 6.13]). Proposition 4.G refines the results of Proposition 4.F — see, e.g., [45, Theorem VII.3.24].

For Proposition 4.H see [45, Exercise VII.5.34]. Proposition 4.I is a consequence of Theorem 4.20 (see, e.g., [45, Theorem VII.3.20]). Proposition 4.J also holds for Banach-space spectral operators of the scalar type (cf. [47, Theorem XV.5.1]), and Proposition 4.K is a particular case of it for the characteristic function in a neighborhood of a clopen set in the σ-algebra of Borel subsets of $\sigma(T)$. Proposition 4.L is the extension of Proposition 3.G from normal to hyponormal operators, which is obtained by the Riesz Decomposition Theorem (Corollary 4.21) — see, for instance, [78, Problem 6.28].

Suggested Readings

Arveson [7]
Brown and Pearcy [23]
Conway [30, 32, 33]
Dowson [39]
Dunford and Schwartz [45, 46, 47]

Hille and Phillips [66]
Radjavi and Rosenthal [98]
Riesz and Sz.-Nagy [100]
Sz.-Nagy and Foiaş [110]
Taylor and Lay [111]

5

Fredholm Theory in Hilbert Space

The central theme of this chapter is the investigation of compact perturbations. Of particular concern will be the properties of the spectrum of an
operator which are invariant under compact perturbations. In other words,
the properties of the spectrum of T which are also possessed by the spectrum
of $T+K$ for every compact operator K. As in Sections 1.8 and 2.6, when
dealing with compact operators, we assume in this chapter that all operators
lie in $\mathcal{B}[\mathcal{H}]$ where $\mathcal{H}$ is a Hilbert space. Although the theory is amenable to
being developed on a Banach space as well, the approaches are quite differ-
ent. Hilbert spaces are endowed with a notably rich structure which allows
significant simplifications. The Banach-space case is considered in Appendix
A, where the differences between the two approaches are examined, and an
account is given of the origins and consequences of such differences.

5.1 Fredholm and Weyl Operators — Fredholm Index

Let $\mathcal{H}$ be a nonzero complex Hilbert space. Let $\mathcal{B}_\infty[\mathcal{H}]$ be the (two-sided) ideal
of compact operators from $\mathcal{B}[\mathcal{H}]$. An operator $T \in \mathcal{B}[\mathcal{H}]$ is *left semi-Fredholm*
if there exist $A \in \mathcal{B}[\mathcal{H}]$ and $K \in \mathcal{B}_\infty[\mathcal{H}]$ for which $AT = I + K$, and *right semi-
Fredholm* if there exist $A \in \mathcal{B}[\mathcal{H}]$ and $K \in \mathcal{B}_\infty[\mathcal{H}]$ for which $TA = I + K$. It is
semi-Fredholm if it is either left or right semi-Fredholm, and *Fredholm* if it is
both left and right semi-Fredholm. Let $\mathcal{F}_\ell[\mathcal{H}]$ and $\mathcal{F}_r[\mathcal{H}]$ be the classes of all
left semi-Fredholm operators and of all right semi-Fredholm operators:

$$\mathcal{F}_\ell[\mathcal{H}] = \{T \in \mathcal{B}[\mathcal{H}]\colon\ AT = I + K \text{ for some } A \in \mathcal{B}[\mathcal{H}] \text{ and some } K \in \mathcal{B}_\infty[\mathcal{H}]\},$$

$$\mathcal{F}_r[\mathcal{H}] = \{T \in \mathcal{B}[\mathcal{H}]\colon\ TA = I + K \text{ for some } A \in \mathcal{B}[\mathcal{H}] \text{ and some } K \in \mathcal{B}_\infty[\mathcal{H}]\}.$$

The classes of all semi-Fredholm and Fredholm operators from $\mathcal{B}[\mathcal{H}]$ will be
denoted by $\mathcal{SF}[\mathcal{H}]$ and $\mathcal{F}[\mathcal{H}]$, respectively:

$$\mathcal{SF}[\mathcal{H}] = \mathcal{F}_\ell[\mathcal{H}] \cup \mathcal{F}_r[\mathcal{H}] \quad \text{and} \quad \mathcal{F}[\mathcal{H}] = \mathcal{F}_\ell[\mathcal{H}] \cap \mathcal{F}_r[\mathcal{H}].$$

According to Proposition 1.W, $K \in \mathcal{B}_\infty[\mathcal{H}]$ if and only if $K^* \in \mathcal{B}_\infty[\mathcal{H}]$. Thus

© Springer Nature Switzerland AG 2020
C. S. Kubrusly, *Spectral Theory of Bounded Linear Operators*,
https://doi.org/10.1007/978-3-030-33149-8_5

$$T \in \mathcal{F}_\ell[\mathcal{H}] \quad \text{if and only if} \quad T^* \in \mathcal{F}_r[\mathcal{H}].$$

Therefore

$$T \in \mathcal{SF}[\mathcal{H}] \quad \text{if and only if} \quad T^* \in \mathcal{SF}[\mathcal{H}],$$

$$T \in \mathcal{F}[\mathcal{H}] \quad \text{if and only if} \quad T^* \in \mathcal{F}[\mathcal{H}].$$

The following results of kernel and range of a Hilbert-space operator and its adjoint (Lemmas 1.4 and 1.5) will often be used throughout this chapter:

$$\mathcal{N}(T^*) = \mathcal{H} \ominus \mathcal{R}(T)^- = \mathcal{R}(T)^\perp,$$

$$\mathcal{R}(T^*) \text{ is closed if and only if } \mathcal{R}(T) \text{ is closed.}$$

Theorem 5.1. (a) *An operator $T \in \mathcal{B}[\mathcal{H}]$ is left semi-Fredholm if and only if $\mathcal{R}(T)$ is closed and $\mathcal{N}(T)$ is finite-dimensional.*

(b) *Hence*

$$\mathcal{F}_\ell[\mathcal{H}] = \{T \in \mathcal{B}[\mathcal{H}]: \mathcal{R}(T) \text{ is closed and } \dim \mathcal{N}(T) < \infty\},$$

$$\mathcal{F}_r[\mathcal{H}] = \{T \in \mathcal{B}[\mathcal{H}]: \mathcal{R}(T) \text{ is closed and } \dim \mathcal{N}(T^*) < \infty\}.$$

Proof. Let A, T, and K be operators on $\mathcal{H}$, where K is compact.

(a_1) If T is left semi-Fredholm, then there are operators A and K such that $AT = I + K$, and so $\mathcal{N}(T) \subseteq \mathcal{N}(AT) = \mathcal{N}(I + K)$ and $\mathcal{R}(AT) = \mathcal{R}(I + K)$. According to the Fredholm Alternative (Corollary 1.20), $\dim \mathcal{N}(I + K) < \infty$ and $\mathcal{R}(I + K)$ is closed. Therefore

(i) $\dim \mathcal{N}(T) < \infty,$

(ii) $\dim T(\mathcal{N}(AT)) < \infty,$

(iii) $\mathcal{R}(AT)$ is closed.

This implies

(iv) $T(\mathcal{N}(AT)^\perp)$ is closed.

Indeed, the restriction $(AT)|_{\mathcal{N}(AT)^\perp} : \mathcal{N}(AT)^\perp \to \mathcal{H}$ is bounded below by (iii) since $(AT)|_{\mathcal{N}(AT)^\perp}$ is injective with range $\mathcal{R}(AT)$ (Theorem 1.2). Thus there exists an $\alpha > 0$ such that $\alpha\|v\| \le \|ATv\| \le \|A\|\,\|Tv\|$ for every $v \in \mathcal{N}(AT)^\perp$. Then $T|_{\mathcal{N}(AT)^\perp} : \mathcal{N}(AT)^\perp \to \mathcal{H}$ is bounded below, and so $T|_{\mathcal{N}(AT)^\perp}$ has a closed range (Theorem 1.2 again), proving (iv). But (ii) and (iv) imply

(v) $\mathcal{R}(T)$ is closed.

In fact, since $\mathcal{H} = \mathcal{N}(AT) + \mathcal{N}(AT)^\perp$ by the Projection Theorem, it follows that $\mathcal{R}(T) = T(\mathcal{H}) = T(\mathcal{N}(AT) + \mathcal{N}(AT)^\perp) = T(\mathcal{N}(AT)) + T(\mathcal{N}(AT)^\perp)$. Thus, by assertions (ii) and (iv), $\mathcal{R}(T)$ is closed (since the sum of a closed linear manifold and a finite-dimensional linear manifold is closed — see Proposition 1.C). This concludes the proof of (v) which, together with (i), concludes half of the claimed result: $T \in \mathcal{F}_\ell[\mathcal{H}] \implies \mathcal{R}(T)^- = \mathcal{R}(T)$ and $\dim \mathcal{N}(T) < \infty$.

(a_2) Conversely, suppose $\dim \mathcal{N}(T)$ is finite and $\mathcal{R}(T)$ is closed. Since $\mathcal{R}(T)$ is closed, the restriction $T|_{\mathcal{N}(T)^\perp} : \mathcal{N}(T)^\perp \to \mathcal{H}$ (which is injective because $\mathcal{N}(T|_{\mathcal{N}(T)^\perp}) = \{0\}$) has a closed range $\mathcal{R}(T|_{\mathcal{N}(T)^\perp}) = \mathcal{R}(T)$. Hence it has a bounded inverse on its range (Theorem 1.2). Let $E \in \mathcal{B}[\mathcal{H}]$ be the orthogonal projection onto $\mathcal{R}(E) = \mathcal{R}(T|_{\mathcal{N}(T)^\perp}) = \mathcal{R}(T)$, and define $A \in \mathcal{B}[\mathcal{H}]$ as follows:

$$A = (T|_{\mathcal{N}(T)^\perp})^{-1} E.$$

If $u \in \mathcal{N}(T)$, then $ATu = 0$ trivially. On the other hand, if $v \in \mathcal{N}(T)^\perp$, then $ATv = (T|_{\mathcal{N}(T)^\perp})^{-1} ET|_{\mathcal{N}(T)^\perp} v = (T|_{\mathcal{N}(T)^\perp})^{-1} T|_{\mathcal{N}(T)^\perp} v = v$. Thus for every $x = u + v$ in $\mathcal{H} = \mathcal{N}(T) + \mathcal{N}(T)^\perp$ we get $ATx = ATu + ATv = v = E'x$, where $E' \in \mathcal{B}[\mathcal{H}]$ is the orthogonal projection onto $\mathcal{N}(T)^\perp$. Therefore

$$AT = I + K,$$

where $-K = I - E' \in \mathcal{B}[\mathcal{H}]$ is the complementary orthogonal projection onto the finite-dimensional space $\mathcal{N}(T)$. Hence K is a finite-rank operator, and so compact (Proposition 1.X). Thus T is left semi-Fredholm.

(b) Since $T \in \mathcal{F}_\ell[\mathcal{H}]$ if and only if $T^* \in \mathcal{F}_r[\mathcal{H}]$, and since $\mathcal{R}(T^*)$ is closed if and only if $\mathcal{R}(T)$ is closed (Lemma 1.5), item (b) follows from item (a). $\quad\square$

Corollary 5.2. *Take an operator* $T \in \mathcal{B}[\mathcal{H}]$.

(a) *T is semi-Fredholm if and only if $\mathcal{R}(T)$ is closed and $\mathcal{N}(T)$ or $\mathcal{N}(T^*)$ is finite-dimensional:*

$$\mathcal{SF}[\mathcal{H}] = \{T \in \mathcal{B}[\mathcal{H}]\colon \mathcal{R}(T) \text{ is closed, } \dim \mathcal{N}(T) < \infty \text{ or } \dim \mathcal{N}(T^*) < \infty\}.$$

(b) *T is Fredholm if and only if $\mathcal{R}(T)$ is closed and both $\mathcal{N}(T)$ and $\mathcal{N}(T^*)$ are finite-dimensional:*

$$\mathcal{F}[\mathcal{H}] = \{T \in \mathcal{B}[\mathcal{H}]\colon \mathcal{R}(T) \text{ is closed, } \dim \mathcal{N}(T) < \infty \text{ and } \dim \mathcal{N}(T^*) < \infty\}.$$

Proof. $\mathcal{SF}[\mathcal{H}] = \mathcal{F}_\ell[\mathcal{H}] \cup \mathcal{F}_r[\mathcal{H}]$ and $\mathcal{F}[\mathcal{H}] = \mathcal{F}_\ell[\mathcal{H}] \cap \mathcal{F}_r[\mathcal{H}]$. Use Theorem 5.1. $\square$

Let $\mathbb{Z}$ be the set of all integers and set $\overline{\mathbb{Z}} = \mathbb{Z} \cup \{-\infty\} \cup \{+\infty\}$, the *extended integers*. Take any T in $\mathcal{SF}[\mathcal{H}]$ so that $\mathcal{N}(T)$ or $\mathcal{N}(T^*)$ is finite-dimensional. The *Fredholm index* $\mathrm{ind}\,(T)$ of T in $\mathcal{SF}[\mathcal{H}]$ is defined in $\overline{\mathbb{Z}}$ by

$$\mathrm{ind}\,(T) = \dim \mathcal{N}(T) - \dim \mathcal{N}(T^*).$$

It is usual to write $\alpha(T) = \dim \mathcal{N}(T)$ and $\beta(T) = \dim \mathcal{N}(T^*)$, and hence $\mathrm{ind}\,(T) = \alpha(T) - \beta(T)$. Since T^* and T lie in $\mathcal{SF}[\mathcal{H}]$ together, we get

$$\mathrm{ind}\,(T^*) = -\mathrm{ind}\,(T).$$

Note: $\beta(T) = \alpha(T^*) = \dim \mathcal{R}(T)^{\perp} = \dim(\mathcal{H} \ominus \mathcal{R}(T)^-) = \dim \mathcal{H}/\mathcal{R}(T)^-$, where the last identity is discussed in the forthcoming Remark 5.3(a).

According to Theorem 5.1, $T \in \mathcal{F}_{\ell}[\mathcal{H}]$ implies $\dim \mathcal{N}(T) < \infty$ and, dually, $T \in \mathcal{F}_r[\mathcal{H}]$ implies $\dim \mathcal{N}(T^*) < \infty$. Therefore

$$T \in \mathcal{F}_{\ell}[\mathcal{H}] \quad \text{implies} \quad \text{ind}\,(T) \neq +\infty,$$

$$T \in \mathcal{F}_r[\mathcal{H}] \quad \text{implies} \quad \text{ind}\,(T) \neq -\infty.$$

In other words, if $T \in \mathcal{SF}[\mathcal{H}] = \mathcal{F}_{\ell}[\mathcal{H}] \cup \mathcal{F}_r[\mathcal{H}]$, then

$$\text{ind}\,(T) = +\infty \quad \text{implies} \quad T \in \mathcal{F}_r[\mathcal{H}] \backslash \mathcal{F}_{\ell}[\mathcal{H}] = \mathcal{F}_r[\mathcal{H}] \backslash \mathcal{F}[\mathcal{H}],$$

$$\text{ind}\,(T) = -\infty \quad \text{implies} \quad T \in \mathcal{F}_{\ell}[\mathcal{H}] \backslash \mathcal{F}_r[\mathcal{H}] = \mathcal{F}_{\ell}[\mathcal{H}] \backslash \mathcal{F}[\mathcal{H}].$$

Remark 5.3. (a) BANACH SPACE. If T is an operator on a Hilbert space $\mathcal{H}$, then $\mathcal{N}(T^*) = \mathcal{R}(T)^{\perp} = \mathcal{H} \ominus \mathcal{R}(T)^-$ by Lemma 1.4. Thus Theorem 5.1 and Corollary 5.2 can be restated with $\mathcal{N}(T^*)$ replaced by $\mathcal{H} \ominus \mathcal{R}(T)^-$. However, consider the quotient space $\mathcal{H}/\mathcal{R}(T)^-$ of $\mathcal{H}$ modulo $\mathcal{R}(T)^-$ consisting of all cosets $x + \mathcal{R}(T)^-$ for each $x \in \mathcal{H}$. The natural mapping of $\mathcal{H}/\mathcal{R}(T)^-$ onto $\mathcal{H} \ominus \mathcal{R}(T)^-$ (viz., $x + \mathcal{R}(T)^- \mapsto Ex$ where E is the orthogonal projection onto $\mathcal{R}(T)^{\perp}$) is an isomorphism. Then $\dim \mathcal{H}/\mathcal{R}(T)^- = \dim(\mathcal{H} \ominus \mathcal{R}(T)^-)$. Therefore we get the following restatement of Theorem 5.1 and Corollary 5.2.

T is right semi-Fredholm if and only if $\mathcal{R}(T)$ is closed and $\mathcal{H}/\mathcal{R}(T)$ is finite-dimensional. It is semi-Fredholm if and only if $\mathcal{R}(T)$ is closed and $\mathcal{N}(T)$ or $\mathcal{H}/\mathcal{R}(T)$ is finite-dimensional. It is Fredholm if and only if $\mathcal{R}(T)$ is closed and both $\mathcal{N}(T)$ and $\mathcal{H}/\mathcal{R}(T)$ are finite-dimensional.

This is how the theory advances in a Banach space. Many properties among those that will be developed in this chapter work smoothly in any Banach space. Some of them, in being translated literally into a Banach-space setting, will behave well if the Banach space is complemented. A normed space $\mathcal{X}$ is *complemented* if every subspace of it has a complementary subspace (i.e., if there is a (closed) subspace $\mathcal{N}$ of $\mathcal{X}$ such that $\mathcal{M} + \mathcal{N} = \mathcal{X}$ and $\mathcal{M} \cap \mathcal{N} = \{0\}$ for every (closed) subspace $\mathcal{M}$ of $\mathcal{X}$). For instance, Proposition 1.D (on complementary subspaces and continuous projections) is an example of a result that acquires its full strength only on complemented Banach spaces. However, *if a Banach space is complemented, then it is isomorphic to a Hilbert space* [91] (i.e., topologically isomorphic, by the Inverse Mapping Theorem). So complemented Banach spaces are identified with Hilbert spaces: *only Hilbert spaces* (up to an isomorphism) *are complemented*. We stick to Hilbert spaces in this chapter. The Banach-space counterpart will be considered in Appendix A.

(b) CLOSED RANGE. As will be verified in Appendix A (cf. Remark A.1 and Corollary A.9), if $\text{codim}\,\mathcal{R}(T) < \infty$, then $\mathcal{R}(T)$ is closed — *codimension* means dimension of any algebraic complement, which coincides with dimension of the quotient space. So, in a Hilbert space, $\dim \mathcal{H}/\mathcal{R}(T) < \infty \iff \operatorname{codim} \mathcal{R}(T) < \infty$ $\implies \operatorname{codim} \mathcal{R}(T)^- < \infty \iff \dim \mathcal{H}/\mathcal{R}(T)^- < \infty \iff \operatorname{codim} \mathcal{N}(T^*)^\perp < \infty \iff$ $\dim \mathcal{N}(T^*) < \infty$ (see also Lemmas 1.4 and 1.5 and the Projection Theorem). Thus, since $\dim \mathcal{H}/\mathcal{R}(T) < \infty \implies \mathcal{R}(T) = \mathcal{R}(T)^-$, then (cf. Theorem 5.1)

$$\{\mathcal{R}(T) = \mathcal{R}(T)^- \text{ and } \dim \mathcal{N}(T^*) < \infty\} \iff \dim \mathcal{H}/\mathcal{R}(T) < \infty.$$

(c) FINITE RANK. The following statement was in fact proved in Theorem 5.1:

  T is left semi-Fredholm if and only if there exists an operator $A \in \mathcal{B}[\mathcal{H}]$
  and a finite-rank operator $K \in \mathcal{B}[\mathcal{H}]$ for which $AT = I + K$.

Since T is right semi-Fredholm if and only if T^* is left semi-Fredholm, and so there exists a finite-rank operator K such that $AT^* = I + K$ or, equivalently, such that $TA^* = I + K^*$, and since K is a finite-rank operator if and only if K^* is (see, e.g., [78, Problem 5.40]), we also get

  T is right semi-Fredholm if and only if there exists an operator $A \in \mathcal{B}[\mathcal{H}]$
  and a finite-rank operator $K \in \mathcal{B}[\mathcal{H}]$ for which $TA = I + K$.

So the definitions of left semi-Fredholm, right semi-Fredholm, semi-Fredholm, and Fredholm operators are equivalently stated if "compact" is replaced with "finite-rank". For Fredholm operators we can even have *the same A when stating that $I - AT$ and $I - TA$ are of finite rank* (cf. Proposition 5.C).

(d) FINITE DIMENSION. Take any operator $T \in \mathcal{B}[\mathcal{H}]$. The rank and nullity identity of linear algebra says $\dim \mathcal{H} = \dim \mathcal{N}(T) + \dim \mathcal{R}(T)$ (cf. proof of Corollary 2.19). Since $\mathcal{H} = \mathcal{R}(T)^- \oplus \mathcal{R}(T)^\perp$ (Projection Theorem), then $\dim \mathcal{H} = \dim \mathcal{R}(T) + \dim \mathcal{R}(T)^\perp$. Thus if $\mathcal{H}$ is finite-dimensional (and so $\mathcal{N}(T)$ and $\mathcal{N}(T^*)$ are finite-dimensional), then $\operatorname{ind}(T) = 0$. Indeed, if $\dim \mathcal{H} < \infty$, then

$$\dim \mathcal{N}(T) - \dim \mathcal{N}(T^*) = \dim \mathcal{N}(T) - \dim \mathcal{R}(T)^\perp$$
$$= \dim \mathcal{N}(T) + \dim \mathcal{R}(T) - \dim \mathcal{H}$$
$$= \dim \mathcal{H} - \dim \mathcal{H} = 0.$$

Since linear manifolds of finite-dimensional spaces are closed, then *on a finite-dimensional space every operator is Fredholm with a null index* (Corollary 5.2):

$$\dim \mathcal{H} < \infty \implies \{T \in \mathcal{F}[\mathcal{H}] \colon \operatorname{ind}(T) = 0\} = \mathcal{B}[\mathcal{H}].$$

(e) FREDHOLM ALTERNATIVE. If $K \in \mathcal{B}_\infty[\mathcal{H}]$ and $\lambda \neq 0$, then $\mathcal{R}(\lambda I - K)$ is closed and $\dim \mathcal{N}(\lambda I - K) = \dim \mathcal{N}(\overline{\lambda} I - K^*) < \infty$. This is the Fredholm Alternative for compact operators of Corollary 1.20, which can be restated in terms of Fredholm indices, according to Corollary 5.2, as follows.

  If $K \in \mathcal{B}_\infty[\mathcal{H}]$ and $\lambda \neq 0$, then $\lambda I - K$ is Fredholm with $\operatorname{ind}(\lambda I - K) = 0$.

However, $\operatorname{ind}(\lambda I - K) = 0$ means $\dim \mathcal{N}(\lambda I - K) = \dim \mathcal{R}(\lambda I - K)^\perp$ (since $\dim \mathcal{N}(\overline{\lambda} I - K^*) = \dim \mathcal{R}(\lambda I - K)^\perp$ by Lemma 1.5), and this implies that

$\mathcal{N}(\lambda I - K) = \{0\}$ if and only if $\mathcal{R}(\lambda I - K)^- = \mathcal{H}$ (equivalently, if and only if $\mathcal{R}(\lambda I - K)^\perp = \{0\}$). Since $\mathcal{R}(\lambda I - K)$ is closed if $\lambda I - K$ is Fredholm, we get still another form of the Fredholm Alternative (cf. Theorems 1.18 and 2.18).

> If $K \in \mathcal{B}_\infty[\mathcal{H}]$ and $\lambda \neq 0$, then $\mathcal{N}(\lambda I - K) = \{0\} \iff \mathcal{R}(\lambda I - K) = \mathcal{H}$

(i.e., if K is a compact operator and λ is a nonzero scalar, then $\lambda I - K$ is injective if and only if it is surjective).

(f) FREDHOLM INDEX. Corollary 5.2 and some of its straightforward consequences can also be naturally rephrased in terms of Fredholm indices. Indeed, $\dim \mathcal{N}(T)$ and $\dim \mathcal{N}(T^*)$ are both finite if and only if $\mathrm{ind}\,(T)$ is finite (reason: $\mathrm{ind}\,(T)$ was defined for semi-Fredholm operators only, and so if one of $\dim \mathcal{N}(T)$ or $\dim \mathcal{N}(T^*)$ is infinite, then the other must be finite). Thus *T is Fredholm if and only if it is semi-Fredholm with a finite index*, and hence

$$T \in \mathcal{SF}[\mathcal{H}] \quad \Longrightarrow \quad \{T \in \mathcal{F}[\mathcal{H}] \iff |\mathrm{ind}\,(T)| < \infty\}.$$

Moreover, since $\mathcal{F}_\ell[\mathcal{H}] \backslash \mathcal{F}_r[\mathcal{H}] = \mathcal{F}_\ell[\mathcal{H}] \backslash \mathcal{F}[\mathcal{H}]$ and $\mathcal{F}_r[\mathcal{H}] \backslash \mathcal{F}_\ell[\mathcal{H}] = \mathcal{F}_r[\mathcal{H}] \backslash \mathcal{F}[\mathcal{H}]$,

$$T \in \mathcal{SF}[\mathcal{H}] \quad \Longrightarrow \quad \{T \in \mathcal{F}_r[\mathcal{H}] \backslash \mathcal{F}_\ell[\mathcal{H}] \iff \mathrm{ind}\,(T) = +\infty\},$$

$$T \in \mathcal{SF}[\mathcal{H}] \quad \Longrightarrow \quad \{T \in \mathcal{F}_\ell[\mathcal{H}] \backslash \mathcal{F}_r[\mathcal{H}] \iff \mathrm{ind}\,(T) = -\infty\}.$$

(g) PRODUCT. Still from Corollary 5.2, a nonzero scalar operator is Fredholm with a null index. This is readily generalized:

> If $T \in \mathcal{F}[\mathcal{H}]$, then $\gamma T \in \mathcal{F}[\mathcal{H}]$ and $\mathrm{ind}\,(\gamma T) = \mathrm{ind}\,(T)$ for every $\gamma \in \mathbb{C} \backslash \{0\}$.

A further generalization leads to the most important property of the index, namely its *logarithmic additivity*: $\mathrm{ind}\,(ST) = \mathrm{ind}\,(S) + \mathrm{ind}\,(T)$ whenever such an addition makes sense.

Theorem 5.4. *Take $S, T \in \mathcal{B}[\mathcal{H}]$.*

(a) *If $S, T \in \mathcal{F}_\ell[\mathcal{H}]$, then $ST \in \mathcal{F}_\ell[\mathcal{H}]$.*
   If $S, T \in \mathcal{F}_r[\mathcal{H}]$, then $ST \in \mathcal{F}_r[\mathcal{H}]$.
   (Therefore, if $S, T \in \mathcal{F}[\mathcal{H}]$, then $ST \in \mathcal{F}[\mathcal{H}]$.)

(b) *If $S, T \in \mathcal{F}_\ell[\mathcal{H}]$ or $S, T \in \mathcal{F}_r[\mathcal{H}]$ (in particular, if $S, T \in \mathcal{F}[\mathcal{H}]$), then*

$$\mathrm{ind}\,(ST) = \mathrm{ind}\,(S) + \mathrm{ind}\,(T).$$

Proof. (a) If S and T are left semi-Fredholm, then there are A, B, K, L in $\mathcal{B}[\mathcal{H}]$, with K, L being compact, such that $BS = I + L$ and $AT = I + K$. Then

$$(AB)(ST) = A(I + L)T = AT + ALT = I + (K + ALT).$$

But $K + ALT$ is compact (because $\mathcal{B}_\infty[\mathcal{H}]$ is an ideal of $\mathcal{B}[\mathcal{H}]$). Thus ST is left semi-Fredholm. Summing up: $S, T \in \mathcal{F}_\ell[\mathcal{H}]$ implies $ST \in \mathcal{F}_\ell[\mathcal{H}]$. Dually,

$S, T \in \mathcal{F}_r[\mathcal{H}]$ if and only if $S^*, T^* \in \mathcal{F}_\ell[\mathcal{H}]$, which (as we saw above) implies $T^* S^* \in \mathcal{F}_\ell[\mathcal{H}]$, which means $ST = (T^* S^*)^* \in \mathcal{F}_r[\mathcal{H}]$.

(b) We split the proof of $\mathrm{ind}\,(ST) = \mathrm{ind}\,(S) + \mathrm{ind}\,(T)$ into three parts.

(b_1) Take $S, T \in \mathcal{F}_\ell[\mathcal{H}]$. First suppose $\mathrm{ind}\,(S)$ and $\mathrm{ind}\,(T)$ are both finite or, equivalently, suppose $S, T \in \mathcal{F}[\mathcal{H}] = \mathcal{F}_\ell[\mathcal{H}] \cap \mathcal{F}_r[\mathcal{H}]$. Consider the surjective transformation $L \colon \mathcal{N}(ST) \to \mathcal{R}(L) \subseteq \mathcal{H}$ defined by

$$Lx = Tx \ \text{ for every } \ x \in \mathcal{N}(ST),$$

which is clearly linear with $\mathcal{N}(L) = \mathcal{N}(T) \cap \mathcal{N}(ST)$ by the definition of L. Since $\mathcal{N}(T) \subseteq \mathcal{N}(ST)$,

$$\mathcal{N}(L) = \mathcal{N}(T).$$

If $y \in \mathcal{R}(L) = L(\mathcal{N}(ST))$, then $y = Lx = Tx$ for some vector $x \in \mathcal{H}$ such that $STx = 0$, and so $Sy = 0$, which implies $y \in \mathcal{R}(T) \cap \mathcal{N}(S)$. Therefore we get $\mathcal{R}(L) \subseteq \mathcal{R}(T) \cap \mathcal{N}(S)$. Conversely, if $y \in \mathcal{R}(T) \cap \mathcal{N}(S)$, then $y = Tx$ for some vector $x \in \mathcal{H}$ and $Sy = 0$, and so $STx = 0$, which implies $x \in \mathcal{N}(ST)$ and $y = Tx = Lx \in \mathcal{R}(L)$. Thus $\mathcal{R}(T) \cap \mathcal{N}(S) \subseteq \mathcal{R}(L)$. Hence

$$\mathcal{R}(L) = \mathcal{R}(T) \cap \mathcal{N}(S).$$

Recalling again the rank and nullity identity of linear algebra, $\dim \mathcal{X} = \dim \mathcal{N}(L) + \dim \mathcal{R}(L)$ for every linear transformation L on a linear space $\mathcal{X}$. Thus, with $\mathcal{X} = \mathcal{N}(ST)$ and since $\dim \mathcal{N}(S) < \infty$,

$$\dim \mathcal{N}(ST) = \dim \mathcal{N}(T) + \dim (\mathcal{R}(T) \cap \mathcal{N}(S))$$
$$= \dim \mathcal{N}(T) + \dim \mathcal{N}(S) + \dim (\mathcal{R}(T) \cap \mathcal{N}(S)) - \dim \mathcal{N}(S).$$

Moreover, since $\mathcal{R}(T)$ is closed, $\mathcal{R}(T) \cap \mathcal{N}(S)$ is a subspace of $\mathcal{N}(S)$. So (by the Projection Theorem) $\mathcal{N}(S) = \mathcal{R}(T) \cap \mathcal{N}(S) \oplus (\mathcal{N}(S) \ominus (\mathcal{R}(T) \cap \mathcal{N}(S)))$, and hence $\dim (\mathcal{N}(S) \ominus (\mathcal{R}(T) \cap \mathcal{N}(S))) = \dim \mathcal{N}(S) - \dim (\mathcal{R}(T) \cap \mathcal{N}(S))$ (because $\dim \mathcal{N}(S) < \infty$). Thus

$$\dim \mathcal{N}(T) + \dim \mathcal{N}(S) = \dim \mathcal{N}(ST) + \dim (\mathcal{N}(S) \ominus (\mathcal{R}(T) \cap \mathcal{N}(S))).$$

Swapping T with S^* and S with T^* (which have finite-dimensional kernels and closed ranges), it follows by Lemma 1.4 and Proposition 1.H(a,b),

$$\dim \mathcal{N}(S^*) + \dim \mathcal{N}(T^*) = \dim \mathcal{N}(T^* S^*) + \dim(\mathcal{N}(T^*) \ominus (\mathcal{R}(S^*) \cap \mathcal{N}(T^*)))$$
$$= \dim \mathcal{N}((ST)^*) + \dim(\mathcal{R}(T)^\perp \ominus (\mathcal{N}(S)^\perp \cap \mathcal{R}(T)^\perp))$$
$$= \dim \mathcal{N}((ST)^*) + \dim(\mathcal{R}(T)^\perp \ominus (\mathcal{R}(T) + \mathcal{N}(S))^\perp)$$
$$= \dim \mathcal{N}((ST)^*) + \dim(\mathcal{N}(S) \ominus (\mathcal{R}(T) \cap \mathcal{N}(S))).$$

Therefore

$$\dim \mathcal{N}(S) - \dim \mathcal{N}(S^*) + \dim \mathcal{N}(T) - \dim \mathcal{N}(T^*)$$
$$= \dim \mathcal{N}(ST) - \dim \mathcal{N}((ST)^*),$$

and so

$$\mathrm{ind}\,(ST) = \mathrm{ind}\,(S) + \mathrm{ind}\,(T).$$

(b$_2$) If $S, T \in \mathcal{F}_\ell[\mathcal{H}]$, then $\mathrm{ind}\,(S), \mathrm{ind}\,(T) \neq +\infty$. Suppose $S, T \in \mathcal{F}_\ell[\mathcal{H}]$ and one of $\mathrm{ind}\,(S)$ or $\mathrm{ind}\,(T)$ is not finite; that is, $\mathrm{ind}\,(S) = -\infty$ or $\mathrm{ind}\,(T) = -\infty$. If $\mathrm{ind}\,(S) = -\infty$, then $\dim \mathcal{N}(S^*) = \infty$, and so $\dim \mathcal{R}(S)^\perp = \dim \mathcal{N}(S^*) = \infty$. Since $\mathcal{R}(ST) \subseteq \mathcal{R}(S)$, we get $\mathcal{R}(S)^\perp \subseteq \mathcal{R}(ST)^\perp$, and hence $\dim \mathcal{N}((TS)^*) = \dim \mathcal{R}(ST)^\perp = \infty$. Since $TS \in \mathcal{F}_\ell[\mathcal{H}]$ by item (a), then $\dim \mathcal{N}(TS) < \infty$ by Theorem 5.1. Thus $\mathrm{ind}\,(TS) = \dim \mathcal{N}(TS) - \dim \mathcal{N}((TS)^*) = -\infty$. On the other hand, if $\mathrm{ind}\,(S) \neq -\infty$, then $\mathrm{ind}\,(T) = -\infty$, and the same argument leads to $\mathrm{ind}\,(ST) = -\infty$. So, in both cases, $\mathrm{ind}\,(ST) = \mathrm{ind}\,(TS)$ and

$$\mathrm{ind}\,(ST) = -\infty = \mathrm{ind}\,(S) + \mathrm{ind}\,(T).$$

(b$_3$) Suppose $S, T \in \mathcal{F}_r[\mathcal{H}]$. Thus $S^*, T^* \in \mathcal{F}_\ell[\mathcal{H}]$, which (as we saw in items (b$_1$) and (b$_2$)) implies that $\mathrm{ind}\,(T^*S^*) = \mathrm{ind}\,(S^*T^*) = \mathrm{ind}\,(S^*) + \mathrm{ind}\,(T^*)$. But $\mathrm{ind}\,(S^*) = -\mathrm{ind}\,(S)$, $\mathrm{ind}\,(T^*) = -\mathrm{ind}\,(T)$, $\mathrm{ind}\,(T^*S^*) = \mathrm{ind}\,((ST)^*) = -\mathrm{ind}\,(ST)$, and $\mathrm{ind}\,(S^*T^*) = \mathrm{ind}\,((TS)^*) = -\mathrm{ind}\,(TS)$. Thus, if $S, T \in \mathcal{F}_r[\mathcal{H}]$,

$$\mathrm{ind}\,(ST) = \mathrm{ind}\,(S) + \mathrm{ind}\,(T). \qquad \square$$

If $S, T \in \mathcal{SF}[\mathcal{H}] \backslash \mathcal{F}[\mathcal{H}]$ (and so $\mathrm{ind}\,(S)$ and $\mathrm{ind}\,(T)$ are both not finite), then the expression $\mathrm{ind}\,(S) + \mathrm{ind}\,(T)$ makes sense as an extended integer in $\overline{\mathbb{Z}}$ if and only if either $\mathrm{ind}\,(S) = \mathrm{ind}\,(T) = +\infty$ or $\mathrm{ind}\,(S) = \mathrm{ind}\,(T) = -\infty$. But this is equivalent to saying that either $S, T \in \mathcal{F}_r[\mathcal{H}] \backslash \mathcal{F}[\mathcal{H}] = \mathcal{F}_r[\mathcal{H}] \backslash \mathcal{F}_\ell[\mathcal{H}]$ or $S, T \in \mathcal{F}_\ell[\mathcal{H}] \backslash \mathcal{F}[\mathcal{H}] = \mathcal{F}_\ell[\mathcal{H}] \backslash \mathcal{F}_r[\mathcal{H}]$. Therefore if $S, T \in \mathcal{SF}[\mathcal{H}]$ with one of them in $\mathcal{F}_r[\mathcal{H}] \backslash \mathcal{F}_\ell[\mathcal{H}]$ and the other in $\mathcal{F}_\ell[\mathcal{H}] \backslash \mathcal{F}_r[\mathcal{H}]$, then the index expression of Theorem 5.4 does not make sense. Moreover, if one of them lies in $\mathcal{F}_r[\mathcal{H}] \backslash \mathcal{F}_\ell[\mathcal{H}]$ and the other lies in $\mathcal{F}_\ell[\mathcal{H}] \backslash \mathcal{F}_r[\mathcal{H}]$, then their product may not be an operator in $\mathcal{SF}[\mathcal{H}]$. For instance, let $\mathcal{H}$ be an infinite-dimensional Hilbert space, and let S_+ be the canonical unilateral shift (of infinite multiplicity) on the Hilbert space $\ell_+^2(\mathcal{H}) = \bigoplus_{k=0}^\infty \mathcal{H}$ (cf. Section 2.7). As is readily verified, $\dim \mathcal{N}(S_+) = \infty$, $\dim \mathcal{N}(S_+^*) = 0$, and $\mathcal{R}(S_+)$ is closed in $\ell_+^2(\mathcal{H})$. In fact, $\mathcal{N}(S_+) = \ell_+^2(\mathcal{H}) \ominus \bigoplus_{k=1}^\infty \mathcal{H} \cong \mathcal{H}$, $\mathcal{N}(S_+^*) = \{0\}$, and $\mathcal{R}(S_+) = \{0\} \oplus \bigoplus_{k=1}^\infty \mathcal{H}$. Therefore $S_+ \in \mathcal{F}_r[\mathcal{H}] \backslash \mathcal{F}_\ell[\mathcal{H}]$ and $S_+^* \in \mathcal{F}_\ell[\mathcal{H}] \backslash \mathcal{F}_r[\mathcal{H}]$ (according to Theorem 5.1). However, $S_+ S_+^* \notin \mathcal{SF}[\mathcal{H}]$. Indeed, $S_+ S_+^* = O \oplus I$ (where O stands for the null operator on $\mathcal{H}$ and I for the identity operator on $\bigoplus_{k=1}^\infty \mathcal{H}$) does not lie in $\mathcal{SF}[\mathcal{H}]$ since it is self-adjoint and $\dim \mathcal{N}(S_+ S_+^*) = \infty$ (cf. Theorem 5.1).

Corollary 5.5. *Take an arbitrary nonnegative integer n. If $T \in \mathcal{F}_\ell[\mathcal{H}]$ (or if $T \in \mathcal{F}_r[\mathcal{H}]$), then $T^n \in \mathcal{F}_\ell[\mathcal{H}]$ (or $T^n \in \mathcal{F}_r[\mathcal{H}]$) and*

$$\mathrm{ind}\,(T^n) = n\,\mathrm{ind}\,(T).$$

Proof. The result holds trivially for $n = 0$ (the identity is Fredholm with index zero), and tautologically for $n = 1$. Thus suppose $n \geq 2$. The result holds for

$n = 2$ by Theorem 5.4, and a trivial induction ensures the claimed result for each $n \geq 2$ by using Theorem 5.4 again. $\qquad\square$

The null operator on an infinite-dimensional space is not Fredholm, and so a compact operator may not be Fredholm. However, the sum of a Fredholm operator and a compact operator is a Fredholm operator with the same index. In other words, the index of a Fredholm operator remains unchanged under compact perturbation, a property which is referred to as *index stability*.

Theorem 5.6. *Take $T \in \mathcal{B}[\mathcal{H}]$ and $K \in \mathcal{B}_\infty[\mathcal{H}]$.*

(a) $T \in \mathcal{F}_\ell[\mathcal{H}] \iff T + K \in \mathcal{F}_\ell[\mathcal{H}]$ *and* $T \in \mathcal{F}_r[\mathcal{H}] \iff T + K \in \mathcal{F}_r[\mathcal{H}]$.

 In particular,

$$T \in \mathcal{F}[\mathcal{H}] \iff T + K \in \mathcal{F}[\mathcal{H}] \quad \text{and} \quad T \in \mathcal{SF}[\mathcal{H}] \iff T + K \in \mathcal{SF}[\mathcal{H}].$$

(b) *Moreover, for every $T \in \mathcal{SF}[\mathcal{H}]$,*

$$\mathrm{ind}\,(T + K) = \mathrm{ind}\,(T).$$

Proof. (a) If $T \in \mathcal{F}_\ell[\mathcal{H}]$, then there exist $A \in \mathcal{B}[\mathcal{H}]$ and $K_1 \in \mathcal{B}_\infty[\mathcal{H}]$ such that $AT = I + K_1$. Hence $A \in \mathcal{F}_r[\mathcal{H}]$. Take any $K \in \mathcal{B}_\infty[\mathcal{H}]$. Set $K_2 = K_1 + AK$, which lies in $\mathcal{B}_\infty[\mathcal{H}]$ because $\mathcal{B}_\infty[\mathcal{H}]$ is an ideal of $\mathcal{B}[\mathcal{H}]$. Since $A(T + K) = I + K_2$, then $T + K \in \mathcal{F}_\ell[\mathcal{H}]$. Therefore $T \in \mathcal{F}_\ell[\mathcal{H}]$ implies $T + K \in \mathcal{F}_\ell[\mathcal{H}]$. The converse also holds because $T = (T + K) - K$. Hence

$$T \in \mathcal{F}_\ell[\mathcal{H}] \iff T + K \in \mathcal{F}_\ell[\mathcal{H}].$$

Dually, if $T \in \mathcal{F}_r[\mathcal{H}]$ and $K \in \mathcal{B}_\infty[\mathcal{H}]$, then $T^* \in \mathcal{F}_\ell[\mathcal{H}]$ and $K^* \in \mathcal{B}_\infty[\mathcal{H}]$, so that $T^* + K^* \in \mathcal{F}_\ell[\mathcal{H}]$, and hence $T + K = (T^* + K^*)^* \in \mathcal{F}_r[\mathcal{H}]$. Therefore

$$T \in \mathcal{F}_r[\mathcal{H}] \iff T + K \in \mathcal{F}_r[\mathcal{H}].$$

(b) First suppose $T \in \mathcal{F}_\ell[\mathcal{H}]$ as above. By the Fredholm Alternative of Corollary 1.20, the operators $I + K_1$ and $I + K_2$ are both Fredholm with index zero (see Remark 5.3(e)). Therefore $AT = I + K_1 \in \mathcal{F}[\mathcal{H}]$ with $\mathrm{ind}\,(AT) = 0$ and $A(T + K) = I + K_2 \in \mathcal{F}[\mathcal{H}]$ with $\mathrm{ind}\,(A(T+K)) = 0$. Recall: since $T \in \mathcal{F}_\ell[\mathcal{H}]$, then $A \in \mathcal{F}_r[\mathcal{H}]$. If $T \in \mathcal{F}[\mathcal{H}]$, then $T + K \in \mathcal{F}[\mathcal{H}]$ by item (a). In this case, applying Theorem 5.4 for operators in $\mathcal{F}_r[\mathcal{H}]$,

$$\mathrm{ind}\,(T + K) + \mathrm{ind}\,(A) = \mathrm{ind}\,(A(T + K)) = 0 = \mathrm{ind}\,(AT) = \mathrm{ind}\,(A) + \mathrm{ind}\,(T).$$

Since $\mathrm{ind}\,(T)$ is finite (whenever $T \in \mathcal{F}[\mathcal{H}]$), then $\mathrm{ind}\,(A) = -\mathrm{ind}\,(T)$ is finite as well, and so we may subtract $\mathrm{ind}\,(A)$ to get

$$\mathrm{ind}\,(T + K) = \mathrm{ind}\,(T).$$

On the other hand, if T lies in $\mathcal{F}_\ell[\mathcal{H}]\backslash\mathcal{F}[\mathcal{H}] = \mathcal{F}_\ell[\mathcal{H}]\backslash\mathcal{F}_r[\mathcal{H}]$, then $T + K$ lies in $\mathcal{F}_\ell[\mathcal{H}]\backslash\mathcal{F}_r[\mathcal{H}]$ by (a). In this case (see Remark 5.3(f)),

$$\mathrm{ind}\,(T + K) = -\infty = \mathrm{ind}\,(T).$$

If $T \in \mathcal{F}_r[\mathcal{H}] \backslash \mathcal{F}[\mathcal{H}]$ and $K \in \mathcal{B}_\infty[\mathcal{H}]$, then $T^* \in \mathcal{F}_\ell[\mathcal{H}] \backslash \mathcal{F}[\mathcal{H}]$ and $K^* \in \mathcal{B}_\infty[\mathcal{H}]$. So

$$\mathrm{ind}\,(T + K) = -\mathrm{ind}\,(T^* + K^*) = -\mathrm{ind}\,(T^*) = \mathrm{ind}\,(T). \qquad \square$$

Remark 5.7. (a) INDEX OF A. We have established the following assertion in the preceding proof.

> If $T \in \mathcal{F}[\mathcal{H}]$ and $A \in \mathcal{B}[\mathcal{H}]$ are such that $AT = I + K$ for some $K \in \mathcal{B}_\infty[\mathcal{H}]$, then $A \in \mathcal{F}[\mathcal{H}]$ and $\mathrm{ind}\,(A) = -\mathrm{ind}\,(T) = \mathrm{ind}\,(T^*)$.

But if $T \in \mathcal{F}[\mathcal{H}]$, then $T \in \mathcal{F}_r[\mathcal{H}]$, so that $TA = I + K$ for some $K \in \mathcal{B}_\infty[\mathcal{H}]$. Thus $A \in \mathcal{F}_\ell[\mathcal{H}]$. And a similar argument leads to the symmetric statement.

> If $T \in \mathcal{F}[\mathcal{H}]$ and $A \in \mathcal{B}[\mathcal{H}]$ are such that $TA = I + K$ for some $K \in \mathcal{B}_\infty[\mathcal{H}]$, then $A \in \mathcal{F}[\mathcal{H}]$ and $\mathrm{ind}\,(A) = -\mathrm{ind}\,(T) = \mathrm{ind}\,(T^*)$.

(b) WEYL OPERATOR. A *Weyl operator* is a Fredholm operator with null index (equivalently, a semi-Fredholm operator with null index). Let

$$\mathcal{W}[\mathcal{H}] = \big\{ T \in \mathcal{F}[\mathcal{H}] \colon \mathrm{ind}\,(T) = 0 \big\}$$

denote the class of all Weyl operators from $\mathcal{B}[\mathcal{H}]$. Since $T \in \mathcal{F}[\mathcal{H}]$ if and only if $T^* \in \mathcal{F}[\mathcal{H}]$ and $\mathrm{ind}\,(T^*) = -\mathrm{ind}\,(T)$, we conclude

$$T \in \mathcal{W}[\mathcal{H}] \quad \Longleftrightarrow \quad T^* \in \mathcal{W}[\mathcal{H}].$$

According to items (d), (e), and (g) in Remark 5.3 we get:

(i) every operator on a finite-dimensional space is a Weyl operator,

$$\dim \mathcal{H} < \infty \quad \Longrightarrow \quad \mathcal{W}[\mathcal{H}] = \mathcal{B}[\mathcal{H}],$$

(ii) the Fredholm Alternative can be rephrased as

$$K \in \mathcal{B}_\infty[\mathcal{H}] \ \text{ and } \ \lambda \neq 0 \quad \Longrightarrow \quad \lambda I - K \in \mathcal{W}[\mathcal{H}],$$

(iii) every nonzero multiple of a Weyl operator is again a Weyl operator,

$$T \in \mathcal{W}[\mathcal{H}] \quad \Longrightarrow \quad \gamma T \in \mathcal{W}[\mathcal{H}] \ \text{ for every } \ \gamma \in \mathbb{C} \backslash \{0\}.$$

In particular, every nonzero scalar operator is a Weyl operator. In fact, the product of two Weyl operators is again a Weyl operator (by Theorem 5.4),

$$S, T \in \mathcal{W}[\mathcal{H}] \quad \Longrightarrow \quad ST \in \mathcal{W}[\mathcal{H}].$$

Thus integral powers of Weyl operators are Weyl operators (Corollary 5.5),

$$T \in \mathcal{W}[\mathcal{H}] \quad \Longrightarrow \quad T^n \in \mathcal{W}[\mathcal{H}] \ \text{ for every } \ n \in \mathbb{N}_0.$$

On an infinite-dimensional space, the identity is Weyl but not compact; and the null operator is compact but not semi-Fredholm. Actually,

$$T \in \mathcal{F}[\mathcal{H}] \cap \mathcal{B}_\infty[\mathcal{H}] \quad \Longleftrightarrow \quad \dim \mathcal{H} < \infty \quad (\Longleftrightarrow \quad T \in \mathcal{W}[\mathcal{H}] \cap \mathcal{B}_\infty[\mathcal{H}]).$$

(Set $T = -K$ in Theorem 5.6, and for the converse see Remark 5.3(d).) Also, if T is normal, then $\mathcal{N}(T) = \mathcal{N}(T^*T) = \mathcal{N}(TT^*) = \mathcal{N}(T^*)$ by Lemma 1.4, and so every normal Fredholm operator is Weyl,

$$T \text{ normal in } \mathcal{F}[\mathcal{H}] \quad \Longrightarrow \quad T \in \mathcal{W}[\mathcal{H}].$$

Since $T \in \mathcal{B}[\mathcal{H}]$ is invertible if and only $\mathcal{N}(T) = \{0\}$ and $\mathcal{R}(T) = \mathcal{H}$ (Theorem 1.1), which happens if and only if $T^* \in \mathcal{B}[\mathcal{H}]$ is invertible (Proposition 1.L), it follows by Corollary 5.2 that every invertible operator is Weyl,

$$T \in \mathcal{G}[\mathcal{H}] \quad \Longrightarrow \quad T \in \mathcal{W}[\mathcal{H}].$$

Moreover, by Theorem 5.6, for every compact $K \in \mathcal{B}_\infty[\mathcal{H}]$,

$$T \in \mathcal{W}[\mathcal{H}] \quad \Longrightarrow \quad T + K \in \mathcal{W}[\mathcal{H}].$$

5.2 Essential (Fredholm) Spectrum and Spectral Picture

An element a in a unital algebra $\mathcal{A}$ is *left invertible* if there is an element a_ℓ in $\mathcal{A}$ (a *left inverse* of a) such that $a_\ell a = 1$ where 1 stands for the identity in $\mathcal{A}$, and it is *right invertible* if there is an element a_r in $\mathcal{A}$ (a *right inverse* of a) such that $a a_r = 1$. An element a in $\mathcal{A}$ is *invertible* if there is an element a^{-1} in $\mathcal{A}$ (the *inverse* of a) such that $a^{-1}a = a a^{-1} = 1$. Thus a in $\mathcal{A}$ is invertible if and only if it has a left inverse a_ℓ in $\mathcal{A}$ and a right inverse a_r in $\mathcal{A}$, which coincide with its inverse a^{-1} in $\mathcal{A}$ (since $a_r = a_\ell a a_r = a_\ell$).

Lemma 5.8. *Take an operator S in the unital Banach algebra $\mathcal{B}[\mathcal{H}]$.*

(a) *S is left invertible if and only if it is injective with a closed range (i.e., $\mathcal{N}(S) = \{0\}$ and $\mathcal{R}(S)^- = \mathcal{R}(S)$).*

(b) *S is left (right) invertible if and only if S^* is right (left) invertible.*

(c) *S is right invertible if and only if it is surjective (i.e., $\mathcal{R}(S) = \mathcal{H}$).*

Proof. An operator $S \in \mathcal{B}[\mathcal{H}]$ has a bounded inverse on its range $\mathcal{R}(S)$ if there exists $S^{-1} \in \mathcal{B}[\mathcal{R}(S), \mathcal{H}]$ such that $S^{-1}S = I \in \mathcal{B}[\mathcal{H}]$, and this implies $SS^{-1} = I \in \mathcal{B}[\mathcal{R}(S)]$ (cf. Section 1.2). Moreover, $S \in \mathcal{B}[\mathcal{H}]$ has a bounded inverse on its range if and only if $\mathcal{N}(S) = \{0\}$ and $\mathcal{R}(S)$ is closed (Theorem 1.2).

Claim. S is left invertible if and only if it has a bounded inverse on its range.

Proof. If $S_\ell \in \mathcal{B}[\mathcal{H}]$ is a left inverse of $S \in \mathcal{B}[\mathcal{H}]$, then $S_\ell S = I \in \mathcal{B}[\mathcal{H}]$. Hence the restriction $S_\ell|_{\mathcal{R}(S)} \in \mathcal{B}[\mathcal{R}(S), \mathcal{H}]$ is such that $S_\ell|_{\mathcal{R}(S)} S = I \in \mathcal{B}[\mathcal{H}]$. Thus

$S^{-1} = S_\ell|_{\mathcal{R}(S)} \in \mathcal{B}[\mathcal{R}(S), \mathcal{H}]$ is a bounded inverse of S on $\mathcal{R}(S)$. Conversely, if $S^{-1} \in \mathcal{B}[\mathcal{R}(S), \mathcal{H}]$ is a bounded inverse of S on $\mathcal{R}(S)$, then $\mathcal{R}(S)$ is closed (cf. Theorem 1.2). Take the orthogonal projection $E \in \mathcal{B}[\mathcal{H}]$ onto the subspace $\mathcal{R}(S)$. Consider the extension $S^{-1}E \in \mathcal{B}[\mathcal{H}]$ of $S^{-1} \in \mathcal{B}[\mathcal{R}(S), \mathcal{H}]$ over the whole space $\mathcal{H}$, as in Proposition 1.G(b). Thus $S^{-1}ES = S^{-1}S = I \in \mathcal{B}[\mathcal{H}]$. Then $S_\ell = S^{-1}E \in \mathcal{B}[\mathcal{H}]$ is a left inverse of $S \in \mathcal{B}[\mathcal{H}]$. This proves the claim.

(a) But S has a bounded inverse on its range if and only if it is injective (i.e., $\mathcal{N}(S) = \{0\}$) and $\mathcal{R}(S)$ is closed, according to Theorem 1.2.

(b) Since $S_\ell S = I$ ($S S_r = I$) if and only if $S^* S_\ell^* = I$ ($S_r^* S^* = I$), it follows that S is left (right) invertible if and only if S^* is right (left) invertible.

(c) Thus, according to items (a) and (b), S is right invertible if and only if $\mathcal{N}(S^*) = \{0\}$ (i.e., $\mathcal{N}(S^*)^\perp = \mathcal{H}$) and $\mathcal{R}(S^*)$ is closed, or equivalently (cf. Lemmas 1.4 and 1.5) $\mathcal{R}(S) = \mathcal{R}(S)^- = \mathcal{H}$, which means $\mathcal{R}(S) = \mathcal{H}$.   □

Let T be an operator in $\mathcal{B}[\mathcal{H}]$. The *left spectrum* $\sigma_\ell(T)$ and the *right spectrum* $\sigma_r(T)$ of $T \in \mathcal{B}[\mathcal{H}]$ are the sets

$$\sigma_\ell(T) = \{\lambda \in \mathbb{C} \colon \lambda I - T \text{ is not left invertible}\},$$

$$\sigma_r(T) = \{\lambda \in \mathbb{C} \colon \lambda I - T \text{ is not right invertible}\},$$

and so the spectrum $\sigma(T)$ of $T \in \mathcal{B}[\mathcal{H}]$ is given by

$$\sigma(T) = \{\lambda \in \mathbb{C} \colon \lambda I - T \text{ is not invertible}\} = \sigma_\ell(T) \cup \sigma_r(T).$$

Corollary 5.9. *Take an arbitrary operator $T \in \mathcal{B}[\mathcal{H}]$.*

$$\sigma_\ell(T) = \{\lambda \in \mathbb{C} \colon \mathcal{R}(\lambda I - T) \text{ is not closed } or \ \mathcal{N}(\lambda I - T) \neq \{0\}\}$$
$$= \sigma(T) \backslash \sigma_{R_1}(T) = \sigma_{AP}(T) \subseteq \sigma(T),$$

$$\sigma_r(T) = \{\lambda \in \mathbb{C} \colon \mathcal{R}(\lambda I - T) \text{ is not closed } or \ \mathcal{N}(\overline{\lambda} I - T^*) \neq \{0\}\}$$
$$= \{\lambda \in \mathbb{C} \colon \mathcal{R}(\lambda I - T) \text{ is not closed } or \ \mathcal{R}(\lambda I - T)^- \neq \mathcal{H}\}$$
$$= \{\lambda \in \mathbb{C} \colon \mathcal{R}(\lambda I - T) \neq \mathcal{H}\}$$
$$= \sigma(T) \backslash \sigma_{P_1}(T) = \sigma_{AP}(T^*)^* \subseteq \sigma(T).$$

Proof. Take $T \in \mathcal{B}[\mathcal{H}]$ and $\lambda \in \mathbb{C}$. Set $S = \lambda I - T$ in $\mathcal{B}[\mathcal{H}]$. By Lemma 5.8, S is not left invertible if and only if $\mathcal{R}(S)$ is not closed or S is not injective (i.e., or $\mathcal{N}(S) \neq 0$). This proves the first expression for $\sigma_\ell(T)$. By Lemma 5.8, S is not right invertible if and only if $\mathcal{R}(S) \neq \mathcal{H}$, which means $\mathcal{R}(S) \neq \mathcal{R}(S)^-$ or $\mathcal{R}(S)^- \neq \mathcal{H}$, which is equivalent to $\mathcal{R}(S) \neq \mathcal{R}(S)^-$ or $\mathcal{N}(S^*) \neq \{0\}$ (Lemma 1.5), thus proving the first expressions for $\sigma_r(T)$. The remaining identities and inclusions follow from Theorem 2.6 (see also the diagram of Section 2.2).   □

Therefore

$$\sigma_\ell(T) \text{ and } \sigma_r(T) \text{ are closed subsets of } \sigma(T).$$

Indeed, the approximate point spectrum $\sigma_{AP}(T)$ is a closed set (cf. Theorem 2.5). Recall from the remarks following Theorems 2.5 and 2.6 that $\sigma_{P_1}(T)$ and $\sigma_{R_1}(T)$ are open sets, and so $\sigma_{AP}(T) = \sigma(T)\backslash\sigma_{R_1}(T) = \sigma(T)\cap(\mathbb{C}\backslash\sigma_{R_1}(T))$ and $\sigma_{AP}(T^*)^* = \sigma(T)\backslash\sigma_{P_1}(T) = \sigma(T)\cap(\mathbb{C}\backslash\sigma_{P_1}(T))$ are closed (and bounded) sets in $\mathbb{C}$, since $\sigma(T)$ is closed (and bounded); and so is the intersection

$$\sigma_\ell(T)\cap\sigma_r(T) = \sigma(T)\backslash\big(\sigma_{P_1}(T)\cup\sigma_{R_1}(T)\big) = \sigma_{AP}(T)\cap\sigma_{AP}(T^*)^*.$$

Actually, as we saw before, $\sigma_{P_1}(T)$ and $\sigma_{R_1}(T)$ are open subsets of $\mathbb{C}$, and so $\sigma_\ell(T)$ and $\sigma_r(T)$ are closed subsets of $\mathbb{C}$ (and this follows because T lies in the unital Banach algebra $\mathcal{B}[\mathcal{H}]$, and therefore $\sigma(T)$ is a compact set, which in fact happens in any unital Banach algebra). Also, since $(S_\ell S)^* = S^* S_\ell^*$ and $(S S_r)^* = S_r^* S^*$, then S_ℓ is a left inverse of S if and only if S_ℓ^* is a right inverse of S^*, and S_r is a right inverse of S if and only if S_r^* is a left inverse of S^*. Therefore (as we can infer from Corollary 5.9 as well)

$$\sigma_\ell(T) = \sigma_r(T^*)^* \quad \text{and} \quad \sigma_r(T) = \sigma_\ell(T^*)^*.$$

Consider the *Calkin algebra* $\mathcal{B}[\mathcal{H}]/\mathcal{B}_\infty[\mathcal{H}]$ — the quotient algebra of $\mathcal{B}[\mathcal{H}]$ modulo the ideal $\mathcal{B}_\infty[\mathcal{H}]$ of all compact operators. If $\dim\mathcal{H} < \infty$, then all operators are compact, and so $\mathcal{B}[\mathcal{H}]/\mathcal{B}_\infty[\mathcal{H}]$ is trivially null. Thus if the Calkin algebra is brought into play, then the space $\mathcal{H}$ is assumed infinite-dimensional (i.e., $\dim\mathcal{H} = \infty$). Since $\mathcal{B}_\infty[\mathcal{H}]$ is a subspace of $\mathcal{B}[\mathcal{H}]$, then $\mathcal{B}[\mathcal{H}]/\mathcal{B}_\infty[\mathcal{H}]$ is a *unital Banach algebra* whenever $\mathcal{H}$ is *infinite-dimensional*. Moreover, consider the *natural map* (or the *natural quotient map*) $\pi\colon\mathcal{B}[\mathcal{H}]\to\mathcal{B}[\mathcal{H}]/\mathcal{B}_\infty[\mathcal{H}]$ which is defined by (see, e.g., [78, Example 2.H])

$$\pi(T) = [T] = \big\{S\in\mathcal{B}[\mathcal{H}]\colon S = T + K \text{ for some } K\in\mathcal{B}_\infty[\mathcal{H}]\big\} = T + \mathcal{B}_\infty[\mathcal{H}]$$

for every T in $\mathcal{B}[\mathcal{H}]$. The origin of the linear space $\mathcal{B}[\mathcal{H}]/\mathcal{B}_\infty[\mathcal{H}]$ is

$$[0] = \mathcal{B}_\infty[\mathcal{H}] \in \mathcal{B}[\mathcal{H}]/\mathcal{B}_\infty[\mathcal{H}],$$

the kernel of the natural map π is

$$\mathcal{N}(\pi) = \big\{T\in\mathcal{B}[\mathcal{H}]\colon \pi(T) = [0]\big\} = \mathcal{B}_\infty[\mathcal{H}] \subseteq \mathcal{B}[\mathcal{H}],$$

and π is a *unital homomorphism*. Indeed, since $\mathcal{B}_\infty[\mathcal{H}]$ is an ideal of $\mathcal{B}[\mathcal{H}]$,

$$\pi(T+T') = (T+T')+\mathcal{B}_\infty[\mathcal{H}] = (T+\mathcal{B}_\infty[\mathcal{H}])+(T'+\mathcal{B}_\infty[\mathcal{H}]) = \pi(T)+\pi(T'),$$
$$\pi(T T') = (T T')+\mathcal{B}_\infty[\mathcal{H}] = (T+\mathcal{B}_\infty[\mathcal{H}])(T'+\mathcal{B}_\infty[\mathcal{H}]) = \pi(T)\pi(T'),$$

for every $T, T'\in\mathcal{B}[\mathcal{H}]$, and $\pi(I) = [I]$ is the identity element of the algebra $\mathcal{B}[\mathcal{H}]/\mathcal{B}_\infty[\mathcal{H}]$. Furthermore, the norm on $\mathcal{B}[\mathcal{H}]/\mathcal{B}_\infty[\mathcal{H}]$ is given by

$$\|[T]\| = \inf_{K\in\mathcal{B}_\infty[\mathcal{H}]} \|T + K\| \leq \|T\|,$$

so that π is a contraction (see, e.g., [78, Section 4.3]).

Theorem 5.10. *Take any operator $T\in\mathcal{B}[\mathcal{H}]$. $T\in\mathcal{F}_\ell[\mathcal{H}]$ (or $T\in\mathcal{F}_r[\mathcal{H}]$) if and only if $\pi(T)$ is left (or right) invertible in the Calkin algebra $\mathcal{B}[\mathcal{H}]/\mathcal{B}_\infty[\mathcal{H}]$.*

Proof. We show the equivalence of the following assertions.

(a) $T \in \mathcal{F}_\ell[\mathcal{H}]$.

(b) There is a pair of operators $\{A, K\}$ with $A \in \mathcal{B}[\mathcal{H}]$ and $K \in \mathcal{B}_\infty[\mathcal{H}]$ such that $AT = I + K$.

(c) There exists a quadruple of operators $\{A, K_1, K_2, K_3\}$ with $A \in \mathcal{B}[\mathcal{H}]$ and $K_1, K_2, K_3 \in \mathcal{B}_\infty[\mathcal{H}]$ such that $(A + K_1)(T + K_2) = I + K_3$.

(d) $\pi(A)\pi(T) = \pi(I)$, identity in $\mathcal{B}[\mathcal{H}]/\mathcal{B}_\infty[\mathcal{H}]$, for some $\pi(A)$ in $\mathcal{B}[\mathcal{H}]/\mathcal{B}_\infty[\mathcal{H}]$.

(e) $\pi(T)$ is left invertible in $\mathcal{B}[\mathcal{H}]/\mathcal{B}_\infty[\mathcal{H}]$.

By definition, (a) and (b) are equivalent, and (b) implies (c) trivially. If (c) holds, then there exist $B \in [A] = \pi(A)$, $S \in [T] = \pi(T)$, and $J \in [I] = \pi(I)$ for which $BS = J$. Therefore $\pi(A)\pi(T) = [A][T] = [B][S] = \pi(B)\pi(S) = \pi(BS) = \pi(J) = [J] = [I] = \pi(I)$, and so (c) implies (d). Now observe that $X \in \pi(A)\pi(T) = \pi(AT)$ if and only if $X = AT + K_1$ for some $K_1 \in \mathcal{B}_\infty[\mathcal{H}]$, and $X \in \pi(I)$ if and only if $X = I + K_2$ for some $K_2 \in \mathcal{B}_\infty[\mathcal{H}]$. Therefore if $\pi(A)\pi(T) = \pi(I)$, then $X = AT + K_1$ if and only if $X = I + K_2$, and hence $AT = I + K$ with $K = K_2 - K_1 \in \mathcal{B}_\infty[\mathcal{H}]$. Thus (d) implies (b). Finally, (d) and (e) are equivalent by the definition of left invertibility.

Outcome: $T \in \mathcal{F}_\ell[\mathcal{H}]$ if and only if $\pi(T)$ is left invertible in $\mathcal{B}[\mathcal{H}]/\mathcal{B}_\infty[\mathcal{H}]$. Dually, $T \in \mathcal{F}_r[\mathcal{H}]$ if and only if $\pi(T)$ is right invertible in $\mathcal{B}[\mathcal{H}]/\mathcal{B}_\infty[\mathcal{H}]$. $\qquad\square$

The *essential spectrum* (or the *Calkin spectrum*) $\sigma_e(T)$ of $T \in \mathcal{B}[\mathcal{H}]$ is the spectrum of $\pi(T)$ in the unital Banach algebra $\mathcal{B}[\mathcal{H}]/\mathcal{B}_\infty[\mathcal{H}]$,

$$\sigma_e(T) = \sigma(\pi(T)),$$

and so $\sigma_e(T)$ is a compact subset of $\mathbb{C}$. Similarly, the *left essential spectrum* $\sigma_{\ell e}(T)$ and the *right essential spectrum* $\sigma_{re}(T)$ of $T \in \mathcal{B}[\mathcal{H}]$ are defined as the left and the right spectrum of $\pi(T)$ in the Calkin algebra $\mathcal{B}[\mathcal{H}]/\mathcal{B}_\infty[\mathcal{H}]$:

$$\sigma_{\ell e}(T) = \sigma_\ell(\pi(T)) \quad \text{and} \quad \sigma_{re}(T) = \sigma_r(\pi(T)).$$

Hence

$$\sigma_e(T) = \sigma_{\ell e}(T) \cup \sigma_{re}(T).$$

By using only the definitions of left and right semi-Fredholm operators, and of left and right essential spectra, we get the following characterization.

Corollary 5.11. *If $T \in \mathcal{B}[\mathcal{H}]$, then*

$$\sigma_{\ell e}(T) = \{\lambda \in \mathbb{C} \colon \lambda I - T \in \mathcal{B}[\mathcal{H}] \backslash \mathcal{F}_\ell[\mathcal{H}]\},$$

$$\sigma_{re}(T) = \{\lambda \in \mathbb{C} \colon \lambda I - T \in \mathcal{B}[\mathcal{H}] \backslash \mathcal{F}_r[\mathcal{H}]\}.$$

Proof. Take $T \in \mathcal{B}[\mathcal{H}]$. According to Theorem 5.10, $\lambda I - T \notin \mathcal{F}_\ell[\mathcal{H}]$ if and only if $\pi(\lambda I - T)$ is not left invertible in $\mathcal{B}[\mathcal{H}]/\mathcal{B}_\infty[\mathcal{H}]$, which means (definition of left spectrum) $\lambda \in \sigma_\ell(\pi(T))$. But $\sigma_{\ell e}(T) = \sigma_\ell(\pi(T))$ — definition of left essential spectrum. Thus $\lambda \in \sigma_{\ell e}(T)$ if and only if $\lambda I - T \notin \mathcal{F}_\ell[\mathcal{H}]$. Dually, $\lambda \in \sigma_{re}(T)$ if and only if $\lambda I - T \notin \mathcal{F}_r[\mathcal{H}]$. $\qquad\square$

Corollary 5.12. (ATKINSON THEOREM).

$$\sigma_e(T) = \{\lambda \in \mathbb{C} \colon \lambda I - T \in \mathcal{B}[\mathcal{H}]\backslash\mathcal{F}[\mathcal{H}]\}.$$

Proof. The expressions for $\sigma_{\ell e}(T)$ and $\sigma_{re}(T)$ in Corollary 5.11 lead to the claimed identity since $\sigma_e(T) = \sigma_{\ell e}(T) \cup \sigma_{re}(T)$ and $\mathcal{F}[\mathcal{H}] = \mathcal{F}_\ell[\mathcal{H}] \cap \mathcal{F}_r[\mathcal{H}]$. $\square$

Thus the essential spectrum is the set of all scalars λ for which $\lambda I - T$ is not Fredholm. The following equivalent version is also frequently used [8].

Corollary 5.13. (ATKINSON THEOREM). *An operator* $T \in \mathcal{B}[\mathcal{H}]$ *is Fredholm if and only if its image* $\pi(T)$ *in the Calkin algebra* $\mathcal{B}[\mathcal{H}]/\mathcal{B}_\infty[\mathcal{H}]$ *is invertible.*

Proof. Straightforward from Theorem 5.10: T is Fredholm if and only if it is both left and right semi-Fredholm, and $\pi(T)$ is invertible in $\mathcal{B}[\mathcal{H}]/\mathcal{B}_\infty[\mathcal{H}]$ if and only if it is both left and right invertible in $\mathcal{B}[\mathcal{H}]/\mathcal{B}_\infty[\mathcal{H}]$. $\qquad\square$

This is usually referred to by saying that T is Fredholm if and only if T is *essentially invertible*. Thus the essential spectrum $\sigma_e(T)$ is the set of all scalars λ for which $\lambda I - T$ is not essentially invertible (i.e., $\lambda I - T$ is not Fredholm), and so the essential spectrum is also called the *Fredholm spectrum*.

As we saw above (and as happens in any unital Banach algebra — in particular, in $\mathcal{B}[\mathcal{H}]$), the sets $\sigma_\ell(T)$ and $\sigma_r(T)$ are closed in $\mathbb{C}$ because $\sigma(T)$ is a compact subset of $\mathbb{C}$. Similarly, in the unital Banach algebra $\mathcal{B}[\mathcal{H}]/\mathcal{B}_\infty[\mathcal{H}]$,

$$\sigma_{\ell e}(T) \text{ and } \sigma_{re}(T) \text{ are closed subsets of } \sigma_e(T)$$

because $\sigma_e(T) = \sigma(\pi(T))$ is a compact set in $\mathbb{C}$. From Corollary 5.11 we get

$$\mathbb{C}\backslash\sigma_{\ell e}(T) = \{\lambda \in \mathbb{C} \colon \lambda I - T \in \mathcal{F}_\ell[\mathcal{H}]\},$$

$$\mathbb{C}\backslash\sigma_{re}(T) = \{\lambda \in \mathbb{C} \colon \lambda I - T \in \mathcal{F}_r[\mathcal{H}]\},$$

which are open subsets of $\mathbb{C}$. Thus, since $\mathcal{SF}[\mathcal{H}] = \mathcal{F}_\ell[\mathcal{H}] \cup \mathcal{F}_r[\mathcal{H}]$,

$$\mathbb{C}\backslash\big(\sigma_{\ell e}(T)\cap\sigma_{re}(T)\big) = \big(\mathbb{C}\backslash\sigma_{\ell e}(T)\big)\cup\big(\mathbb{C}\backslash\sigma_{re}(T)\big) = \{\lambda \in \mathbb{C} \colon \lambda I - T \in \mathcal{SF}[\mathcal{H}]\},$$

which is again an open subset of $\mathbb{C}$. Therefore the intersection

$$\sigma_{\ell e}(T) \cap \sigma_{re}(T) = \{\lambda \in \mathbb{C} \colon \lambda I - T \in \mathcal{B}[\mathcal{H}]\backslash\mathcal{SF}[\mathcal{H}]\}$$

is a closed subset of $\mathbb{C}$. Since $\mathcal{F}[\mathcal{H}] = \mathcal{F}_\ell[\mathcal{H}] \cap \mathcal{F}_r[\mathcal{H}]$, the complement of the union $\sigma_{\ell e}(T) \cup \sigma_{re}(T)$ is given by (cf. Corollary 5.12)

$$\mathbb{C}\backslash\sigma_e(T) = \mathbb{C}\backslash(\sigma_{\ell e}(T) \cup \sigma_{re}(T))$$
$$= (\mathbb{C}\backslash\sigma_{\ell e}(T)) \cap (\mathbb{C}\backslash\sigma_{re}(T)) = \{\lambda \in \mathbb{C} \colon \lambda I - T \in \mathcal{F}[\mathcal{H}]\},$$

which is an open subset of $\mathbb{C}$.

Corollary 5.14. *If $T \in \mathcal{B}[\mathcal{H}]$, then*

$$\sigma_{\ell e}(T) = \{\lambda \in \mathbb{C} \colon \mathcal{R}(\lambda I - T) \text{ is not closed } \text{ or } \dim \mathcal{N}(\lambda I - T) = \infty\} \subseteq \sigma(T),$$

$$\sigma_{re}(T) = \{\lambda \in \mathbb{C} \colon \mathcal{R}(\lambda I - T) \text{ is not closed } \text{ or } \dim \mathcal{N}(\overline{\lambda} I - T^*) = \infty\} \subseteq \sigma(T).$$

Proof. Theorem 5.1 and Corollary 5.11. For the inclusions in $\sigma(T)$ see the diagram of Section 2.2. (See also Remark 5.3(b).)   □

By Corollaries 5.9 and 5.14,

$$\sigma_{\ell e}(T) \subseteq \sigma_{\ell}(T) \quad \text{and} \quad \sigma_{re}(T) \subseteq \sigma_r(T).$$

Thus

$$\sigma_e(T) \subseteq \sigma(T).$$

Also, still by Corollary 5.14,

$$\sigma_{\ell e}(T) = \sigma_{re}(T^*)^* \quad \text{and} \quad \sigma_{re}(T) = \sigma_{\ell e}(T^*)^*,$$

and hence

$$\sigma_e(T) = \sigma_e(T^*)^*.$$

Moreover, by the previous results and the diagram of Section 2.2,

$$\sigma_C(T) \subseteq \sigma_{\ell e}(T) \subseteq \sigma_{\ell}(T) = \sigma(T)\backslash\sigma_{R_1}(T) = \sigma_{AP}(T),$$

$$\sigma_C(T) = \sigma_C(T^*)^* \subseteq \sigma_{\ell e}(T^*)^* = \sigma_{re}(T) \subseteq \sigma_r(T) = \sigma(T)\backslash\sigma_{P_1}(T) = \sigma_{AP}(T^*)^*,$$

and so

$$\sigma_{\ell e}(T) \cap \sigma_{re}(T) \subseteq \sigma(T)\backslash(\sigma_{P_1}(T) \cup \sigma_{R_1}(T)),$$

which is an inclusion of closed sets. Therefore

$$\sigma_{P_1}(T) \cup \sigma_{R_1}(T) \subseteq \sigma(T)\backslash(\sigma_{\ell e}(T) \cap \sigma_{re}(T)) \subseteq \mathbb{C}\backslash(\sigma_{\ell e}(T) \cap \sigma_{re}(T)),$$

which is an inclusion of open sets.

Remark 5.15. (a) FINITE DIMENSION. For any $T \in \mathcal{B}[\mathcal{H}]$,

$$\sigma_e(T) \neq \varnothing \quad \Longleftrightarrow \quad \dim \mathcal{H} = \infty.$$

Actually, since $\sigma_e(T)$ was defined as the spectrum of $\pi(T)$ in the Calkin algebra $\mathcal{B}[\mathcal{H}]/\mathcal{B}_\infty[\mathcal{H}]$, which is a unital Banach algebra if and only if $\mathcal{H}$ is infinite-dimensional (in a finite-dimensional space all operators are compact), and since spectra are nonempty in a unital Banach algebra, we get

$$\dim \mathcal{H} = \infty \quad \Longrightarrow \quad \sigma_e(T) \neq \varnothing.$$

However, if we take the equivalent expression of the essential spectrum in Corollary 5.12, namely $\sigma_e(T) = \{\lambda \in \mathbb{C}: \lambda I - T \in \mathcal{B}[\mathcal{H}]\backslash\mathcal{F}[\mathcal{H}]\}$, then the converse holds by Remark 5.3(d):

$$\dim \mathcal{H} < \infty \quad \Longrightarrow \quad \sigma_e(T) = \varnothing.$$

(b) ATKINSON THEOREM. An argument similar to the one in the proof of Corollary 5.13, using each expression for $\sigma_{\ell e}(T)$ and $\sigma_{re}(T)$ in Corollary 5.11 separately, leads to the following result.

An operator $T \in \mathcal{B}[\mathcal{H}]$ lies in $\mathcal{F}_\ell[\mathcal{H}]$ (or in $\mathcal{F}_r[\mathcal{H}]$) if and only if its image $\pi(T)$ in the Calkin algebra $\mathcal{B}[\mathcal{H}]/\mathcal{B}_\infty[\mathcal{H}]$ is left (right) invertible.

(c) COMPACT PERTURBATION. For every $K \in \mathcal{B}_\infty[\mathcal{H}]$,

$$\sigma_e(T + K) = \sigma_e(T).$$

Indeed, take an arbitrary compact operator $K \in \mathcal{B}_\infty[\mathcal{H}]$. Since $\pi(T + K) = \pi(T)$, we get $\sigma_e(T) = \sigma(\pi(T)) = \sigma(\pi(T + K)) = \sigma_e(T + K)$ by the definition of $\sigma_e(T)$ in $\mathcal{B}[\mathcal{H}]/\mathcal{B}_\infty[\mathcal{H}]$. Similarly, by the definitions of $\sigma_{\ell e}(T)$ and $\sigma_{re}(T)$ in $\mathcal{B}[\mathcal{H}]/\mathcal{B}_\infty[\mathcal{H}]$ we get $\sigma_{\ell e}(T) = \sigma_\ell(\pi(T)) = \sigma_\ell(\pi(T + K)) = \sigma_{\ell e}(T + K)$ and $\sigma_{re}(T) = \sigma_r(\pi(T)) = \sigma_r(\pi(T + K)) = \sigma_{re}(T + K)$. Thus

$$\sigma_{\ell e}(T + K) = \sigma_{\ell e}(T) \quad \text{and} \quad \sigma_{re}(T + K) = \sigma_{re}(T).$$

Such invariance is a major feature of Fredholm theory.

Now take any operator T in $\mathcal{B}[\mathcal{H}]$. For each $k \in \overline{\mathbb{Z}}\backslash\{0\}$ set

$$\sigma_k(T) = \{\lambda \in \mathbb{C}: \lambda I - T \in \mathcal{SF}[\mathcal{H}] \text{ and } \mathrm{ind}\,(\lambda I - T) = k\}.$$

As we have seen before (Section 2.7), a *component* of a set in a topological space is any maximal connected subset of it. A *hole* of a set in a topological space is any bounded component of its complement. If a set has an open complement (i.e., if it is closed), then a hole of it must be open. We show that each $\sigma_k(T)$ with finite index is a *hole of* $\sigma_e(T)$ which lies in $\sigma(T)$ (i.e., if $k \in \mathbb{Z}\backslash\{0\}$, then $\sigma_k(T) \subseteq \sigma(T)\backslash\sigma_e(T)$ is a bounded component of the open set $\mathbb{C}\backslash\sigma_e(T)$), and hence it is an open set. Moreover, we also show that $\sigma_{+\infty}(T)$ and $\sigma_{-\infty}(T)$ are holes of $\sigma_{re}(T)$ and of $\sigma_{\ell e}(T)$ which lie in $\sigma_{\ell e}(T)$ and $\sigma_{re}(T)$ (in fact, $\sigma_{+\infty}(T) = \sigma_{\ell e}(T)\backslash\sigma_{re}(T)$ is a bounded component of the open set $\mathbb{C}\backslash\sigma_{re}(T)$, and $\sigma_{-\infty}(T) = \sigma_{re}(T)\backslash\sigma_{\ell e}(T)$ is a bounded component of the open set $\mathbb{C}\backslash\sigma_{\ell e}(T)$), and hence they are open sets. The sets $\sigma_{+\infty}(T)$ and $\sigma_{-\infty}(T)$ — which are holes of $\sigma_{re}(T)$ and of $\sigma_{\ell e}(T)$ but are not holes of $\sigma_e(T)$ — are called *pseudoholes* of $\sigma_e(T)$.

Let $\sigma_{PF}(T)$ denote the set of all eigenvalues of T of finite multiplicity,

$$\sigma_{PF}(T) = \{\lambda \in \sigma_P(T) \colon \dim \mathcal{N}(\lambda I - T) < \infty\}$$
$$= \{\lambda \in \mathbb{C} \colon 0 < \dim \mathcal{N}(\lambda I - T) < \infty\}.$$

According to the diagram of Section 2.2 and Corollary 5.14,

$$\sigma_{AP}(T) = \sigma_{\ell e}(T) \cup \sigma_{PF}(T).$$

Theorem 5.16. *Take an arbitrary operator* $T \in \mathcal{B}[\mathcal{H}]$. *For each* $k \in \overline{\mathbb{Z}} \backslash \{0\}$,

(a) $$\sigma_k(T) = \sigma_{-k}(T^*)^*.$$

If $k \in \mathbb{Z} \backslash \{0\}$, *then*

(b) $$\sigma_k(T) = \{\lambda \in \mathbb{C} \colon \lambda I - T \in \mathcal{F}[\mathcal{H}] \ \text{and} \ \mathrm{ind}\,(\lambda I - T) = k\}$$
$$\subseteq \begin{cases} \sigma_{PF}(T), & 0 < k, \\ \sigma_{PF}(T^*)^*, & k < 0, \end{cases}$$

so that $\sigma_k(T) \subseteq \sigma_P(T) \cup \sigma_R(T) \subseteq \sigma(T)$, *and*

$$\bigcup_{k \in \mathbb{Z} \backslash \{0\}} \sigma_k(T) = \{\lambda \in \mathbb{C} \colon \lambda I - T \in \mathcal{F}[\mathcal{H}] \ \text{and} \ \mathrm{ind}\,(\lambda I - T) \neq 0\}$$
$$= \{\lambda \in \mathbb{C} \colon \lambda I - T \in \mathcal{F}[\mathcal{H}] \backslash \mathcal{W}[\mathcal{H}]\}$$
$$\subseteq \{\lambda \in \mathbb{C} \colon \lambda I - T \in \mathcal{F}[\mathcal{H}]\} = \mathbb{C} \backslash \sigma_e(T).$$

If $k = \pm\infty$, *then*

(c) $$\sigma_{+\infty}(T) \subseteq \sigma_P(T) \backslash \sigma_{PF}(T) \quad \text{and} \quad \sigma_{-\infty}(T) \subseteq \sigma_P(T^*)^* \backslash \sigma_{PF}(T^*)^*.$$

Thus

$$\bigcup_{k \in \overline{\mathbb{Z}} \backslash \{0\}} \sigma_k(T) \subseteq \sigma_P(T) \cup \sigma_R(T) \subseteq \sigma(T).$$

Moreover,

$$\sigma_{+\infty}(T) = \{\lambda \in \mathbb{C} \colon \lambda I - T \in \mathcal{SF}[\mathcal{H}] \ \text{and} \ \mathrm{ind}\,(\lambda I - T) = +\infty\}$$
$$= \{\lambda \in \mathbb{C} \colon \lambda I - T \in \mathcal{F}_r[\mathcal{H}] \backslash \mathcal{F}_\ell[\mathcal{H}]\}$$
$$= \sigma_{\ell e}(T) \backslash \sigma_{re}(T) = \sigma_e(T) \backslash \sigma_{re}(T) \subseteq \mathbb{C} \backslash \sigma_{re}(T),$$

$$\sigma_{-\infty}(T) = \{\lambda \in \mathbb{C} \colon \lambda I - T \in \mathcal{SF}[\mathcal{H}] \ \text{and} \ \mathrm{ind}\,(\lambda I - T) = -\infty\}$$
$$= \{\lambda \in \mathbb{C} \colon \lambda I - T \in \mathcal{F}_\ell[\mathcal{H}] \backslash \mathcal{F}_r[\mathcal{H}]\}$$
$$= \sigma_{re}(T) \backslash \sigma_{\ell e}(T) = \sigma_e(T) \backslash \sigma_{\ell e}(T) \subseteq \mathbb{C} \backslash \sigma_{\ell e}(T),$$

which are all open subsets of $\mathbb{C}$, *and hence*

$$\sigma_{+\infty}(T) \subseteq \sigma_e(T) \quad \text{and} \quad \sigma_{-\infty}(T) \subseteq \sigma_e(T).$$

Furthermore, with

(d) $\overline{\mathbb{Z}}^{+}\backslash\{0\} = \overline{\mathbb{N}} = \mathbb{N}\cup\{+\infty\}$ *(the set of all extended positive integers), and* $\overline{\mathbb{Z}}^{-}\backslash\{0\} = -\overline{\mathbb{N}} = -\mathbb{N}\cup\{-\infty\}$ *(the set of all extended negative integers), we get*

$$\sigma_{P_1}(T) \subseteq \bigcup_{k\in\overline{\mathbb{Z}}^{+}\backslash\{0\}} \sigma_k(T)$$
$$= \{\lambda\in\mathbb{C}: \lambda I - T \in \mathcal{SF}[\mathcal{H}] \text{ and } 0 < \text{ind}(\lambda I - T)\},$$

$$\sigma_{R_1}(T) \subseteq \bigcup_{k\in\overline{\mathbb{Z}}^{-}\backslash\{0\}} \sigma_k(T)$$
$$= \{\lambda\in\mathbb{C}: \lambda I - T \in \mathcal{SF}[\mathcal{H}] \text{ and } \text{ind}(\lambda I - T) < 0\},$$

and so

$$\sigma_{P_1}(T)\cup\sigma_{R_1}(T) \subseteq \bigcup_{k\in\overline{\mathbb{Z}}\backslash\{0\}} \sigma_k(T)$$
$$= \{\lambda\in\mathbb{C}: \lambda I - T \in \mathcal{SF}[\mathcal{H}] \text{ and } \text{ind}(\lambda I - T) \neq 0\}$$
$$\subseteq \{\lambda\in\mathbb{C}: \lambda I - T \in \mathcal{SF}[\mathcal{H}]\} = \mathbb{C}\backslash(\sigma_{\ell e}(T)\cap\sigma_{re}(T)),$$

which are all open subsets of $\mathbb{C}$.

(e) *For any $k\in\overline{\mathbb{Z}}\backslash\{0\}$ the set $\sigma_k(T)$ is a hole of $\sigma_{\ell e}(T)\cap\sigma_{re}(T)$. If k is finite (i.e., if $k\in\mathbb{Z}\backslash\{0\}$), then $\sigma_k(T)$ is a hole of $\sigma_e(T) = \sigma_{\ell e}(T)\cup\sigma_{re}(T)$. Otherwise, $\sigma_{+\infty}(T)$ is a hole of $\sigma_{re}(T)$ and $\sigma_{-\infty}(T)$ is a hole of $\sigma_{\ell e}(T)$. Thus $\sigma_k(T)$ is an open set for every $k\in\overline{\mathbb{Z}}\backslash\{0\}$.*

Proof. (a) $\lambda I - T \in \mathcal{SF}[\mathcal{H}]$ if and only if $\overline{\lambda}I - T^* \in \mathcal{SF}[\mathcal{H}]$, and $\text{ind}(\lambda I - T) = -\text{ind}(\overline{\lambda}I - T^*)$ (cf. Section 5.1). Thus $\lambda\in\sigma_k(T)$ if and only if $\overline{\lambda}\in\sigma_{-k}(T^*)$.

(b) The expression for $\sigma_k(T)$ — with $\mathcal{SF}[\mathcal{H}]$ replaced with $\mathcal{F}[\mathcal{H}]$ — holds since for finite k all semi-Fredholm operators are Fredholm (cf. Remark 5.3(f)). Take any $k\in\mathbb{Z}\backslash\{0\}$. If $k > 0$ and $\lambda\in\sigma_k(T)$, then $0 < \text{ind}(\lambda I - T) < \infty$. Hence

$$0 \leq \dim\mathcal{N}(\overline{\lambda}I - T^*) < \dim\mathcal{N}(\lambda I - T) < \infty,$$

and so $0 < \dim\mathcal{N}(\lambda I - T) < \infty$. Thus λ is an eigenvalue of T of finite multiplicity. Dually, if $k < 0$ (i.e., $-k > 0$) and $\lambda\in\sigma_k(T) = \sigma_{-k}(T^*)^*$, then $\overline{\lambda}$ is an eigenvalue of T^* of finite multiplicity. Also,

$$\mathbb{C}\backslash\sigma_e(T) = \{\lambda\in\mathbb{C}: \lambda I - T \in \mathcal{F}[\mathcal{H}]\}$$

by Corollary 5.12. This closes the proof of (b).

(c) According to Corollary 5.2 and by the definitions of $\sigma_{+\infty}(T)$ and $\sigma_{-\infty}(T)$, $\lambda\in\mathbb{C}$ lies in $\sigma_{+\infty}(T)$ if and only if $\mathcal{R}(\lambda I - T)$ is closed, $\dim\mathcal{N}(\overline{\lambda}I - T^*)$ is finite, and $\dim\mathcal{N}(\lambda I - T)$ is infinite. Thus if $\lambda\in\sigma_{+\infty}(T)$, then

$$0 \leq \dim\mathcal{N}(\overline{\lambda}I - T^*) < \dim\mathcal{N}(\lambda I - T) = \infty,$$

and so λ is an eigenvalue of T of infinite multiplicity. Dually, if $\lambda\in\sigma_{-\infty}(T) = \sigma_{+\infty}(T^*)^*$, then $\overline{\lambda}$ is an eigenvalue of T^* of infinite multiplicity. Thus the expressions for $\sigma_{+\infty}(T)$ and $\sigma_{-\infty}(T)$ follow from Remark 5.3(f) and Corollary 5.11, since $\sigma_e(T) = \sigma_{\ell e}(T) \cup \sigma_{re}(T)$. Also, $\sigma_{+\infty}(T)$ and $\sigma_{-\infty}(T)$ are open sets in $\mathbb{C}$ because $\sigma_{\ell e}(T)$, $\sigma_{re}(T)$ and so $\sigma_e(T)$ are all closed sets in $\mathbb{C}$.

(d) Recall: $\mathcal{M}^- = \mathcal{H}$ if and only if $\mathcal{M}^\perp = \{0\}$ (Section 1.3). So $\mathcal{R}(\lambda I - T)^- = \mathcal{H}$ if and only if $\mathcal{N}(\overline{\lambda} I - T^*) = \{0\}$ (Lemma 1.4) or, equivalently, if and only if $\dim \mathcal{N}(\overline{\lambda} I - T^*) = 0$. Thus, according to the diagram of Section 2.2,

$$\sigma_{P_1}(T) = \big\{\lambda \in \mathbb{C}\colon \mathcal{R}(\lambda I - T) \text{ is closed}, \, 0 = \dim \mathcal{N}(\overline{\lambda} I - T^*) \neq \dim \mathcal{N}(\lambda I - T)\big\}$$
$$= \big\{\lambda \in \mathbb{C}\colon \mathcal{R}(\lambda I - T) \text{ is closed}, \, 0 = \dim \mathcal{N}(\overline{\lambda} I - T^*), \, \mathrm{ind}\,(\lambda I - T) > 0\big\}.$$

Therefore, by Corollary 5.2,

$$\sigma_{P_1}(T) = \big\{\lambda \in \mathbb{C}\colon \lambda I - T \in \mathcal{SF}[\mathcal{H}], \, 0 = \dim \mathcal{N}(\overline{\lambda} I - T^*), \, \mathrm{ind}\,(\lambda I - T) > 0\big\}$$
$$\subseteq \big\{\lambda \in \mathbb{C}\colon \lambda I - T \in \mathcal{SF}[\mathcal{H}], \, \mathrm{ind}\,(\lambda I - T) > 0\big\} = \bigcup\nolimits_{k \in \overline{\mathbb{Z}}^+ \backslash \{0\}} \sigma_k(T).$$

Dually (cf. item (a) and Theorem 2.6),

$$\sigma_{R_1}(T) = \sigma_{P_1}(T^*)^* \subseteq \Big(\bigcup\nolimits_{k \in \overline{\mathbb{Z}}^+ \backslash \{0\}} \sigma_k(T^*)\Big)^*$$
$$= \Big(\bigcup\nolimits_{k \in \overline{\mathbb{Z}}^+ \backslash \{0\}} \sigma_{-k}(T)^*\Big)^* = \bigcup\nolimits_{k \in \overline{\mathbb{Z}}^- \backslash \{0\}} \sigma_k(T).$$

Finally, as we had verified before,

$$\mathbb{C} \backslash \sigma_{\ell e}(T) \cap \sigma_{re}(T) = \big\{\lambda \in \mathbb{C}\colon \lambda I - T \in \mathcal{SF}[\mathcal{H}]\big\}$$

according to Corollary 5.11.

(e) By item (d) we get

$$\bigcup\nolimits_{k \in \overline{\mathbb{Z}} \backslash \{0\}} \sigma_k(T) \subseteq \mathbb{C} \backslash \sigma_{\ell e}(T) \cap \sigma_{re}(T) = \big\{\lambda \in \mathbb{C}\colon \lambda I - T \in \mathcal{SF}[\mathcal{H}]\big\}.$$

Take an arbitrary $\lambda \in \bigcup_{k \in \overline{\mathbb{Z}} \backslash \{0\}} \sigma_k(T)$ so that $\lambda I - T \in \mathcal{SF}[\mathcal{H}]$. By Proposition 5.D, $\mathrm{ind}\,(\zeta I - T)$ is constant for every ζ in a punctured disk $B_\varepsilon(\lambda) \backslash \{\lambda\}$ centered at λ for some positive radius ε. Thus, for every pair of distinct extended integers $j, k \in \overline{\mathbb{Z}} \backslash \{0\}$, the sets $\sigma_j(T)$ and $\sigma_k(T)$ are not only disjoint, but they are (bounded) components (i.e., maximal connected subsets) of the open set $\mathbb{C} \backslash \sigma_{\ell e}(T) \cap \sigma_{re}(T)$, and hence they are holes of the closed set $\sigma_{\ell e}(T) \cap \sigma_{re}(T)$. By item (b), if j, k are finite (i.e., if $j, k \in \mathbb{Z} \backslash \{0\}$), then $\sigma_j(T)$ and $\sigma_k(T)$ are (bounded) components of the open set $\mathbb{C} \backslash \sigma_e(T) = \mathbb{C} \backslash \sigma_{\ell e}(T) \cup \sigma_{re}(T)$ and, in this case, they are holes of the closed set $\sigma_e(T)$. By item (c), $\sigma_{+\infty}(T)$ and $\sigma_{-\infty}(T)$ are (bounded) components of the open sets $\mathbb{C} \backslash \sigma_{re}(T)$ and $\mathbb{C} \backslash \sigma_{\ell e}(T)$, and so $\sigma_{+\infty}(T)$ and $\sigma_{-\infty}(T)$ are holes of the closed sets $\sigma_{re}(T)$ and $\sigma_{\ell e}(T)$, respectively. Thus in any case $\sigma_k(T)$ is an open subset of $\mathbb{C}$. Therefore $\bigcup_{k \in \overline{\mathbb{Z}} \backslash \{0\}} \sigma_k(T)$ is open as claimed in (d). $\qquad \square$

The following figure illustrates a simple instance of holes and pseudoholes of the essential spectrum. Observe that the right and left essential spectra $\sigma_{\ell e}(T)$ and $\sigma_{re}(T)$ differ only by pseudoholes $\sigma_{\pm\infty}(T)$.

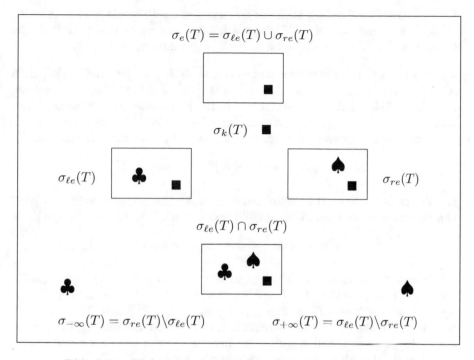

Fig. § 5.2. Holes and pseudoholes of the essential spectrum

Now for $k = 0$ we define the set $\sigma_0(T)$ as the following subset of $\sigma(T)$.

$$\sigma_0(T) = \{\lambda \in \sigma(T): \lambda I - T \in \mathcal{SF}[\mathcal{H}] \text{ and } \mathrm{ind}\,(\lambda I - T) = 0\}$$
$$= \{\lambda \in \sigma(T): \lambda I - T \in \mathcal{F}[\mathcal{H}] \text{ and } \mathrm{ind}\,(\lambda I - T) = 0\}$$
$$= \{\lambda \in \sigma(T): \lambda I - T \in \mathcal{W}[\mathcal{H}]\}.$$

Corollary 5.2, the diagram of Section 2.2, and Theorem 2.6 give an equivalent description for the set $\sigma_0(T) \subseteq \sigma(T)$:

$$\sigma_0(T) = \{\lambda \in \sigma(T): \mathcal{R}(\lambda I - T) \text{ is closed and}$$
$$\dim \mathcal{N}(\lambda I - T) = \dim \mathcal{N}(\overline{\lambda} I - T^*) < \infty\}$$
$$= \{\lambda \in \sigma_P(T): \mathcal{R}(\lambda I - T) = \mathcal{R}(\lambda I - T)^- \neq \mathcal{H} \text{ and}$$
$$\dim \mathcal{N}(\lambda I - T) = \dim \mathcal{N}(\overline{\lambda} I - T^*) < \infty\}$$
$$= \{\lambda \in \sigma_{P_4}(T): \dim \mathcal{N}(\lambda I - T) = \dim \mathcal{N}(\overline{\lambda} I - T^*) < \infty\},$$

where $\sigma_{P_4}(T) = \{\lambda \in \sigma_P(T): \mathcal{R}(\lambda I - T)^- = \mathcal{R}(\lambda I - T) \neq \mathcal{H}\}$. In fact,

$$\sigma_0(T) \subseteq \sigma_{PF}(T) \cap \sigma_{PF}(T^*)^* \subseteq \sigma_P(T) \cap \sigma_P(T^*)^*.$$

Thus

$$\sigma_0(T) = \sigma_0(T^*)^*.$$

Observe that if $\lambda \in \sigma_0(T)$, then the definition of $\sigma_0(T)$ forces λ to be in $\sigma(T)$, while in the definition of $\sigma_k(T)$ for $k \neq 0$ this happens naturally. In fact, the definition of $\sigma_0(T)$ forces λ to be in $\sigma_P(T)$ (specifically, in $\sigma_{P_4}(T)$).

Remark 5.17. (a) FREDHOLM ALTERNATIVE. If $K \in \mathcal{B}_\infty[\mathcal{H}]$ and $\lambda \neq 0$, then $\lambda I - K$ is Fredholm with $\mathrm{ind}(\lambda I - K) = 0$ (i.e., $\lambda I - K$ is Weyl; in symbols, $\lambda I - K \in \mathcal{W}[\mathcal{H}]$). This is a restatement of the Fredholm Alternative as in Remark 5.3(e). Since $\sigma_0(K) = \{\lambda \in \sigma(K) : \lambda I - K \in \mathcal{W}[\mathcal{H}]\}$, the Fredholm Alternative can be restated once again as follows (compare with Theorem 2.18):

$$K \in \mathcal{B}_\infty[\mathcal{H}] \quad \Longrightarrow \quad \sigma(K) \backslash \{0\} = \sigma_0(K) \backslash \{0\}.$$

(b) FINITE DIMENSION. If $\mathcal{H}$ is finite-dimensional, then every operator is Weyl (i.e., if $\dim \mathcal{H} < \infty$, then $\mathcal{W}[\mathcal{H}] = \mathcal{B}[\mathcal{H}]$ — cf. Remark 5.3(d) or 5.7(c)). Thus,

$$\dim \mathcal{H} < \infty \quad \Longrightarrow \quad \sigma(T) = \sigma_0(T).$$

(c) COMPACT PERTURBATION. The expressions for $\sigma_k(T)$, $\sigma_{+\infty}(T)$, and $\sigma_{-\infty}(T)$, viz., $\sigma_k(T) = \{\lambda \in \mathbb{C} : \lambda I - T \in \mathcal{F}[\mathcal{H}]$ and $\mathrm{ind}(\lambda I - T) = k\}$ for $k \in \mathbb{Z}\backslash\{0\}$, $\sigma_{+\infty}(T) = \sigma_{\ell e}(T)\backslash\sigma_{re}(T)$, and $\sigma_{-\infty}(T) = \sigma_{re}(T)\backslash\sigma_{\ell e}(T)$ as in Theorem 5.16, together with the results of Theorem 5.4, ensure that the sets $\sigma_k(T)$ for each $k \in \mathbb{Z}\backslash\{0\}$, $\sigma_{+\infty}(T)$, and $\sigma_{-\infty}(T)$, are invariant under compact perturbation as well (see Remark 5.15(c)). In other words, for every $K \in \mathcal{B}_\infty[\mathcal{H}]$

$$\sigma_k(T + K) = \sigma_k(T) \quad \text{for every} \ \ k \in \overline{\mathbb{Z}}\backslash\{0\}.$$

Such invariance, however, does not apply to $\sigma_0(T)$. Example: if $\dim \mathcal{H} < \infty$ (where all operators are compact), and if $T = I$, then $\sigma_0(T) = \sigma(T) = \{1\}$ and $\sigma_0(T - T) = \sigma(T - T) = \{0\}$. Thus there is a compact $K = -T$ for which

$$\{0\} = \sigma_0(T + K) \neq \sigma_0(T) = \{1\}.$$

This will be generalized in the proof Theorem 5.24, where it is proved that *a nonempty $\sigma_0(T)$ is never invariant under compact perturbation.*

The partition of the spectrum $\sigma(T)$ obtained in the next corollary is called the *Spectral Picture* of T [97].

Corollary 5.18. (SPECTRAL PICTURE). *If $T \in \mathcal{B}[\mathcal{H}]$, then*

$$\sigma(T) = \sigma_e(T) \cup \bigcup_{k \in \mathbb{Z}} \sigma_k(T)$$

and

$$\sigma_e(T) = \big(\sigma_{\ell e}(T) \cap \sigma_{re}(T)\big) \cup \sigma_{+\infty}(T) \cup \sigma_{-\infty}(T),$$

where

$$\sigma_e(T) \cap \bigcup_{k \in \mathbb{Z}} \sigma_k(T) = \varnothing,$$

$$\sigma_k(T) \cap \sigma_j(T) = \varnothing \quad \text{for every } j, k \in \overline{\mathbb{Z}} \text{ such that } j \neq k,$$

and

$$\big(\sigma_{\ell e}(T) \cap \sigma_{re}(T)\big) \cap \big(\sigma_{+\infty}(T) \cup \sigma_{+\infty}\big) = \varnothing.$$

Proof. The collection $\{\sigma_k(T)\}_{k \in \overline{\mathbb{Z}}}$ is clearly pairwise disjoint. By Theorem 5.16(b) and the definition of $\sigma_0(T)$,

$$\bigcup_{k \in \mathbb{Z}} \sigma_k(T) = \{\lambda \in \sigma(T) \colon \lambda I - T \in \mathcal{F}[\mathcal{H}]\}.$$

Since $\sigma_e(T) \subseteq \sigma(T)$ we get

$$\sigma_e(T) = \{\lambda \in \sigma(T) \colon \lambda I - T \notin \mathcal{F}[\mathcal{H}]\}$$

according to Corollary 5.12. This leads to the following partition of the spectrum: $\sigma(T) = \sigma_e(T) \cup \bigcup_{k \in \mathbb{Z}} \sigma_k(T)$ with $\sigma_e(T) \cap \bigcup_{k \in \mathbb{Z}} \sigma_k(T) = \varnothing$. Moreover, since $\sigma_e(T) = \sigma_{\ell e}(T) \cup \sigma_{re}(T) \subseteq \sigma(T)$ we also get

$$\sigma_{\ell e}(T) \cap \sigma_{re}(T) = \{\lambda \in \sigma(T) \colon \lambda I - T \notin \mathcal{SF}[\mathcal{H}]\} \subseteq \sigma_e(T)$$

from Corollary 5.11. Furthermore, according to Theorem 5.16(c),

$$\sigma_{+\infty}(T) \cup \sigma_{-\infty}(T) = \{\lambda \in \sigma(T) \colon \lambda I - T \in \mathcal{SF}[\mathcal{H}] \backslash \mathcal{F}[\mathcal{H}]\}$$

and $\sigma_{+\infty}(T) \cap \sigma_{-\infty}(T) = \varnothing$. Thus we obtain the following partition of the essential spectrum as well: $\sigma_e(T) = \big(\sigma_{\ell e}(T) \cap \sigma_{re}(T)\big) \cup \sigma_{+\infty}(T) \cup \sigma_{-\infty}(T)$ with $\big(\sigma_{\ell e}(T) \cap \sigma_{re}(T)\big) \cap \big(\sigma_{+\infty}(T) \cup \sigma_{+\infty}\big) = \varnothing$. $\qquad\square$

As we saw above, the collection $\{\sigma_k(T)\}_{k \in \mathbb{Z} \backslash \{0\}}$ consists of pairwise disjoint *open* sets (cf. Theorem 5.16), which are subsets of $\sigma(T) \backslash \sigma_e(T)$. These are the *holes of the essential spectrum* $\sigma_e(T)$. Moreover, $\sigma_{\pm\infty}(T)$ also are *open* sets (cf. Theorem 5.16), which are subsets of $\sigma_e(T)$. These are the *pseudoholes of* $\sigma_e(T)$ (they are holes of $\sigma_{re}(T)$ and $\sigma_{\ell e}(T)$). Thus the spectral picture of T consists of the essential spectrum $\sigma_e(T)$, the holes $\sigma_k(T)$ and pseudoholes $\sigma_{\pm\infty}(T)$ (to each is associated a nonzero index k in $\mathbb{Z} \backslash \{0\}$), and the set $\sigma_0(T)$. It is worth noticing that any spectral picture can be attained [28] by an operator in $\mathcal{B}[\mathcal{H}]$ (see also Proposition 5.G).

5.3 Riesz Points and Weyl Spectrum

The set $\sigma_0(T)$ will play a rather important role in the sequel. It consists of an open set $\tau_0(T)$ and the set $\pi_0(T)$ of isolated points of $\sigma(T)$ the Riesz idempotents associated with which have finite rank. The next proposition says that

$\pi_0(T)$ is precisely the set of all isolated points of $\sigma(T)$ which lie in $\sigma_0(T)$. Let $E_\lambda \in \mathcal{B}[\mathcal{H}]$ be the Riesz idempotent associated with an *isolated* point λ of the spectrum $\sigma(T)$ of an operator $T \in \mathcal{B}[\mathcal{H}]$.

Theorem 5.19. *If λ is an* isolated *point of $\sigma(T)$, then the following assertions are pairwise equivalent.*

(a) $\lambda \in \sigma_0(T)$.

(b) $\lambda \notin \sigma_e(T)$.

(c) $\lambda I - T \in \mathcal{F}[\mathcal{H}]$.

(d) $\mathcal{R}(\lambda I - T)$ *is closed and* $\dim \mathcal{N}(\lambda I - T) < \infty$.

(e) $\lambda \notin \sigma_{\ell e}(T) \cap \sigma_{re}(T)$.

(f) $\dim \mathcal{R}(E_\lambda) < \infty$.

Proof. Let T be an operator on an infinite-dimensional complex Hilbert space.

(a)$\Rightarrow$(b)$\Rightarrow$(c). By definition, we have $\sigma_0(T) = \{\lambda \in \sigma(T)\colon \lambda I - T \in \mathcal{W}[\mathcal{H}]\} = \{\lambda \in \sigma(T)\colon \lambda I - T \in \mathcal{F}[\mathcal{H}]$ and $\mathrm{ind}(\lambda I - T) = 0\}$. Thus (a) implies (c) tautologically, and (c) is equivalent to (b) by Corollary 5.12.

(c)$\Rightarrow$(d). By Corollary 5.2, $\lambda I - T \in \mathcal{F}[\mathcal{H}]$ if and only if $\mathcal{R}(\lambda I - T)$ is closed, $\dim \mathcal{N}(\lambda I - T) < \infty$, and $\dim \mathcal{N}(\bar\lambda I - T^*) < \infty$. Thus (c) implies (d) trivially.

(d)$\Rightarrow$(e). By Corollary 5.14, $\sigma_{\ell e}(T) \cap \sigma_{re}(T) = \{\lambda \in \mathbb{C}\colon \mathcal{R}(\lambda I - T)$ is not closed or $\dim \mathcal{N}(\lambda I - T) = \dim \mathcal{N}(\bar\lambda I - T^*) = \infty\}$. Thus (d) implies (e).

From now on suppose λ is an isolated point of the spectrum $\sigma(T)$ of T.

(e)$\Rightarrow$(a). If $\lambda \in \sigma(T)\backslash\sigma_0(T)$, then $\lambda \in \sigma_e(T) \cup \bigcup_{k \in \mathbb{Z}\backslash\{0\}} \sigma_k(T)$ by Corollary 5.18, where $\bigcup_{k \in \mathbb{Z}\backslash\{0\}} \sigma_k(T)$ is an open subset of $\mathbb{C}$ according to Theorem 5.16, and so it has no isolated point. Hence if λ is an isolated pont of $\sigma(T)\backslash\sigma_0(T)$, then λ lies in $\sigma_e(T) = (\sigma_{\ell e}(T) \cap \sigma_{re}(T)) \cup (\sigma_{+\infty}(T) \cup \sigma_{-\infty}(T))$, where the set $\sigma_{+\infty}(T) \cup \sigma_{-\infty}(T)$ is open in $\mathbb{C}$, and hence it likewise has no isolated point (cf. Theorem 5.16 and Corollary 5.18 again). Therefore $\lambda \in \sigma_{\ell e}(T) \cap \sigma_{re}(T)$. Outcome: if λ is an isolated point of $\sigma(T)$ and $\lambda \notin \sigma_{\ell e}(T) \cap \sigma_{re}(T)$, then $\lambda \in \sigma_0(T)$. Thus (e) implies (a).

So assertions (a), (b), (c), (d), and (e) are pairwise equivalent. As we will see next, (f) is also equivalent to them.

(f)$\Rightarrow$(d). If λ is an isolated point of $\sigma(T)$, then $\Delta = \{\lambda\}$ is a spectral set for the operator T. Thus (f) implies (d) by Corollary 4.22.

(a)$\Rightarrow$(f). Let $\pi\colon \mathcal{B}[\mathcal{H}] \to \mathcal{B}[\mathcal{H}]/\mathcal{B}_\infty[\mathcal{H}]$ be the natural map of $\mathcal{B}[\mathcal{H}]$ into the Calkin algebra $\mathcal{B}[\mathcal{H}]/\mathcal{B}_\infty[\mathcal{H}]$, which is a unital homomorphism of the unital Banach algebra $\mathcal{B}[\mathcal{H}]$ to the unital Banach algebra $\mathcal{B}[\mathcal{H}]/\mathcal{B}_\infty[\mathcal{H}]$, whenever the complex Hilbert space $\mathcal{H}$ is infinite-dimensional.

Claim. If $T \in \mathcal{B}[\mathcal{H}]$ and $\zeta \in \rho(T)$, then $\zeta \in \rho(\pi(T))$ and

$$\pi\big((\zeta I - T)^{-1}\big) = \big(\zeta \pi(I) - \pi(T)\big)^{-1}.$$

Proof. Since $\mathcal{B}_\infty[\mathcal{H}]$ is an ideal of $\mathcal{B}[\mathcal{H}]$,

$$
\begin{aligned}
\pi\big((\zeta I - T)^{-1}\big)\,\pi(\zeta I - T) &= \big((\zeta I - T)^{-1} + \mathcal{B}_\infty[\mathcal{H}]\big)\big((\zeta I - T) + \mathcal{B}_\infty[\mathcal{H}]\big) \\
&= (\zeta I - T)^{-1}(\zeta I - T) + \mathcal{B}_\infty[\mathcal{H}] \\
&= I + \mathcal{B}_\infty[\mathcal{H}] = \pi(I) = [I],
\end{aligned}
$$

where $\pi(I) = [I]$ is the identity in $\mathcal{B}[\mathcal{H}]/\mathcal{B}_\infty[\mathcal{H}]$. Therefore

$$\pi\big((\zeta I - T)^{-1}\big) = \big(\pi(\zeta I - T)\big)^{-1} = \big(\zeta I - T + \mathcal{B}_\infty[\mathcal{H}]\big)^{-1} = \big(\zeta \pi(I) - \pi(T)\big)^{-1},$$

completing the proof of the claimed result.

Suppose an isolated point λ of $\sigma(T)$ lies in $\sigma_0(T)$. According to Corollary 5.18, $\lambda \notin \sigma_e(T)$, which means (by definition) $\lambda \notin \sigma(\pi(T))$, and hence $\lambda \in \rho(\pi(T))$. Thus $\lambda \pi(I) - \pi(T)$ is invertible in $\mathcal{B}[\mathcal{H}]/\mathcal{B}_\infty[\mathcal{H}]$. Since the resolvent function of an element in a unital complex Banach algebra is analytic on the resolvent set (cf. proof of Claim 1 in the proof of Lemma 4.13), it then follows that $(\zeta \pi(I) - \pi(T))^{-1} \colon \Lambda \to \mathcal{B}[\mathcal{H}]/\mathcal{B}_\infty[\mathcal{H}]$ is analytic on any nonempty open subset Λ of $\mathbb{C}$ for which $(\text{ins}\,\Upsilon_\lambda)^- \subset \Lambda$, where Υ_λ is any simple closed rectifiable positively oriented curve (e.g., any positively oriented circle) enclosing λ but no other point of $\sigma(T)$. Hence, by the Cauchy Theorem (cf. Claim 2 in the proof of Lemma 4.13), we get $\int_{\Upsilon_\lambda}(\zeta \pi(I) - \pi(T))^{-1}d\zeta = [O]$, where $[O] = \pi(O) = \mathcal{B}_\infty[\mathcal{H}]$ is the origin in the Calkin algebra $\mathcal{B}[\mathcal{H}]/\mathcal{B}_\infty[\mathcal{H}]$. Then, with $E_\lambda = \frac{1}{2\pi i}\int_{\Upsilon_\lambda}(\zeta I - T)^{-1}d\zeta$ standing for the Riesz idempotent associated with λ, and since $\pi \colon \mathcal{B}[\mathcal{H}] \to \mathcal{B}[\mathcal{H}]/\mathcal{B}_\infty[\mathcal{H}]$ is linear and bounded,

$$\pi(E_\lambda) = \tfrac{1}{2\pi i}\int_{\Upsilon_\lambda} \pi\big((\zeta I - T)^{-1}\big)\,d\zeta = \tfrac{1}{2\pi i}\int_{\Upsilon_\lambda}\big(\zeta \pi(I) - \pi(T)\big)^{-1}d\zeta = [O].$$

Therefore $E_\lambda \in \mathcal{B}_\infty[\mathcal{H}]$. But the restriction of a compact operator to an invariant subspace is again compact (cf. Proposition 1.V), and so the identity $I = E_\lambda|_{\mathcal{R}(E_\lambda)} \colon \mathcal{R}(E_\lambda) \to \mathcal{R}(E_\lambda)$ on $\mathcal{R}(E_\lambda)$ is compact. Then $\dim \mathcal{R}(E_\lambda) < \infty$ (cf. Proposition 1.Y). Thus (a) implies (f). $\qquad\square$

A *Riesz point* of an operator T is an isolated point λ of $\sigma(T)$ for which the Riesz idempotent E_λ has finite rank (i.e., for which $\dim \mathcal{R}(E_\lambda) < \infty$). Let $\sigma_{\mathrm{iso}}(T)$ denote the set of all isolated points of the spectrum $\sigma(T)$,

$$\sigma_{\mathrm{iso}}(T) = \big\{\lambda \in \sigma(T)\colon \lambda \text{ is an isolated point of } \sigma(T)\big\}.$$

So its complement in $\sigma(T)$,

$$\sigma_{\mathrm{acc}}(T) = \sigma(T)\backslash\sigma_{\mathrm{iso}}(T),$$

is the set of all accumulation points of $\sigma(T)$. Let $\pi_0(T)$ denote the set of all isolated points of $\sigma(T)$ lying in $\sigma_0(T)$:

$$\pi_0(T) = \sigma_{\text{iso}}(T) \cap \sigma_0(T).$$

Since $\sigma_0(T) \subseteq \sigma_{P_4}(T) \subseteq \sigma_P(T)$, it follows by Theorem 5.19 that

$$\pi_0(T) = \{\lambda \in \sigma_{\text{iso}}(T) \colon \dim \mathcal{R}(E_\lambda) < \infty\},$$

and so

$$\pi_0(T) \text{ is precisely the set of all Riesz points of } T,$$

and these are eigenvalues of T. Indeed,

$$\sigma_{\text{iso}}(T) \backslash \sigma_e(T) = \sigma_{\text{iso}}(T) \cap \sigma_0(T) \subseteq \sigma_{P_4}(T) \subseteq \sigma_P(T).$$

Summing up (cf. Theorem 5.19 and the diagram of Section 2.2):

$$
\begin{aligned}
\pi_0(T) &= \{\lambda \in \sigma_P(T) \colon \lambda \in \sigma_{\text{iso}}(T) \text{ and } \lambda \in \sigma_0(T)\} = \sigma_{\text{iso}}(T) \cap \sigma_0(T) \\
&= \{\lambda \in \sigma_P(T) \colon \lambda \in \sigma_{\text{iso}}(T) \text{ and } \lambda \notin \sigma_e(T)\} = \sigma_{\text{iso}}(T) \backslash \sigma_e(T) \\
&= \{\lambda \in \sigma_P(T) \colon \lambda \in \sigma_{\text{iso}}(T) \text{ and } \dim \mathcal{R}(E_\lambda) < \infty\} \\
&= \{\lambda \in \sigma_P(T) \colon \lambda \in \sigma_{\text{iso}}(T) \text{ and } \lambda \notin (\sigma_{\ell e}(T) \cap \sigma_{re}(T))\}
\end{aligned}
$$
$$= \{\lambda \in \sigma_P(T) \colon \lambda \in \sigma_{\text{iso}}(T), \ \dim \mathcal{N}(\lambda I - T) < \infty, \text{ and } \mathcal{R}(\lambda I - T) \text{ is closed}\}$$
$$= \{\lambda \in \sigma_{P_4}(T) \colon \lambda \in \sigma_{\text{iso}}(T) \text{ and } \dim \mathcal{N}(\lambda I - T) = \dim \mathcal{N}(\bar{\lambda} I - T^*) < \infty\}.$$

Let $\tau_0(T)$ denote the complement of $\pi_0(T)$ in $\sigma_0(T)$,

$$\tau_0(T) = \sigma_0(T) \backslash \pi_0(T) \subseteq \sigma_{P_4}(T),$$

and so $\{\pi_0(T), \tau_0(T)\}$ is a partition of $\sigma_0(T)$. That is, $\tau_0(T) \cap \pi(T)_0 = \varnothing$ and

$$\sigma_0(T) = \tau_0(T) \cup \pi_0(T).$$

Corollary 5.20. $\tau_0(T)$ is an open subset of $\mathbb{C}$.

Proof. First, $\tau_0(T) = \sigma_0(T) \backslash \pi_0(T) \subseteq \sigma(T)$ has no isolated point. Moreover, if $\lambda \in \tau_0(T)$, then $\lambda I - T \in \mathcal{W}[\mathcal{H}]$ and $\lambda \in \mathbb{C} \backslash \sigma_{\ell e}(T) \cap \sigma_{re}(T)$ (by Corollary 5.18). Thus (since λ is not an isolated point), there exists a nonempty open ball $B_\varepsilon(\lambda)$ centered at λ and included in $\tau_0(T)$ (cf. Proposition 5.D). Therefore $\tau_0(T)$ is an open subset of $\mathbb{C}$. $\square$

As usual, let $\partial\sigma(T)$ denote the boundary of $\sigma(T)$.

Corollary 5.21. *If* $\lambda \in \partial\sigma(T)$, *then either* λ *is an isolated point of* $\sigma(T)$ *in* $\pi_0(T)$ *or* $\lambda \in \sigma_{\ell e}(T) \cap \sigma_{re}(T)$.

Proof. If $\lambda \in \partial\sigma(T)$, then $\lambda \notin \left(\bigcup_{k \in \mathbb{Z} \backslash \{0\}} \sigma_k(T) \cup \sigma_{+\infty}(T) \cup \sigma_{-\infty}(T)\right)$ (since this set is open by Theorem 5.16 and is included in $\sigma(T)$, which is closed). Thus

$\lambda \in (\sigma_{\ell e}(T) \cap \sigma_{re}(T)) \cup \sigma_0(T)$ by Corollary 5.18. But $\sigma_0(T) = \tau_0(T) \cup \pi_0(T)$, where $\pi_0(T) = \sigma_{\mathrm{iso}}(T) \cap \sigma_0(T)$, and so $\pi_0(T)$ consists of isolated points only. However $\lambda \notin \tau_0(T)$ because $\tau_0(T)$ is again an open set (Corollary 5.20) included in $\sigma(T)$, which is closed. Thus $\lambda \in (\sigma_{\ell e}(T) \cap \sigma_{re}(T)) \cup \pi_0(T)$. $\qquad\square$

Let $\pi_{00}(T)$ denote the set of all isolated eigenvalues of finite multiplicity,

$$\pi_{00}(T) = \sigma_{\mathrm{iso}}(T) \cap \sigma_{PF}(T).$$

Since $\sigma_0(T) \subseteq \sigma_{PF}(T)$, then $\pi_0(T) \subseteq \pi_{00}(T)$. Indeed,

$$\pi_0(T) = \sigma_{\mathrm{iso}}(T) \cap \sigma_0(T) \subseteq \sigma_{\mathrm{iso}}(T) \cap \sigma_{PF}(T) = \pi_{00}(T).$$

Corollary 5.22. $\pi_0(T) = \{\lambda \in \pi_{00}(T) \colon \mathcal{R}(\lambda I - T) \text{ is closed}\}.$

Proof. If $\lambda \in \pi_0(T)$, then $\lambda \in \pi_{00}(T)$ and $\mathcal{R}(\lambda I - T)$ is closed (since $\lambda \in \sigma_0(T)$). Conversely, if $\mathcal{R}(\lambda I - T)$ is closed and $\lambda \in \pi_{00}(T)$, then $\mathcal{R}(\lambda I - T)$ is closed and $\dim \mathcal{N}(\lambda I - T) < \infty$ (since $\lambda \in \sigma_{PF}(T)$), which means $\lambda \in \sigma_0(T)$ by Theorem 5.19 (since $\lambda \in \sigma_{\mathrm{iso}}(T)$). Thus $\lambda \in \sigma_{\mathrm{iso}}(T) \cap \sigma_0(T) = \pi_0(T)$. $\qquad\square$

The set $\pi_0(T)$ of all Riesz points of T is sometimes referred to as the set of isolated eigenvalues of T of *finite algebraic multiplicity* (sometimes also called *normal eigenvalues* of T [64, p. 5], but we reserve this terminology for eigenvalues satisfying the inclusion in Lemma 1.13(a)) — see also the first paragraph of Section 2.7. The set $\pi_{00}(T)$ of all isolated eigenvalues of T of finite multiplicity is sometimes referred to as the set of isolated eigenvalues of T of *finite geometric multiplicity*.

The *Weyl spectrum* of an operator $T \in \mathcal{B}[\mathcal{H}]$ is the set

$$\sigma_w(T) = \bigcap_{K \in \mathcal{B}_\infty[\mathcal{H}]} \sigma(T + K),$$

which is the largest part of $\sigma(T)$ that remains unchanged under compact perturbations. In other words, $\sigma_w(T)$ is the largest part of $\sigma(T)$ such that

$$\sigma_w(T + K) = \sigma_w(T) \quad \text{for every} \quad K \in \mathcal{B}_\infty[\mathcal{H}].$$

Another characterization of $\sigma_w(T)$ will be given in Theorem 5.24.

Lemma 5.23. (a) *If $T \in \mathcal{SF}[\mathcal{H}]$ with $\mathrm{ind}\,(T) \leq 0$, then $T \in \mathcal{F}_\ell[\mathcal{H}]$ and there exists a compact operator $K' \in \mathcal{B}_\infty[\mathcal{H}]$ for which $T + K'$ is left invertible (i.e., such that there exists an operator $A' \in \mathcal{B}[\mathcal{H}]$ for which $A'(T + K') = I$).*

(b) *If $T \in \mathcal{SF}[\mathcal{H}]$ with $\mathrm{ind}\,(T) \geq 0$, then $T \in \mathcal{F}_r[\mathcal{H}]$ and there exists a compact operator $K'' \in \mathcal{B}_\infty[\mathcal{H}]$ for which $T + K''$ is right invertible (i.e., such that there exists an operator $A'' \in \mathcal{B}[\mathcal{H}]$ for which $(T + K'')A'' = I$).*

(c) *If $T \in \mathcal{SF}[\mathcal{H}]$, then $\mathrm{ind}\,(T) = 0$ if and only if there exists a compact operator $K \in \mathcal{B}_\infty[\mathcal{H}]$ for which $T + K$ is invertible (i.e., such that there exists an operator $A \in \mathcal{B}[\mathcal{H}]$ for which $A(T + K) = (T + K)A = I$).*

Proof. Take an operator T in $\mathcal{B}[\mathcal{H}]$.

Claim. If $T \in \mathcal{SF}[\mathcal{H}]$ with $\mathrm{ind}\,(T) \leq 0$, then $T \in \mathcal{F}_\ell[\mathcal{H}]$ and there exists a compact (actually, a finite-rank) operator $K \in \mathcal{B}_\infty[\mathcal{H}]$ for which $\mathcal{N}(T+K) = \{0\}$.

Proof. Take $T \in \mathcal{SF}[\mathcal{H}]$. If $\mathrm{ind}\,(T) \leq 0$, then $\dim \mathcal{N}(T) - \dim \mathcal{N}(T^*) \leq 0$. Thus $\dim \mathcal{N}(T) \leq \dim \mathcal{N}(T^*)$ and $\dim \mathcal{N}(T) < \infty$ (and so $T \in \mathcal{F}_\ell[\mathcal{H}]$ by Theorem 5.1). Let $\{e_i\}_{i=1}^n$ be an orthonormal basis for $\mathcal{N}(T)$, and let B be an orthonormal basis for $\mathcal{N}(T^*) = \mathcal{R}(T)^\perp$ (cf. Lemma 1.4) whose cardinality is not less than n (since $\dim \mathcal{N}(T) \leq \dim \mathcal{N}(T^*)$). Take any orthonormal set $\{f_k\}_{k=1}^n \subseteq B$ and define a map $K \colon \mathcal{H} \to \mathcal{H}$ by

$$Kx = \sum_{j=1}^n \langle x\,;e_j\rangle f_j \quad \text{for every } x \in \mathcal{H}.$$

This is clearly linear and bounded (a contraction, actually). In fact, the map K is a finite-rank operator (thus compact — K lies in $\mathcal{B}_0[\mathcal{H}] \subseteq \mathcal{B}_\infty[\mathcal{H}]$) whose range is included in $\mathcal{R}(T)^\perp$. Indeed,

$$\mathcal{R}(K) \subseteq \bigvee \{f_j\}_{j=1}^n \subseteq \bigvee B = \mathcal{N}(T^*) = \mathcal{R}(T)^\perp.$$

Take any $x \in \mathcal{N}(T)$ and consider its Fourier series expansion with respect to the orthonormal basis $\{e_i\}_{i=1}^n$, namely $x = \sum_{i=1}^n \langle x\,;e_i\rangle e_i$. Thus

$$\|x\|^2 = \sum_{j=1}^n |\langle x\,;e_i\rangle|^2 = \|Kx\|^2 \quad \text{for every } x \in \mathcal{N}(T).$$

Now, if $x \in \mathcal{N}(T+K)$, then $Tx = -Kx$, and hence $Tx \in \mathcal{R}(T) \cap \mathcal{R}(K) \subseteq \mathcal{R}(T) \cap \mathcal{R}(T)^\perp = \{0\}$, so that $Tx = 0$ (i.e., $x \in \mathcal{N}(T)$). Thus $\|x\| = \|Kx\| = \|Tx\| = 0$, and hence $x = 0$. Therefore we get the claimed result:

$$\mathcal{N}(T+K) = \{0\}.$$

(a) If $T \in \mathcal{SF}[\mathcal{H}]$ with $\mathrm{ind}\,(T) \leq 0$, then $T \in \mathcal{F}_\ell[\mathcal{H}]$ and there exists $K \in \mathcal{B}_\infty[\mathcal{H}]$ such that $\mathcal{N}(T+K) = \{0\}$ by the preceding claim. Now $T + K \in \mathcal{F}_\ell[\mathcal{H}]$ according to Theorem 5.6, and so $\mathcal{R}(T+K)$ is closed by Theorem 5.1. Thus $T + K$ is left invertible by Lemma 5.8.

(b) If $T \in \mathcal{SF}[\mathcal{H}]$ has $\mathrm{ind}\,(T) \geq 0$, then $T^* \in \mathcal{SF}[\mathcal{H}]$ has $\mathrm{ind}\,(T^*) = -\mathrm{ind}\,(T) \leq 0$. Thus according to item (a) $T^* \in \mathcal{F}_\ell[\mathcal{H}]$ and so $T \in \mathcal{F}_r[\mathcal{H}]$ and there exists $K \in \mathcal{B}_\infty[\mathcal{H}]$ for which $T^* + K$ is left invertible. Hence $T + K^* = (T^* + K)^*$ is right invertible with $K^* \in \mathcal{B}_\infty[\mathcal{H}]$ (cf. Proposition 1.W in Section 1.9).

(c) If $T \in \mathcal{SF}[\mathcal{H}]$ with $\mathrm{ind}\,(T) = 0$, then $T \in \mathcal{F}[\mathcal{H}]$ and there exists $K \in \mathcal{B}_\infty[\mathcal{H}]$ such that $\mathcal{N}(T+K) = \{0\}$ by the preceding claim. Since $\mathrm{ind}\,(T+K) = \mathrm{ind}\,(T) = 0$ (Theorem 5.6), we get $\mathcal{N}((T+K)^*) = \{0\}$. So $\mathcal{R}(T+K)^\perp = \{0\}$ (Lemma 1.4), and hence $\mathcal{R}(T+K) = \mathcal{H}$. Thus $T + K$ is invertible (Theorem 1.1). Conversely, if there exists $K \in \mathcal{B}_\infty[\mathcal{H}]$ such that $T + K$ is invertible, then $T + K$ is Weyl (since every invertible operator is Weyl), and so is T (Theorem 5.6); that is, $T \in \mathcal{F}[\mathcal{H}]$ with $\mathrm{ind}\,(T) = 0$. $\qquad \square$

According to the claim in the preceding proof, the statement of Lemma 5.23 holds if "compact" is replaced with "finite-rank" (see Remark 5.3(c)).

Theorem 5.24. (SCHECHTER THEOREM). *If* $T \in \mathcal{B}[\mathcal{H}]$, *then*

$$\sigma_w(T) = \sigma_e(T) \cup \bigcup_{k \in \mathbb{Z} \setminus \{0\}} \sigma_k(T) = \sigma(T) \setminus \sigma_0(T).$$

Proof. Take an operator T in $\mathcal{B}[\mathcal{H}]$.

Claim. If $\lambda \in \sigma_0(T)$, then there is a $K \in \mathcal{B}_\infty[\mathcal{H}]$ for which $\lambda \notin \sigma(T + K)$.

Proof. If $\lambda \in \sigma_0(T)$, then $\lambda I - T \in \mathcal{SF}[\mathcal{H}]$ with $\mathrm{ind}\,(\lambda I - T) = 0$. Thus according to Lemma 5.23(c) $\lambda I - (T + K)$ is invertible for some $K \in \mathcal{B}_\infty[\mathcal{H}]$, and therefore $\lambda \in \rho(T + K)$. This completes the proof of the claimed result.

From Corollary 5.18,

$$\sigma(T) = \sigma_e(T) \cup \bigcup_{k \in \mathbb{Z} \setminus \{0\}} \sigma_k(T) \cup \sigma_0(T),$$

where the above sets are all pairwise disjoint, and so

$$\sigma_0(T) = \sigma(T) \setminus \left(\sigma_e(T) \cup \bigcup_{k \in \mathbb{Z} \setminus \{0\}} \sigma_k(T) \right).$$

Take an arbitrary $\lambda \in \sigma(T)$. If λ lies in $\sigma_e(T) \cup \bigcup_{k \in \mathbb{Z} \setminus \{0\}} \sigma_k(T)$, then λ lies in $\sigma_e(T + K) \cup \bigcup_{k \in \mathbb{Z} \setminus \{0\}} \sigma_k(T + K)$ for every $K \in \mathcal{B}_\infty[\mathcal{H}]$ since

$$\sigma_e(T + K) = \sigma_e(T) \quad \text{and} \quad \sigma_k(T + K) = \sigma_k(T) \text{ for every } k \subset \mathbb{Z} \setminus \{0\}$$

for every $K \in \mathcal{B}_\infty[\mathcal{H}]$ (cf. Remarks 5.15(c) and 5.17(c)). On the other hand, if $\lambda \in \sigma_0(T)$, then the preceding claim ensures the existence of a $K \in \mathcal{B}_\infty[\mathcal{H}]$ for which $\lambda \notin \sigma(T + K)$. As the largest part of $\sigma(T)$ that remains invariant under compact perturbation is $\sigma_w(T)$, we get

$$\sigma_w(T) = \sigma_e(T) \cup \bigcup_{k \in \mathbb{Z} \setminus \{0\}} \sigma_k(T) = \sigma(T) \setminus \sigma_0(T). \qquad \square$$

Since $\sigma_w(T) \subseteq \sigma(T)$ by the very definition of $\sigma_w(T)$,

$$\sigma_e(T) \subseteq \sigma_w(T) \subseteq \sigma(T).$$

Chronologically, Theorem 5.24 precedes the spectral picture of Corollary 5.18 [105]. Since $\sigma_k(T) \subseteq \sigma(T) \setminus \sigma_e(T)$ for all $k \in \mathbb{Z}$, we also get

$$\sigma_w(T) \setminus \sigma_e(T) = \bigcup_{k \in \mathbb{Z} \setminus \{0\}} \sigma_k(T)$$

which is the collection of all holes of the essential spectrum. Thus

$$\sigma_e(T) = \sigma_w(T) \quad \Longleftrightarrow \quad \bigcup_{k \in \mathbb{Z} \setminus \{0\}} \sigma_k(T) = \varnothing$$

(i.e., if and only if the essential spectrum has no holes). Since $\sigma_w(T)$ and $\sigma_0(T)$ are both subsets of $\sigma(T)$, it follows by Theorem 5.24 that $\sigma_0(T)$ is the complement of $\sigma_w(T)$ in $\sigma(T)$,

$$\sigma_0(T) = \sigma(T)\backslash\sigma_w(T),$$

and therefore $\{\sigma_w(T), \sigma_0(T)\}$ forms a partition of the spectrum $\sigma(T)$:

$$\sigma(T) = \sigma_w(T) \cup \sigma_0(T) \quad \text{and} \quad \sigma_w(T) \cap \sigma_0(T) = \varnothing.$$

Thus

$$\sigma_w(T) = \sigma(T) \quad \Longleftrightarrow \quad \sigma_0(T) = \varnothing,$$

and so, by Theorem 5.24 again,

$$\sigma_e(T) = \sigma_w(T) = \sigma(T) \quad \Longleftrightarrow \quad \bigcup_{k\in\mathbb{Z}} \sigma_k(T) = \varnothing.$$

Moreover, $\sigma_w(T)$ is always compact (it is the intersection $\bigcap_{K\in\mathcal{B}_\infty[\mathcal{H}]} \sigma(T + K)$ of compact sets in $\mathbb{C}$), and so $\sigma_0(T)$ is closed in $\mathbb{C}$ if and only if $\sigma_0(T) = \pi_0(T) \subseteq \sigma_{\text{iso}}(T)$ (since $\{\sigma_w(T), \sigma_0(T)\}$ is a partition of $\sigma(T)$ and $\{\pi_0(T), \tau_0(T)\}$ is a partition of $\sigma_0(T)$). Here is a third characterization of the Weyl spectrum.

Corollary 5.25. *For every operator* $T \in \mathcal{B}[\mathcal{H}]$,

$$\sigma_w(T) = \{\lambda \in \mathbb{C}: \lambda I - T \in \mathcal{B}[\mathcal{H}]\backslash\mathcal{W}[\mathcal{H}]\}.$$

Proof. If $\lambda \in \rho(T)$, then $\lambda I - T$ is invertible, and so $\lambda I - T \in \mathcal{W}[\mathcal{H}]$ (since invertible operators are Weyl). Therefore if $\lambda I - T \notin \mathcal{W}[\mathcal{H}]$, then $\lambda \in \sigma(T)$, and the result follows from Theorem 5.24 and the definition of $\sigma_0(T)$:

$$\sigma_w(T) = \sigma(T)\backslash\sigma_0(T) \quad \text{and} \quad \sigma_0(T) = \{\lambda \in \sigma(T): \lambda I - T \in \mathcal{W}[\mathcal{H}]\}. \quad \square$$

Then the Weyl spectrum $\sigma_w(T)$ is the set of all scalars λ for which $\lambda I - T$ is not a Weyl operator (i.e., for which $\lambda I - T$ is not a Fredholm operator of index zero). Since an operator lies in $\mathcal{W}[\mathcal{H}]$ together with its adjoint, it follows by Corollary 5.25 that $\lambda \in \sigma_w(T)$ if and only if $\overline{\lambda} \in \sigma_w(T^*)$:

$$\sigma_w(T) = \sigma_w(T^*)^*.$$

Theorem 5.24 also gives us another characterization of the set $\mathcal{W}[\mathcal{H}]$ of all Weyl operators from $\mathcal{B}[\mathcal{H}]$ in terms of the set $\sigma_0(T) = \sigma(T)\backslash\sigma_w(T)$.

Corollary 5.26.

$$\mathcal{W}[\mathcal{H}] = \{T \in \mathcal{B}[\mathcal{H}]: 0 \in \rho(T) \cup \sigma_0(T)\} = \{T \in \mathcal{F}[\mathcal{H}]: 0 \in \rho(T) \cup \sigma_0(T)\}.$$

Proof. Take $T \in \mathcal{B}[\mathcal{H}]$. If $T \in \mathcal{W}[\mathcal{H}]$, then $T \in \mathcal{F}[\mathcal{H}]$ and $T + K$ is invertible for some $K \in \mathcal{B}_\infty[\mathcal{H}]$ by Lemma 5.23(c), which means $0 \in \rho(T + K)$ or, equivalently, $0 \notin \sigma(T + K)$, and so $0 \notin \sigma_w(T + K)$. Thus $0 \notin \sigma_w(T)$ (cf. definition

of the Weyl spectrum preceding Lemma 5.23), and hence $0 \in \rho(T) \cup \sigma_0(T)$
by Theorem 5.24. The converse is (almost) trivial: if $0 \in \rho(T) \cup \sigma_0(T)$, then
either $0 \in \rho(T)$, so that $T \in \mathcal{G}[\mathcal{H}] \subset \mathcal{W}[\mathcal{H}]$ by Remark 5.7(b); or $0 \in \sigma_0(T)$,
which means $0 \in \sigma(T)$ and $T \in \mathcal{W}[\mathcal{H}]$ by the very definition of $\sigma_0(T)$. This
proves the first identity, which implies the second one since $\mathcal{W}[\mathcal{H}] \subseteq \mathcal{F}[\mathcal{H}]$. $\square$

Remark 5.27. (a) FINITE DIMENSION. Take any operator $T \in \mathcal{B}[\mathcal{H}]$. Since
$\sigma_e(T) \subseteq \sigma_w(T)$, then by Remark 5.15(a)

$$\sigma_w(T) = \varnothing \quad \Longrightarrow \quad \dim \mathcal{H} < \infty$$

and, by Remark 5.3(d) and Corollary 5.25,

$$\dim \mathcal{H} < \infty \quad \Longrightarrow \quad \sigma_w(T) = \varnothing.$$

Thus

$$\sigma_w(T) \neq \varnothing \quad \Longleftrightarrow \quad \dim \mathcal{H} = \infty.$$

(b) MORE ON FINITE DIMENSION. Let $T \in \mathcal{B}[\mathcal{H}]$ be an operator on a finite-
dimensional space. Then $\mathcal{B}[\mathcal{H}] = \mathcal{W}[\mathcal{H}]$ by Remark 5.3(d) and so $\sigma_0(T) = \sigma(T)$
by the definition of $\sigma_0(T)$. Moreover, according to Corollary 2.9, $\sigma(T) =
\sigma_{\mathrm{iso}}(T)$ (since $\#\sigma(T) < \infty$), and $\sigma(T) = \sigma_{PF}(T)$ (since $\dim \mathcal{H} < \infty$). Hence

$$\dim \mathcal{H} < \infty \quad \Longrightarrow \quad \sigma(T) = \sigma_{PF}(T) = \sigma_{\mathrm{iso}}(T) = \sigma_0(T) = \pi_0(T) = \pi_{00}(T)$$

and so, either by Theorem 5.24 or by item (a) and Remark 5.15(a),

$$\dim \mathcal{H} < \infty \quad \Longrightarrow \quad \sigma_e(T) = \sigma_w(T) = \varnothing.$$

(c) FREDHOLM ALTERNATIVE. The Fredholm Alternative has been restated
again and again in Corollary 1.20, Theorem 2.18, Remark 5.3(e), Remark
5.7(b), and Remark 5.17(a). Here is an ultimate restatement of it. If K is com-
pact (i.e., if $K \in \mathcal{B}_\infty[\mathcal{H}]$), then $\sigma(K)\backslash\{0\} = \sigma_0(K)\backslash\{0\}$ by Remark 5.17(a).
But $\sigma(K)\backslash\{0\} \subseteq \sigma_{\mathrm{iso}}(K)$ by Corollary 2.20, and $\pi_0(K) = \sigma_{\mathrm{iso}}(K) \cap \sigma_0(K)$.
Therefore (compare with Theorem 2.18 and Remark 5.17(a)),

$$K \in \mathcal{B}_\infty[\mathcal{H}] \quad \Longrightarrow \quad \sigma(K)\backslash\{0\} = \pi_0(K)\backslash\{0\}.$$

As $\pi_0(K) = \sigma_{\mathrm{iso}}(K) \cap \sigma_0(K) \subseteq \sigma_{\mathrm{iso}}(K) \cap \sigma_{PF}(K) = \pi_{00}(K) \subseteq \sigma_{PF}(T) \subseteq \sigma_P(T)$,

$$\sigma(K)\backslash\{0\} = \sigma_{PF}(K)\backslash\{0\} = \pi_{00}(K)\backslash\{0\} = \pi_0(K)\backslash\{0\}.$$

(d) MORE ON COMPACT OPERATORS. Take a compact operator $K \in \mathcal{B}_\infty[\mathcal{H}]$.
By Remark 5.3(e) (Fredholm Alternative) if $\lambda \neq 0$, then $\lambda I - K \in \mathcal{W}[\mathcal{H}]$.
Thus, according to Corollary 5.25, $\sigma_w(K)\backslash\{0\} = \varnothing$. If $\dim \mathcal{H} = \infty$, then
$\sigma_w(K) = \{0\}$ by item (a). Since $\sigma_e(K) \subseteq \sigma_w(K)$, it follows by Remark 5.15(a)
that if $\dim \mathcal{H} = \infty$, then $\varnothing \neq \sigma_e(K) \subseteq \{0\}$. Hence

$$K \in \mathcal{B}_\infty[\mathcal{H}] \text{ and } \dim \mathcal{H} = \infty \quad \Longrightarrow \quad \sigma_e(K) = \sigma_w(K) = \{0\}.$$

Therefore, by the Fredholm Alternative of item (c),

$$K \in \mathcal{B}_{\infty}[\mathcal{H}] \ \text{ and } \ \dim \mathcal{H} = \infty \quad \Longrightarrow \quad \sigma(K) = \pi_0(K) \cup \{0\}, \quad \sigma_w(K) = \{0\}.$$

(e) NORMAL OPERATORS. Recall: T is normal if and only if $\lambda I - T$ is normal. Since every normal Fredholm operator is Weyl (see Remark 5.7(b)), we may invoke Corollaries 5.12 and 5.25 to get

$$T \text{ normal} \quad \Longrightarrow \quad \sigma_e(T) = \sigma_w(T).$$

Take any $T \in \mathcal{B}[\mathcal{H}]$ and consider the following expressions (cf. Theorem 5.24 or Corollary 5.25 and definitions of $\sigma_0(T)$, $\pi_0(T)$, and $\pi_{00}(T)$).

$$\sigma_0(T) = \sigma(T) \backslash \sigma_w(T), \quad \pi_0(T) = \sigma_{\mathrm{iso}}(T) \cap \sigma_0(T), \quad \pi_{00}(T) = \sigma_{\mathrm{iso}}(T) \cap \sigma_{PF}(T).$$

The equivalent assertions in the next corollary define an important class of operators. An operator for which any of those equivalent assertions holds is said to satisfy *Weyl's Theorem*. This will be discussed later in Section 5.5.

Corollary 5.28. *The assertions below are equivalent.*

(a) $\sigma_0(T) = \pi_{00}(T)$.

(b) $\sigma(T) \backslash \pi_{00}(T) = \sigma_w(T)$.

Moreover, if

$$\sigma_0(T) = \sigma_{\mathrm{iso}}(T)$$

(i.e., if $\sigma_w(T) = \sigma_{\mathrm{acc}}(T)$), then the above equivalent assertions hold true. Conversely, if either of the above equivalent assertions holds true, then

$$\pi_0(T) = \pi_{00}(T).$$

Proof. By Theorem 5.24, $\sigma(T) \backslash \sigma_0(T) = \sigma_w(T)$, and so (a) implies (b). If (b) holds, then $\sigma_0(T) = \sigma(T) \backslash \sigma_w(T) = \sigma(T) \backslash (\sigma(T) \backslash \pi_{00}(T)) = \pi_{00}(T)$ because $\pi_{00}(T) \subseteq \sigma(T)$, and so (b) implies (a). Moreover, if $\sigma_0(T) = \sigma_{\mathrm{iso}}(T)$ (or, if their complements in $\sigma(T)$ coincide), then (a) holds because $\sigma_0(T) \subseteq \sigma_{PF}(T)$. Conversely, if (a) holds, then $\pi_0(T) = \sigma_{\mathrm{iso}}(T) \cap \sigma_0(T) = \sigma_{\mathrm{iso}}(T) \cap \pi_{00}(T) = \pi_{00}(T)$ because $\pi_{00}(T) \subseteq \sigma_{\mathrm{iso}}(T)$. $\square$

5.4 Browder Operators and Browder Spectrum

As usual, take $T \in \mathcal{B}[\mathcal{H}]$ (or, more generally, take any linear transformation of a linear space into itself). In fact, part of this section is purely algebraic, and the entire section (as the whole chapter) has a natural counterpart in a Banach space. Let $\mathbb{N}_0$ be the set of all nonnegative integers and consider the nonnegative integral powers T^n of T. As is readily verified,

$$\mathcal{N}(T^n) \subseteq \mathcal{N}(T^{n+1}) \quad \text{and} \quad \mathcal{R}(T^{n+1}) \subseteq \mathcal{R}(T^n)$$

for every $n \in \mathbb{N}_0$. Thus $\{\mathcal{N}(T^n)\}$ and $\{\mathcal{R}(T^n)\}$ are nondecreasing and nonincreasing (in the inclusion ordering) sequences of subsets of $\mathcal{H}$, respectively.

Lemma 5.29. *Let n_0 be an arbitrary integer in $\mathbb{N}_0$.*

(a) *If* $\mathcal{N}(T^{n_0+1}) = \mathcal{N}(T^{n_0})$, *then* $\mathcal{N}(T^{n+1}) = \mathcal{N}(T^n)$ *for every* $n \geq n_0$.

(b) *If* $\mathcal{R}(T^{n_0+1}) = \mathcal{R}(T^{n_0})$, *then* $\mathcal{R}(T^{n+1}) = \mathcal{R}(T^n)$ *for every* $n \geq n_0$.

Proof. (a) Rewrite the statements in (a) as follows.

$$\mathcal{N}(T^{n_0+1}) = \mathcal{N}(T^{n_0}) \implies \mathcal{N}(T^{n_0+k+1}) = \mathcal{N}(T^{n_0+k}) \text{ for every } k \geq 0.$$

The claimed result holds trivially for $k = 0$. Suppose it holds for some $k \geq 0$. Take an arbitrary $x \in \mathcal{N}(T^{n_0+k+2})$ so that $T^{n_0+k+1}(Tx) = T^{n_0+k+2}x = 0$. Thus $Tx \in \mathcal{N}(T^{n_0+k+1}) = \mathcal{N}(T^{n_0+k})$, and so $T^{n_0+k+1}x = T^{n_0+k}(Tx) = 0$, which implies $x \in \mathcal{N}(T^{n_0+k+1})$. Hence $\mathcal{N}(T^{n_0+k+2}) \subseteq \mathcal{N}(T^{n_0+k+1})$. However, $\mathcal{N}(T^{n_0+k+1}) \subseteq \mathcal{N}(T^{n_0+k+2})$ since $\{\mathcal{N}(T^n)\}$ is nondecreasing, and therefore $\mathcal{N}(T^{n_0+k+2}) = \mathcal{N}(T^{n_0+k+1})$. Then the claimed result holds for $k + 1$ whenever it holds for k, which completes the proof of (a) by induction.

(b) Rewrite the statements in (b) as follows.

$$\mathcal{R}(T^{n_0+1}) = \mathcal{R}(T^{n_0}) \implies \mathcal{R}(T^{n_0+k+1}) = \mathcal{R}(T^{n_0+k}) \text{ for every } k \geq 1.$$

The claimed result holds trivially for $k = 0$. Suppose it holds for some integer $k \geq 0$. Take an arbitrary $y \in \mathcal{R}(T^{n_0+k+1})$ so that $y = T^{n_0+k+1}x = T(T^{n_0+k}x)$ for some $x \in \mathcal{H}$, and hence $y = Tu$ for some $u \in \mathcal{R}(T^{n_0+k})$. If $\mathcal{R}(T^{n_0+k}) = \mathcal{R}(T^{n_0+k+1})$, then $u \in \mathcal{R}(T^{n_0+k+1})$, and so $y = T(T^{n_0+k+1}v)$ for some $v \in \mathcal{H}$. Thus $y \in \mathcal{R}(T^{n_0+k+2})$. Therefore $\mathcal{R}(T^{n_0+k+1}) \subseteq \mathcal{R}(T^{n_0+k+2})$. Since the sequence $\{\mathcal{R}(T^n)\}$ is nonincreasing we get $\mathcal{R}(T^{n_0+k+2}) \subseteq \mathcal{R}(T^{n_0+k+1})$. Hence $\mathcal{R}(T^{n_0+k+2}) = \mathcal{R}(T^{n_0+k+1})$. Thus the claimed result holds for $k + 1$ whenever it holds for k, which completes the proof of (b) by induction. $\quad\square$

Let $\overline{\mathbb{N}}_0 = \mathbb{N}_0 \cup \{+\infty\}$ denote the set of all *extended nonnegative integers* with its natural (extended) ordering. The *ascent* and *descent* of an operator $T \in \mathcal{B}[\mathcal{H}]$ are defined as follows. The *ascent* of T, asc(T), is the least (extended) nonnegative integer for which $\mathcal{N}(T^{n+1}) = \mathcal{N}(T^n)$:

$$\operatorname{asc}(T) = \min\{n \in \overline{\mathbb{N}}_0 : \mathcal{N}(T^{n+1}) = \mathcal{N}(T^n)\},$$

and the *descent* of T, dsc(T), is the least (extended) nonnegative integer for which $\mathcal{R}(T^{n+1}) = \mathcal{R}(T^n)$:

$$\operatorname{dsc}(T) = \min\{n \in \overline{\mathbb{N}}_0 : \mathcal{R}(T^{n+1}) = \mathcal{R}(T^n)\}.$$

The ascent of T is null if and only if T is injective, and the descent of T is null if and only if T is surjective. Indeed, since $\mathcal{N}(I) = \{0\}$ and $\mathcal{R}(I) = \mathcal{H}$,

$$\operatorname{asc}(T) = 0 \iff \mathcal{N}(T) = \{0\},$$
$$\operatorname{dsc}(T) = 0 \iff \mathcal{R}(T) = \mathcal{H}.$$

Again, the notion of ascent and descent holds for every linear transformation of a linear space into itself, and so does the next result.

Lemma 5.30. *Take any operator $T \in \mathcal{B}[\mathcal{H}]$.*

(a) *If $\mathrm{asc}\,(T) < \infty$ and $\mathrm{dsc}\,(T) = 0$, then $\mathrm{asc}\,(T) = 0$.*

(b) *If $\mathrm{asc}\,(T) < \infty$ and $\mathrm{dsc}\,(T) < \infty$, then $\mathrm{asc}\,(T) = \mathrm{dsc}\,(T)$.*

Proof. (a) Suppose $\mathrm{dsc}\,(T) = 0$ (i.e., suppose $\mathcal{R}(T) = \mathcal{H}$). If $\mathrm{asc}\,(T) \neq 0$ (i.e., if $\mathcal{N}(T) \neq \{0\}$), then take $0 \neq x_1 \in \mathcal{N}(T) \cap \mathcal{R}(T)$ and x_2, x_3 in $\mathcal{R}(T) = \mathcal{H}$ as follows: $x_1 = Tx_2$ and $x_2 = Tx_3$, and so $x_1 = T^2 x_3$. Continuing this way, we can construct a sequence $\{x_n\}_{n \geq 1}$ of vectors in $\mathcal{H} = \mathcal{R}(T)$ such that $x_n = Tx_{n+1}$ and $0 \neq x_1 = T^n x_{n+1}$ lies in $\mathcal{N}(T)$, and so $T^{n+1} x_{n+1} = 0$. Therefore $x_{n+1} \in \mathcal{N}(T^{n+1}) \backslash \mathcal{N}(T^n)$ for each $n \geq 1$, and hence $\mathrm{asc}\,(T) = \infty$ by Lemma 5.29. Summing up: If $\mathrm{dsc}\,(T) = 0$, then $\mathrm{asc}\,(T) \neq 0$ implies $\mathrm{asc}\,(T) = \infty$.

(b) Set $m = \mathrm{dsc}\,(T)$, so that $\mathcal{R}(T^m) = \mathcal{R}(T^{m+1})$, and set $S = T|_{\mathcal{R}(T^m)}$. By Proposition 1.E(a), $\mathcal{R}(T^m)$ is T-invariant, and so $S \in \mathcal{B}[\mathcal{R}(T^m)]$. Also $\mathcal{R}(S) = S(\mathcal{R}(T^m)) = \mathcal{R}(ST^m) = \mathcal{R}(T^{m+1}) = \mathcal{R}(T^m)$. Thus S is surjective, which means $\mathrm{dsc}\,(S) = 0$. Moreover, since $\mathrm{asc}\,(S) < \infty$ (because $\mathrm{asc}\,(T) < \infty$), we get $\mathrm{asc}\,(S) = 0$ by (a), which means $\mathcal{N}(S) = \{0\}$. Take $x \in \mathcal{N}(T^{m+1})$ and set $y = T^m x$ in $\mathcal{R}(T^m)$. Thus $Sy = T^{m+1} x = 0$, and so $y = 0$. Therefore $x \in \mathcal{N}(T^m)$. Then $\mathcal{N}(T^{m+1}) \subseteq \mathcal{N}(T^m)$ which implies $\mathcal{N}(T^{m+1}) = \mathcal{N}(T^m)$ since $\{\mathcal{N}(T^n)\}$ is nondecreasing. Consequently, $\mathrm{asc}\,(T) \leq m$ by Lemma 5.29. To prove the reverse inequality, suppose $m \neq 0$ (otherwise apply (a)) and take a vector z in $\mathcal{R}(T^{m-1}) \backslash \mathcal{R}(T^m)$ so that $z = T^{m-1} u$ and $Tz = T(T^{m-1} u) = T^m u$ lies in $\mathcal{R}(T^m)$ for some u in $\mathcal{H}$. Since $T^m(\mathcal{R}(T^m)) = \mathcal{R}(T^{2m}) = \mathcal{R}(T^m)$, we get $Tz = T^m v$ for some v in $\mathcal{R}(T^m)$. Now observe that $T^m(u - v) = Tz - Tz = 0$ and $T^{m-1}(u - v) = z - T^{m-1} v \neq 0$ (reason: $T^{m-1} v \in \mathcal{R}(T^{2m-1}) = \mathcal{R}(T^m)$ since $v \in \mathcal{R}(T^m)$, and $z \notin \mathcal{R}(T^m)$). Therefore $(u - v) \in \mathcal{N}(T^m) \backslash \mathcal{N}(T^{m-1})$, and so $\mathrm{asc}\,(T) \geq m$. Outcome: If $\mathrm{dsc}\,(T) = m$ and $\mathrm{asc}\,(T) < \infty$, then $\mathrm{asc}\,(T) = m$. $\square$

Lemma 5.31. *If $T \in \mathcal{F}[\mathcal{H}]$, then $\mathrm{asc}\,(T) = \mathrm{dsc}\,(T^*)$ and $\mathrm{dsc}\,(T) = \mathrm{asc}\,(T^*)$.*

Proof. Take any $T \in \mathcal{B}[\mathcal{H}]$. We will use freely the relations between range and kernel involving adjoints and orthogonal complements of Lemma 1.4.

Claim 1. $\mathrm{asc}\,(T) < \infty$ if and only if $\mathrm{dsc}\,(T^*) < \infty$.

$\mathrm{dsc}\,(T) < \infty$ if and only if $\mathrm{asc}\,(T^*) < \infty$.

Proof. Take an arbitrary $n \in \mathbb{N}_0$. If $\mathrm{asc}\,(T) = \infty$, then $\mathcal{N}(T^n) \subset \mathcal{N}(T^{n+1})$, and so $\mathcal{N}(T^{n+1})^\perp \subset \mathcal{N}(T^n)^\perp$. Equivalently, $\mathcal{R}(T^{*(n+1)})^- \subset \mathcal{R}(T^{*n})^-$, which ensures the proper inclusion $\mathcal{R}(T^{*(n+1)}) \subset \mathcal{R}(T^{*n})$, and hence $\mathrm{dsc}\,(T^*) = \infty$. Thus $\mathrm{asc}\,(T) = \infty$ implies $\mathrm{dsc}\,(T^*) = \infty$. Dually (since $T^{**} = T$), $\mathrm{asc}\,(T^*) = \infty$ implies $\mathrm{dsc}\,(T) = \infty$:

$$\mathrm{asc}\,(T) = \infty \implies \mathrm{dsc}\,(T^*) = \infty \quad \text{and} \quad \mathrm{asc}\,(T^*) = \infty \implies \mathrm{dsc}\,(T) = \infty.$$

Conversely, if $\mathrm{dsc}\,(T) = \infty$, then $\mathcal{R}(T^{n+1}) \subset \mathcal{R}(T^n)$. Therefore $\mathcal{R}(T^n)^{\perp} \subset \mathcal{R}(T^{n+1})^{\perp}$. Equivalently, $\mathcal{N}(T^{*n}) \subset \mathcal{N}(T^{*(n+1)})$, and so $\mathrm{asc}\,(T^*) = \infty$. Thus $\mathrm{dsc}\,(T) = \infty$ implies $\mathrm{asc}\,(T^*) = \infty$. Dually, $\mathrm{dsc}\,(T^*) = \infty$ implies $\mathrm{asc}\,(T) = \infty$:

$$\mathrm{dsc}\,(T) = \infty \implies \mathrm{asc}\,(T^*) = \infty \quad \text{and} \quad \mathrm{dsc}\,(T^*) = \infty \implies \mathrm{asc}\,(T) = \infty.$$

Summing up: $\mathrm{asc}\,(T) = \infty$ if and only if $\mathrm{dsc}\,(T^*) = \infty$ and $\mathrm{dsc}\,(T) = \infty$ if and only if $\mathrm{asc}\,(T^*) = \infty$, which completes the proof of Claim 1.

Claim 2. If $\mathrm{dsc}\,(T) < \infty$, then $\mathrm{asc}\,(T^*) \leq \mathrm{dsc}\,(T)$.

       If $\mathrm{dsc}\,(T^*) < \infty$, then $\mathrm{asc}\,(T) \leq \mathrm{dsc}\,(T^*)$.

Proof. If $\mathrm{dsc}\,(T) < \infty$, then set $n_0 = \mathrm{dsc}\,(T) \in \mathbb{N}_0$ and take any integer $n \geq n_0$. Thus $\mathcal{R}(T^n) = \mathcal{R}(T^{n_0})$, and hence $\mathcal{R}(T^n)^- = \mathcal{R}(T^{n_0})^-$ so that $\mathcal{N}(T^{*n})^{\perp} = \mathcal{N}(T^{*n_0})^{\perp}$, which implies $\mathcal{N}(T^{*n}) = \mathcal{N}(T^{*n_0})$. Therefore $\mathrm{asc}\,(T^*) \leq n_0$, and so $\mathrm{asc}\,(T^*) \leq \mathrm{dsc}\,(T)$. Dually, if $\mathrm{dsc}\,(T^*) < \infty$, then $\mathrm{asc}\,(T) \leq \mathrm{dsc}\,(T^*)$. This completes the proof of Claim 2.

Claim 3. If $\mathrm{asc}\,(T) < \infty$ and $\mathcal{R}(T^n)$ is closed for every $n \geq \mathrm{asc}\,(T)$,

       then $\mathrm{dsc}\,(T^*) \leq \mathrm{asc}\,(T)$.

       If $\mathrm{asc}\,(T^*) < \infty$ and $\mathcal{R}(T^n)$ is closed for every $n \geq \mathrm{asc}\,(T^*)$,

       then $\mathrm{dsc}\,(T) \leq \mathrm{asc}\,(T^*)$.

Proof. If $\mathrm{asc}\,(T) < \infty$, then set $n_0 = \mathrm{asc}\,(T) \in \mathbb{N}_0$ and take an arbitrary integer $n \geq n_0$. Thus $\mathcal{N}(T^n) = \mathcal{N}(T^{n_0})$ or, equivalently, $\mathcal{R}(T^{*n})^{\perp} = \mathcal{R}(T^{*n_0})^{\perp}$, so that $\mathcal{R}(T^{*n_0})^- = \mathcal{R}(T^{*n})^-$. Suppose $\mathcal{R}(T^n)$ is closed. Thus $\mathcal{R}(T^{*n}) = \mathcal{R}(T^{n*})$ is closed (Lemma 1.5). Then $\mathcal{R}(T^{*n}) = \mathcal{R}(T^{*n_0})$, and therefore $\mathrm{dsc}\,(T^*) \leq n_0$, so that $\mathrm{dsc}\,(T^*) \leq \mathrm{asc}\,(T)$. Dually, if $\mathrm{asc}\,(T^*) < \infty$ and $\mathcal{R}(T^n)$ is closed (equivalently, $\mathcal{R}(T^{*n})$ is closed — Lemma 1.5) for every integer $n \geq \mathrm{asc}\,(T^*)$, then $\mathrm{dsc}\,(T) \leq \mathrm{asc}\,(T^*)$, completing the proof of Claim 3.

Outcome: If $\mathcal{R}(T^n)$ is closed for every $n \in \mathbb{N}_0$ (so that we can apply Claim 3), then it follows by Claims 1, 2, and 3 that

$$\mathrm{asc}\,(T) = \mathrm{dsc}\,(T^*) \quad \text{and} \quad \mathrm{dsc}\,(T) = \mathrm{asc}\,(T^*).$$

In particular, if $T \in \mathcal{F}[\mathcal{H}]$, then $T^n \in \mathcal{F}[\mathcal{H}]$ by Corollary 5.5, and therefore $\mathcal{R}(T^n)$ is closed for every integer $n \in \mathbb{N}_0$ by Corollary 5.2. Thus if T is Fredholm, then $\mathrm{asc}\,(T) = \mathrm{dsc}\,(T^*)$ and $\mathrm{dsc}\,(T) = \mathrm{asc}\,(T^*)$. $\qquad\square$

A *Browder operator* is a Fredholm operator with finite ascent and finite descent. Let $\mathcal{B}_r[\mathcal{H}]$ denote the class of all Browder operators from $\mathcal{B}[\mathcal{H}]$:

$$\mathcal{B}_r[\mathcal{H}] = \{T \in \mathcal{F}[\mathcal{H}]: \mathrm{asc}\,(T) < \infty \text{ and } \mathrm{dsc}\,(T) < \infty\};$$

equivalently, according to Lemma 5.30(b),

$$\mathcal{B}_r[\mathcal{H}] = \{T \in \mathcal{F}[\mathcal{H}]: \mathrm{asc}\,(T) = \mathrm{dsc}\,(T) < \infty\}$$

(i.e., $\mathcal{B}_r[\mathcal{H}] = \{T \in \mathcal{F}[\mathcal{H}]: \mathrm{asc}\,(T) = \mathrm{dsc}\,(T) = m$ for some $m \in \mathbb{N}_0\}$). Thus

$$\mathcal{F}[\mathcal{H}]\backslash\mathcal{B}_r[\mathcal{H}] = \{T \in \mathcal{F}[\mathcal{H}]: \mathrm{asc}\,(T) = \infty \ \text{ or }\ \mathrm{dsc}\,(T) = \infty\}$$

and, by Lemma 5.31,

$$T \in \mathcal{B}_r[\mathcal{H}] \quad \text{if and only if} \quad T^* \in \mathcal{B}_r[\mathcal{H}].$$

Two linear manifolds of a linear space $\mathcal{X}$, say $\mathcal{R}$ and $\mathcal{N}$, are said to be *algebraic complements* of each other (or *complementary linear manifolds*) if they sum the whole space and are algebraically disjoint; that is, if

$$\mathcal{R} + \mathcal{N} = \mathcal{X} \quad \text{and} \quad \mathcal{R} \cap \mathcal{N} = \{0\}.$$

A pair of subspaces (i.e., closed linear manifolds) of a normed space that are algebraic complements of each other are called *complementary subspaces*.

Lemma 5.32. *If $T \in \mathcal{B}[\mathcal{H}]$ with $\mathrm{asc}\,(T) = \mathrm{dsc}\,(T) = m$ for some $m \in \mathbb{N}_0$, then $\mathcal{R}(T^m)$ and $\mathcal{N}(T^m)$ are algebraic complements of each other.*

Proof. Take any $T \in \mathcal{B}[\mathcal{H}]$. (In fact, as will become clear during the proof, this is a purely algebraic result which holds for every linear transformation of a linear space into itself with coincident finite ascent and descent.)

Claim 1. If $\mathrm{asc}\,(T) = m$, then $\mathcal{R}(T^m) \cap \mathcal{N}(T^m) = \{0\}$.

Proof. If $y \in \mathcal{R}(T^m) \cap \mathcal{N}(T^m)$, then $y = T^m x$ for some $x \in \mathcal{H}$ and $T^m y = 0$, so that $T^{2m}x = T^m(T^m x) = T^m y = 0$; that is, $x \in \mathcal{N}(T^{2m}) = \mathcal{N}(T^m)$ since $\mathrm{asc}\,(T) = m$. Hence $y = T^m x = 0$, proving Claim 1.

Claim 2. If $\mathrm{dsc}\,(T) = m$, then $\mathcal{R}(T^m) + \mathcal{N}(T^m) = \mathcal{H}$.

Proof. If $\mathrm{dsc}\,(T) = m$, then $\mathcal{R}(T^m) = \mathcal{R}(T^{m+1})$. Set $S = T|_{\mathcal{R}(T^m)}$. According to Proposition 1.E(a), $\mathcal{R}(T^m)$ is T-invariant, and so $S \in \mathcal{B}[\mathcal{R}(T^m)]$ with $\mathcal{R}(S) = S(\mathcal{R}(T^m)) = \mathcal{R}(ST^m) = \mathcal{R}(T^{m+1}) = \mathcal{R}(T^m)$. Then S is surjective, and hence S^m is surjective too (surjectiveness of S implies $\mathrm{dsc}\,(S) = 0$) Therefore $\mathcal{R}(S^m) = \mathcal{R}(T^m)$. Take an arbitrary vector $x \in \mathcal{H}$. Thus $T^m x \in \mathcal{R}(T^m) = \mathcal{R}(S^m)$, and so there exists a vector $u \in \mathcal{R}(T^m)$ for which $S^m u = T^m x$. Since $S^m u = T^m u$ (because $S^m = (T|_{\mathcal{R}(T^m)})^m = T^m|_{\mathcal{R}(T^m)}$), we get $T^m u = T^m x$ and hence $v = x - u$ is in $\mathcal{N}(T^m)$. Then $x = u + v$ lies in $\mathcal{R}(T^m) + \mathcal{N}(T^m)$. Consequently $\mathcal{H} \subseteq \mathcal{R}(T^m) + \mathcal{N}(T^m)$, proving Claim 2. $\square$

Theorem 5.33. *Take $T \in \mathcal{B}[\mathcal{H}]$ and consider the following assertions.*

(a) *T is Browder but not invertible   (i.e., $T \in \mathcal{B}_r[\mathcal{H}]/\mathcal{G}[\mathcal{H}]$).*

(b) *$T \in \mathcal{F}[\mathcal{H}]$ is such that $\mathcal{R}(T^m)$ and $\mathcal{N}(T^m)$ are complementary subspaces for some $m \in \mathbb{N}$.*

(c) *$T \in \mathcal{W}[\mathcal{H}]$ (i.e., T is Weyl, which means T is Fredholm of index zero).*

Claim: (a) $\Longrightarrow$ (b) $\Longrightarrow$ (c).

Proof. (a)$\Rightarrow$(b). An operator $T \in \mathcal{B}[\mathcal{H}]$ is invertible (equivalently, $\mathcal{R}(T) = \mathcal{H}$ and $\mathcal{N}(T) = \{0\}$) if and only if $T \in \mathcal{G}[\mathcal{H}]$ (i.e., $T \in \mathcal{B}[\mathcal{H}]$ has a bounded inverse) by Theorem 1.1. Moreover, every invertible operator is Fredholm (by the definition of Fredholm operators), and $\mathcal{R}(T) = \mathcal{H}$ and $\mathcal{N}(T) = \{0\}$ (i.e., $T \in \mathcal{G}[\mathcal{H}]$) if and only if $\operatorname{asc}(T) = \operatorname{dsc}(T) = 0$ (by the definition of ascent and descent). Thus invertible operators are Browder, which in turn are Fredholm,

$$\mathcal{G}[\mathcal{H}] \subset \mathcal{B}_r[\mathcal{H}] \subseteq \mathcal{F}[\mathcal{H}].$$

(The last inclusion is proper only in infinite-dimensional spaces.) Therefore if $T \in \mathcal{B}_r[\mathcal{H}]$ and $T \notin \mathcal{G}[\mathcal{H}]$, then $\operatorname{asc}(T) = \operatorname{dsc}(T) = m$ for some integer $m \geq 1$, and so $\mathcal{N}(T^m)$ and $\mathcal{R}(T^m)$ are complementary linear manifolds of $\mathcal{H}$ by Lemma 5.32. If $T \in \mathcal{B}_r[\mathcal{H}]$, then $T \in \mathcal{F}[\mathcal{H}]$, so that $T^m \in \mathcal{F}[\mathcal{H}]$ by Corollary 5.5. Hence $\mathcal{R}(T^m)$ is closed by Corollary 5.2. But $\mathcal{N}(T^m)$ is closed since T^m is bounded. Thus $\mathcal{R}(T^m)$ and $\mathcal{N}(T^m)$ are subspaces of $\mathcal{H}$.

(b)$\Rightarrow$(c). If (b) holds, then $T \in \mathcal{F}[\mathcal{H}]$. Thus $T^n \in \mathcal{F}[\mathcal{H}]$ by Corollary 5.5, and so $\mathcal{N}(T^n)$ and $\mathcal{N}(T^{n*})$ are finite-dimensional and $\mathcal{R}(T^n)$ is closed by Corollary 5.2, for every $n \in \mathbb{N}$. In addition, suppose there exists $m \in \mathbb{N}$ such that

$$\mathcal{R}(T^m) + \mathcal{N}(T^m) = \mathcal{H} \quad \text{and} \quad \mathcal{R}(T^m) \cap \mathcal{N}(T^m) = \{0\}.$$

Since $\mathcal{R}(T^m)$ is closed, $\mathcal{R}(T^m) + \mathcal{R}(T^m)^\perp = \mathcal{H}$ (Projection Theorem), where $\mathcal{R}(T^m)^\perp = \mathcal{N}(T^{m*})$ (Lemma 1.4). Then

$$\mathcal{R}(T^m) + \mathcal{N}(T^{m*}) = \mathcal{H} \quad \text{and} \quad \mathcal{R}(T^m) \cap \mathcal{N}(T^{m*}) = \{0\}.$$

Thus $\mathcal{N}(T^m)$ and $\mathcal{N}(T^{m*})$ are both algebraic complements of $\mathcal{R}(T^m)$, and so they have the same (finite) dimension (see, e.g., [78, Theorem 2.18] — it is in fact the codimension of $\mathcal{R}(T^m)$). Then $\operatorname{ind}(T^m) = 0$. Since $T \in \mathcal{F}[\mathcal{H}]$, we get $m \operatorname{ind}(T) = \operatorname{ind}(T^m) = 0$ (see Corollary 5.5). Therefore (recalling that $m \neq 0$) $\operatorname{ind}(T) = 0$, and so $T \in \mathcal{W}[\mathcal{H}]$. $\qquad\square$

Since $\mathcal{G}[\mathcal{H}] \subset \mathcal{W}[\mathcal{H}]$ (cf. Remark 5.7(b)),

$$\mathcal{G}[\mathcal{H}] \subset \mathcal{B}_r[\mathcal{H}] \subseteq \mathcal{W}[\mathcal{H}] \subseteq \mathcal{F}[\mathcal{H}] \subseteq \mathcal{B}[\mathcal{H}].$$

(The last three inclusions are proper only in infinite-dimensional spaces — Remarks 5.3(d), 5.7(b) and 5.36(a)). The inclusion $\mathcal{B}_r[\mathcal{H}] \subseteq \mathcal{W}[\mathcal{H}]$ can be otherwise verified by Proposition 5.J. Also, the assertion "T is Browder and not invertible" (i.e., $T \in \mathcal{B}_r[\mathcal{H}] \backslash \mathcal{G}[\mathcal{H}]$) means "$T \in \mathcal{B}_r[\mathcal{H}]$ and $0 \in \sigma(T)$". According to Theorem 5.34 below, in this case 0 must be an isolated point of $\sigma(T)$. Precisely, $T \in \mathcal{B}_r[\mathcal{H}]$ and $0 \in \sigma(T)$ *if and only if* $T \in \mathcal{F}[\mathcal{H}]$ and $0 \in \sigma_{\mathrm{iso}}(T)$.

Theorem 5.34. *Take an operator* $T \in \mathcal{B}[\mathcal{H}]$.

(a) *If* $T \in \mathcal{B}_r[\mathcal{H}]$ *and* $0 \in \sigma(T)$, *then* $0 \in \pi_0(T)$.

(b) *If* $T \in \mathcal{F}[\mathcal{H}]$ *and* $0 \in \sigma_{\mathrm{iso}}(T)$, *then* $T \in \mathcal{B}_r[\mathcal{H}]$.

Proof. (a) As we have seen before, $\sigma_0(T) = \{\lambda \in \sigma(T) \colon \lambda I - T \in \mathcal{W}[\mathcal{H}]\}$ and $\mathcal{B}_r[\mathcal{H}] \subset \mathcal{W}[\mathcal{H}]$. So if $T \in \mathcal{B}_r[\mathcal{H}]$ and $0 \in \sigma(T)$, then

$$0 \in \sigma_0(T).$$

Moreover, Theorem 5.33 ensures the existence of an integer $m \geq 1$ for which

$$\mathcal{R}(T^m) + \mathcal{N}(T^m) = \mathcal{H} \quad \text{and} \quad \mathcal{R}(T^m) \cap \mathcal{N}(T^m) = \{0\}.$$

Equivalently,

$$\mathcal{R}(T^m) \oplus \mathcal{N}(T^m) = \mathcal{H} \quad \text{and} \quad \mathcal{R}(T^m) \cap \mathcal{N}(T^m) = \{0\}$$

(where the direct sum is not necessarily orthogonal and equality in fact means equivalence — i.e., equality means equality up to a topological isomorphism). Since both $\mathcal{R}(T^m)$ and $\mathcal{N}(T^m)$ are T^m-invariant,

$$T^m = T^m|_{\mathcal{R}(T^m)} \oplus T^m|_{\mathcal{N}(T^m)}.$$

Since $\mathcal{N}(T^m) \neq \{0\}$ (otherwise T^m would be invertible) and $T^m|_{\mathcal{N}(T^m)} = O$ acts on the nonzero space $\mathcal{N}(T^m)$, we get $T^m = T^m|_{\mathcal{R}(T^m)} \oplus O$. Hence (see Proposition 2.F even though the direct sum is not orthogonal)

$$\sigma(T)^m = \sigma(T^m) = \sigma(T^m|_{\mathcal{R}(T^m)}) \cup \{0\}$$

by Theorem 2.7 (Spectral Mapping). Since $T^m|_{\mathcal{R}(T^m)} \colon \mathcal{R}(T^m) \to \mathcal{R}(T^m)$ is surjective and injective, and since $\mathcal{R}(T^m)$ is closed because $T^m \in \mathcal{F}[\mathcal{H}]$, we also get $T^m|_{\mathcal{R}(T^m)} \in \mathcal{G}[\mathcal{R}(T^m)]$. Thus $0 \in \rho(T^m|_{\mathcal{R}(T^m)})$ and so $0 \notin \sigma(T^m|_{\mathcal{R}(T^m)})$. Then $\sigma(T^m)$ is disconnected, and therefore 0 is an isolated point of $\sigma(T)^m = \sigma(T^m)$. Then 0 is an isolated point of $\sigma(T)$. That is,

$$0 \in \sigma_{\mathrm{iso}}(T).$$

Outcome: $0 \in \pi_0(T) = \sigma_{\mathrm{iso}}(T) \cap \sigma_0(T)$.

(b) Take an arbitrary $m \in \mathbb{N}$. If $0 \in \sigma_{\mathrm{iso}}(T)$, then $0 \in \sigma_{\mathrm{iso}}(T^m)$. Indeed, if 0 is an isolated pont of $\sigma(T)$, then $0 \in \sigma(T)^m = \sigma(T^m)$ and hence 0 is an isolated pont of $\sigma(T)^m = \sigma(T^m)$. Therefore

$$\sigma(T^m) = \{0\} \cup \Delta,$$

where Δ is a compact subset of $\mathbb{C}$ which does not contain 0. Then $\{0\}$ and Δ are complementary spectral sets for the operator T^m (i.e., they form a disjoint partition for $\sigma(T^m)$). Let $E_0 = E_{\{0\}} \in \mathcal{B}[\mathcal{H}]$ be the Riesz idempotent associated with 0. By Corollary 4.22

$$\mathcal{N}(T^m) \subseteq \mathcal{R}(E_0).$$

Claim. If $T \in \mathcal{F}[\mathcal{H}]$ and $0 \in \sigma_{\mathrm{iso}}(T)$, then $\dim \mathcal{R}(E_0) < \infty$.

Proof. Theorem 5.19.

Thus if $T \in \mathcal{F}[\mathcal{H}]$ and $0 \in \sigma_{\mathrm{iso}}(T)$, then $\dim \mathcal{N}(T^m) \leq \dim \mathcal{R}(E_0) < \infty$. Since $\{\mathcal{N}(T^n)\}$ is a nondecreasing sequence of subspaces, $\{\dim \mathcal{N}(T^n)\}$ is a nondecreasing sequence of positive integers, which is bounded by the (finite) integer $\dim \mathcal{R}(E_0)$. This ensures

$$T \in \mathcal{F}[\mathcal{H}] \quad \text{and} \quad 0 \in \sigma_{\mathrm{iso}}(T) \quad \Longrightarrow \quad \mathrm{asc}(T) < \infty.$$

Since $\sigma(T)^* = \sigma(T^*)$ we get $0 \in \sigma_{\mathrm{iso}}(T)$ if and only if $0 \in \sigma_{\mathrm{iso}}(T^*)$, and so

$$T \in \mathcal{F}[\mathcal{H}] \quad \text{and} \quad 0 \in \sigma_{\mathrm{iso}}(T) \quad \Longrightarrow \quad \mathrm{asc}(T^*) < \infty,$$

because $T \in \mathcal{F}[\mathcal{H}]$ if and only if $T^* \in \mathcal{F}[\mathcal{H}]$. Therefore, according to Lemma 5.31, $\mathrm{asc}(T) < \infty$ and $\mathrm{dsc}(T) < \infty$ which means (by definition) $T \in \mathcal{B}_r[\mathcal{H}]$. $\square$

Theorem 5.34 also gives us another characterization of the set $\mathcal{B}_r[\mathcal{H}]$ of all Browder operators from $\mathcal{B}[\mathcal{H}]$ (compare with Corollary 5.26).

Corollary 5.35.

$$\mathcal{B}_r[\mathcal{H}] = \big\{T \in \mathcal{F}[\mathcal{H}] : 0 \in \rho(T) \cup \sigma_{\mathrm{iso}}(T)\big\} = \big\{T \in \mathcal{F}[\mathcal{H}] : 0 \in \rho(T) \cup \pi_0(T)\big\}.$$

Proof. Consider the set $\pi_0(T) = \sigma_{\mathrm{iso}}(T) \cap \sigma_0(T)$ of Riesz points of T.

Claim. If $T \in \mathcal{F}[\mathcal{H}]$, then $0 \in \sigma_{\mathrm{iso}}(T)$ if and only if $0 \in \pi_0(T)$.

Proof. Theorem 5.19.

Thus the expressions for $\mathcal{B}_r[\mathcal{H}]$ follow from (a) and (b) in Theorem 5.34. $\square$

Remark 5.36. (a) FINITE DIMENSION. Since $\{\mathcal{N}(T^n)\}$ is nondecreasing and $\{\mathcal{R}(T^n)\}$ is nonincreasing, we get $\dim \mathcal{N}(T^n) \leq \dim \mathcal{N}(T^{n+1}) \leq \dim \mathcal{H}$ and $\dim \mathcal{R}(T^{n+1}) \leq \dim \mathcal{R}(T^n) \leq \dim \mathcal{H}$ for every $n \geq 0$. Thus if $\mathcal{H}$ is finite-dimensional, then $\mathrm{asc}(T) < \infty$ and $\mathrm{dsc}(T) < \infty$. Therefore (see Remark 5.3(d))

$$\dim \mathcal{H} < \infty \quad \Longrightarrow \quad \mathcal{B}_r[\mathcal{H}] = \mathcal{B}[\mathcal{H}].$$

(b) COMPLEMENTS. By definition $\sigma(T) \backslash \sigma_{\mathrm{iso}}(T) = \sigma_{\mathrm{acc}}(T)$ and

$$\begin{aligned}
\sigma(T) \backslash \pi_0(T) &= \sigma(T) \backslash \big(\sigma_{\mathrm{iso}}(T) \cap \sigma_0(T)\big) \\
&= \big(\sigma(T) \backslash \sigma_{\mathrm{iso}}(T)\big) \cup \big(\sigma(T) \backslash \sigma_0(T)\big) = \sigma_{\mathrm{acc}}(T) \cup \sigma_w(T).
\end{aligned}$$

Thus, since $\mathcal{G}[\mathcal{H}] \subset \mathcal{B}_r[\mathcal{H}] \subseteq \mathcal{W}[\mathcal{H}] \subseteq \mathcal{F}[\mathcal{H}]$, we get by Corollary 5.35

$$\mathcal{F}[\mathcal{H}] \backslash \mathcal{B}_r[\mathcal{H}] = \big\{T \in \mathcal{F}[\mathcal{H}] : 0 \in \sigma_{\mathrm{acc}}(T)\big\} = \big\{T \in \mathcal{F}[\mathcal{H}] : 0 \in \big(\sigma_{\mathrm{acc}}(T) \cup \sigma_w(T)\big)\big\}.$$

Since $\mathcal{W}[\mathcal{H}] \backslash \mathcal{B}_r[\mathcal{H}] = \mathcal{W}[\mathcal{H}] \cap \big(\mathcal{F}[\mathcal{H}] \backslash \mathcal{B}_r[\mathcal{H}]\big)$, then $\mathcal{W}[\mathcal{H}] \backslash \mathcal{B}_r[\mathcal{H}] = \{T \in \mathcal{W}[\mathcal{H}] : 0 \in \sigma_{\mathrm{acc}}(T)\}$. Moreover, since $\sigma_0(T) = \tau_0(T) \cup \pi_0(T)$ where $\tau_0(T)$ is an open subset of $\mathbb{C}$ (according to Corollary 5.20), and since $\pi_0(T) = \sigma_{\mathrm{iso}}(T) \cap \sigma_0(T)$, we get $\sigma_0(T) \backslash \sigma_{\mathrm{iso}}(T) = \tau_0(T) \subseteq \sigma_{\mathrm{acc}}(T)$. Hence $\mathcal{W}[\mathcal{H}] \backslash \mathcal{B}_r[\mathcal{H}] = \{T \in \mathcal{W}[\mathcal{H}] : 0 \in \tau_0(T)\}$ by Corollaries 5.26 and 5.35. Summing up:

$$W[\mathcal{H}]\backslash\mathcal{B}_r[\mathcal{H}] = \{T \in W[\mathcal{H}]\colon 0 \in \sigma_{\mathrm{acc}}(T)\} = \{T \in W[\mathcal{H}]\colon 0 \in \tau_0(T)\}.$$

Furthermore, since $\sigma(T)\backslash\sigma_0(T) = \sigma_w(T)$ we get by Corollary 5.26

$$\mathcal{F}[\mathcal{H}]\backslash W[\mathcal{H}] = \{T \in \mathcal{F}[\mathcal{H}]\colon 0 \in \sigma_w(T)\}.$$

(c) ANOTHER CHARACTERIZATION. *An operator $T \in \mathcal{B}[\mathcal{H}]$ is Browder if and only if it is Fredholm and $\lambda I - T$ is invertible for sufficiently small $\lambda \neq 0$.*

Indeed, by the identity $\mathcal{B}_r[\mathcal{H}] = \{T \in \mathcal{F}[\mathcal{H}]\colon 0 \in \rho(T) \cup \sigma_{\mathrm{iso}}(T)\}$ in Corollary 5.35, $T \in \mathcal{B}_r[\mathcal{H}]$ if and only if $T \in \mathcal{F}[\mathcal{H}]$ and either $0 \in \rho(T)$ or 0 is an isolated point of $\sigma(T) = \mathbb{C}\backslash\rho(T)$. Since the set $\rho(T)$ is open in $\mathbb{C}$, this means $T \in \mathcal{B}_r[\mathcal{H}]$ if and only if $T \in \mathcal{F}[\mathcal{H}]$ and there exists an $\varepsilon > 0$ for which $B_\varepsilon(0)\backslash\{0\} \subset \rho(T)$, where $B_\varepsilon(0)$ is the open ball centered at 0 with radius ε. This can be restated as follows: $T \in \mathcal{B}_r[\mathcal{H}]$ if and only if $T \in \mathcal{F}[\mathcal{H}]$ and $\lambda \in \rho(T)$ whenever $0 \neq |\lambda| < \varepsilon$ for some $\varepsilon > 0$. But this is precisely the claimed assertion.

The *Browder spectrum* of an operator $T \in \mathcal{B}[\mathcal{H}]$ is the set $\sigma_b(T)$ of all complex numbers λ for which $\lambda I - T$ is not Browder,

$$\sigma_b(T) = \{\lambda \in \mathbb{C}\colon \lambda I - T \in \mathcal{B}[\mathcal{H}]\backslash\mathcal{B}_r[\mathcal{H}]\}$$

(compare with Corollary 5.25). Since an operator lies in $\mathcal{B}_r[\mathcal{H}]$ together with its adjoint, it follows at once that $\lambda \in \sigma_b(T)$ if and only if $\overline{\lambda} \in \sigma_b(T^*)$:

$$\sigma_b(T) = \sigma_b(T^*)^*.$$

Moreover, by the preceding definition if $\lambda \in \sigma_b(T)$, then $\lambda I - T \notin \mathcal{B}_r[\mathcal{H}]$, and so either $\lambda I - T \notin \mathcal{F}[\mathcal{H}]$ or $\lambda I - T \in \mathcal{F}[\mathcal{H}]$ and $\mathrm{asc}\,(\lambda I - T) = \mathrm{dsc}\,(\lambda I - T) = \infty$ (cf. definition of Browder operators). In the former case $\lambda \in \sigma_e(T) \subseteq \sigma(T)$ according to Corollary 5.12. In the latter case $\lambda I - T$ is not invertible (because $\mathcal{N}(\lambda I - T) = \{0\}$ if and only if $\mathrm{asc}\,(\lambda I - T) = 0$ — also $\mathcal{R}(\lambda I - T) = \mathcal{H}$ if and only if $\mathrm{dsc}\,(\lambda I - T) = 0$), and so $\lambda \notin \rho(T)$; that is, $\lambda \in \sigma(T)$. Therefore

$$\sigma_b(T) \subseteq \sigma(T).$$

Since $\mathcal{B}_r[\mathcal{H}] \subset W[\mathcal{H}]$, we also get $\sigma_w(T) \subseteq \sigma_b(T)$ by Corollary 5.25. Hence

$$\sigma_e(T) \subseteq \sigma_w(T) \subseteq \sigma_b(T) \subseteq \sigma(T).$$

Equivalently, $\mathcal{G}[\mathcal{H}] \subseteq \mathcal{B}_r[\mathcal{H}] \subseteq W[\mathcal{H}] \subseteq \mathcal{F}[\mathcal{H}]$.

Corollary 5.37. *For every operator $T \in \mathcal{B}[\mathcal{H}]$,*

$$\sigma_b(T) = \sigma_e(T) \cup \sigma_{\mathrm{acc}}(T) = \sigma_w(T) \cup \sigma_{\mathrm{acc}}(T).$$

Proof. Since $\sigma(T) = \{\lambda\} - \sigma(\lambda I - T)$ by Theorem 2.7 (Spectral Mapping), $0 \in \sigma(\lambda I - T)\backslash\sigma_{\mathrm{iso}}(\lambda I - T)$ if and only if $\lambda \in \sigma(T)\backslash\sigma_{\mathrm{iso}}(T)$. Thus by the definition of $\sigma_b(T)$ and Corollaries 5.12 and 5.35,

$$\sigma_b(T) = \{\lambda \in \mathbb{C} \colon \lambda I - T \notin \mathcal{F}[\mathcal{H}] \text{ or } 0 \notin \rho(\lambda I - T) \cup \sigma_{\mathrm{iso}}(\lambda I - T)\}$$
$$= \{\lambda \in \mathbb{C} \colon \lambda \in \sigma_e(T) \text{ or } 0 \in \sigma(\lambda I - T) \backslash \sigma_{\mathrm{iso}}(\lambda I - T)\}$$
$$= \{\lambda \in \mathbb{C} \colon \lambda \in \sigma_e(T) \text{ or } \lambda \in \sigma(T) \backslash \sigma_{\mathrm{iso}}(T)\}$$
$$= \sigma_e(T) \cup \big(\sigma(T) \backslash \sigma_{\mathrm{iso}}(T)\big) = \sigma_e(T) \cup \sigma_{\mathrm{acc}}(T).$$

However, by Theorem 5.24 $\sigma_w(T) = \sigma_e(T) \cup \kappa(T)$ where

$$\kappa(T) = \bigcup\nolimits_{k \in \mathbb{Z} \backslash \{0\}} \sigma_k(T) \subseteq \sigma(T)$$

is the collection of all holes of $\sigma_e(T)$, thus a union of open sets in $\mathbb{C}$ by Theorem 5.16, and therefore is itself an open set in $\mathbb{C}$. Hence $\kappa(T) \backslash \sigma_{\mathrm{acc}}(T) = \varnothing$ and so

$$\sigma_w(T) \backslash \sigma_{\mathrm{acc}}(T) = \big(\sigma_e(T) \cup \kappa(T)\big) \backslash \sigma_{\mathrm{acc}}(T)$$
$$= \big(\sigma_e(T) \backslash \sigma_{\mathrm{acc}}(T)\big) \cup \big(\kappa(T) \backslash \sigma_{\mathrm{acc}}(T)\big) = \sigma_e(T) \backslash \sigma_{\mathrm{acc}}(T). \qquad \square$$

The next result is particularly important: it gives another partition of $\sigma(T)$.

Corollary 5.38. *For every operator* $T \in \mathcal{B}[\mathcal{H}]$,

$$\sigma_b(T) = \sigma(T) \backslash \pi_0(T).$$

Proof. By the Spectral Picture (Corollary 5.18) and Corollary 5.37,

$$\sigma(T) \backslash \sigma_b(T) = \sigma(T) \backslash \big(\sigma_e(T) \cup \sigma_{\mathrm{acc}}(T)\big) = \big(\sigma(T) \backslash \sigma_e(T)\big) \cap \big(\sigma(T) \backslash \sigma_{\mathrm{acc}}(T)\big)$$
$$= \Big(\bigcup\nolimits_{k \in \mathbb{Z}} \sigma_k(T)\Big) \cap \sigma_{\mathrm{iso}}(T) = \sigma_0(T) \cap \sigma_{\mathrm{iso}}(T) = \pi_0(T)$$

according to the definition of $\pi_0(T)$, since $\sigma_k(T)$ is an open set in $\mathbb{C}$ for every $0 \neq k \in \mathbb{Z}$ (Theorem 5.16) and so it has no isolated point. $\qquad \square$

Since $\sigma_b(T) \cup \pi_0(T) \subseteq \sigma(T)$, we get from Corollary 5.38

$$\sigma_b(T) \cup \pi_0(T) = \sigma(T) \quad \text{and} \quad \sigma_b(T) \cap \pi_0(T) = \varnothing,$$

and so $\{\sigma_b(T), \pi_0(T)\}$ forms a partition of the spectrum $\sigma(T)$. Thus

$$\pi_0(T) = \sigma(T) \backslash \sigma_b(T) = \{\lambda \in \sigma(T) \colon \lambda I - T \in \mathcal{B}_r[\mathcal{H}]\},$$

and

$$\sigma_b(T) = \sigma(T) \iff \pi_0(T) = \varnothing.$$

Corollary 5.39. *For every operator* $T \in \mathcal{B}[\mathcal{H}]$,

$$\pi_0(T) = \sigma_{\mathrm{iso}}(T) \backslash \sigma_b(T) = \sigma_{\mathrm{iso}}(T) \backslash \sigma_w(T) = \sigma_{\mathrm{iso}}(T) \backslash \sigma_e(T)$$
$$= \pi_{00}(T) \backslash \sigma_b(T) = \pi_{00}(T) \backslash \sigma_w(T) = \pi_{00}(T) \backslash \sigma_e(T).$$

Proof. The identities involving $\sigma_b(T)$ are readily verified. In fact, recall that $\pi_0(T) = \sigma_{\mathrm{iso}}(T) \cap \sigma_0(T) \subseteq \sigma_{\mathrm{iso}}(T) \cap \sigma_{PF}(T) = \pi_{00}(T) \subseteq \sigma_{\mathrm{iso}}(T) \subseteq \sigma(T)$. Since $\sigma_b(T) \cap \pi_0(T) = \varnothing$ and $\sigma(T) \backslash \pi_0(T) = \sigma_b(T)$ (Corollary 5.38), we get

$$\pi_0(T) = \pi_0(T) \backslash \sigma_b(T) \subseteq \pi_{00}(T) \backslash \sigma_b(T) \subseteq \sigma_{\mathrm{iso}}(T) \backslash \sigma_b(T) \subseteq \sigma(T) \backslash \sigma_b(T) = \pi_0(T).$$

To prove the identities involving $\sigma_w(T)$ and $\sigma_e(T)$, proceed as follows. Since $\sigma_0(T) = \sigma(T) \backslash \sigma_w(T)$ (Theorem 5.24), we get

$$\pi_0(T) = \sigma_{\mathrm{iso}}(T) \cap \sigma_0(T) = \sigma_{\mathrm{iso}}(T) \cap \big(\sigma(T) \backslash \sigma_w(T)\big) = \sigma_{\mathrm{iso}}(T) \backslash \sigma_w(T).$$

Moreover, as we saw in the proof of Corollary 5.37, $\sigma_w(T) = \sigma_e(T) \cup \kappa(T)$ where $\kappa(T) \subseteq \sigma(T)$ is open in $\mathbb{C}$ (union of open sets). Hence

$$\sigma_{\mathrm{iso}}(T) \backslash \sigma_w(T) = \sigma_{\mathrm{iso}}(T) \backslash (\sigma_e(T) \cup \kappa(T)) = (\sigma_{\mathrm{iso}}(T) \backslash \sigma_e(T)) \cap (\sigma_{\mathrm{iso}}(T) \backslash \kappa(T)).$$

But $\sigma_{\mathrm{iso}}(T) \cap \kappa(T) = \varnothing$ because $\kappa(T)$ is a subset of $\sigma(T)$ which is open in $\mathbb{C}$. Then $\sigma_{\mathrm{iso}}(T) \backslash \kappa(T) = \sigma_{\mathrm{iso}}(T)$ and so

$$\sigma_{\mathrm{iso}}(T) \backslash \sigma_w(T) = \sigma_{\mathrm{iso}}(T) \backslash \sigma_e(T).$$

Since $\pi_0(T) \subseteq \pi_{00}(T) \subseteq \sigma_{\mathrm{iso}}(T)$, $\pi_0(T) \subseteq \sigma_0(T) = \sigma(T) \backslash \sigma_w(T)$, $\sigma_e(T) \subseteq \sigma_w(T)$ (cf. Theorem 5.24), and as we have just verified $\sigma_{\mathrm{iso}}(T) \backslash \sigma_w(T) = \pi_0(T)$ and $\sigma_{\mathrm{iso}}(T) \backslash \sigma_e(T) = \sigma_{\mathrm{iso}}(T) \backslash \sigma_w(T)$, then

$$\pi_{00}(T) \backslash \sigma_w(T) \subseteq \pi_{00}(T) \backslash \sigma_e(T) \subseteq \sigma_{\mathrm{iso}}(T) \backslash \sigma_e(T)$$
$$= \sigma_{\mathrm{iso}}(T) \backslash \sigma_w(T) = \pi_0(T) = \pi_0(T) \backslash \sigma_w(T) \subseteq \pi_{00}(T) \backslash \sigma_w(T),$$

and therefore

$$\pi_0(T) = \pi_{00}(T) \backslash \sigma_w(T) = \pi_{00}(T) \backslash \sigma_e(T). \qquad \square$$

Remark 5.40. (a) Finite Dimension. Take any $T \in \mathcal{B}[\mathcal{H}]$. Since every operator on a finite-dimensional space is Browder (cf. Remark 5.36(a)), and according to the definition of the Browder spectrum $\sigma_b(T)$,

$$\dim \mathcal{H} < \infty \quad \Longrightarrow \quad \sigma_b(T) = \varnothing.$$

The converse holds by Remark 5.27(a) since $\sigma_w(T) \subseteq \sigma_b(T)$:

$$\dim \mathcal{H} = \infty \quad \Longrightarrow \quad \sigma_b(T) \neq \varnothing.$$

Thus

$$\sigma_b(T) \neq \varnothing \quad \Longleftrightarrow \quad \dim \mathcal{H} = \infty,$$

and $\sigma_b(T) = \sigma(T) \backslash \pi_0(T)$ (cf. Corollary 5.38) is a compact set in $\mathbb{C}$ because $\sigma(T)$ is compact and $\pi_0(T) \subseteq \sigma_{\mathrm{iso}}(T)$. Moreover, by Remark 5.27(b),

$$\dim \mathcal{H} < \infty \quad \Longrightarrow \quad \sigma_e(T) = \sigma_w(T) = \sigma_b(T) = \varnothing.$$

(b) Compact Operators. Since $\sigma_{\mathrm{acc}}(K) \subseteq \{0\}$ if K is compact (cf. Corollary 2.20), it follows by Remark 5.27(d), Corollary 5.37, and item (a) that

$$K \in \mathcal{B}_\infty[\mathcal{H}] \text{ and } \dim \mathcal{H} = \infty \quad \Longrightarrow \quad \sigma_e(K) = \sigma_w(K) = \sigma_b(K) = \{0\}.$$

(c) Finite Algebraic and Geometric Multiplicities. We saw in Corollary 5.28 that the set of Riesz points $\pi_0(T) = \sigma_{\text{iso}}(T) \cap \sigma_0(T)$ and the set of isolated eigenvalues of finite multiplicity $\pi_{00}(T) = \sigma_{\text{iso}}(T) \cap \sigma_{PF}(T)$ coincide if any of the equivalent assertions of Corollary 5.28 holds true. The previous corollary supplies a collection of conditions equivalent to $\pi_0(T) = \pi_{00}(T)$. In fact, by Corollary 5.39 the following four assertions are pairwise equivalent.

(i) $\pi_0(T) = \pi_{00}(T)$.

(ii) $\sigma_e(T) \cap \pi_{00}(T) = \varnothing$ (i.e., $\sigma_e(T) \cap \sigma_{\text{iso}}(T) \cap \sigma_{PF}(T) = \varnothing$).

(iii) $\sigma_w(T) \cap \pi_{00}(T) = \varnothing$ (i.e., $\sigma_w(T) \cap \sigma_{\text{iso}}(T) \cap \sigma_{PF}(T) = \varnothing$).

(iv) $\sigma_b(T) \cap \pi_{00}(T) = \varnothing$ (i.e., $\sigma_b(T) \cap \sigma_{\text{iso}}(T) \cap \sigma_{PF}(T) = \varnothing$).

Take any $T \in \mathcal{B}[\mathcal{H}]$ and consider again the following relations:

$$\sigma_0(T) = \sigma(T) \backslash \sigma_w(T), \quad \pi_0(T) = \sigma_{\text{iso}}(T) \cap \sigma_0(T), \quad \pi_{00}(T) = \sigma_{\text{iso}}(T) \cap \sigma_{PF}(T).$$

We have defined a class of operators for which the equivalent assertions of Corollary 5.28 hold true (namely, operators that satisfy Weyl's Theorem). The following equivalent assertions define another important class of operators. An operator for which any of these equivalent assertions holds is said to satisfy *Browder's Theorem*. This will be discussed later in Section 5.5.

Corollary 5.41. *The assertions below are pairwise equivalent.*

(a) $\sigma_0(T) = \pi_0(T)$.

(b) $\sigma(T) \backslash \pi_0(T) = \sigma_w(T)$.

(c) $\sigma(T) = \sigma_w(T) \cup \pi_{00}(T)$.

(d) $\sigma(T) = \sigma_w(T) \cup \sigma_{\text{iso}}(T)$.

(e) $\sigma_0(T) \subseteq \sigma_{\text{iso}}(T)$.

(f) $\sigma_0(T) \subseteq \pi_{00}(T)$.

(g) $\sigma_{\text{acc}}(T) \subseteq \sigma_w(T)$.

(h) $\sigma_w(T) = \sigma_b(T)$.

Proof. Consider the partition $\{\sigma_w(T), \sigma_0(T)\}$ of $\sigma(T)$ and the inclusions $\pi_0(T) \subseteq \pi_{00}(T) \subseteq \sigma(T)$. Thus (a) and (b) are equivalent and, if (a) holds,

$$\sigma(T) = \sigma_w(T) \cup \sigma_0(T) = \sigma_w(T) \cup \pi_0(T) \subseteq \sigma_w(T) \cup \pi_{00}(T) \subseteq \sigma(T),$$

and so (a) implies (c); and (c) implies (d) because $\pi_{00}(T) \subseteq \sigma_{\text{iso}}(T)$:

$$\sigma(T) = \sigma_w(T) \cup \pi_{00}(T) \subseteq \sigma_w(T) \cup \sigma_{\text{iso}}(T) \subseteq \sigma(T).$$

If (d) holds, then (since $\sigma_0(T) = \sigma(T) \backslash \sigma_w(T)$)

$$\sigma_0(T) = \sigma(T) \backslash \sigma_w(T) = (\sigma_w(T) \cup \sigma_{\text{iso}}(T)) \backslash \sigma_w(T) \subseteq \sigma_{\text{iso}}(T).$$

Thus (d) implies (e). Since $\pi_0(T) = \sigma_{\mathrm{iso}}(T) \cap \sigma_0(T)$, the inclusion in (e) is equivalent to the identity in (a). Since $\pi_{00}(T) = \sigma_{\mathrm{iso}}(T) \cap \sigma_{PF}(T)$ and since $\sigma_0(T) \subseteq \sigma_{PF}(T)$, the inclusions in (e) and (f) are equivalent. Since $\sigma_{\mathrm{acc}}(T)$ and $\sigma_{\mathrm{iso}}(T)$ are complements of each other in $\sigma(T)$, and $\sigma_w(T)$ and $\sigma_0(T)$ are complements of each other in $\sigma(T)$, the inclusions in (e) and (g) are equivalent. Similarly, since $\sigma_w(T)$ and $\sigma_0(T)$ are complements of each other in $\sigma(T)$, and $\sigma_b(T)$ and $\pi_0(T)$ are complements of each other in $\sigma(T)$ (cf. Theorem 5.24 and Corollary 5.38), the identities in (h) and (a) are equivalent as well. $\square$

The Browder spectrum is not invariant under compact perturbation. In fact, if $T \in \mathcal{B}[\mathcal{H}]$ is such that $\sigma_b(T) \neq \sigma_w(T)$ (i.e., if T is such that any of the equivalent assertions in Corollary 5.41 fails — and so all of them fail), then there exists a compact $K \in \mathcal{B}_\infty[\mathcal{H}]$ for which

$$\sigma_b(T + K) \neq \sigma_b(T).$$

Indeed, if $\sigma_b(T) \neq \sigma_w(T)$, then $\sigma_w(T) \subset \sigma_b(T)$, and so $\sigma_b(T)$ is not invariant under compact perturbation because $\sigma_w(T)$ is the largest part of $\sigma(T)$ that remains unchanged under compact perturbation. However, as we will see in the forthcoming Theorem 5.44 (whose proof is based on Lemmas 5.42 and 5.43 below), $\sigma_b(T)$ is invariant under perturbation by compact operators that commute with T. We follow the framework set forth in [79].

Set $\mathcal{A}[\mathcal{H}] = \mathcal{B}[\mathcal{H}]$ where $\mathcal{H}$ is an infinite-dimensional complex Hilbert space. Let $\mathcal{A}'[\mathcal{H}]$ be a unital closed subalgebra of the *unital complex Banach algebra* $\mathcal{A}[\mathcal{H}]$, thus a unital complex Banach algebra itself. Take an arbitrary operator T in $\mathcal{A}'[\mathcal{H}]$. Let $\mathcal{B}_\infty'[\mathcal{H}] = \mathcal{B}_\infty[\mathcal{H}] \cap \mathcal{A}'[\mathcal{H}]$ denote the collection of all compact operators from $\mathcal{A}'[\mathcal{H}]$, and let $\mathcal{F}'[\mathcal{H}]$ denote the class of all Fredholm operators in the unital complex Banach algebra $\mathcal{A}'[\mathcal{H}]$. That is (Proposition 5.C),

$$\mathcal{F}'[\mathcal{H}] = \big\{T \in \mathcal{A}'[\mathcal{H}]\colon\ AT - I \text{ and } TA - I \text{ are in } \mathcal{B}_\infty'[\mathcal{H}] \text{ for some } A \in \mathcal{A}'[\mathcal{H}]\big\}.$$

Therefore $\mathcal{F}'[\mathcal{H}] \subseteq \mathcal{F}[\mathcal{H}] \cap \mathcal{A}'[\mathcal{H}]$. The inclusion is proper in general (see, e.g., [11, p. 86]) but it becomes an identity if $\mathcal{A}'[\mathcal{H}]$ is a $*$-subalgebra of the C*-algebra $\mathcal{A}[\mathcal{H}]$; that is, if the unital closed subalgebra $\mathcal{A}'[\mathcal{H}]$ is a $*$-algebra (see, e.g., [11, Theorem A.1.3]). However, if the inclusion becomes an identity, then by Corollary 5.12 the essential spectra in $\mathcal{A}[\mathcal{H}]$ and in $\mathcal{A}'[\mathcal{H}]$ coincide:

$$\mathcal{F}'[\mathcal{H}] = \mathcal{F}[\mathcal{H}] \cap \mathcal{A}'[\mathcal{H}] \quad \Longrightarrow \quad \sigma_e'(T) = \sigma_e(T),$$

where

$$\sigma_e'(T) = \big\{\lambda \in \sigma'(T)\colon\ \lambda I - T \in \mathcal{A}'[\mathcal{H}] \backslash \mathcal{F}'[\mathcal{H}]\big\}$$

is the essential spectrum of $T \in \mathcal{A}'[\mathcal{H}]$ with respect to $\mathcal{A}'[\mathcal{H}]$, with $\sigma'(T)$ denoting the spectrum of $T \in \mathcal{A}'[\mathcal{H}]$ with respect to the unital complex Banach algebra $\mathcal{A}'[\mathcal{H}]$. Similarly, let $\mathcal{W}'[\mathcal{H}]$ denote the class of Weyl operators in $\mathcal{A}'[\mathcal{H}]$,

$$\mathcal{W}'[\mathcal{H}] = \big\{T \in \mathcal{F}'[\mathcal{H}]\colon\ \mathrm{ind}\,(T) = 0\big\},$$

and let
$$\sigma'_w(T) = \{\lambda \in \sigma'(T) \colon \lambda I - T \in \mathcal{A}'[\mathcal{H}]\backslash\mathcal{W}'[\mathcal{H}]\}$$

be the Weyl spectrum of $T \in \mathcal{A}'[\mathcal{H}]$ with respect to $\mathcal{A}'[\mathcal{H}]$. If $T \in \mathcal{A}'[\mathcal{H}]$, set
$$\sigma'_0(T) = \sigma'(T)\backslash\sigma'_w(T) = \{\lambda \in \sigma'(T) \colon \lambda I - T \in \mathcal{W}'[\mathcal{H}]\}.$$

Moreover, let $\mathcal{B}'_r[\mathcal{H}]$ stand for the class of Browder operators in $\mathcal{A}'[\mathcal{H}]$,
$$\mathcal{B}'_r[\mathcal{H}] = \{T \in \mathcal{F}'[\mathcal{H}] \colon 0 \in \rho'(T) \cup \sigma'_{\mathrm{iso}}(T)\}$$

according to Corollary 5.35, where $\sigma'_{\mathrm{iso}}(T)$ denotes the set of all isolated points of $\sigma'(T)$, and $\rho'(T) = \mathbb{C}\backslash\sigma'(T)$ is the resolvent set of $T \in \mathcal{A}'[\mathcal{H}]$ with respect to the unital complex Banach algebra $\mathcal{A}'[\mathcal{H}]$. Let
$$\sigma'_b(T) = \{\lambda \in \sigma'(T) \colon \lambda I - T \in \mathcal{A}'[\mathcal{H}]\backslash\mathcal{B}'_r[\mathcal{H}]\}$$

be the Browder spectrum of $T \in \mathcal{A}'[\mathcal{H}]$ with respect to $\mathcal{A}'[\mathcal{H}]$.

Lemma 5.42. *Consider the preceding setup. If $T \in \mathcal{A}'[\mathcal{H}]$, then*
$$\sigma'(T) = \sigma(T) \quad \Longrightarrow \quad \sigma'_b(T) = \sigma_b(T).$$

Proof. Take an arbitrary operator T in $\mathcal{A}'[\mathcal{H}]$. Suppose
$$\sigma'(T) = \sigma(T),$$

so that $\rho'(T) = \rho(T)$ and $\sigma'_{\mathrm{iso}}(T) = \sigma_{\mathrm{iso}}(T)$. Therefore if $\lambda \in \sigma'_{\mathrm{iso}}(T)$, then the Riesz idempotent associated with λ,
$$E_\lambda = \tfrac{1}{2\pi i} \int_{\Upsilon_\lambda} (\zeta I - T)^{-1}\, d\zeta,$$

lies in $\mathcal{A}'[\mathcal{H}]$ (because $\sigma'_{\mathrm{iso}}(T) = \sigma_{\mathrm{iso}}(T)$, and also $(\zeta I - T)^{-1} \in \mathcal{A}'[\mathcal{H}]$ whenever $(\zeta I - T)^{-1} \in \mathcal{A}[\mathcal{H}] = \mathcal{B}[\mathcal{H}]$ since $\rho'(T) = \rho(T)$). Thus by Theorem 5.19,
$$\lambda \in \sigma'_0(T) \quad \Longleftrightarrow \quad \dim \mathcal{R}(E_\lambda) < \infty \quad \Longleftrightarrow \quad \lambda \in \sigma_0(T)$$

whenever $\lambda \in \sigma'_{\mathrm{iso}}(T)$, since $\sigma'_{\mathrm{iso}}(T) = \sigma_{\mathrm{iso}}(T)$. Hence
$$\pi'_0(T) = \sigma'_0(T) \cap \sigma'_{\mathrm{iso}}(T) = \sigma_0(T) \cap \sigma_{\mathrm{iso}}(T) = \pi_0(T).$$

Then by Corollary 5.38,
$$\sigma_b(T) = \sigma(T)\backslash\pi_0(T) = \sigma'(T)\backslash\pi'_0(T) = \sigma'_b(T). \qquad \square$$

Let $\{T\}'$ denote the *commutant* of $T \in \mathcal{A}[\mathcal{H}]$ (i.e., the collection of all operators in $\mathcal{B}[\mathcal{H}]$ that commute with $T \in \mathcal{B}[\mathcal{H}]$). This is a unital subalgebra of the unital complex Banach algebra $\mathcal{A}[\mathcal{H}] = \mathcal{B}[\mathcal{H}]$, which is weakly closed in $\mathcal{B}[\mathcal{H}]$ (see, e.g., [75, Problem 3.7]), and hence uniformly closed (i.e., closed in the operator norm topology of $\mathcal{B}[\mathcal{H}]$). Therefore $\mathcal{A}'[\mathcal{H}] = \{T\}'$ is a unital closed subalgebra of the unital complex Banach algebra $\mathcal{A}[\mathcal{H}] = \mathcal{B}[\mathcal{H}]$.

Lemma 5.43. *Consider the preceding setup with* $\mathcal{A}'[\mathcal{H}] = \{T\}'$. *If* $0 \in \sigma'(T)$ *and* $T \in \mathcal{W}'[\mathcal{H}]$, *then* $0 \in \sigma'_{\mathrm{iso}}(T)$. *Equivalently,*

$$\mathcal{A}'[\mathcal{H}] = \{T\}' \ \ and \ \ 0 \in \sigma'_0(T) \ \implies \ 0 \in \sigma'_{\mathrm{iso}}(T).$$

Proof. Take $T \in \mathcal{B}[\mathcal{H}]$. Let $\mathcal{A}'[\mathcal{H}]$ be a unital closed subalgebra of $\mathcal{A}[\mathcal{H}] = \mathcal{B}[\mathcal{H}]$ that includes T. Suppose $0 \in \sigma'_0(T)$, which means $0 \in \sigma'(T)$ and $T \in \mathcal{W}'[\mathcal{H}]$. Since $T \in \mathcal{W}'[\mathcal{H}]$, there is a compact $K \in \mathcal{B}'_\infty[\mathcal{H}] = \mathcal{B}_\infty[\mathcal{H}] \cap \mathcal{A}'[\mathcal{H}]$, actually a finite-rank operator, and an operator $A \in \mathcal{A}'[\mathcal{H}]$ such that (cf. Lemma 5.23(c))

$$A(T + K) = (T + K)A = I.$$

So $A \in \mathcal{A}'[\mathcal{H}]$ is invertible with inverse $A^{-1} = T + K \in \mathcal{A}'[\mathcal{H}]$. If $\mathcal{A}'[\mathcal{H}] = \{T\}'$, then $AK = KA$ and so $A^{-1}K = KA^{-1}$. Let $\mathcal{A}''[\mathcal{H}]$ be the unital closed commutative subalgebra of $\mathcal{A}'[\mathcal{H}]$ generated by A, A^{-1}, and K. Since $T = A^{-1} - K$, $\mathcal{A}''[\mathcal{H}]$ includes T. Let $\sigma'_{\mathrm{iso}}(T)$ and $\sigma''_{\mathrm{iso}}(T)$ stand for the sets of isolated points of the spectra $\sigma'(T)$ and $\sigma''(T)$ of T with respect to the Banach algebras $\mathcal{A}'[\mathcal{H}]$ and $\mathcal{A}''[\mathcal{H}]$, respectively. Since $\mathcal{A}''[\mathcal{H}] \subseteq \mathcal{A}'[\mathcal{H}]$, it follows by Proposition 2.Q(a) that $\sigma'(T) \subseteq \sigma''(T)$. Hence $0 \in \sigma''(T)$ because $0 \in \sigma'_0(T) \subseteq \sigma'(T)$.

Claim. $0 \in \sigma''_{\mathrm{iso}}(T)$.

Proof. Let $\widehat{\mathcal{A}}''[\mathcal{H}]$ denote the collection of all algebra homomorphisms of $\mathcal{A}''[\mathcal{H}]$ into $\mathbb{C}$. From Proposition 2.Q(b) we get

$$\sigma''(A^{-1}) = \{\Phi(A^{-1}) \in \mathbb{C} \colon \Phi \in \widehat{\mathcal{A}}''[\mathcal{H}]\},$$

which is bounded away from zero (since $0 \in \rho''(A^{-1})$), and

$$\sigma''(K) = \{\Phi(K) \in \mathbb{C} \colon \Phi \in \widehat{\mathcal{A}}''[\mathcal{H}]\} = \{0\} \cup \{\Phi(K) \in \mathbb{C} \colon \Phi \in \widehat{\mathcal{A}}''_F[\mathcal{H}]\},$$

where $\widehat{\mathcal{A}}''_F[\mathcal{H}] \subseteq \widehat{\mathcal{A}}''[\mathcal{H}]$ is a set of nonzero homomorphisms, which is *finite* since K is a finite-rank operator, and so it has a finite spectrum (Corollary 2.19). Also, $0 \in \sigma''(T) = \sigma''(A^{-1} - K)$ if and only if $0 = \Phi(A^{-1} - K) = \Phi(A^{-1}) - \Phi(K)$ for some $\Phi \in \widehat{\mathcal{A}}''[\mathcal{H}]$. If $\Phi \in \widehat{\mathcal{A}}''[\mathcal{H}] \backslash \widehat{\mathcal{A}}''_F[\mathcal{H}]$, then $\Phi(K) = 0$. So $\Phi(A^{-1}) = 0$, which is a contradiction (since $0 \notin \sigma''(A^{-1})$). Thus $\Phi \in \widehat{\mathcal{A}}''_F[\mathcal{H}]$, and therefore $\Phi(A^{-1} - K) = \Phi(A^{-1}) - \Phi(K) = 0$. Then $\Phi(A^{-1}) = \Phi(K)$ for at most a finite number of homomorphisms Φ in $\widehat{\mathcal{A}}''_F[\mathcal{H}]$. Hence, since the set $\{\Phi(A^{-1}) \in \mathbb{C} \colon \Phi \in \widehat{\mathcal{A}}''[\mathcal{H}]\} = \sigma''(A^{-1})$ is bounded away from zero, and since the set

$$\{\Phi \in \widehat{\mathcal{A}}''_F[\mathcal{H}] \colon \Phi(A^{-1}) = \Phi(K)\} = \{\Phi \in \widehat{\mathcal{A}}''_F[\mathcal{H}] \colon 0 \in \sigma''(A^{-1} - K)\}$$

is finite, it follows that 0 is an isolated point of $\sigma''(A^{-1} - K) = \sigma''(T)$, which concludes the proof of the claimed result.

Since $0 \in \sigma'(T) \subseteq \sigma''(T)$, then $0 \in \sigma'_{\mathrm{iso}}(T)$. $\square$

The next characterization of the Browder spectrum is the counterpart of the very definition of the Weyl spectrum. It says that $\sigma_b(T)$ is the largest part of $\sigma(T)$ that remains unchanged under compact perturbations in the commutant of T. That is, $\sigma_b(T)$ is the largest part of $\sigma(T)$ such that [88, 70]

$$\sigma_b(T + K) = \sigma_b(T) \quad \text{for every} \quad K \in \mathcal{B}_\infty[\mathcal{H}] \cap \{T\}'.$$

Theorem 5.44. *For every* $T \in \mathcal{B}[\mathcal{H}]$,

$$\sigma_b(T) = \bigcap_{K \in \mathcal{B}_\infty[\mathcal{H}] \cap \{T\}'} \sigma(T + K).$$

Proof. Take $T \in \mathcal{B}[\mathcal{H}]$. Let $\mathcal{A}'[\mathcal{H}]$ be a unital closed subalgebra of the unital complex Banach algebra $\mathcal{A}[\mathcal{H}] = \mathcal{B}[\mathcal{H}]$ (thus a unital complex Banach algebra itself). If $T \in \mathcal{A}'[\mathcal{H}]$, then let $\sigma'(T)$, $\sigma_b'(T)$, and $\sigma_w'(T)$ be the spectrum, the Browder spectrum, and the Weyl spectrum of T with respect to $\mathcal{A}'[\mathcal{H}]$.

Claim 1. If $\mathcal{A}'[\mathcal{H}] = \{T\}'$, then $\sigma'(T) = \sigma(T)$.

Proof. Suppose $\mathcal{A}'[\mathcal{H}] = \{T\}'$. Trivially, $T \in \mathcal{A}'[\mathcal{H}]$. Let $\mathcal{P}[\mathcal{H}] = \mathcal{P}(T)$ be the collection of all polynomials $p(T)$ in T with complex coefficients, which is a unital commutative subalgebra of $\mathcal{A}[\mathcal{H}] = \mathcal{B}[\mathcal{H}]$. Consider the collection $\mathcal{T}$ of all unital commutative subalgebras of $\mathcal{A}[\mathcal{H}]$ containing T. Every element of $\mathcal{T}$ is included in $\mathcal{A}'[\mathcal{H}]$, and $\mathcal{T}$ is partially ordered (in the inclusion ordering) and nonempty (e.g., $\mathcal{P}[\mathcal{H}] \in \mathcal{T}$— and so $\mathcal{P}[\mathcal{H}] \subseteq \mathcal{A}'[\mathcal{H}]$). Moreover, every chain in $\mathcal{T}$ has an upper bound in $\mathcal{T}$ (the union of all subalgebras in a given chain of subalgebras in $\mathcal{T}$ is again a subalgebra in $\mathcal{T}$). Thus according to Zorn's Lemma, $\mathcal{T}$ has a maximal element, say $\mathcal{A}''[\mathcal{H}] = \mathcal{A}''(T) \in \mathcal{T}$. Hence there is a maximal commutative subalgebra $\mathcal{A}''[\mathcal{H}]$ of $\mathcal{B}[\mathcal{H}]$ containing T (which is unital and closed — see the paragraph that precedes Proposition 2.Q). Therefore

$$\mathcal{A}''[\mathcal{H}] \subseteq \mathcal{A}'[\mathcal{H}] \subseteq \mathcal{A}[\mathcal{H}].$$

Let $\sigma''(T)$ denote the spectrum of T with respect to $\mathcal{A}''[\mathcal{H}]$. If the preceding inclusions hold true, and since $\mathcal{A}''[\mathcal{H}]$ is a maximal commutative subalgebra of $\mathcal{A}[\mathcal{H}]$, then Proposition 2.Q(a,b) ensures the following relations:

$$\sigma(T) \subseteq \sigma'(T) \subseteq \sigma''(T) = \sigma(T).$$

Claim 2. If $\mathcal{A}'[\mathcal{H}] = \{T\}'$, then $\sigma_b(T) = \sigma_b'(T)$.

Proof. Claim 1 and Lemma 5.42.

Claim 3. If $\mathcal{A}'[\mathcal{H}] = \{T\}'$, then $\sigma_b'(T) = \sigma_w'(T)$.

Proof. By definition $\lambda \in \sigma_0'(T)$ if and only if $\lambda \in \sigma'(T)$ and $\lambda I - T \in \mathcal{W}'[\mathcal{H}]$. But $\lambda \in \sigma'(T)$ if and only if $0 \in \sigma'(\lambda I - T)$ by the Spectral Mapping Theorem (Theorem 2.7). Hence

$$\lambda \in \sigma_0'(T) \quad \Longleftrightarrow \quad 0 \in \sigma_0'(\lambda I - T).$$

Since $\mathcal{A}'[\mathcal{H}] = \{T\}' = \{\lambda I - T\}'$ we get by Lemma 5.43

$$0 \in \sigma_0'(\lambda I - T) \implies 0 \in \sigma_{\mathrm{iso}}'(\lambda I - T).$$

However, applying the Spectral Mapping Theorem again,

$$0 \in \sigma_{\mathrm{iso}}'(\lambda I - T) \iff \lambda \in \sigma_{\mathrm{iso}}'(T).$$

Therefore

$$\sigma_0'(T) \subseteq \sigma_{\mathrm{iso}}'(T),$$

which means (cf. Corollary 5.41)

$$\sigma_w'(T) = \sigma_b'(T).$$

Claim 4. $\sigma_w'(T) = \bigcap_{K \in \mathcal{B}_\infty[\mathcal{H}] \cap \mathcal{A}'[\mathcal{H}]} \sigma(T + K).$

Proof. This is the definition of the Weyl spectrum of T with respect to $\mathcal{A}'[\mathcal{H}]$: $\sigma_w'(T) = \bigcap_{K \in \mathcal{B}_\infty'[\mathcal{H}]} \sigma(T + K)$, where $\mathcal{B}_\infty'[\mathcal{H}] = \mathcal{B}_\infty[\mathcal{H}] \cap \mathcal{A}'[\mathcal{H}]$.

By Claims 2, 3, and 4 we have

$$\sigma_b(T) = \bigcap_{K \in \mathcal{B}_\infty[\mathcal{H}] \cap \{T\}'} \sigma(T + K). \qquad \square$$

5.5 Remarks on Browder and Weyl Theorems

This final section consist of a brief survey on Weyl's and Browder's Theorems. As such, and unlike the previous sections, some of the assertions discussed here, instead of being fully proved, will be accompanied by a reference to indicate where the proof can be found.

Take any operator $T \in \mathcal{B}[\mathcal{H}]$ and consider the partitions $\{\sigma_w(T), \sigma_0(T)\}$ and $\{\sigma_b(T), \pi_0(T)\}$ of the spectrum of T in terms of Weyl and Browder spectra $\sigma_w(T)$ and $\sigma_b(T)$ and their complements $\sigma_0(T)$ and $\pi_0(T)$ in $\sigma(T)$ as in Theorem 5.24 and Corollary 5.38, so that

$$\sigma_w(T) = \sigma(T) \backslash \sigma_0(T) \quad \text{and} \quad \sigma_b(T) = \sigma(T) \backslash \pi_0(T),$$

where

$$\sigma_0(T) = \{\lambda \in \sigma(T) \colon \lambda I - T \in \mathcal{W}[\mathcal{H}]\} \quad \text{and} \quad \pi_0(T) = \sigma_{\mathrm{iso}}(T) \cap \sigma_0(T).$$

Although $\sigma_0(T) \subseteq \sigma_{PF}(T)$ and $\sigma_{\mathrm{acc}}(T) \subseteq \sigma_b(T)$ (i.e., $\pi_0(T) \subseteq \sigma_{\mathrm{iso}}(T)$) for every T, in general $\sigma_0(T)$ may not be included in $\sigma_{\mathrm{iso}}(T)$; equivalently, $\sigma_{\mathrm{acc}}(T)$ may not be included in $\sigma_w(T)$. Recall the definition of $\pi_{00}(T)$:

$$\pi_{00}(T) = \sigma_{\mathrm{iso}}(T) \cap \sigma_{PF}(T).$$

Definition 5.45. An operator T is said to *satisfy Weyl's Theorem* (or *Weyl's Theorem holds* for T) if any of the equivalent assertions of Corollary 5.28 holds true. Therefore *an operator T satisfies Weyl's Theorem if*

$$\sigma_0(T) = \sigma_{\mathrm{iso}}(T) \cap \sigma_{PF}(T).$$

In other words, if $\sigma_0(T) = \pi_{00}(T)$ or, equivalently, if $\sigma(T) \backslash \pi_{00}(T) = \sigma_w(T)$.

Further necessary and sufficient conditions for an operator to satisfy Weyl's Theorem can be found in [51]. See also [69, Chapter 11].

Definition 5.46. An operator T is said to *satisfy Browder's Theorem* (or *Browder's Theorem holds* for T) if any of the equivalent assertions of Corollary 5.41 holds true. Therefore *an operator T satisfies Browder's Theorem if*

$$\sigma_0(T) \subseteq \sigma_{\mathrm{iso}}(T) \cap \sigma_{PF}(T).$$

That is, $\sigma_0(T) \subseteq \pi_{00}(T)$ or, equivalently,

$$\sigma_0(T) = \pi_0(T), \quad \text{or} \quad \sigma_0(T) \subseteq \sigma_{\mathrm{iso}}(T), \quad \text{or}$$

$$\sigma_{\mathrm{acc}}(T) \subseteq \sigma_w(T), \quad \text{or} \quad \sigma_w(T) = \sigma_b(T).$$

A word on terminology. The expressions "T satisfies Weyl's Theorem" and "T satisfies Browder's Theorem" have become standard and are often used in current literature and we will stick to them, although saying T "satisfies Weyl's (or Browder's) *property*" rather than "satisfies Weyl's (or Browder's) *Theorem*" would perhaps sound more appropriate.

Remark 5.47. (a) SUFFICIENCY FOR WEYL'S.

$$\sigma_w(T) = \sigma_{\mathrm{acc}}(T) \quad \Longrightarrow \quad T \text{ satisfies Weyl's Theorem,}$$

according to Corollary 5.28.

(b) BROWDER'S NOT WEYL'S. Consider Definitions 5.45 and 5.46. If T satisfies Weyl's Theorem, then it obviously satisfies Browder's Theorem. An operator T satisfies Browder's but not Weyl's Theorem if and only if the proper inclusion $\sigma_0(T) \subset \sigma_{\mathrm{iso}}(T) \cap \sigma_{PF}(T)$ holds true, and so there exists an isolated eigenvalue of finite multiplicity not in $\sigma_0(T)$ (i.e., there exists an isolated eigenvalue of finite multiplicity in $\sigma_w(T) = \sigma(T) \backslash \sigma_0(T)$). Outcome:

If T satisfies Browder's Theorem but not Weyl's Theorem, then
$$\sigma_w(T) \cap \sigma_{\mathrm{iso}}(T) \cap \sigma_{PF}(T) \neq \varnothing.$$

(c) EQUIVALENT CONDITION. The preceding result can be extended as follows (cf. Remark 5.40(c)). Consider the equivalent assertions of Corollary 5.28 and of Corollary 5.41. If Browder's Theorem holds and $\pi_0(T) = \pi_{00}(T)$, then Weyl's Theorem holds (i.e., if $\sigma_0(T) = \pi_0(T)$ and $\pi_0(T) = \pi_{00}(T)$, then $\sigma_0(T) = \pi_{00}(T)$ tautologically). Conversely, if Weyl's Theorem holds, then $\pi_0(T) = \pi_{00}(T)$ (see Corollary 5.28), and Browder's Theorem holds trivially. Summing up:

Weyl's Theorem holds $\iff$ *Browder's Theorem holds and $\pi_0(T) = \pi_{00}(T)$.*

That is, Weyl's Theorem holds if and only if Browder's Theorem and any of the equivalent assertions of Remark 5.40(c) hold true.

(d) TRIVIAL CASES. By Remark 5.27(a), if $\dim \mathcal{H} < \infty$, then $\sigma_0(T) = \sigma(T)$. Thus, according to Corollary 2.19, $\dim \mathcal{H} < \infty \implies \sigma_0(T) = \pi_{00}(T) = \sigma(T)$:

> *Every operator on a finite-dimensional space satisfies Weyl's Theorem.*

This extends to finite-rank but not to compact operators (see examples in [44]). On the other hand, if $\sigma_P(T) = \varnothing$, then $\sigma_0(T) = \pi_{00}(T) = \varnothing$:

> *Every operator without eigenvalues satisfies Weyl's Theorem.*

Since $\sigma_0(T) \subseteq \sigma_{PF}(T)$, this extends to operators with $\sigma_{PF}(T) = \varnothing$.

Investigating quadratic forms with compact difference, Hermann Weyl proved in 1909 that Weyl's Theorem holds for self-adjoint operators (i.e., *every self-adjoint operator satisfies Weyl's Theorem*) [114]. We present next a contemporary proof of Weyl's original result. First recall from Sections 1.5 and 1.6 the following elementary (proper) inclusion of classes of operators:

$$\text{Self-Adjoint} \ \subset \ \text{Normal} \ \subset \ \text{Hyponormal}.$$

Theorem 5.48. (WEYL'S THEOREM). *If $T \in \mathcal{B}[\mathcal{H}]$ is self-adjoint, then*

$$\sigma_0(T) = \sigma_{\mathrm{iso}}(T) \cap \sigma_{PF}(T).$$

Proof. Take an arbitrary operator $T \in \mathcal{B}[\mathcal{H}]$.

Claim 1. If T is self-adjoint, then $\sigma_0(T) = \pi_0(T)$.

Proof. If T is self-adjoint, then $\sigma(T) \subset \mathbb{R}$ (Proposition 2.A). Thus no subset of $\sigma(T)$ is open in $\mathbb{C}$, and hence $\sigma_0(T) = \tau_0(T) \cup \pi_0(T) = \pi_0(T)$ because $\tau_0(T)$ is open in $\mathbb{C}$ (Corollary 5.20).

Claim 2. If T is hyponormal and $\lambda \in \pi_{00}(T)$, then $\mathcal{R}(\lambda I - T)$ is closed.

Proof. $\lambda \in \pi_{00}(T) = \sigma_{\mathrm{iso}}(T) \cap \sigma_{PF}(T)$ if and only if λ is an isolated point of $\sigma(T)$ and $0 < \dim \mathcal{N}(\lambda I - T) < \infty$. Thus, if $\lambda \in \pi_{00}(T)$ and T is hyponormal, then $\dim \mathcal{R}(E_\lambda) < \infty$ by Proposition 4.L, where E_λ is the Riesz idempotent associated with λ. Then $\mathcal{R}(\lambda I - T)$ is closed by Corollary 4.22.

Claim 3. If T is hyponormal, then $\pi_0(T) = \pi_{00}(T)$.

Proof. By Claim 2 if T is hyponormal, then (cf. Corollary 5.22)

$$\pi_0(T) = \{\lambda \in \pi_{00}(T) : \mathcal{R}(\lambda I - T) \text{ is closed}\} = \pi_{00}(T).$$

In particular, if T is self-adjoint, then $\pi_0(T) = \pi_{00}(T)$ and so, by Claim 1

$$\sigma_0(T) = \pi_0(T) = \pi_{00}(T) = \sigma_{\mathrm{iso}}(T) \cap \sigma_{PF}(T). \qquad \square$$

What the preceding proof says is this: if T is self-adjoint, then it satisfies Browder's Theorem, and if T is hyponormal, then $\pi_0(T) = \pi_{00}(T)$. Thus if T is self-adjoint, then T satisfies Weyl's Theorem by Remark 5.47(c). Theorem 5.48 was extended to normal operators in [107]. This can be verified by Proposition 5.E, according to which if T is normal, then $\sigma(T) \backslash \sigma_e(T) = \pi_{00}(T)$. But Corollary 5.18 says $\sigma(T) \backslash \sigma_e(T) = \bigcup_{k \in \mathbb{Z} \backslash \{0\}} \sigma_k(T) \cup \sigma_0(T)$, where $\bigcup_{k \in \mathbb{Z} \backslash \{0\}} \sigma_k(T)$ is open in $\mathbb{C}$ (cf. Theorem 5.16) and $\pi_{00}(T) = \sigma_{\mathrm{iso}}(T) \cap \sigma_{PF}(T)$ is closed in $\mathbb{C}$. Thus $\sigma_0(T) = \pi_{00}(T)$. Therefore *every normal operator satisfies Weyl's Theorem*. Moreover, Theorem 5.48 was further extended to hyponormal operators in [27] and to seminormal operators in [16]. In other words:

If T or T^* is hyponormal, then T satisfies Weyl's Theorem.

Some additional definitions and terminology will be needed. An operator is *isoloid* if every isolated point of its spectrum is an eigenvalue; that is, T is isoloid if $\sigma_{\mathrm{iso}}(T) \subseteq \sigma_P(T)$. A Hilbert-space operator T is said to be *dominant* if $\mathcal{R}(\lambda I - T) \subseteq \mathcal{R}(\overline{\lambda} I - T^*)$ for every $\lambda \in \mathbb{C}$ or, equivalently, if for each $\lambda \in \mathbb{C}$ there is an $M_\lambda \geq 0$ for which $(\lambda I - T)(\lambda I - T)^* \leq M_\lambda (\lambda I - T)^*(\lambda I - T)$ [37]. Therefore *every hyponormal operator is dominant* (with $M_\lambda = 1$) *and isoloid* (Proposition 4.L). Recall that a subspace $\mathcal{M}$ of a Hilbert space $\mathcal{H}$ is invariant for an operator $T \in \mathcal{B}[\mathcal{H}]$ (or T-invariant) if $T(\mathcal{M}) \subseteq \mathcal{M}$, and reducing if it is invariant for both T and T^*. A *part* of an operator is a restriction of it to an invariant subspace, and a *direct summand* is a restriction of it to a reducing subspace. The main result in [16] reads as follows (see also [17]).

Theorem 5.49. *If each finite-dimensional eigenspace of an operator on a Hilbert space is reducing, and if every direct summand of it is isoloid, then it satisfies Weyl's Theorem.*

This is a fundamental result that includes many of the previous results along this line, and has also been frequently applied to yield further results, mainly through the following corollary.

Corollary 5.50. *If an operator on a Hilbert space is dominant and every direct summand of it is isoloid, then it satisfies Weyl's Theorem.*

Proof. Take any $T \in \mathcal{B}[\mathcal{H}]$. The announced result can be restated as follows. If $\mathcal{R}(\lambda I - T) \subseteq \mathcal{R}(\overline{\lambda} I - T^*)$ *for every* $\lambda \in \mathbb{C}$, *and the restriction* $T|_\mathcal{M}$ *of T to each reducing subspace $\mathcal{M}$ is such that* $\sigma_{\mathrm{iso}}(T|_\mathcal{M}) \subseteq \sigma_P(T|_\mathcal{M})$, *then T satisfies Weyl's Theorem*. To prove it we need the following elementary result, which extends Lemma 1.13(a) from hyponormal to dominant operators. Take an arbitrary scalar $\lambda \in \mathbb{C}$.

Claim. If $\mathcal{R}(\lambda I - T) \subseteq \mathcal{R}(\overline{\lambda} I - T^*)$, then $\mathcal{N}(\lambda I - T) \subseteq \mathcal{N}(\overline{\lambda} I - T^*)$.

Proof. Take any $S \in \mathcal{B}[\mathcal{H}]$. If $\mathcal{R}(S) \subseteq \mathcal{R}(S^*)$, then $\mathcal{R}(S^*)^\perp \subseteq \mathcal{R}(S)^\perp$, which is equivalent to $\mathcal{N}(S) \subseteq \mathcal{N}(S^*)$ by Lemma 1.4, thus proving the claimed result.

If $\mathcal{R}(\lambda I - T) \subseteq \mathcal{R}(\overline{\lambda} I - T^*)$, then $\mathcal{N}(\lambda I - T)$ reduces T by the preceding Claim and Lemma 1.14(b). Therefore if $\mathcal{R}(\lambda I - T) \subseteq \mathcal{R}(\overline{\lambda} I - T^*)$ for every $\lambda \in \mathbb{C}$, then every eigenspace of T is reducing and so is, in particular, every finite-dimensional eigenspace of T. Thus the stated result is a straightforward consequence of Theorem 5.49. □

Since every hyponormal operator is dominant, every direct summand of a hyponormal operator is again hyponormal (in fact, every part of a hyponormal operator is again hyponormal — Proposition 1.P), and every hyponormal operator is isoloid (Proposition 4.L), then Corollary 5.50 offers another proof that *every hyponormal operator satisfies Weyl's Theorem*.

We need a few more definitions and terminologies. An operator $T \in \mathcal{B}[\mathcal{H}]$ is *paranormal* if $\|Tx\|^2 \leq \|T^2x\| \, \|x\|$ for every $x \in \mathcal{H}$, and *totally hereditarily normaloid* (THN) if all parts of it are normaloid, as well as the inverse of all invertible parts. (Tautologically, totally hereditarily normaloid operators are normaloid.) Hyponormal operators are paranormal and dominant, but paranormal operators are not necessarily dominant. These classes are related by proper inclusion (see [43, Remark 1] and [44, Proposition 2]):

$$\text{Hyponormal} \ \subset \ \text{Paranormal} \ \subset \ \text{THN} \ \subset \ (\text{Normaloid} \cap \text{Isoloid}).$$

Weyl's Theorem has been extended to classes of nondominant operators that properly include the hyponormal operators. For instance, it was extended to paranormal operators in [112] and beyond to totally hereditarily normaloid operators in [42]. In fact [42, Lemma 2.5]:

If T is totally hereditarily normaloid, then both

T and T^* satisfy Weyl's Theorem

(variations along this line in [40, Theorem 3.9], [41, Corollary 2.16]). So Weyl's Theorem holds for paranormal operators and their adjoints and, in particular, Weyl's Theorem holds for hyponormal operators and their adjoints.

Let T and S be arbitrary operators acting on Hilbert spaces. First we consider their (orthogonal) direct sum. The spectrum of a direct sum coincides with the union of the spectra of the summands by Proposition 2.F(b),

$$\sigma(T \oplus S) = \sigma(T) \cup \sigma(S).$$

For the Weyl spectra, only inclusion is ensured. In fact, the Weyl spectrum of a direct sum is included in the union of the Weyl spectra of the summands,

$$\sigma_w(T \oplus S) \subseteq \sigma_w(T) \cup \sigma_w(S),$$

but equality does not hold in general [61]. However, the equality holds if $\sigma_w(T) \cap \sigma_w(S)$ has empty interior [90],

$$\left(\sigma_w(T) \cap \sigma_w(S)\right)^\circ = \varnothing \quad \Longrightarrow \quad \sigma_w(T \oplus S) = \sigma_w(T) \cup \sigma_w(S).$$

In general, Weyl's Theorem does not transfer from T and S to their direct sum $T \oplus S$. The above identity involving the Weyl spectrum of a direct sum, viz., $\sigma_w(T \oplus S) = \sigma_w(T) \cup \sigma_w(S)$, when it holds, plays an important role in establishing sufficient conditions for a direct sum to satisfy Weyl's Theorem. This was recently investigated in [89] and [44]. As for the problem of transferring Browder's Theorem from T and S to their direct sum $T \oplus S$, the following necessary and sufficient condition was proved in [61, Theorem 4].

Theorem 5.51. *If both operators T and S satisfy Browder's Theorem, then the direct sum $T \oplus S$ satisfies Browder's Theorem if and only if*

$$\sigma_w(T \oplus S) = \sigma_w(T) \cup \sigma_w(S).$$

Now consider the tensor product $T \otimes S$ of a pair of Hilbert-space operators T and S (for an expository paper on tensor products which will suffice for our needs see [76]). As is known from [22], the spectrum of a tensor product coincides with the product of the spectra of the factors,

$$\sigma(T \otimes S) = \sigma(T) \cdot \sigma(S).$$

For the Weyl spectrum the following inclusion was proved in [68]:

$$\sigma_w(T \otimes S) \subseteq \sigma_w(T) \cdot \sigma(S) \cup \sigma(T) \cdot \sigma_w(S).$$

However, it remained an open question whether such an inclusion might be an identity; that is, it was not known if there existed a pair of operators T and S for which the above inclusion was proper. This question was solved recently, as we will see later. Sufficient conditions ensuring that the equality holds were investigated in [83]. For instance, if

$$\sigma_e(T)\backslash\{0\} = \sigma_w(T)\backslash\{0\} \quad \text{and} \quad \sigma_e(S)\backslash\{0\} = \sigma_w(S)\backslash\{0\}$$

(which holds, in particular, for compact operators T and S), or if

$$\sigma_w(T \otimes S) = \sigma_b(T \otimes S)$$

(the tensor product satisfies Browder's Theorem), then [83, Proposition 6]

$$\sigma_w(T \otimes S) = \sigma_w(T) \cdot \sigma(S) \cup \sigma(T) \cdot \sigma_w(S).$$

Again, Weyl's Theorem does not necessarily transfer from T and S to their tensor product $T \otimes S$. The preceding identity involving the Weyl spectrum of a tensor product, namely, $\sigma_w(T \otimes S) = \sigma_w(T) \cdot \sigma(S) \cup \sigma(T) \cdot \sigma_w(S)$, when it holds, plays a crucial role in establishing sufficient conditions for a tensor product to satisfy Weyl's Theorem, as was recently investigated in [108], [83], and [84]. As for the problem of transferring Browder's Theorem from T and S

to their tensor product $T \otimes S$, the following necessary and sufficient condition was proved in [83, Corollary 6].

Theorem 5.52. *If both operators T and S satisfy Browder's Theorem, then the tensor product $T \otimes S$ satisfies Browder's Theorem if and only if*

$$\sigma_w(T \otimes S) = \sigma_w(T) \cdot \sigma(S) \cup \sigma(T) \cdot \sigma_w(S).$$

According to Theorem 5.52, if there exist operators T and S satisfying Browder's Theorem for which $T \otimes S$ does not satisfy Browder's Theorem, then the Weyl spectrum identity, viz., $\sigma_w(T \otimes S) = \sigma_w(T) \cdot \sigma(S) \cup \sigma(T) \cdot \sigma_w(S)$, does *not* hold for such a pair of operators. An example of a pair of operators that satisfy Weyl's Theorem (and thus satisfy Browder's Theorem) but whose tensor product does not satisfy Browder's Theorem was recently supplied in [73]. Therefore, [73] and [83] together ensure that *there exists a pair of operators T and S for which the inclusion*

$$\sigma_w(T \otimes S) \subset \sigma_w(T) \cdot \sigma(S) \cup \sigma(T) \cdot \sigma_w(S)$$

is proper; that is, for which the Weyl spectrum identity fails.

5.6 Supplementary Propositions

Proposition 5.A. *Both classes of left and right semi-Fredholm operators $\mathcal{F}_\ell[\mathcal{H}]$ and $\mathcal{F}_r[\mathcal{H}]$ are open sets in $\mathcal{B}[\mathcal{H}]$, and so are $\mathcal{SF}[\mathcal{H}]$ and $\mathcal{F}[\mathcal{H}]$.*

Proposition 5.B. *The mapping $\mathrm{ind}(\cdot) \colon \mathcal{SF}[\mathcal{H}] \to \overline{\mathbb{Z}}$ is continuous, where the topology on $\mathcal{SF}[\mathcal{H}]$ is the topology inherited from $\mathcal{B}[\mathcal{H}]$, and the topology on $\overline{\mathbb{Z}}$ is the discrete topology.*

Proposition 5.C. *Take $T \in \mathcal{B}[\mathcal{H}]$. The following assertions are equivalent.*

(a) *$T \in \mathcal{F}[\mathcal{H}]$.*

(b) *There exists $A \in \mathcal{B}[\mathcal{H}]$ such that $I - AT$ and $I - TA$ are compact.*

(c) *There exists $A \in \mathcal{B}[\mathcal{H}]$ such that $I - AT$ and $I - TA$ are finite-rank.*

Proposition 5.D. *Take an operator T in $\mathcal{B}[\mathcal{H}]$. If $\lambda \in \mathbb{C} \backslash \sigma_{\ell e}(T) \cap \sigma_{re}(T)$, then there exists an $\varepsilon > 0$ such that $\dim \mathcal{N}(\zeta I - T)$ and $\dim \mathcal{N}(\overline{\zeta} I - T^*)$ are constant in the punctured disk $B_\varepsilon(\lambda) \backslash \{\lambda\} = \{\zeta \in \mathbb{C} \colon 0 < |\zeta - \lambda| < \varepsilon\}$.*

Proposition 5.E. *If $T \in \mathcal{B}[\mathcal{H}]$ is normal, then $\sigma_{\ell e}(T) = \sigma_{re}(T) = \sigma_e(T)$ and*

$$\sigma(T) \backslash \sigma_e(T) = \sigma_{\mathrm{iso}}(T) \cap \sigma_{PF}(T) \qquad (\textit{i.e.,} \ \ \sigma(T) \backslash \sigma_e(T) = \pi_{00}).$$

(*Thus $\sigma_e(T) = \sigma_w(T) = \sigma_b(T)$ and $\sigma_0(T) = \pi_{00}(T) = \pi_0(T)$ — no holes or pseudoholes — cf. Theorems 5.18, 5.24, 5.38, 5.48 and Remark 5.27(e).*)

Proposition 5.F. *Let* $\mathbb{D}$ *denote the open unit disk centered at the origin of the complex plane, and let* $\mathbb{T} = \partial\mathbb{D}$ *be the unit circle. If* $S_+ \in \mathcal{B}[\ell_+^2]$ *is a unilateral shift on* $\ell_+^2 = \ell_+^2(\mathbb{C})$, *then*

$$\sigma_{\ell e}(S_+) = \sigma_{re}(S_+) = \sigma_e(S_+) = \sigma_C(S_+) = \partial\sigma(S_+) = \mathbb{T},$$

and $\mathrm{ind}\,(\lambda I - S_+) = -1$ *if* $|\lambda| < 1$ *(i.e., if* $\lambda \in \mathbb{D} = \sigma(S_+)\backslash\partial\sigma(S_+))$.

Proposition 5.G. *If* $\varnothing \neq \Omega \subset \mathbb{C}$ *is a compact set,* $\{\Lambda_k\}_{k\in\mathbb{Z}}$ *is a collection of open sets included in* Ω *whose nonempty sets are pairwise disjoint, and* $\Delta \subseteq \Omega$ *is a discrete set (i.e., containing only isolated points), then there exists* $T \in \mathcal{B}[\mathcal{H}]$ *such that* $\sigma(T) = \Omega$, $\sigma_k(T) = \Lambda_k$ *for each* $k \neq 0$, $\sigma_0(T) = \Lambda_0 \cup \Delta$, *and for each* $\lambda \in \Delta$ *there is a positive integer* $n_\lambda \in \mathbb{N}$ *such that* $\dim \mathcal{R}(E_\lambda) = n_\lambda$ *(where* E_λ *is the Riesz idempotent associated with the isolated point* $\lambda \in \Delta$).

The above proposition shows that *every spectral picture is attainable* [28].

If $K \in \mathcal{B}_\infty[\mathcal{H}]$ is compact, then by Remark 5.27(b,d) either $\sigma_w(K) = \varnothing$ if $\dim \mathcal{H} < \infty$, or $\sigma_w(K) = \{0\}$ if $\dim \mathcal{H} = \infty$. According to the next result, *on an infinite-dimensional* separable *Hilbert space, every compact operator is a commutator.* (An operator T is a *commutator* if there are operators A and B such that $T = AB - BA$.)

Proposition 5.H. *If* $T \in \mathcal{B}[\mathcal{H}]$ *is an operator on an infinite-dimensional* separable *Hilbert space* $\mathcal{H}$, *then*

$$0 \in \sigma_w(T) \quad \Longrightarrow \quad T \text{ is a commutator.}$$

Proposition 5.I. *Take any operator* T *in* $\mathcal{B}[\mathcal{H}]$.

(a) *If* $\dim \mathcal{N}(T) < \infty$ *or* $\dim \mathcal{N}(T^*) < \infty$, *then*

    (a_1) $\mathrm{asc}\,(T) < \infty \implies \dim \mathcal{N}(T) \leq \dim \mathcal{N}(T^*)$,

    (a_2) $\mathrm{dsc}\,(T) < \infty \implies \dim \mathcal{N}(T^*) \leq \dim \mathcal{N}(T)$.

(b) *If* $\dim \mathcal{N}(T) = \dim \mathcal{N}(T^*) < \infty$, *then* $\mathrm{asc}\,(T) < \infty \iff \mathrm{dsc}\,(T) < \infty$.

Proposition 5.J. *Suppose* $T \in \mathcal{B}[\mathcal{H}]$ *is a Fredholm operator (i.e.,* $T \in \mathcal{F}[\mathcal{H}])$.

(a) *If* $\mathrm{asc}\,(T) < \infty$ *and* $\mathrm{dsc}\,(T) < \infty$, *then* $\mathrm{ind}\,(T) = 0$.

(b) *If* $\mathrm{ind}\,(T) = 0$, *then* $\mathrm{asc}\,(T) < \infty$ *if and only if* $\mathrm{dsc}\,(T) < \infty$.

Therefore

$$\mathcal{B}_r[\mathcal{H}] \subseteq \{T \in \mathcal{W}[\mathcal{H}]\colon \mathrm{asc}\,(T) < \infty \iff \mathrm{dsc}\,(T) < \infty\},$$

$$\mathcal{W}[\mathcal{H}]\backslash\mathcal{B}_r[\mathcal{H}] = \{T \in \mathcal{W}[\mathcal{H}]\colon \mathrm{asc}\,(T) = \mathrm{dsc}\,(T) = \infty\}.$$

The next result is an extension of the Fredholm Alternative of Remark 5.7(b), and also of the ultimate form of it in Remark 5.27(c).

Proposition 5.K. *Take any compact operator K in $\mathcal{B}_\infty[\mathcal{H}]$.*

(a) *If $\lambda \neq 0$, then $\lambda I - K \in \mathcal{B}_r[\mathcal{H}]$.*

(b) $\sigma(K)\backslash\{0\} = \pi_0(K)\backslash\{0\} = \sigma_{\mathrm{iso}}(K)\backslash\{0\}$.

Proposition 5.L. *Take $T \in \mathcal{B}[\mathcal{H}]$. If $\sigma_w(T)$ is simply connected (so it has no holes), then $T + K$ satisfies Browder's Theorem for every $K \in \mathcal{B}_\infty[\mathcal{H}]$.*

Notes: Propositions 5.A to 5.C are standard results on Fredholm and semi-Fredholm operators. For instance, see [97, Proposition 1.25] or [30, Proposition XI.2.6] for Proposition 5.A, and [97, Proposition 1.17] or [30, Proposition XI.3.13] for Proposition 5.B. For Proposition 5.C see [7, Remark 3.3.3] or [58, Problem 181]. The locally constant dimension of kernels is considered in Proposition 5.D (see, e.g., [30, Theorem XI.6.7]), and a finer analysis of the spectra of normal operators and of unilateral shifts is discussed in Propositions 5.E and 5.F (see, e.g., [30, Proposition XI.4.6] and [30, Example XI.4.10]). Every spectral picture is attainable [28], and this is described in Proposition 5.G — see [30, Proposition XI.6.13]. For Proposition 5.H see [17, §7]. The results of Proposition 5.I are from [26, p. 57] (see also [36]), and Proposition 5.J is an immediate consequence of Proposition 5.I. Regarding the Fredholm Alternative version of Proposition 5.K, for item (a) see [96, Theorem 1.4.6], and item (b) follows from Corollary 5.35, which goes back to Corollary 2.20. The compact perturbation result of Proposition 5.L is from [10, Theorem 11].

Suggested Readings

Aiena [3]

Arveson [7]

Barnes, Murphy, Smyth, and West [11]

Caradus, Pfaffenberger, and Yood [26]

Conway [30, 32]

Douglas [38]

Halmos [58]

Harte [60]

Istrățescu [69]

Kato [72]

Müller [95]

Murphy [96]

Pearcy [97]

Schechter [106]

Sunder [109]

Taylor and Lay [111]

Appendices

Appendix A
Aspects of Fredholm Theory in Banach Space

Appendix B
A Glimpse at Multiplicity Theory

A

Aspects of Fredholm Theory in Banach Space

Chapter 5, whose framework was borrowed from [77], dealt with Fredholm theory in Hilbert space. The Hilbert-space geometry yields a tremendously rich structure leading to remarkable simplifications. In a Banach space, not all subspaces are complemented. This drives the need for a finer analysis of semi-Fredholm operators on Banach spaces where, in addition to the notions of left and right semi-Fredholmness of Section 5.1, the concepts of upper and lower semi-Fredholmness are also required. Semi-Fredholm operators can be defined either as the union of the class of all left semi-Fredholm and the class of all right semi-Fredholm operators (as in Section 5.1) on the one hand or, on the other hand, as the union of the class of all upper semi-Fredholm and the class of all lower semi-Fredholm operators (which will be defined in Section A.4). These two ways of handling semi-Fredholm operators are referred to as *left-right* and *upper-lower* approaches. In a Hilbert space these approaches coincide, and the reason they do coincide is precisely the fact that every subspace of a Hilbert space is complemented (as a consequence of the Projection Theorem of Section 1.3). Such complementation may, however, fail in a Banach space (as was briefly commented in Remark 5.3(a)). And this is the source of the difference between these two approaches, which arises when investigating semi-Fredholm operators on a Banach space. This appendix draws a parallel between the left-right and upper-lower ways of handling semi-Fredholm operators, and provides an analysis of some consequences of the lack of complementation in Banach spaces. The framework here follows [85].

A.1 Quotient Space

Let $\mathcal{X}$ be a linear space over a field $\mathbb{F}$ and let $\mathcal{M}$ be a linear manifold of $\mathcal{X}$. An *algebraic complement* of $\mathcal{M}$ is any linear manifold $\mathcal{N}$ of $\mathcal{X}$ such that $\mathcal{M} + \mathcal{N} = \mathcal{X}$ and $\mathcal{M} \cap \mathcal{N} = \{0\}$. Every linear manifold has an algebraic complement, and two linear manifolds $\mathcal{M}$ and $\mathcal{N}$ satisfying the above identities are referred to as a pair of *complementary linear manifolds*. See Section 1.1.

Take a linear manifold $\mathcal{M}$ of a linear space $\mathcal{X}$. The *quotient space* $\mathcal{X}/\mathcal{M}$ of $\mathcal{X}$ modulo $\mathcal{M}$ is the collection of all *translations* of the linear manifold $\mathcal{M}$

© Springer Nature Switzerland AG 2020
C. S. Kubrusly, *Spectral Theory of Bounded Linear Operators*,
https://doi.org/10.1007/978-3-030-33149-8

(also called *affine spaces* or *linear varieties*). For each vector $x \in X$ the set $[x] = x + M \in X/M$ (i.e., the translate of M by x) is referred to as the *coset* of x modulo M (which is an equivalence class $[x] \subset X$ of vectors in X, where the equivalence relation $\sim$ is given by $x' \sim x$ if $x - x' \in M$). The quotient space X/M is the collection of all cosets $[x]$ modulo a given M for all $x \in X$, which becomes a linear space under the natural definitions of vector addition and scalar product in X/M (viz., $[x] + [y] = [x + y]$ and $\alpha[x] = [\alpha x]$ for $x, y \in X$ and $\alpha \in \mathbb{F}$ — see, e.g., [78, Example 2.H]). The *natural map* (or the *natural quotient map*) $\pi \colon X \to X/M$ of X onto X/M is defined by

$$\pi(x) = [x] = x + M \quad \text{for every } x \in X.$$

The origin $[0]$ of the linear space X/M is

$$[0] = M \in X/M \qquad (\text{i.e., } [y] = [0] \iff y \in M).$$

The kernel $N(\pi)$ and range $R(\pi)$ of the natural map π are given by

$$N(\pi) = \pi^{-1}([0]) = \{x \in X \colon \pi(x) = [0]\}$$
$$= \{x \in X \colon [x] = [0]\} = \{x \in X \colon x \in M\} = M \subseteq X,$$
$$R(\pi) = \pi(X) = \{[y] \in X/M \colon [y] = \pi(x) \text{ for some } x \in X\}$$
$$= \{[y] \in X/M \colon y + M = x + M \text{ for some } x \in X\} = X/M,$$

so that π is in fact surjective. The map π is a linear transformation between the linear spaces X and X/M. If S is an arbitrary linear manifold of X, then

$$\pi(S) = \{[y] \in X/M \colon [y] = \pi(s) \text{ for some } s \in S\}$$
$$= \{[y] \in X/M \colon [y] = [s] \text{ for some } s \in S\}$$
$$= \{[y] \in X/M \colon [y] - [s] = [0] \text{ for some } s \in S\}$$
$$= \{[y] \in X/M \colon [y - s] = [0] \text{ for some } s \in S\}$$
$$= \{[y] \in X/M \colon y - s \in M \text{ for some } s \in S\}$$
$$= \{[y] \in X/M \colon y \in M + s \text{ for some } s \in S\}$$
$$= \{[y] \in X/M \colon y \in M + S\}.$$

(In particular, $\pi(X) = \{[y] \in X/M \colon y \in X\} = X/M$ since $M + X = X$; that is, again, π is surjective). Hence the inverse image of $\pi(S)$ under π is

$$\pi^{-1}(\pi(S)) = \{x \in X \colon \pi(x) \in \pi(S)\} = \{x \in X \colon [x] \in \pi(S)\}$$
$$= \{x \in X \colon x \in M + S\} = M + S.$$

If N is a linear manifold of X for which $X = M + N$, then

$$\pi(N) = \{[y] \in X/M \colon y \in M + N\} = \{[y] \in X/M \colon y \in X\} = X/M,$$

and in this case the restriction $\pi|_N \colon N \to X/M$ of π to N is surjective as well. Moreover, if $M \cap N = \varnothing$, then

$$\mathcal{N}(\pi|_{\mathcal{N}}) = \pi|_{\mathcal{N}}^{-1}([0]) = \{v \in \mathcal{N} : \pi(v) = [0]\}$$
$$= \{v \in \mathcal{N} : [v] = [0]\} = \{v \in \mathcal{N} : v \in \mathcal{M}\} = \{0\},$$

and in this case the restriction $\pi|_{\mathcal{N}} : \mathcal{N} \to \mathcal{X}/\mathcal{M}$ of π to $\mathcal{N}$ is injective. Since π is linear, its restriction to a linear manifold is again linear.

Remark. A word on terminology. By an *isomorphism* (or an *algebraic isomorphism*, or a *linear-space isomorphism*) we mean a linear invertible transformation between linear spaces. A *topological isomorphism* is a continuous invertible linear transformation with a continuous inverse between topological vector spaces (i.e., an isomorphism and a homeomorphism). When actin between normed spaces this is also called a *normed-space isomorphism*. Dimension is preserved by (plain) isomorphisms, closedness is preserved by topological isomorphisms. An *isometric isomorphism* is an isomorphism and an isometry between normed spaces (a particular case of topological isomorphism). A *unitary transformation* is an isometric isomorphism between inner product spaces.

Thus if $\mathcal{N}$ is an algebraic complement of $\mathcal{M}$, then $\pi|_{\mathcal{N}} : \mathcal{N} \to \mathcal{X}/\mathcal{M}$ is an algebraic isomorphism (sometimes referred to as the *natural quotient isomorphism*) between the linear spaces $\mathcal{N}$ and $\mathcal{X}/\mathcal{M}$. Hence every algebraic complement of $\mathcal{M}$ is isomorphic to the quotient space $\mathcal{X}/\mathcal{M}$. Thus, with $\cong$ standing for "algebraically isomorphic to",

$$\mathcal{N} \cong \mathcal{X}/\mathcal{M} \text{ for every algebraic complement } \mathcal{N} \text{ of } \mathcal{M}.$$

Then every algebraic complement of $\mathcal{M}$ has the same (constant) dimension:

$$\dim \mathcal{N} = \dim \mathcal{X}/\mathcal{M} \text{ for every algebraic complement } \mathcal{N} \text{ of } \mathcal{M},$$

and so two algebraic complements of $\mathcal{M}$ are algebraically isomorphic. Such an invariant (dimension of any algebraic complement) is the *codimension* of $\mathcal{M}$ (notation: $\operatorname{codim} \mathcal{M}$ — see, e.g., [78, Lemma 2.17, Theorem 2.18]). Thus

$$\operatorname{codim} \mathcal{M} = \dim \mathcal{X}/\mathcal{M}.$$

Remark A.1. Let $\mathcal{M}$, $\mathcal{N}$, and $\mathcal{R}$ be linear manifolds of a linear space $\mathcal{X}$.

(a) CODIMENSION. *The codimension of $\mathcal{M}$ is the dimension of any algebraic complement of $\mathcal{M}$.* In other words, the codimension of $\mathcal{M}$ is the dimension of any linear manifold $\mathcal{N}$ for which $\mathcal{M} + \mathcal{N} = \mathcal{X}$ and $\mathcal{M} \cap \mathcal{N} = \{0\}$. However,

$$\mathcal{M} + \mathcal{R} = \mathcal{X} \text{ and } \dim \mathcal{R} < \infty \implies \operatorname{codim} \mathcal{M} < \infty$$

even if $\mathcal{R}$ is not algebraically disjoint from $\mathcal{M}$ (i.e., even if $\mathcal{M} \cap \mathcal{R} \neq \{0\}$). In fact, if $\mathcal{M} + \mathcal{R} = \mathcal{X}$, then the linear manifold $\mathcal{N} = \mathcal{M} \cap \mathcal{R} \subseteq \mathcal{R}$ is an algebraic complement of $\mathcal{M}$. Thus $\dim \mathcal{R} < \infty$ implies $\dim \mathcal{N} < \infty$ so that $\operatorname{codim} \mathcal{M} < \infty$.

(b) ANOTHER PROOF. Another way to show

$$\dim \mathcal{N} = \dim \mathcal{X}/\mathcal{M}$$

if $\mathcal{N}$ is an algebraic complement of $\mathcal{M}$ without using the natural map π goes as follows. Let $\mathcal{M}$ be a linear manifold of $\mathcal{X}$ and let $\mathcal{N}$ be any algebraic complement of $\mathcal{M}$. Let Δ be an index set and consider the sets $\{e_\delta\}_{\delta \in \Delta} \subseteq \mathcal{N}$ and $\{[e_\delta]\}_{\delta \in \Delta} \subseteq \mathcal{X} \backslash \mathcal{M}$, where $[e_\delta] = e_\delta + \mathcal{M}$ for each $\delta \in \Delta$. Thus

$$\{e_\delta\}_{\delta \in \Delta} \text{ is linearly independent} \quad \Longleftrightarrow \quad \{[e_\delta]\}_{\delta \in \Delta} \text{ is linearly independent.}$$

Indeed, take an arbitrary e_{δ_0} in $\{e_\delta\}_{\delta \in \Delta}$, consider the respective arbitrary $[e_{\delta_0}]$ in $\{[e_\delta]\}_{\delta \in \Delta}$, and let $\{\alpha_\delta\}_{\delta \in \Delta}$ be an arbitrary set of scalars. If

$$e_{\delta_0} = \sum\nolimits_{\delta \neq \delta_0} \alpha_\delta \, e_\delta, \qquad (*)$$

then $[e_{\delta_0}] = \left[\sum_{\delta \neq \delta_0} \alpha_\delta \, e_\delta \right] = \sum_{\delta \neq \delta_0} [\alpha_\delta \, e_\delta] = \sum_{\delta \neq \delta_0} \alpha_\delta \, [e_\delta]$. Conversely, if

$$[e_{\delta_0}] = \sum\nolimits_{\delta \neq \delta_0} \alpha_\delta \, [e_\delta], \qquad (**)$$

then $[e_{\delta_0}] = \sum_{\delta \neq \delta_0} [\alpha_\delta \, e_\delta] = \left[\sum_{\delta \neq \delta_0} \alpha_\delta \, e_\delta \right]$ and hence $e_{\delta_0} - \sum_{\delta \neq \delta_0} \alpha_\delta \, e_\delta \in \mathcal{M}$ so that $e_{\delta_0} = \sum_{\delta \neq \delta_0} \alpha_\delta \, e_\delta$ because $\mathcal{M} \cap \mathcal{N} = \{0\}$. Thus $(*)$ holds true if and only if $(**)$ holds true. However, $\{e_\delta\}_{\delta \in \Delta}$ is linearly independent if and only if the only set of scalars for which $(*)$ holds is $\alpha_\delta = 0$ for all $\delta \in \Delta$, and $\{[e_\delta]\}_{\delta \in \Delta}$ is linearly independent if and only if the only set of scalars for which $(**)$ holds is $\alpha_\delta = 0$ for all $\delta \in \Delta$. This proves the above equivalence. Moreover,

$$\{e_\delta\}_{\delta \in \Delta} \text{ spans } \mathcal{N} \quad \Longleftrightarrow \quad \{[e_\delta]\}_{\delta \in \Delta} \text{ spans } \mathcal{X} \backslash \mathcal{M}.$$

In fact, take an arbitrary $[x] \in \mathcal{X} \backslash \mathcal{M}$, where $x = u + v \in \mathcal{M} + \mathcal{N}$ is an arbitrary vector in $\mathcal{X} = \mathcal{M} + \mathcal{N}$, with u being an arbitrary vector in $\mathcal{M}$ and v being an arbitrary vector in $\mathcal{N}$. If $\{e_\delta\}_{\delta \in \Delta}$ spans $\mathcal{N}$, then there exists a sequence of scalars $\{\alpha_\delta(v)\}_{\delta \in \Delta}$ such that $v = \sum_{\delta \in \Delta} \alpha_\delta(v) \, e_\delta$. Hence

$$[x] = x + \mathcal{M} = u + v + \mathcal{M} = v + \mathcal{M} = [v]$$
$$= \sum\nolimits_{\delta \in \Delta} \alpha_\delta(v) e_\delta + \mathcal{M} = \sum\nolimits_{\delta \in \Delta} \alpha_\delta(v)(e_\delta + \mathcal{M}) = \sum\nolimits_{\delta \in \Delta} \alpha_\delta(v)[e_\delta],$$

and so $\{[e_\delta]\}_{\delta \in \Delta}$ spans $\mathcal{X} \backslash \mathcal{M}$. Conversely, if $\{[e_\delta]\}_{\delta \in \Delta}$ spans $\mathcal{X} \backslash \mathcal{M}$, then there exists a sequence of scalars $\{\alpha_\delta([x])\}_{\delta \in \Delta}$ such that

$$v + \mathcal{M} = v + u + \mathcal{M} = x + \mathcal{M} = [x]$$
$$= \sum\nolimits_{\delta \in \Delta} \alpha_\delta([x]) \, [e_\delta] = \sum\nolimits_{\delta \in \Delta} \alpha_\delta([x]) \, (e_\delta + \mathcal{M}) = \sum\nolimits_{\delta \in \Delta} \alpha_\delta([x]) \, e_\delta + \mathcal{M}.$$

Thus $v = \sum_{\delta \in \Delta} \alpha_\delta([x]) \, e_\delta$ (because $\mathcal{M} \cap \mathcal{N} = \{0\}$), and so $\{e_\delta\}_{\delta \in \Delta}$ spans $\mathcal{N}$. Hence the above equivalence also holds true. By both equivalences

$$\{e_\delta\}_{\delta \in \Delta} \text{ is a Hamel basis for } \mathcal{N} \quad \Longleftrightarrow \quad \{[e_\delta]\}_{\delta \in \Delta} \text{ is a Hamel basis for } \mathcal{X} \backslash \mathcal{M},$$

and so $\dim \mathcal{N} = \dim \mathcal{X} / \mathcal{M}$ as claimed.

(c) COMPLEMENTARY DIMENSION. Let $\mathcal{M}$ and $\mathcal{N}$ be algebraic complements in a linear space $\mathcal{X}$ (i.e., $\mathcal{M}$ and $\mathcal{N}$ are complementary linear manifolds). Then

$$\dim \mathcal{X} = \dim \mathcal{M} + \dim \mathcal{N}$$

(see, e.g., [111, Theorem 6.2]). In other words, for any linear manifold, dimension plus codimension coincides with the dimension of the linear space:

$$\dim \mathcal{X} = \dim \mathcal{M} + \dim \mathcal{X}/\mathcal{M} = \dim \mathcal{M} + \operatorname{codim} \mathcal{M}.$$

This is another consequence of the rank and nullity identity of linear algebra,

$$\dim \mathcal{X} = \dim \mathcal{R}(L) + \dim \mathcal{N}(L)$$

for every linear transformation $L\colon \mathcal{X} \to \mathcal{Y}$ between linear spaces $\mathcal{X}$ and $\mathcal{Y}$. (In fact, if $\mathcal{M}$ and $\mathcal{N}$ are complementary linear manifolds of a linear space $\mathcal{X}$, then there is a projection $E\colon \mathcal{X} \to \mathcal{X}$ with $\mathcal{R}(E) = \mathcal{M}$ and $\mathcal{N}(E) = \mathcal{R}(I - E) = \mathcal{N}$ — as will be revisited below in the first paragraph of Section A.2 — and $\mathcal{R}(E) \cong \mathcal{X}/\mathcal{N}(E)$ — as will be seen in Remark A.2(a) below.)

Recall: a *subspace* is a *closed* linear manifold of a normed space. Let $\mathcal{M}$ be a linear manifold of a normed space $\mathcal{X}$. Consider the map $\|\cdot\|\colon \mathcal{X}/\mathcal{M} \to \mathbb{R}$ defined for each coset $[x]$ in the quotient space $\mathcal{X}/\mathcal{M}$ by

$$\big\|[x]\big\| = \big\|x + \mathcal{M}\big\| = \inf_{u \in \mathcal{M}} \|x + u\| = d(x, \mathcal{M}) \leq \|x\|.$$

This gauges the distance of x to $\mathcal{M}$ (with is invariant for all representatives in the equivalence class $[x]$). It is a seminorm in $\mathcal{X}/\mathcal{M}$ which becomes a norm (the *quotient norm*) in $\mathcal{X}/\mathcal{M}$ if $\mathcal{M}$ is closed. So if $\mathcal{M}$ is a subspace, then equip the quotient space $\mathcal{X}/\mathcal{M}$ with its quotient norm. When we refer to the normed space $\mathcal{X}/\mathcal{M}$, it is understood that $\mathcal{M}$ is a subspace and the quotient space $\mathcal{X}/\mathcal{M}$ is equipped with the quotient norm. Thus if $\mathcal{M}$ is a subspace of a normed space $\mathcal{X}$, then *the natural quotient map* $\pi\colon \mathcal{X} \to \mathcal{X}/\mathcal{M}$ *is a* (linear) *contraction* (i.e., $\big\|\pi(x)\big\| \leq \|x\|$ for every $x \in \mathcal{X}$), and so π is continuous.

Let $\mathcal{X}$ be a normed space. Take the normed algebra $\mathcal{B}[\mathcal{X}]$. Let $\mathcal{X}^* = \mathcal{B}[\mathcal{X}, \mathbb{F}]$ be the dual space of $\mathcal{X}$ (so that $\mathcal{X}^*$ is a Banach space), and let $T^* \in \mathcal{B}[\mathcal{X}^*]$ stand for the *normed-space adjoint* of an operator $T \in \mathcal{B}[\mathcal{X}]$, defined as the unique bounded linear operator on $\mathcal{X}^*$ for which $f \circ T = T^* f \in \mathcal{X}^*$ for every $f \in \mathcal{X}^*$ (i.e., $f(Tx) = (T^* f)(x)$ for every $f \in \mathcal{X}^*$ and every $x \in \mathcal{X}$; see, e.g., [93, Section 3.1] or [106, Section 3.2]). The usual basic properties of Hilbert-space adjoints are transferred to normed-space adjoints. For instance, I^* is the identity in $\mathcal{B}[\mathcal{X}^*]$ (where I is the identity in $\mathcal{B}[\mathcal{X}]$), $\|T^*\| = \|T\|$, $(T+S)^* = T^* + S^*$, $(TS)^* = S^* T^*$ and, slightly different, $(\alpha T)^* = \alpha T^*$ for $\alpha \in \mathbb{F}$ and $T, S \in \mathcal{B}[\mathcal{X}]$ (see, e.g., [93, Proposition 3.1.2, 3.1.4, 3.1.6, 3.1.10]).

Remark A.2. Let $\mathcal{X}$ and $\mathcal{Y}$ be linear spaces over the same field.

(a) THE FIRST ISOMORPHISM THEOREM. If $L\colon \mathcal{X} \to \mathcal{Y}$ is a linear transformation,

$$\mathcal{R}(L) \cong \mathcal{X}/\mathcal{N}(L).$$

Here $\cong$ means an algebraic isomorphism (see, e.g., [101, Theorem 3.5]). This holds tautologically for bounded linear transformations between normed spaces. In particular, if $T \in \mathcal{B}[\mathcal{X}]$ where $\mathcal{X}$ is a Banach space and $\mathcal{R}(T)$ is closed, then

$$\mathcal{R}(T) \cong \mathcal{X}/\mathcal{N}(T).$$

Now $\cong$ means a topological isomorphism (see, e.g., [93, Theorem 1.7.14]).

(b) BASIC DUAL RESULTS. Let $\mathcal{X}$ be a normed space, and let $T^* \in \mathcal{B}[\mathcal{X}^*]$ be the normed-space adjoint of $T \in \mathcal{B}[\mathcal{X}]$. If $\mathcal{X}$ is a Banach space, then the range of T^* is closed in $\mathcal{X}^*$ if and only if the range of T is closed in $\mathcal{X}$ (see, e.g., [93, Lemma 3.1.21] — compare with Lemma 1.5):

$$\mathcal{R}(T^*)^- = \mathcal{R}(T^*) \quad \Longleftrightarrow \quad \mathcal{R}(T)^- = \mathcal{R}(T).$$

The dual of the result in (a) holds for normed-space operators,

$$\mathcal{N}(T^*) \cong \big(\mathcal{X}/\mathcal{R}(T)\big)^*$$

(see, e.g., [93, Theorems 1.10.17, Lemma 3.1.16]). Here $\cong$ means an isometric isomorphism. Moreover, if $\mathcal{X}$ is a normed space, then (see, e.g., [93, 1.10.8]),

$$\dim \mathcal{X} < \infty \quad \Longleftrightarrow \quad \dim \mathcal{X}^* < \infty.$$

In this finite-dimensional case we have

$$\dim \mathcal{X}^* = \dim \mathcal{X},$$

since if $\dim \mathcal{X} < \infty$, then $\mathcal{X}^* = \mathcal{B}[\mathcal{X}, \mathbb{F}] = \mathcal{L}[\mathcal{X}, \mathbb{F}]$ with $\mathcal{L}[\mathcal{X}, \mathbb{F}]$ denoting the linear space of all linear functionals on $\mathcal{X}$ (see, e.g,. [101, Corollary 3.13]).

(c) RANGE AND KERNEL DUALITY. For a normed space $\mathcal{X}$ we get from (b)

$$\dim \mathcal{N}(T^*) < \infty \quad \Longleftrightarrow \quad \operatorname{codim} \mathcal{R}(T) < \infty.$$

Similarly, if $\mathcal{X}$ is a Banach space, then

$$\dim \mathcal{N}(T) < \infty \quad \Longleftrightarrow \quad \operatorname{codim} \mathcal{R}(T^*) < \infty.$$

This needs an explanation since T^{**} may not be identified with T for normed-space adjoints (on nonreflexive spaces). The second equivalence, however, comes from the first one by using the identity (which holds in a Banach space)

$$\dim \mathcal{N}(T^{**}) = \dim \mathcal{N}(T).$$

To verify the above identity proceed as follows. Let $\mathcal{X}$ be a normed space and let Θ be the *natural map* (the *natural second dual map*). That is, for each $x \in \mathcal{X}$ take $\varphi_x \in \mathcal{X}^{**}$ such that $\varphi_x(f) = f(x)$ for every $f \in \mathcal{X}^*$ and define $\Theta : \mathcal{X} \to \mathcal{R}(\Theta) \subseteq \mathcal{X}^{**}$ by $\Theta(x) = \varphi_x$. This is an isometric isomorphism of $\mathcal{X}$ onto $\mathcal{R}(\Theta)$, and Θ has a closed range if $\mathcal{X}$ is Banach. If $T^{**} \in \mathcal{B}[\mathcal{X}^{**}]$ is the normed-space adjoint of $T^* \in \mathcal{B}[\mathcal{X}^*]$, then $T^{**}(\Theta(\mathcal{X})) \subseteq \Theta(\mathcal{X})$ (i.e., $\mathcal{R}(\Theta) \subseteq \mathcal{X}^{**}$ is T^{**}-invariant) [93, Proposition 3.1.13]. As $\Theta \in \mathcal{B}[\mathcal{X}, \mathcal{X}^{**}]$ is also injective, let $\Theta^{-1} \in \mathcal{B}[\mathcal{R}(\Theta), \mathcal{X}]$ be its inverse on its closed range. So [93, Proposition 3.1.13]

$$\Theta^{-1} T^{**} \Theta = T.$$

Therefore $x \in \mathcal{N}(T)$ if and only if $\Theta(x) \in \mathcal{N}(T^{**})$. (In fact, by the above identity, $x \in \mathcal{N}(T) \Rightarrow \Theta^{-1}(T^{**}\Theta x) = 0 \Rightarrow (T^{**}\Theta x) = 0$ (since Θ^{-1} is injective) $\Rightarrow \Theta x \in \mathcal{N}(T^{**}) \Rightarrow Tx = 0 \Rightarrow x \in \mathcal{N}(T)$.) This implies

$$\Theta(\mathcal{N}(T)) = \mathcal{N}(T^{**}).$$

Indeed, since $x \in \mathcal{N}(T) \Rightarrow \Theta x \in \mathcal{N}(T^{**})$, then $\Theta(\mathcal{N}(T)) \subseteq \mathcal{N}(T^{**}) \cap \mathcal{R}(\Theta)$; equivalently, $\mathcal{N}(T) \subseteq \Theta^{-1}(\mathcal{N}(T^{**}))$. On the other hand, since $\Theta x \in \mathcal{N}(T^{**}) \Rightarrow x \in \mathcal{N}(T) \cap \mathcal{R}(\Theta)$, then $\Theta^{-1}(\mathcal{N}(T^{**})) \subseteq \mathcal{N}(T)$. So $\mathcal{N}(T) = \Theta^{-1}(\mathcal{N}(T^{**}))$, proving the identity. Now consider the restriction of Θ to $\mathcal{N}(T)$,

$$\Theta|_{\mathcal{N}(T)} : \mathcal{N}(T) \to \mathcal{R}(\Theta|_{\mathcal{N}(T)}) \subseteq \mathcal{R}(\Theta) \subseteq \mathcal{X}^{**},$$

which is linear and injective since Θ is. Actually, $\mathcal{N}(\Theta|_{\mathcal{N}(T)}) \subseteq \mathcal{N}(\Theta) = \{0\}$ and so $\dim \mathcal{N}(\Theta|_{\mathcal{N}(T)}) = 0$. Moreover, by the above identity, $\mathcal{R}(\Theta|_{\mathcal{N}(T)}) = \mathcal{N}(T^{**})$ and so $\dim \mathcal{R}(\Theta|_{\mathcal{N}(T)}) = \dim \mathcal{N}(T^{**})$. Thus by linearity

$$\dim \mathcal{N}(T^{**}) = \dim \mathcal{R}(\Theta|_{\mathcal{N}(T)}) + \dim \mathcal{N}(\Theta|_{\mathcal{N}(T)}) = \dim \mathcal{N}(T).$$

(d) FREDHOLM ALTERNATIVE. Let $\mathcal{X}$ be a Banach space. If $\lambda \neq 0$ and $K \in \mathcal{B}_\infty[\mathcal{X}]$ (i.e., if K is compact), then $\mathcal{R}(\lambda I - K)$ is closed and

$$\dim \mathcal{N}(\lambda I - K) = \dim \mathcal{N}(\lambda I - K^*) < \infty.$$

This is the Fredholm Alternative in Corollary 1.20 (which as commented in Section 1.8 still holds in a Banach space — see, e.g., [30, Theorem VII.7.9] or [93, Lemma 3.4.21]). Thus with the help of (a) and (b) we get

$$\operatorname{codim} \mathcal{R}(\lambda I - K^*) = \dim \mathcal{N}(\lambda I - K) = \dim \mathcal{N}(\lambda I - K^*) = \operatorname{codim} \mathcal{R}(\lambda I - K).$$

Moreover, this leads to the Fredholm Alternative version in Theorem 2.18 according to the diagram of Section 2.2:

$$\sigma(K) \backslash \{0\} = \sigma_P(K) \backslash \{0\} = \sigma_{P_4}(K) \backslash \{0\}.$$

(e) THREE-SPACE PROPERTY. If $\mathcal{M}$ is a subspace of a Banach space $\mathcal{X}$ (and so $\mathcal{M}$ is a Banach space), then the quotient space $\mathcal{X}/\mathcal{M}$ is a Banach space (see, e.g., [78, Proposition 4.10], [30, Theorem III.4.2(b)], [93, Theorem 1.7.7]). Conversely, if $\mathcal{M}$ is a Banach space (i.e., if $\mathcal{M}$ is complete in a normed space $\mathcal{X}$, and so $\mathcal{M}$ is closed) and the quotient space $\mathcal{X}/\mathcal{M}$ is a Banach space, then $\mathcal{X}$ is a Banach space (see, e.g., [78, Problem 4.13], [30, Exercise III.4.5]). If $\mathcal{X}/\mathcal{M}$ is a normed space equipped with the quotient norm, then $\mathcal{M}$ must be a subspace of $\mathcal{X}$; if $\mathcal{X}$ is Banach, then so is $\mathcal{M}$. Thus if any two of $\mathcal{M}$, $\mathcal{X}$ and $\mathcal{X}/\mathcal{M}$ are Banach spaces, then so is the other (see also [93, Theorem 1.7.9]).

A.2 Complementation

If $E\colon \mathcal{X} \to \mathcal{X}$ is a projection (i.e., an idempotent linear transformation on a linear space $\mathcal{X}$), then $\mathcal{R}(E)$ and $\mathcal{N}(E)$ are complementary linear manifolds, and conversely if $\mathcal{M}$ and $\mathcal{N}$ are complementary linear manifolds, then there is a unique projection $E\colon \mathcal{X} \to \mathcal{X}$ with $\mathcal{R}(E) = \mathcal{M} = \mathcal{N}(I - E)$ and $\mathcal{N}(E) = \mathcal{N} = \mathcal{R}(I - E)$, where $I - E\colon \mathcal{X} \to \mathcal{X}$ is the *complementary projection* of E.

Proposition A.3. *On a Banach space, a projection with a closed range and a closed kernel is continuous.*

Proof. Let $E\colon \mathcal{X} \to \mathcal{X}$ be a projection on a normed space $\mathcal{X}$ such that $\mathcal{M} = \mathcal{R}(E)$ and $\mathcal{N} = \mathcal{N}(E)$ are subspaces (i.e., complementary closed linear manifolds) of $\mathcal{X}$. Take the *natural mapping* $\Pi\colon \mathcal{M} \oplus \mathcal{N} \to \mathcal{M} + \mathcal{N}$ *of the direct sum* $\mathcal{M} \oplus \mathcal{N}$ (equipped with any of its usual p-norms $\| \cdot \|_p$) *onto the ordinary sum* $\mathcal{X} = \mathcal{M} + \mathcal{N}$, which is given by $\Pi(u, v) = u + v$ for every (u, v) in $\mathcal{M} \oplus \mathcal{N}$. Consider the mapping $P\colon \mathcal{M} \oplus \mathcal{N} \to \mathcal{M} \subseteq \mathcal{X}$ defined by $P(u, v) = u$ for every (u, v) in $\mathcal{M} \oplus \mathcal{N}$. Since P is a linear contraction (as $\|P(u, v)\|^p = \|u\|^p \leq \|u\|^p + \|v\|^p = \|(u, v)\|_p^p$ for every $(u, v) \in \mathcal{M} \oplus \mathcal{N}$ and any $p \geq 1$), since Π is a topological isomorphism whenever $\mathcal{X}$ is a Banach space (by Theorem 1.1 — see, e.g., [78, Problem 4.34]), and since $E = P \circ \Pi^{-1}\colon \mathcal{M} + \mathcal{N} = \mathcal{X} \to \mathcal{X}$ is the composition of two continuous functions, then E is continuous. $\square$

A subspace $\mathcal{M}$ of a normed space $\mathcal{X}$ is *complemented* if it has a subspace as an algebraic complement; that is, a *closed* linear manifold $\mathcal{M}$ of a normed space $\mathcal{X}$ is complemented if there is a *closed* linear manifold $\mathcal{N}$ of $\mathcal{X}$ such that $\mathcal{M}$ and $\mathcal{N}$ are algebraic complements. In this case, $\mathcal{M}$ and $\mathcal{N}$ are *complementary subspaces*.

Remark A.4. Complementary Subspace and Continuous Projection. If $\mathcal{X}$ is a normed space and $E\colon \mathcal{X} \to \mathcal{X}$ is a continuous projection, then $\mathcal{R}(E)$ and $\mathcal{N}(E)$ are complementary subspaces of $\mathcal{X}$. This is straightforward since $\mathcal{R}(E)$ and $\mathcal{N}(E)$ arc complementary linear manifolds (whenever E is an idempotent linear transformation) and $\mathcal{R}(E) = \mathcal{N}(I - E)$. Conversely, if $\mathcal{M}$ and $\mathcal{N}$ are complementary subspaces of a Banach space $\mathcal{X}$, then the (unique) projection $E\colon \mathcal{X} \to \mathcal{X}$ with $\mathcal{R}(E) = \mathcal{M}$ and $\mathcal{N}(E) = \mathcal{N}$ is continuous by Proposition A.3. Therefore if $\mathcal{X}$ is a Banach space, then the assertions below are equivalent.

(a$_1$) A subspace $\mathcal{M}$ of $\mathcal{X}$ is complemented.

(a$_2$) There exists a projection $E \in \mathcal{B}[\mathcal{X}]$ with $\mathcal{R}(E) = \mathcal{M}$.

(a$_3$) There exists a projection $I - E \in \mathcal{B}[\mathcal{X}]$ with $\mathcal{N}(I - E) = \mathcal{M}$.

Since in a finite-dimensional normed space every linear manifold is closed, then *in a finite-dimensional normed space every subspace is complemented.*

(b) If $\mathcal{M}$ and $\mathcal{N}$ are complementary subspaces of a Banach space $\mathcal{X}$, then $\mathcal{M} \cong \mathcal{X}/\mathcal{N}$ (topologically isomorphic).

(Let $E \in \mathcal{B}[\mathcal{X}]$ be the continuous projection with $\mathcal{R}(E) = \mathcal{M}$ and $\mathcal{N}(E) = \mathcal{N}$. By the First Isomorphism Theorem of Remark A.2(a), $\mathcal{R}(E) \cong \mathcal{X}/\mathcal{N}(E)$).

Linear manifolds of a Banach space with finite dimension are complemented subspaces. On the other hand, not all linear manifolds of a Banach space with finite codimension are subspaces. If a linear manifold of a Banach space is, however, the range of an operator with finite codimension, then it is a subspace. These facts are discussed next.

Remark A.5. FINITE-DIMENSIONAL COMPLEMENTED SUBSPACES.

(a) Finite-dimensional subspaces of a Banach space are complemented.

In fact, if $\mathcal{M}$ is a finite-dimensional subspace of a Banach space $\mathcal{X}$, then we show next the existence of a continuous projection $E\colon \mathcal{X} \to \mathcal{X}$ with $\mathcal{R}(E) = \mathcal{M}$. Thus $\mathcal{M}$ is complemented by Remark A.4 (since $\mathcal{X}$ is Banach and $\mathcal{M}$ is closed).

To verify the existence of such a *continuous* projection proceed as follows. Suppose $\mathcal{M}$ is nonzero to avoid trivialities. Let $\{e_j\}_{j=1}^m$ be any Hamel basis of unit vectors ($\|e_j\| = 1$) for $\mathcal{M}$ where $m = \dim \mathcal{M}$. Thus every $u \in \mathcal{M}$ is uniquely written as $u = \sum_{j=1}^m \beta_j(u)e_j$. For each j consider the functional $\beta_j \colon \mathcal{M} \to \mathbb{F}$ of the vector expansion in $\mathcal{M}$. By uniqueness these functionals are linear, thus continuous since $\mathcal{M}$ is finite-dimensional. By the Hahn–Banach Theorem (see, e.g., [78, Theorem 4.62]) let $\widehat{\beta}_j \colon \mathcal{X} \to \mathbb{F}$ be a norm-preserving ($\|\widehat{\beta}_j\| = \|\beta_j\|$) bounded linear extension of each β_j over the whole space $\mathcal{X}$ ($\widehat{\beta}_j(u) = \beta_j(u)$ for $u \in \mathcal{M}$). Take the transformation $E\colon \mathcal{X} \to \mathcal{X}$ defined by $Ex = \sum_{j=1}^m \widehat{\beta}_j(x)e_j$ for every $x \in \mathcal{X}$. Then $\mathcal{R}(E) \subseteq \mathcal{M}$ and $Eu = \sum_{j=1}^m \beta_j(u)e_j = u$ for every $u \in \mathcal{M}$, and so $\mathcal{M} \subseteq \mathcal{R}(E)$. Therefore $\mathcal{R}(E) = \mathcal{M}$. This E is linear since the functionals $\widehat{\beta}_j \colon \mathcal{X} \to \mathbb{F}$ are linear. It is bounded since these functionals are bounded,

$$\|Ex\| = \left\| \sum_{j=1}^m \widehat{\beta}_j(x)e_j \right\| \leq \left(\sum_{j=1}^m \|\beta_j\| \right) \|x\|,$$

and idempotent since E is linear,

$$E^2 x = E\left(\sum_{j=1}^m \widehat{\beta}_j(x)e_j \right) = \sum_{j=1}^m \widehat{\beta}_j(x)Ee_j = \sum_{j=1}^m \widehat{\beta}_j(x)e_j = Ex,$$

for every $x \in \mathcal{X}$. Thus E is a bounded (i.e., continuous) projection. By the way, since $Ex = 0$ if and only if $\widehat{\beta}_j(x) = 0$ for all j, the resulting complementary subspace of $\mathcal{M} = \mathcal{R}(E)$ is $\mathcal{N} = \mathcal{N}(E) = \bigcap_{j=1}^m \mathcal{N}(\widehat{\beta}_j)$.

Here is another way to state the same result. Suppose $\mathcal{X}$ is a Banach space.

(a′) Every finite-dimensional subspace of $\mathcal{X}$ has a closed algebraic complement,

which does not mean that *every* algebraic complement of a finite-dimensional subspace of a Banach space is closed. Equivalently, these do not say that if a linear manifold has a finite codimension, then it is a complemented *subspace*.

(a″) There are nonclosed linear manifolds of $\mathcal{X}$ with finite codimension.

Example 1. Let $f\colon \mathcal{X} \to \mathbb{F}$ be a nonzero linear functional on a nonzero normed space $\mathcal{X}$. Then $\dim \mathcal{R}(f) = 1$ (i.e., $\mathcal{R}(f) = \mathbb{F}$). Also $\mathcal{R}(f) \cong \mathcal{X}/\mathcal{N}(f)$ by the algebraic version of the First Isomorphism Theorem of Remark A.2(a) (where $\cong$ means algebraic isomorphism) and so

$$\operatorname{codim} \mathcal{N}(f) = \dim \mathcal{X}/\mathcal{N}(f) = \dim \mathcal{R}(f) = 1.$$

Moreover (see, e.g., [93, Proposition 1.7.16, Theorem 1.7.15]),

$$\mathcal{N}(f)^- \neq \mathcal{N}(f) \iff f \text{ is unbounded} \iff \mathcal{N}(f)^- = \mathcal{X}.$$

Therefore, if f is unbounded, then $\mathcal{N}(f)$ is a nonclosed (and dense) linear manifold of the normed space $\mathcal{X}$ (in particular, of a Banach space) with a finite codimension (i.e., with a one-dimensional complementary subspace).

Note. On the other hand, as we will see in Corollary A.9, for the special case where the linear manifold is the range of an operator (bounded and linear) on a Banach space, if it has a finite codimension, then it must be closed: *If T is a Banach-space operator, and if $\operatorname{codim} \mathcal{R}(T) < \infty$, then $\mathcal{R}(T)$ is closed.*

The above example does not show that assertion (a) fails if $\mathcal{X}$ is not Banach.

(a‴) If a normed space is incomplete, then there may exist an uncomplemented (nonzero) finite-dimensional subspace of it.

Example 2. Take the normed space $C[0,1]$ of all (equivalence classes of) scalar-valued continuous functions on the interval $[0,1]$ equipped with the norm $\|\cdot\|_1$. Let $v\colon [0,1] \to \mathbb{R}$ be a discontinuous function defined by $v(t) = 0$ for $t \in [0, \frac{1}{2})$ and $v(t) = 1$ for $t \in [\frac{1}{2}, 1]$. Take the normed space $\mathcal{X} = C[0,1] + \langle v \rangle$ equipped with the norm $\|\cdot\|_1$ (with $\langle v \rangle = \operatorname{span}\{v\}$). This $\mathcal{X}$ is not Banach (since $C[0,1]$ is dense in $L^1[0,1]$ and so is $\mathcal{X} \subsetneq L^1[0,1]$). Also, $C[0,1]$ is a linear manifold of $\mathcal{X}$, it is an algebraic complement of $\langle v \rangle$ in $\mathcal{X}$, and it is not closed in $\mathcal{X}$: for any scalar α there is a sequence $\{u_n\}$ with $u_n \in C[0,1]$ such that $u_n \to \alpha v$ (i.e., for each $\varepsilon > 0$ there is a $k_\varepsilon \in \mathbb{N}$ such that if $k \geq k_\varepsilon$ then $\|u_k - \alpha v\|_1 < \varepsilon$).

Claim. Every algebraic complement $\mathcal{M}$ of $\langle v \rangle$ in $\mathcal{X}$ is dense in $\mathcal{X}$.

Proof. Since $C[0,1] \subseteq \mathcal{M} + \langle v \rangle = \mathcal{X}$, then $u_k = m_k + \alpha_k v$ with $m_k \in \mathcal{M}$. Thus $d(m_k, \langle v \rangle) = \inf_{w \in \langle v \rangle} \|u_k - \alpha_k v - w\|_1 = \inf_{w \in \langle v \rangle} \|u_k - w\|_1 \leq \|u_k - \alpha v\|_1 < \varepsilon$ for every $k \geq k_\varepsilon$. Hence $d(\mathcal{M}, \langle v \rangle) = 0$ and so $\mathcal{M}^- = \mathcal{X} = \mathcal{M} + \langle v \rangle$, proving the claimed assertion. Then every algebraic complement of $\langle v \rangle$ is not closed in $\mathcal{X}$.

(b) Subspaces of a normed space with finite codimension are complemented.

This goes along the same line of the previous statements, it does not require completeness, and it is trivially verified once closedness is assumed *a priori*. Indeed, if $\mathcal{M}$ is a linear manifold of a normed space $\mathcal{X}$ which is already closed, and if it has a finite-dimensional algebraic complement $\mathcal{N}$, then $\mathcal{N}$ is closed, and so $\mathcal{M}$ and $\mathcal{N}$ are complementary subspaces. But again this does not mean that a linear manifold (even if $\mathcal{X}$ were a Banach space) of finite codimension is a subspace (i.e., is closed), as we saw in Example 1 above.

An ordinary sum of closed linear manifolds may not be closed (even if they are algebraically disjoint; even in a Hilbert space if they are not orthogonal). But if one of them is finite-dimensional, then the sum is closed.

Proposition A.6. *Suppose $\mathcal{M}$ and $\mathcal{N}$ are subspaces of a normed space. If $\dim \mathcal{N} < \infty$, then $\mathcal{M} + \mathcal{N}$ is closed.*

Proof. Let $\mathcal{M}$ and $\mathcal{N}$ be subspaces of a normed space $\mathcal{X}$. Suppose $\mathcal{M}$ and $\mathcal{N}$ are nontrivial to avoid trivialities. Since $\mathcal{M}$ is a subspace, then $\mathcal{X}/\mathcal{M}$ is a normed space equipped with the quotient norm. Take the natural map $\pi \colon \mathcal{X} \to \mathcal{X}/\mathcal{M}$ and consider its restriction $\pi|_{\mathcal{N}} \colon \mathcal{N} \to \pi(\mathcal{N}) \subseteq \mathcal{X}/\mathcal{M}$ to $\mathcal{N}$, which is linear. Thus $\dim \mathcal{N}(\pi|_{\mathcal{N}}) + \dim \mathcal{R}(\pi|_{\mathcal{N}}) = \dim \mathcal{N}$. If $\mathcal{N}$ is finite-dimensional, then $\dim \pi(\mathcal{N}) = \dim \mathcal{R}(\pi|_{\mathcal{N}}) \leq \dim \mathcal{N} < \infty$. Being finite-dimensional, the linear manifold $\pi(\mathcal{N})$ is in fact a subspace of $\mathcal{X}$ (i.e., it is closed in $\mathcal{X}/\mathcal{M}$). So the inverse image $\pi^{-1}(\pi(\mathcal{N}))$ of $\pi(\mathcal{N})$ under π is closed as well because π is continuous. But $\pi^{-1}(\pi(\mathcal{N})) = \mathcal{N} + \mathcal{M}$ as we saw in Section A.1. $\qquad \square$

Proposition A.7. *Suppose $\mathcal{M}$ and $\mathcal{N}$ are subspaces of a Banach space. If $\dim \mathcal{N} < \infty$ and $\mathcal{M}$ is complemented, then $\mathcal{M} + \mathcal{N}$ is a complemented subspace.*

Proof. Let $\mathcal{M}$ and $\mathcal{N}$ be subspaces of a Banach space $\mathcal{X}$. Again, suppose $\mathcal{M}$ and $\mathcal{N}$ are nontrivial to avoid trivialities. If $\mathcal{M}$ is complemented, then there exists a subspace $\mathcal{M}'$ of $\mathcal{X}$ (a complement of $\mathcal{M}$ in $\mathcal{X}$) such that

$$\mathcal{X} = \mathcal{M} + \mathcal{M}' \quad \text{with} \quad \mathcal{M} \cap \mathcal{M}' = \{0\}.$$

Then $\mathcal{M} + \mathcal{N} \subseteq \mathcal{M} + \mathcal{M}'$. Set $\mathcal{S} = \mathcal{N} \cap \mathcal{M}'$, again a subspace of $\mathcal{X}$, and so

$$\mathcal{M} + \mathcal{N} = \mathcal{M} + \mathcal{S} \quad \text{with} \quad \mathcal{M} \cap \mathcal{S} - \{0\}.$$

If $\dim \mathcal{N} < \infty$, then $\dim \mathcal{S} \leq \dim \mathcal{N} < \infty$ (as $\mathcal{S} \subseteq \mathcal{N}$). Thus $\mathcal{S}$ is complemented (Remark A.5(a)). But $\mathcal{S}$ is also included in $\mathcal{M}'$. So there is a subspace $\mathcal{S}'$ of $\mathcal{X}$ including the complement $\mathcal{M}$ of $\mathcal{M}'$, which is a complement of $\mathcal{S}$ in $\mathcal{X}$,

$$\mathcal{X} = \mathcal{S} + \mathcal{S}' \quad \text{with} \quad \mathcal{S} \cap \mathcal{S}' = \{0\}, \quad \mathcal{S} \subseteq \mathcal{M}' \quad \text{and} \quad \mathcal{M} \subseteq \mathcal{S}'.$$

Since $\mathcal{M}$ is complemented in $\mathcal{X}$ with complement $\mathcal{M}'$, there exists a projection $E \in \mathcal{B}[\mathcal{X}]$ for which $\mathcal{R}(E) = \mathcal{M}$ and $\mathcal{N}(E) = \mathcal{M}'$ (Remark A.4). Similarly, since $\mathcal{S}$ is complemented in $\mathcal{X}$ with complement $\mathcal{S}'$, there exists a projection $P \in \mathcal{B}[\mathcal{X}]$ for which $\mathcal{R}(P) = \mathcal{S}$ and $\mathcal{N}(P) = \mathcal{S}'$. Since $\mathcal{R}(E) = \mathcal{M} \subseteq \mathcal{S}' = \mathcal{N}(P)$ and $\mathcal{R}(P) = \mathcal{S} \subseteq \mathcal{M}' = \mathcal{N}(E)$, then $P + E \in \mathcal{B}[\mathcal{X}]$ is a projection. Moreover, $\mathcal{R}(P + E) = \mathcal{R}(E) + \mathcal{R}(P) = \mathcal{M} + \mathcal{S}$ (since $\mathcal{R}(E) \cap \mathcal{R}(P) = \mathcal{M} \cap \mathcal{S} = \varnothing$), where $\mathcal{M} + \mathcal{S}$ is a subspace of $\mathcal{X}$ (Proposition A.6). Therefore the subspace $\mathcal{M} + \mathcal{N} = \mathcal{M} + \mathcal{S}$ is complemented (Remark A.4). $\qquad \square$

Consider the converse to Proposition A.6. Is $\mathcal{M}$ closed whenever $\mathcal{M} + \mathcal{N}$ is closed and $\dim \mathcal{N} < \infty$? If $\mathcal{M}$ is the range of a bounded linear transformation between Banach spaces $\mathcal{X}$ and $\mathcal{Y}$, then the answer is yes.

Proposition A.8. *Suppose $\mathcal{X}$ and $\mathcal{Y}$ are Banach spaces, take any $T \in \mathcal{B}[\mathcal{X}, \mathcal{Y}]$, and let $\mathcal{N}$ be a finite-dimensional subspace of $\mathcal{Y}$. If $\mathcal{R}(T) + \mathcal{N}$ is closed, then $\mathcal{R}(T)$ is closed as well.*

Proof. Take $T \in \mathcal{B}[\mathcal{X}, \mathcal{Y}]$ where $\mathcal{X}$ and $\mathcal{Y}$ are Banach spaces. Let $\mathcal{N}$ be a finite-dimensional subspace of $\mathcal{Y}$. Let $\mathcal{S}$ be an algebraic complement of $\mathcal{R}(T) \cap \mathcal{N}$ in $\mathcal{N}$ (i.e., $(\mathcal{R}(T) \cap \mathcal{N}) + \mathcal{S} = \mathcal{N}$ and $(\mathcal{R}(T) \cap \mathcal{N}) \cap \mathcal{S} = \{0\}$). Then $\mathcal{R}(T) + \mathcal{S} = \mathcal{R}(T) + \mathcal{N}$. (In fact, $\mathcal{R}(T) + \mathcal{N} = \mathcal{R}(T) + (\mathcal{R}(T) \cap \mathcal{N}) + \mathcal{S} \subseteq \mathcal{R}(T) + \mathcal{S} \subseteq \mathcal{R}(T) + \mathcal{N}$). Since $\mathcal{N}(T)$ is a subspace, $\mathcal{X}/\mathcal{N}(T)$ is a Banach space. Thus, since $\mathcal{S}$ is finite-dimensional, $\mathcal{X}/\mathcal{N}(T) \oplus \mathcal{S}$ is a Banach space. Consider the transformation $\theta \colon \mathcal{X}/\mathcal{N}(T) \oplus \mathcal{S} \to \mathcal{Y}$ given by $\theta(x + \mathcal{N}(T), s) = T(x) + s$ for every $x \in \mathcal{X}$ and $s \in \mathcal{S}$. This θ is linear, bounded, and injective (as $\mathcal{R}(T) \cap \mathcal{S} = \{0\}$). Moreover, $\mathcal{R}(\theta) = \mathcal{R}(T) + \mathcal{S} = \mathcal{R}(T) + \mathcal{N}$. Hence if $\mathcal{R}(T) + \mathcal{N}$ is closed, then $\mathcal{R}(\theta) \subseteq \mathcal{Y}$ is closed. Therefore the linear, bounded, and injective θ with a closed range acting between Banach spaces must have a bounded inverse on its range (according to Theorem 1.2). Since $\mathcal{X}/\mathcal{N}(T) \oplus \{0\}$ is closed in $\mathcal{X}/\mathcal{N}(T) \oplus \mathcal{S}$, then $(\theta^{-1})^{-1}(\mathcal{X}/\mathcal{N}(T) \oplus \{0\}) = \theta(\mathcal{X}/\mathcal{N}(T) \oplus \{0\}) = \mathcal{R}(T)$ is closed in $\mathcal{Y}$. $\square$

Corollary A.9. *Let $\mathcal{X}$ and $\mathcal{Y}$ be Banach spaces. Take any $T \in \mathcal{B}[\mathcal{X}, \mathcal{Y}]$. If $\operatorname{codim} \mathcal{R}(T) < \infty$, then $\mathcal{R}(T)$ is closed and so it is a complemented subspace.*

Proof. If $\operatorname{codim} \mathcal{R}(T) < \infty$, then $\dim \mathcal{N} < \infty$ for any algebraic complement $\mathcal{N}$ of $\mathcal{R}(T)$. Since $\mathcal{R}(T) + \mathcal{N} = \mathcal{Y}$ is closed, then so is $\mathcal{R}(T)$ by Proposition A.8. $\square$

A finite-rank linear transformation between Banach spaces does not need to be continuous (but if it is, then it must be compact). Finite-rank projections, however, are continuous (thus compact).

Corollary A.10. *A finite-rank projection on a Banach space is continuous.*

Proof. Let $E \colon \mathcal{X} \to \mathcal{X}$ be a finite-rank projection on a normed space $\mathcal{X}$. Set $\mathcal{M} = \mathcal{R}(E)$ so that $\dim \mathcal{M} < \infty$ and hence $\mathcal{M}$ is closed (i.e., a subspace of $\mathcal{X}$). Take the linear manifold $\mathcal{N}(E)$ of $\mathcal{X}$, which is an algebraic complement of $\mathcal{M}$. Suppose $\mathcal{X}$ is a Banach space. By Remark A.5(a) the linear manifold $\mathcal{N}(E)$ is closed in $\mathcal{X}$ and so E is continuous by Proposition A.3. $\square$

Remark. However, $\mathcal{R}(T)$ is not necessarily closed if $\operatorname{codim} \mathcal{R}(T)^- < \infty$. There are $T \in \mathcal{B}[\mathcal{X}, \mathcal{Y}]$ between Banach spaces $\mathcal{X}$ and $\mathcal{Y}$ for which $\operatorname{codim} \mathcal{R}(T)^- = 0$ and $\mathcal{R}(T)$ is not closed. Example: if $T = \operatorname{diag}\{\frac{1}{k}\}$ on $\mathcal{X} = \ell^2$, then $0 \in \sigma_C(T)$ (0 is in the continuous spectrum of T), and so $\mathcal{R}(T)$ is not closed but is dense in ℓ^2. Thus the unique algebraic complement of $\mathcal{R}(T)^- = \ell^2$ is $\{0\}$, and hence $\operatorname{codim} \mathcal{R}(T)^- = 0$. Therefore $\operatorname{codim} \mathcal{R}(T)^- = 0$ and $\mathcal{R}(T)$ is not closed.

A.3 Range-Kernel Complementation

A normed space is *complemented* if every subspace has a complementary subspace (i.e., if every subspace of it is complemented — recall: by a subspace we mean a *closed* linear manifold). According to the Projection Theorem of Section 1.3, every Hilbert space is complemented. However, as we saw in Remark 5.3(a), if a Banach space is complemented, then it is isomorphic (i.e., topologically isomorphic) to a Hilbert space [91] (see also [71]). Thus complemented Banach spaces are identified with Hilbert spaces — only Hilbert spaces (up to an isomorphism) are complemented.

Definition A.11. Let $\mathcal{X}$ and $\mathcal{Y}$ be normed spaces and define the following classes of transformations from $\mathcal{B}[\mathcal{X},\mathcal{Y}]$.

$$\Gamma_R[\mathcal{X},\mathcal{Y}] = \big\{T \in \mathcal{B}[\mathcal{X},\mathcal{Y}] \colon \mathcal{R}(T)^- \text{ is a complemented subspace of } \mathcal{Y}\big\},$$

$$\Gamma_N[\mathcal{X},\mathcal{Y}] = \big\{T \in \mathcal{B}[\mathcal{X},\mathcal{Y}] \colon \mathcal{N}(T) \text{ is a complemented subspace of } \mathcal{X}\big\},$$

$$\Gamma[\mathcal{X},\mathcal{Y}] = \Gamma_R[\mathcal{X},\mathcal{Y}] \cap \Gamma_N[\mathcal{X},\mathcal{Y}]$$
$$= \big\{T \in \mathcal{B}[\mathcal{X},\mathcal{Y}] \colon \mathcal{R}(T)^- \text{ and } \mathcal{N}(T) \text{ are complemented}\big\}.$$

If $\mathcal{X} = \mathcal{Y}$, then write

$$\Gamma_R[\mathcal{X}] = \Gamma_R[\mathcal{X},\mathcal{X}], \qquad \Gamma_N[\mathcal{X}] = \Gamma_N[\mathcal{X},\mathcal{X}], \qquad \Gamma[\mathcal{X}] = \Gamma_R[\mathcal{X}] \cap \Gamma_N[\mathcal{X}],$$

which are subsets of $\mathcal{B}[\mathcal{X}]$. The class $\Gamma_R[\mathcal{X}]$ has been denoted by $\tilde{\zeta}(\mathcal{X})$ in [25].

A Banach space $\mathcal{X}$ is *range complemented* if $\Gamma_R[\mathcal{X}] = \mathcal{B}[\mathcal{X}]$, *kernel complemented* if $\Gamma_N[\mathcal{X}] = \mathcal{B}[\mathcal{X}]$, and *range-kernel complemented* if $\Gamma[\mathcal{X}] = \mathcal{B}[\mathcal{X}]$ (i.e., if $\Gamma_R[\mathcal{X}] = \Gamma_N[\mathcal{X}] = \mathcal{B}[\mathcal{X}]$). Hilbert spaces are complemented, and consequently they are range-kernel complemented. If a Banach space $\mathcal{X}$ is complemented (i.e., if $\mathcal{X}$ is essentially a Hilbert space), then it is trivially range-kernel complemented. Is the converse true? At the very least, for an arbitrary Banach space the set $\Gamma[\mathcal{X}] = \Gamma_R[\mathcal{X}] \cap \Gamma_N[\mathcal{X}]$ is algebraically and topologically large in the sense that it includes nonempty open groups. For instance, the group $\mathcal{G}[\mathcal{X}]$ of all invertible operators from $\mathcal{B}[\mathcal{X}]$ (i.e., of all operators from $\mathcal{B}[\mathcal{X}]$ with a bounded inverse) is open in $\mathcal{B}[\mathcal{X}]$ and trivially included in $\Gamma[\mathcal{X}]$.

Proposition A.12. *Let $\mathcal{X}$ and $\mathcal{Y}$ be a Banach spaces.*

(a) *If $T \in \mathcal{B}[\mathcal{X},\mathcal{Y}]$ is has a finite-dimensional range, then it lies in $\Gamma_R[\mathcal{X},\mathcal{Y}]$. If $T \in \mathcal{B}[\mathcal{X},\mathcal{Y}]$ has a finite-dimensional kernel, then it lies in $\Gamma_N[\mathcal{X},\mathcal{Y}]$.*

(b) *If range of $T \in \mathcal{B}[\mathcal{X},\mathcal{Y}]$ has a finite codimension, then T lies in $\Gamma_R[\mathcal{X},\mathcal{Y}]$. If kernel of $T \in \mathcal{B}[\mathcal{X},\mathcal{Y}]$ has a finite codimension, then T lies in $\Gamma_N[\mathcal{X},\mathcal{Y}]$.*

(c) *Every invertible operator from $\mathcal{B}[\mathcal{X},\mathcal{Y}]$ has complemented closed range and kernel, and so lies in $\Gamma[\mathcal{X},\mathcal{Y}] = \Gamma_R[\mathcal{X},\mathcal{Y}] \cap \Gamma_N[\mathcal{X},\mathcal{Y}]$. In particular $\mathcal{G}[\mathcal{X}] \subseteq \Gamma[\mathcal{X}] = \Gamma_R[\mathcal{X}] \cap \Gamma_N[\mathcal{X}]$, and $\mathcal{G}[\mathcal{X}]$ is open in $\mathcal{B}[\mathcal{X}]$.*

(d) *$T \in \mathcal{B}[\mathcal{X},\mathcal{Y}]$ is left invertible (with a bounded left inverse) if and only if it is injective with a closed and complemented range.*

In other words, $T \in \Gamma_R[\mathcal{X}, \mathcal{Y}]$, $\mathcal{N}(T) = \{0\}$, and $\mathcal{R}(T)^- = \mathcal{R}(T)$ (or T has a complemented range and satisfies any of the equivalent conditions in Theorem 1.2) if and only if T has a left inverse in $\mathcal{B}[\mathcal{Y}, \mathcal{X}]$. (Compare with the Hilbert-space counterpart in Lemma 5.8(a).)

(e) $T \in \mathcal{B}[\mathcal{X}, \mathcal{Y}]$ *has a right inverse in $\mathcal{B}[\mathcal{Y}, \mathcal{X}]$ if and only if it is surjective and lies in $\Gamma_N[\mathcal{X}, \mathcal{Y}]$.*

Proof. Assertion (a) is a particular case of Remark A.5(a), and assertion (b) comes from Corollary A.9 and Remark A.5(b) since $\mathcal{N}(T)$ is a subspace.

(c) If $T \in \mathcal{G}[\mathcal{X}, \mathcal{Y}]$, then $\mathcal{R}(T) = \mathcal{Y}$ and $\mathcal{N}(T) = \{0\}$. Moreover, $\mathcal{G}[\mathcal{X}]$ is open in the uniform topology of $\mathcal{B}[\mathcal{X}]$ (see, e.g., [78, Problem 4.48]).

(d) Suppose $T \in \mathcal{B}[\mathcal{X}, \mathcal{Y}]$ has a left inverse in $\mathcal{B}[\mathcal{Y}, \mathcal{X}]$. In other words, suppose there exists $S \in \mathcal{B}[\mathcal{Y}, \mathcal{X}]$ such that $ST = I \in \mathcal{B}[\mathcal{X}]$, which trivially implies $\mathcal{N}(T) = \{0\}$ ($Tx = 0 \Rightarrow x = Ix = STx = 0$); that is, T is injective. Since $(TS)^2 = TSTS = TS$ in $\mathcal{B}[\mathcal{Y}]$ and $\mathcal{R}(T) = \mathcal{R}(TST) \subseteq \mathcal{R}(TS) \subseteq \mathcal{R}(T)$, then $TS \in \mathcal{B}[\mathcal{Y}]$ is a (continuous) projection with $\mathcal{R}(TS) = \mathcal{R}(T) \subseteq \mathcal{Y}$, and so $\mathcal{R}(T)$ is closed and complemented in $\mathcal{Y}$ by Remark A.4. Conversely, suppose $T \in \mathcal{B}[\mathcal{X}, \mathcal{Y}]$ is injective with a closed range. Thus T has a bounded inverse on its range according to Theorem 1.2, which means there exists $S' \in \mathcal{B}[\mathcal{R}(T), \mathcal{X}]$ for which $S'T = I \in \mathcal{B}[\mathcal{X}]$. Moreover, if the closed range is complemented in $\mathcal{Y}$, then there exists a (continuous) projection $E \in \mathcal{B}[\mathcal{Y}]$ with $\mathcal{R}(E) = \mathcal{R}(T)$ by Remark A.4. Thus take the operator $S = S'E \in \mathcal{B}[\mathcal{Y}, \mathcal{X}]$ so that $ST = S'ET = S'T = I \in \mathcal{B}[\mathcal{X}]$; that is, $T \in \mathcal{B}[\mathcal{X}, \mathcal{Y}]$ has a left inverse S in $\mathcal{B}[\mathcal{Y}, \mathcal{X}]$.

(e) Suppose $T \in \mathcal{B}[\mathcal{X}, \mathcal{Y}]$ has a right inverse in $\mathcal{B}[\mathcal{Y}, \mathcal{X}]$. In other words, suppose there exists $S \in \mathcal{B}[\mathcal{Y}, \mathcal{X}]$ such that $TS = I \in \mathcal{B}[\mathcal{Y}]$, which trivially implies $\mathcal{R}(T) = \mathcal{Y}$ ($y \in \mathcal{Y} \Rightarrow y = Iy = TSy \in \mathcal{R}(T)$); that is, T is surjective. Since $(ST)^2 = STST = ST$ in $\mathcal{B}[\mathcal{X}]$, then $ST \in \mathcal{B}[\mathcal{X}]$ is a (continuous) projection and $\mathcal{N}(ST) = \mathcal{N}(T)$ ($Tx = 0 \Rightarrow STx = 0 \Rightarrow Tx = TSTx = 0$). Hence $\mathcal{N}(T) = \mathcal{N}(ST) = \mathcal{R}(I - ST)$ is complemented by Remark A.4. Conversely, suppose T lies in $\Gamma_N[\mathcal{X}, \mathcal{Y}]$ so that $\mathcal{N}(T)$ is a complemented subspace of $\mathcal{X}$, which means there exists a subspace $\mathcal{M}$ of $\mathcal{X}$ for which $\mathcal{X} = \mathcal{N}(T) + \mathcal{M}$ and $\mathcal{N}(T) \cap \mathcal{M} = \{0\}$. Thus the restriction $T|_{\mathcal{M}} \in \mathcal{B}[\mathcal{M}, \mathcal{Y}]$ is injective. If T is surjective, then so is $T|_{\mathcal{M}}$ because $\mathcal{R}(T|_{\mathcal{M}}) = \mathcal{R}(T) = \mathcal{Y}$. Hence $T|_{\mathcal{M}} \in \mathcal{B}[\mathcal{M}, \mathcal{Y}]$ has an inverse $(T|_{\mathcal{M}})^{-1} \in \mathcal{B}[\mathcal{Y}, \mathcal{M}]$. So $T(T|_{\mathcal{M}})^{-1} = T|_{\mathcal{M}}(T|_{\mathcal{M}})^{-1} = I \in \mathcal{B}[\mathcal{Y}]$. Since $\mathcal{N}(T)$ and $\mathcal{M}$ are complementary, let $J \in \mathcal{B}[\mathcal{M}, \mathcal{X}]$ be the *natural embedding of $\mathcal{M}$ into $\mathcal{X}$* (i.e., $J(u) = 0 + u \in \mathcal{N}(T) + \mathcal{M} = \mathcal{X}$). Therefore we get $TJ(T|_{\mathcal{M}})^{-1}y = T(0 + (T|_{\mathcal{M}})^{-1}y) = T(T|_{\mathcal{M}})^{-1}y = y$ for every $y \in \mathcal{Y}$. Then $TJ(T|_{\mathcal{M}})^{-1} = I \in \mathcal{B}[\mathcal{Y}]$ and so $J(T|_{\mathcal{M}})^{-1} \in \mathcal{B}[\mathcal{Y}, \mathcal{X}]$ is a right inverse of T. $\square$

Proposition A.13. *Take $T \in \mathcal{B}[\mathcal{X}]$ on a Banach space $\mathcal{X}$.*

(a_1) *If $T \in \Gamma_R[\mathcal{X}]$, then $T^* \in \Gamma_N[\mathcal{X}^*]$.*

(a_2) *If $\mathcal{X}$ is reflexive and $T^* \in \Gamma_N[\mathcal{X}^*]$, then $T \in \Gamma_R[\mathcal{X}]$.*

(b$_1$) *If $\mathcal{X}$ is reflexive and $T^* \in \Gamma_R[\mathcal{X}^*]$, then $T \in \Gamma_N[\mathcal{X}]$.*

(b$_2$) *If $\mathcal{R}(T)$ is closed and $T \in \Gamma_N[\mathcal{X}]$, then $T^* \in \Gamma_R[\mathcal{X}^*]$.*

Proof. See [82, Theorem 3.1]. $\qquad\qquad\qquad\qquad\qquad\qquad\qquad\qquad\square$

A normed space is *reflexive* if the natural second dual map is an isometric isomorphism. Every reflexive normed space is a Banach space.

Theorem A.14. *If $T \in \mathcal{B}[\mathcal{X}]$ is a compact operator on a reflexive Banach space $\mathcal{X}$ with a Schauder basis, then*

$$T \in \Gamma_R[\mathcal{X}] \cap \Gamma_N[\mathcal{X}] \quad and \quad T^* \in \Gamma_R[\mathcal{X}^*] \cap \Gamma_N[\mathcal{X}^*].$$

Proof. (We borrow the proof of [82, Corollary 5.1].) Suppose $T \in \mathcal{B}[\mathcal{X}]$ is compact and $\mathcal{X}$ has a Schauder basis. Thus there is a sequence $\{T_n\}$ of finite-rank operators $T_n \in \mathcal{B}[\mathcal{X}]$ such that $T_n \xrightarrow{u} T$ (i.e., $\{T_n\}$ converges uniformly to T) and $\mathcal{R}(T_n) \subseteq \mathcal{R}(T_{n+1}) \subseteq \mathcal{R}(T)^-$ (see, e.g., [78, Problem 4.58]). Since each T_n is finite-rank (i.e., $\dim \mathcal{R}(T_n) < \infty$), we get $\mathcal{R}(T_n) = \mathcal{R}(T_n)^-$. Moreover, finite-dimensional subspaces of a Banach space are complemented (Remark A.5(a), also Proposition A.12(a)). Then $T_n \in \Gamma_R[\mathcal{X}]$; equivalently, there exist continuous projections $E_n \in \mathcal{B}[\mathcal{X}]$ and $I - E_n \in \mathcal{B}[\mathcal{X}]$ (which as such have closed ranges) with $\mathcal{R}(E_n) = \mathcal{R}(T_n)$ — Remark A.4(a,b), where $\{\mathcal{R}(E_n)\}$ is an increasing sequence of subspaces. Since $\{\mathcal{R}(E_n)\}$ is a monotone sequence of subspaces, $\lim_n \mathcal{R}(E_n)$ exists in the following sense:

$$\lim_n \mathcal{R}(E_n) = \bigcap_{n \geq 1} \bigvee_{k \geq n} \mathcal{R}(E_k) = \left(\bigcup_{n \geq 1} \bigcap_{k \geq n} \mathcal{R}(E_k) \right)^-,$$

where $\bigvee_{k \geq n} \mathcal{R}(E_k)$ is the closure of the span of $\{\bigcup_{k \geq n} \mathcal{R}(E_k)\}$ [25, Definition 1]. Thus regarding the complementary projections $I - E_n$, $\lim_n \mathcal{R}(I - E_n)$ also exists. Moreover, $\lim_n \mathcal{R}(E_n)$ is a subspace of $\mathcal{X}$ included in $\mathcal{R}(T)^-$ ($\lim_n \mathcal{R}(E_n) \subseteq \mathcal{R}(T)^-$ because $\mathcal{R}(T_n) \subseteq \mathcal{R}(T)^-$). Since $\mathcal{X}$ has a Schauder basis and T is compact, the sequence of operators $\{E_n\}$ converges strongly (see, e.g., [78, Hint to Problem 4.58]) and so $\{E_n\}$ is a bounded sequence. Thus since (i) $\mathcal{X}$ is reflexive, (ii) $T_n \xrightarrow{s} T$ (i.e., $\{T_n\}$ converges strongly because it converges uniformly), (iii) $T_n \in \Gamma_R[\mathcal{X}]$, (iv) $\mathcal{R}(E_n) = \mathcal{R}(T_n)$, (v) $\{\|E_n\|\}$ is bounded, (vi) $\lim_n \mathcal{R}(T_n) \subseteq \mathcal{R}(T)^-$, and (vii) $\lim_n \mathcal{R}(I - E_n)$ exists, then

$$T \in \Gamma_R[\mathcal{X}]$$

[25, Theorem 2]. Hence $\mathcal{R}(T)^-$ is complemented if T is compact and $\mathcal{X}$ is reflexive with a Schauder basis. Since reflexivity for $\mathcal{X}$ is equivalent to reflexivity for $\mathcal{X}^*$ (see, e.g., [30, Theorem V.4.2]), since $\mathcal{X}^*$ has a Schauder basis whenever $\mathcal{X}$ has (see, e.g., [93, Theorem 4.4.1]), and since T is compact if and only if T^* is compact (see, e.g., [93, Theorem 3.4.15]), then $\mathcal{R}(T^*)^-$ is also complemented:

$$T^* \in \Gamma_R[\mathcal{X}^*].$$

Therefore, since $\mathcal{X}$ is reflexive, Proposition A.13 ensures

$$T \in \Gamma_R[\mathcal{X}] \implies T^* \in \Gamma_N[\mathcal{X}^*] \quad and \quad T^* \in \Gamma_R[\mathcal{X}^*] \implies T \in \Gamma_N[\mathcal{X}]. \quad \square$$

A.4 Upper-Lower and Left-Right Approaches

Consider the following classes $\Phi_+[\mathcal{X}]$, $\Phi_-[\mathcal{X}]$, $\Phi[\mathcal{X}]$ of operators from $\mathcal{B}[\mathcal{X}]$.

Definition A.15. Let $\mathcal{X}$ be a Banach space. The set

$$\Phi_+[\mathcal{X}] = \{T \in \mathcal{B}[\mathcal{X}] \colon \mathcal{R}(T) \text{ is closed and } \dim \mathcal{N}(T) < \infty\}$$

is the class of *upper semi-Fredholm* operators from $\mathcal{B}[\mathcal{X}]$, while

$$\Phi_-[\mathcal{X}] = \{T \in \mathcal{B}[\mathcal{X}] \colon \mathcal{R}(T) \text{ is closed and } \dim \mathcal{X}/\mathcal{R}(T) < \infty\}$$

is the class of *lower semi-Fredholm* operators from $\mathcal{B}[\mathcal{X}]$, and their intersection

$$\Phi[\mathcal{X}] = \Phi_+[\mathcal{X}] \cap \Phi_-[\mathcal{X}]$$

is the class of *Fredholm* operators from $\mathcal{B}[\mathcal{X}]$. Since $\operatorname{codim} \mathcal{R}(T) = \dim \mathcal{X}/\mathcal{R}(T)$ and since $\mathcal{R}(T)$ is closed whenever $\operatorname{codim} \mathcal{R}(T) < \infty$ by Corollary A.9, then the class of lower semi-Fredholm operators can be equivalently written as

$$\Phi_-[\mathcal{X}] = \{T \in \mathcal{B}[\mathcal{X}] \colon \operatorname{codim} \mathcal{R}(T) < \infty\}$$

(cf. Remark 5.3(b)), and so the class of Fredholm operators can be written as

$$\Phi[\mathcal{X}] = \{T \in \mathcal{B}[\mathcal{X}] \colon \dim \mathcal{N}(T) < \infty \text{ and } \operatorname{codim} \mathcal{R}(T) < \infty\}$$
$$= \{T \in \mathcal{B}[\mathcal{X}] \colon \mathcal{R}(T) \text{ is closed, } \dim \mathcal{N}(T) < \infty, \dim \mathcal{X}/\mathcal{R}(T) < \infty\}.$$

The sets $\Phi_+[\mathcal{X}]$ and $\Phi_-[\mathcal{X}]$ are open (and $\Phi_+[\mathcal{X}]\backslash\Phi_-[\mathcal{X}]$ and $\Phi_-[\mathcal{X}]\backslash\Phi_+[\mathcal{X}]$ are closed) in $\mathcal{B}[\mathcal{X}]$ (see, e.g., [95, Proposition 16.11, Proof of Corollary 18.2]).

Proposition A.16. *Let $\mathcal{X}$ be a Banach space.*

(a) $\Phi_+[\mathcal{X}] \subseteq \Gamma_N[\mathcal{X}]$.

(b) $\Phi_-[\mathcal{X}] \subseteq \Gamma_R[\mathcal{X}]$.

(c) $\Phi[\mathcal{X}] \subseteq \Gamma[\mathcal{X}]$.

Proof. If $T \in \Phi_+[\mathcal{X}]$, then $\dim \mathcal{N}(T) < \infty$ and so $\mathcal{N}(T)$ is complemented by Proposition A.12(a). Thus $\Phi_+[\mathcal{X}] \subseteq \Gamma_N[\mathcal{X}]$. On the other hand, if $T \in \Phi_-[\mathcal{X}]$, then $\operatorname{codim} \mathcal{R}(T) < \infty$, and hence $\mathcal{R}(T)$ is complemented by Proposition A.12(b). Thus $\Phi_-[\mathcal{X}] \subseteq \Gamma_R[\mathcal{X}]$. Consequently, $\Phi[\mathcal{X}] \subseteq \Gamma[\mathcal{X}]$. $\qquad\square$

Proposition A.17. *Let $T \in \mathcal{B}[\mathcal{X}]$ be an operator on a Banach space $\mathcal{X}$ and let $T^* \in \mathcal{B}[\mathcal{X}^*]$ be its normed-space adjoint on the Banach space $\mathcal{X}^*$. Then*

(a) $T \in \Phi_+[\mathcal{X}]$ *if and only if* $T^* \in \Phi_-[\mathcal{X}^*]$,

(b) $T \in \Phi_-[\mathcal{X}]$ *if and only if* $T^* \in \Phi_+[\mathcal{X}^*]$,

(c) $T \in \Phi[\mathcal{X}]$ *if and only if* $T^* \in \Phi[\mathcal{X}^*]$.

Proof. This follows from Definition A.15 since, according to Remark A.2(b,c),

$$\mathcal{R}(T^*)^- = \mathcal{R}(T^*) \quad \Longleftrightarrow \quad \mathcal{R}(T)^- = \mathcal{R}(T),$$
$$\dim \mathcal{N}(T^*) < \infty \quad \Longleftrightarrow \quad \text{codim}\,\mathcal{R}(T) < \infty,$$
$$\dim \mathcal{N}(T) \quad = \quad \dim \mathcal{N}(T^{**}). \qquad \square$$

Proposition A.18. *Let $A, T \in \mathcal{B}[\mathcal{X}]$ be operators on a Banach space $\mathcal{X}$.*

(a) *If $A, T \in \Phi_-[\mathcal{X}]$, then $AT \in \Phi_-[\mathcal{X}]$.*

(b) *If $A, T \in \Phi_+[\mathcal{X}]$, then $AT \in \Phi_+[\mathcal{X}]$.*

(c) *If $A, T \in \Phi[\mathcal{X}]$, then $AT \in \Phi[\mathcal{X}]$.*

Proof. (a) If $A, T \in \Phi_-[\mathcal{X}]$, then $\text{codim}\,\mathcal{R}(A) < \infty$ and $\text{codim}\,\mathcal{R}(T) < \infty$ (Definition A.15). Thus there exist subspaces $\mathcal{M}$ and $\mathcal{N}$ of $\mathcal{X}$ such that $\dim \mathcal{M} < \infty$ and $\dim \mathcal{N} < \infty$ for which $\mathcal{X} = \mathcal{R}(A) + \mathcal{M}$ and $\mathcal{X} = \mathcal{R}(T) + \mathcal{N}$. Therefore

$$\mathcal{X} = A(\mathcal{X}) + \mathcal{M} = A(\mathcal{R}(T) + \mathcal{N}) + \mathcal{M}$$
$$= A(\mathcal{R}(T)) + A(\mathcal{N}) + \mathcal{M} = \mathcal{R}(AT) + A(\mathcal{N}) + \mathcal{M} = \mathcal{R}(AT) + \mathcal{R}$$

with $\mathcal{R} = A(\mathcal{N}) + \mathcal{M}$. Since $\dim \mathcal{N} < \infty$, then $\dim A(\mathcal{N}) < \infty$ because A is linear. Thus $\dim \mathcal{R} < \infty$ because $\dim \mathcal{M} < \infty$. Hence $\text{codim}\,\mathcal{R}(AT) < \infty$ by Remark A.1(a), and so $AT \in \Phi_-[\mathcal{X}]$ by Definition A.15.

(b) Dually, if $A, T \in \Phi_+[\mathcal{X}]$, then $A^*, T^* \subset \Phi_-[\mathcal{X}^*]$ by Proposition A.17 so that $T^*A^* \in \Phi_-[\mathcal{X}^*]$ by item (a) or, equivalently, $(AT)^* \in \Phi_-[\mathcal{X}^*]$ and so $AT \in \Phi_+[\mathcal{X}]$ applying Proposition A.17 again.

(c) Assertions (a) and (b) imply assertion (c) according to Definition A.15. $\square$

Proposition A.19. *Let $A, T \in \mathcal{B}[\mathcal{X}]$ be operators on a Banach space $\mathcal{X}$.*

(a) *If $AT \in \Phi_-[\mathcal{X}]$, then $A \in \Phi_-[\mathcal{X}]$.*

(b) *If $AT \in \Phi_+[\mathcal{X}]$, then $T \in \Phi_+[\mathcal{X}]$.*

(c) *If $AT \in \Phi[\mathcal{X}]$, then $A \in \Phi_-[\mathcal{X}]$ and $T \in \Phi_+[\mathcal{X}]$.*

Proof. (a) If $AT \in \Phi_-[\mathcal{X}]$, then $\text{codim}\,\mathcal{R}(AT) < \infty$ by Definition A.15. Since $\mathcal{R}(AT) \subseteq \mathcal{R}(A)$ we get $\dim \mathcal{R}(AT) \leq \dim \mathcal{R}(A)$ and therefore $\text{codim}\,\mathcal{R}(A) \leq \text{codim}\,\mathcal{R}(AT)$. Hence $\text{codim}\,\mathcal{R}(A) < \infty$ and so $A \in \Phi_-[\mathcal{X}]$ by Definition A.15.

(b) Dually, if $A, T \in \Phi_+[\mathcal{X}]$, then $AT \in \Phi_+[\mathcal{X}]$ by Proposition A.18. Hence $T^*A^* = (AT)^* \in \Phi_-[\mathcal{X}^*]$ by Proposition A.17. Thus $T^* \in \Phi_-[\mathcal{X}^*]$ by item (a), and so $T \in \Phi_+[\mathcal{X}]$ by Proposition A.17.

(c) Apply assertions (a) and (b) to get (c) by the definition of $\Phi[\mathcal{X}]$. $\square$

Next consider the classes $\mathcal{F}_\ell[\mathcal{X}]$, $\mathcal{F}_r[\mathcal{X}]$, $\mathcal{F}[\mathcal{X}]$ of operators from $\mathcal{B}[\mathcal{X}]$ as defined in Section 5.1, now for Banach-space operators.

Definition A.20. Let $\mathcal{X}$ be a Banach space. The set

$$\mathcal{F}_\ell[\mathcal{X}] = \{T \in \mathcal{B}[\mathcal{X}] \colon T \text{ is left essentially invertible}\}$$
$$= \{T \in \mathcal{B}[\mathcal{X}] \colon AT = I + K \text{ for some } A \in \mathcal{B}[\mathcal{X}] \text{ and some } K \in \mathcal{B}_\infty[\mathcal{X}]\}$$

is the class of *left semi-Fredholm* operators from $\mathcal{B}[\mathcal{X}]$, while

$$\mathcal{F}_r[\mathcal{X}] = \{T \in \mathcal{B}[\mathcal{X}] \colon T \text{ is right essentially invertible}\}$$
$$= \{T \in \mathcal{B}[\mathcal{X}] \colon TA = I + K \text{ for some } A \in \mathcal{B}[\mathcal{X}] \text{ and some } K \in \mathcal{B}_\infty[\mathcal{H}]\}$$

is the class of *right semi-Fredholm* operators from $\mathcal{B}[\mathcal{X}]$, and their intersection

$$\mathcal{F}[\mathcal{X}] = \mathcal{F}_\ell[\mathcal{X}] \cap \mathcal{F}_r[\mathcal{X}] = \{T \in \mathcal{B}[\mathcal{X}] \colon T \text{ is essentially invertible}\}$$

is the class of *Fredholm* operators from $\mathcal{B}[\mathcal{X}]$.

Definitions of $\mathcal{F}_\ell[\mathcal{X}]$ and $\mathcal{F}_r[\mathcal{X}]$ can be equivalently stated if "compact" is replaced with "finite-rank" above (see, e.g., [7, Remark 3.3.3] or [95, Theorems 16.14 and 16.15] — cf. Remark 5.3(c) and Proposition 5.C).

The sets $\mathcal{F}_\ell[\mathcal{X}]$ and $\mathcal{F}_r[\mathcal{X}]$ are open in $\mathcal{B}[\mathcal{X}]$ since they are inverse images under the quotient (continuous) map $\pi \colon \mathcal{B}[\mathcal{X}] \to \mathcal{B}[\mathcal{X}]/\mathcal{B}_\infty[\mathcal{X}]$ of the left and right invertible elements in the Calkin algebra $\mathcal{B}[\mathcal{X}]/\mathcal{B}_\infty[\mathcal{X}]$ of $\mathcal{B}[\mathcal{X}]$ modulo the ideal $\mathcal{B}_\infty[\mathcal{X}]$ (see, e.g., [30, Proposition XI.2.6]) — cf. Proposition 5.A.

Proposition A.21. *Let $T \in \mathcal{B}[\mathcal{X}]$ be an operator on a Banach space $\mathcal{X}$ and let $T^* \in \mathcal{B}[\mathcal{X}^*]$ be its normed-space adjoint on the Banach space $\mathcal{X}^*$.*

(a) *If $T \in \mathcal{F}_r[\mathcal{X}]$, then $T^* \in \mathcal{F}_\ell[\mathcal{X}^*]$.*

(b) *If $T \in \mathcal{F}_\ell[\mathcal{X}]$, then $T^* \in \mathcal{F}_r[\mathcal{X}^*]$.*

(c) *If $T \in \mathcal{F}[\mathcal{X}]$, then $T^* \in \mathcal{F}[\mathcal{X}^*]$.*

The converses hold if $\mathcal{X}$ is reflexive.

Proof. The implications follow from Definition A.20 since $(AT)^* = T^*A^*$, $(TA)^* = A^*T^*$, $(I+K)^* = I^* + K^*$, and K^* is compact whenever K is (this is Schauder's Theorem — see, e.g., [93, Theorem 3.4.15]). The converses hold if $\mathcal{X}$ is reflexive, when $T^{**} = \Theta T \Theta^{-1}$ where Θ is the *natural embedding of $\mathcal{X}$ into its second dual* $\mathcal{X}^{**}$ (see, e.g., [93, Proposition 3.1.13]), which is surjective and thus an isometric isomorphism of $\mathcal{X}$ onto $\mathcal{X}^{**}$ whenever $\mathcal{X}$ is reflexive, and $\mathcal{B}_\infty[\mathcal{X}]$ is an ideal of $\mathcal{B}[\mathcal{X}]$. $\qquad\square$

Left and right and upper and lower semi-Fredholm operators are linked by range and kernel complementation as follows (see, e.g., [95, Theorems 16.14 and 16.15]): $T \in \mathcal{F}_\ell[\mathcal{X}]$ *if and only if* $T \in \Phi_+[\mathcal{X}]$ *and* $\mathcal{R}(T)$ *is complemented*; $T \in \mathcal{F}_r[\mathcal{X}]$ *if and only if* $T \in \Phi_-[\mathcal{X}]$ *and* $\mathcal{N}(T)$ *is complemented*.

Theorem A.22. *Let $\mathcal{X}$ be a Banach space. Then*

$$\mathcal{F}_\ell[\mathcal{X}] = \Phi_+[\mathcal{X}] \cap \Gamma_R[\mathcal{X}]$$
$$= \{T \in \Phi_+[\mathcal{X}]: \mathcal{R}(T) \text{ is a complemented subspace of } \mathcal{X}\},$$
$$\mathcal{F}_r[\mathcal{X}] = \Phi_-[\mathcal{X}] \cap \Gamma_N[\mathcal{X}]$$
$$= \{T \in \Phi_-[\mathcal{X}]: \mathcal{N}(T) \text{ is a complemented subspace of } \mathcal{X}\}.$$

Proof. (a) Suppose $T \in \Phi_+[\mathcal{X}]$. Then $\mathcal{R}(T)$ is closed and $\dim \mathcal{N}(T) < \infty$. Suppose $T \in \Gamma_R[\mathcal{X}]$ and take the projection $E \in \mathcal{B}[\mathcal{X}]$ with $\mathcal{R}(E) = \mathcal{R}(T)$ (Remark A.4). But $\dim \mathcal{N}(T) < \infty$ implies $T \in \Gamma_N[\mathcal{X}]$ (Remark A.5(a), also Proposition A.12(a)). Then $\mathcal{X} = \mathcal{N}(T) + \mathcal{M}$ for some *subspace* $\mathcal{M}$ of $\mathcal{X}$ for which $\mathcal{N}(T) \cap \mathcal{M} = \{0\}$, and the restriction $T|_\mathcal{M} \in \mathcal{B}[\mathcal{M}, \mathcal{R}(T)]$ is injective and surjective. Thus take its inverse $(T|_\mathcal{M})^{-1} \in \mathcal{B}[\mathcal{R}(T), \mathcal{M}]$ and set $A = (T|_\mathcal{M})^{-1}E$ in $\mathcal{B}[\mathcal{X}]$ (with $\mathcal{R}(A) = \mathcal{M}$). Hence $AT = (T|_\mathcal{M})^{-1}ET = (T|_\mathcal{M})^{-1}E|_{\mathcal{R}(T)}T = (T|_\mathcal{M})^{-1}T: \mathcal{X} \to \mathcal{M}$ so that $AT|_\mathcal{M} = I \in \mathcal{B}[\mathcal{M}]$. Then $(I - AT)|_\mathcal{M} = O$ and so $K = (I - AT) \in \mathcal{B}[\mathcal{X}]$ is finite-rank (since $\dim \mathcal{N}(T) < \infty$), thus compact. Therefore $AT = I + (-K)$, which means $T \in \mathcal{F}_\ell[\mathcal{X}]$ by Definition A.20. Hence

$$\Phi_+[\mathcal{X}] \cap \Gamma_R[\mathcal{X}] \subseteq \mathcal{F}_\ell[\mathcal{X}].$$

(b) Suppose $T \in \mathcal{F}_\ell$. Thus there exists $A \in \mathcal{B}[\mathcal{X}]$ such that $AT = I - K$ for some $K \in \mathcal{B}_\infty[\mathcal{X}]$ (Definition A.20). By the Fredholm Alternative in Remark A.2(d) we get $\dim \mathcal{N}(I - K) = \operatorname{codim} \mathcal{R}(I - K) < \infty$ and so $AT \in \Phi[\mathcal{X}]$ (Definition A.15). Hence (Proposition A.19(c))

$$T \in \Phi_+[\mathcal{X}].$$

Actually, the operator $I - K$ is Weyl as in Remark 5.7(b,ii) — see, e.g., [111, Theorem V.7.11]. However, as in Proposition 5.K (an extension of the Fredholm Alternative), $I - K$ is in fact Browder (recall: Browder operators are Weyl operators), meaning $\operatorname{asc}(I - K) = \operatorname{dsc}(I - K) < \infty$, which clearly survives from Hilbert to Banach spaces — as happens with the entire Section 5.4 (see, e.g., [111, Theorem V.7.9]). Therefore there exists an integer $n \geq 0$ such that $\mathcal{R}((AT)^n) = \mathcal{R}((AT)^{n+1})$. Thus set

$$\mathcal{M} = \mathcal{R}((AT)^n) \quad \text{so that} \quad AT(\mathcal{M}) = \mathcal{M}.$$

Since $AT \in \Phi[\mathcal{X}]$ we get $(AT)^n \in \Phi[\mathcal{X}]$ (Proposition A.18(c)), and so

$$\mathcal{M} \text{ is a complemented subspace of } \mathcal{X} \text{ with } \operatorname{codim} \mathcal{M} < \infty$$

(Corollary A.9). Hence $\mathcal{X} = \mathcal{M} + \mathcal{N}$ with $\mathcal{M} \cap \mathcal{N} = \{0\}$ for some subspace $\mathcal{N}$ of $\mathcal{X}$ with $\dim \mathcal{N} < \infty$. Let $E \in \mathcal{B}[\mathcal{X}]$ be the (continuous) projection for which $\mathcal{R}(E) = \mathcal{M}$ and $\mathcal{N}(E) = \mathcal{N}$ (Remark A.4) and consider the operator

$$EAT|_\mathcal{M} = E(I - K)|_\mathcal{M} = (E|_\mathcal{M} - EK|_\mathcal{M}) = I - K' \in \mathcal{B}[\mathcal{M}]$$

since $\mathcal{R}(E) = \mathcal{M}$, where the first identity operator acts on $\mathcal{X}$ and the second on $\mathcal{M}$, with $K' = EK|_\mathcal{M}$ (which is compact since $\mathcal{B}_\infty[\mathcal{M}]$ is an ideal of $\mathcal{B}[\mathcal{M}]$ and restriction of compact is compact). Moreover, since $AT(\mathcal{M}) = \mathcal{M} = \mathcal{R}(E)$,

$$\mathcal{R}(I - K') = \mathcal{M}.$$

By the Fredholm Alternative (Remark A.2(d)) we get $1 \in \rho(K)$ and so $I - K'$ is invertible (indeed, if $1 \in \sigma_{P_4}(K)$, then $\mathcal{R}(I - K') \neq \mathcal{M}$). Therefore

$$I = (I - K')^{-1}(I - K') = (I - K')^{-1}EAT|_{\mathcal{M}}.$$

Thus $T|_{\mathcal{M}} \in \mathcal{B}[\mathcal{X}]$ is left invertible, and so $T(\mathcal{M}) = \mathcal{R}(T|_{\mathcal{M}})$ is a complemented subspace of $\mathcal{X}$ (Proposition A.12(d)). Moreover, since $\mathcal{X} = \mathcal{M} + \mathcal{N}$ with $\mathcal{M} \cap \mathcal{N} = \{0\}$, every $x \in \mathcal{X}$ has a unique representation as $x = u + v$ with $u \in \mathcal{M}$ and $v \in \mathcal{N}$, and so $\mathcal{R}(T) = T(\mathcal{X}) = T(\mathcal{M}) + T(\mathcal{N})$. Furthermore, since $\dim \mathcal{N} < \infty$, then $\dim T(\mathcal{N}) < \infty$. Summing up:

$$\mathcal{R}(T) = T(\mathcal{M}) + T(\mathcal{N})$$

is the sum of a complemented subspace and a finite-dimensional subspace. Since $T(\mathcal{M})$ is a subspace and $T(\mathcal{N})$ is finite-dimensional, then $\mathcal{R}(T)$ is a subspace (Proposition A.6). Moreover, since $T(\mathcal{M})$ is complemented, and $T(\mathcal{N})$ is finite-dimensional, then $\mathcal{R}(T)$ is complemented (Proposition A.7). Therefore

$$\mathcal{F}_\ell[\mathcal{X}] \subseteq \Phi_+[\mathcal{X}] \cap \Gamma_R[\mathcal{X}].$$

(c) By the inclusions in (a) and (b) we get the first identity. The proof of the second identity is similar, following a symmetric argument. (However, for reflexive Banach spaces the second identity follows at once from the first one through Proposition A.13 and Propositions A.17 and A.21.) $\square$

In view of Theorem A.22, operators in $\mathcal{F}_\ell[\mathcal{X}]$ and $\mathcal{F}_r[\mathcal{X}]$ are also referred to as *Atkinson operators* [50, Theorem 2.3] (left and right respectively).

Corollary A.23. *Let $\mathcal{X}$ be a Banach space. Then*

$$\mathcal{F}[\mathcal{X}] = \Phi[\mathcal{X}].$$

Proof. $\mathcal{F}[\mathcal{X}] = \Phi[\mathcal{X}] \cap \Gamma[\mathcal{X}] = \Phi[\mathcal{X}]$ (Theorem A.22 and Proposition A.16). $\square$

Remark A.24. FURTHER PROPERTIES. Corollary A.23 justifies the same terminology, *Fredholm operators*, for the classes $\mathcal{F}[\mathcal{X}]$ and $\Phi[\mathcal{X}]$. Further immediate consequences of Theorem A.22 and Proposition A.16 read as follows.

(a) $\mathcal{F}_\ell[\mathcal{X}] \cup \mathcal{F}_r[\mathcal{X}] = (\Phi_+[\mathcal{X}] \cup \Phi_-[\mathcal{X}]) \cap \Gamma[\mathcal{X}],$

$\quad \mathcal{F}_\ell[\mathcal{X}] \backslash \mathcal{F}_r[\mathcal{X}] = (\Phi_+[\mathcal{X}] \backslash \Phi_-[\mathcal{X}]) \cap \Gamma_R[\mathcal{X}] = (\Phi_+[\mathcal{X}] \backslash \Phi_-[\mathcal{X}]) \cap \Gamma[\mathcal{X}],$

$\quad \mathcal{F}_r[\mathcal{X}] \backslash \mathcal{F}_\ell[\mathcal{X}] = (\Phi_-[\mathcal{X}] \backslash \Phi_+[\mathcal{X}]) \cap \Gamma_N[\mathcal{X}] = (\Phi_-[\mathcal{X}] \backslash \Phi_+[\mathcal{X}]) \cap \Gamma[\mathcal{X}].$

(b) If $\Phi_+[\mathcal{X}] = \Phi_-[\mathcal{X}]$, then $\mathcal{F}_\ell[\mathcal{X}] = \mathcal{F}_r[\mathcal{X}] = \Phi_+[\mathcal{X}] = \Phi_-[\mathcal{X}]$.

(c) If $\Phi_+[\mathcal{X}] \subseteq \Gamma_R[\mathcal{X}]$ and $\Phi_-[\mathcal{X}] \subseteq \Gamma_N[\mathcal{X}]$,

$\qquad\qquad$ then $\Phi_+[\mathcal{X}] = \mathcal{F}_\ell[\mathcal{X}]$ and $\Phi_-[\mathcal{X}] = \mathcal{F}_r[\mathcal{X}]$.

In particular,

(d) if $\Gamma_R[\mathcal{X}] = \Gamma_N[\mathcal{X}]$, then $\Phi_+[\mathcal{X}] = \mathcal{F}_\ell[\mathcal{X}]$ and $\Phi_-[\mathcal{X}] = \mathcal{F}_r[\mathcal{X}]$.

In this case the classes $\Gamma_R[\mathcal{X}]$ and $\Gamma_N[\mathcal{X}]$ play no role in Theorem A.22.

Corollary A.25. *Let $\mathcal{X}$ be a Banach space and take $T \in \mathcal{B}[\mathcal{X}]$.*

(a) *If $T \in \Gamma_R[\mathcal{X}]$ is bounded below (or T satisfies any of the equivalent conditions in Theorem 1.2) and $K \in \mathcal{B}[\mathcal{X}]$ is compact, then*

$$\mathcal{R}(T + K) \text{ is closed} \quad \text{and} \quad T + K \in \Gamma[\mathcal{X}] = \Gamma_R[\mathcal{X}] \cap \Gamma_N[\mathcal{X}].$$

(b) *In particular, if $G \in \mathcal{G}[\mathcal{X}]$ and $K \in \mathcal{B}[\mathcal{X}]$ is compact, then*

$$\mathcal{R}(G + K) \text{ is closed} \quad \text{and} \quad G + K \in \Gamma[\mathcal{X}] = \Gamma_R[\mathcal{X}] \cap \Gamma_N[\mathcal{X}].$$

(c) *More particularly, for any compact $K \in \mathcal{B}[\mathcal{X}]$ and any nonzero scalar λ,*

$$\mathcal{R}(\lambda I - K) \text{ is closed} \quad \text{and} \quad \lambda I - K \in \Gamma[\mathcal{X}] = \Gamma_R[\mathcal{X}] \cap \Gamma_N[\mathcal{X}].$$

Proof. (a) Suppose $T \in \mathcal{B}[\mathcal{X}]$ is bounded below. Equivalently, suppose T is injective with a closed range (Theorem 1.2). Moreover, suppose $T \in \Gamma_R[\mathcal{X}]$. Thus T has a left inverse in $\mathcal{B}[\mathcal{X}]$ (Proposition A.12(d)), and therefore $T \in \mathcal{F}_\ell[\mathcal{X}]$ (Definition A.20). But $\mathcal{F}_\ell[\mathcal{X}]$ is invariant under compact perturbation (as in Theorem 5.6). Thus $T + K \in \mathcal{F}_\ell[\mathcal{X}]$ for every compact $K \in \mathcal{B}[\mathcal{X}]$, and so $T + K \in \Gamma_R[\mathcal{X}] \cap \Phi_+[\mathcal{X}]$ (Theorem A.22). Since $\Phi_+[\mathcal{X}] \subseteq \Gamma_N[\mathcal{X}]$ (Proposition A.16), $T + K \in \Gamma_R[\mathcal{X}] \cap \Gamma_N[\mathcal{X}]$. Since $T + K \in \Phi_+[\mathcal{X}]$, $\mathcal{R}(T + K)$ is closed (Definition A.15). (For a proof without using Theorem A.22, see [67, Theorem 2].)

(b) Invertible operators are trivially bounded below with complemented range.

(c) Set $G = \lambda I \neq O$ in (b), or use the Fredholm Alternative of Remark A.2(d) together with Proposition A.12(a,b). (For $\lambda = 0$ the result in (c) still holds if the Banach space $\mathcal{X}$ is reflexive with a Schauder basis, by Theorem A.14.) □

A.5 Essential Spectrum and Spectral Picture Again

Let T be an operator on a Banach space $\mathcal{X}$. Consider the characterizations in Corollaries 5.11 and 5.12, which equally hold in a Banach-space setting.

Definition A.26. For each $T \in \mathcal{B}[\mathcal{X}]$ let

$$\sigma_{\ell e}(T) = \{\lambda \in \mathbb{C} \colon \lambda I - T \in \mathcal{B}[\mathcal{X}] \backslash \mathcal{F}_\ell[\mathcal{X}]\}$$
$$= \{\lambda \in \mathbb{C} \colon \lambda I - T \text{ is not left essentially invertible}\}$$

be the *left essential spectrum* (or *left semi-Fredholm spectrum*) of T, and

$$\sigma_{re}(T) = \{\lambda \in \mathbb{C} \colon \lambda I - T \in \mathcal{B}[\mathcal{X}] \backslash \mathcal{F}_r[\mathcal{X}]\}$$
$$= \{\lambda \in \mathbb{C} \colon \lambda I - T \text{ is not right essentially invertible}\}$$

be the *right essential spectrum* (or *right semi-Fredholm spectrum*) of T. Let

$$\sigma_e(T) = \sigma_{\ell e}(T) \cup \sigma_{re}(T)$$
$$= \{\lambda \in \mathbb{C} \colon \lambda I - T \in \mathcal{B}[\mathcal{X}] \backslash (\mathcal{F}_\ell[\mathcal{X}] \cap \mathcal{F}_r[\mathcal{X}])\}$$
$$= \{\lambda \in \mathbb{C} \colon \lambda I - T \in \mathcal{B}[\mathcal{X}] \backslash \mathcal{F}[\mathcal{X}]\}$$

be the *essential spectrum* (or *Fredholm spectrum*) of T.

As in Section 5.2, $\sigma_{\ell e}(T)$ and $\sigma_{re}(T)$ are the left and right spectra of the natural image $\pi(T)$ of T in the Calkin algebra $\mathcal{B}[\mathcal{X}]/\mathcal{B}_\infty[\mathcal{X}]$, and so $\sigma_{\ell e}(T)$, $\sigma_{re}(T)$, and $\sigma_e(T)$ are compact subsets of the spectrum $\sigma(T)$. Similarly, corresponding to the classes $\Phi_+[\mathcal{X}]$, $\Phi_-[\mathcal{X}]$, and $\Phi[\mathcal{X}]$, define analogous sets.

Definition A.27. For each $T \in \mathcal{B}[\mathcal{X}]$ let

$$\sigma_{e_+}(T) = \{\lambda \in \mathbb{C}: \lambda I - T \in \mathcal{B}[\mathcal{X}]\backslash\Phi_+[\mathcal{X}]\}$$
$$= \{\lambda \in \mathbb{C}: \mathcal{R}(\lambda I - T) \text{ is not closed or } \dim \mathcal{N}(\lambda I - T) = \infty\}$$

be the *upper semi-Fredholm spectrum* of T, and

$$\sigma_{e_-}(T) = \{\lambda \in \mathbb{C}: \lambda I - T \in \mathcal{B}[\mathcal{X}]\backslash\Phi_-[\mathcal{X}]\}$$
$$= \{\lambda \in \mathbb{C}: \mathcal{R}(\lambda I - T) \text{ is not closed or } \dim \mathcal{X}/\mathcal{R}(\lambda I - T) = \infty\}$$
$$= \{\lambda \in \mathbb{C}: \operatorname{codim} \mathcal{R}(\lambda I - T) = \infty\}$$

be the *lower semi-Fredholm spectrum* of T. Let

$$\sigma_e(T) = \sigma_{e_+}(T) \cup \sigma_{e_-}(T)$$
$$= \{\lambda \in \mathbb{C}: \lambda I - T \in \mathcal{B}[\mathcal{X}]\backslash(\Phi_+[\mathcal{X}] \cap \Phi_-[\mathcal{X}])\}$$
$$= \{\lambda \in \mathbb{C}: \lambda I - T \in \mathcal{B}[\mathcal{X}]\backslash\Phi[\mathcal{X}]\}$$

be the *essential spectrum* (or *Fredholm spectrum*) of T.

Since $\mathcal{F}[\mathcal{X}] = \Phi[\mathcal{X}]$ (Corollary A.23), there is no ambiguity in the definition of the essential spectrum. Indeed,

$$\sigma_e(T) = \sigma_{\ell e}(T) \cup \sigma_{re}(T) = \sigma_{e_+}(T) \cup \sigma_{e_-}(T).$$

Also $\sigma_{e_+}(T)$ and $\sigma_{e_-}(T)$ are compact subsets of $\sigma(T)$ (see, e.g., [95, Proposition 6.2 and Theorems 16.7 and 16.11]). Corresponding to the classes $\Gamma_R[\mathcal{X}]$, $\Gamma_N[\mathcal{X}]$, and $\Gamma[\mathcal{X}]$, define the following subsets of $\mathbb{C}$ (actually, subsets of $\sigma(T)$).

Definition A.28. For each operator $T \in \mathcal{B}[\mathcal{X}]$, consider the following sets.

$$\vartheta_R(T) = \{\lambda \in \mathbb{C}: \lambda I - T \in \mathcal{B}[\mathcal{X}]\backslash\Gamma_R[\mathcal{X}]\}$$
$$= \{\lambda \in \mathbb{C}: \mathcal{R}(\lambda I - T)^- \text{ is not complemented}\},$$

$$\vartheta_N(T) = \{\lambda \in \mathbb{C}: \lambda I - T \in \mathcal{B}[\mathcal{X}]\backslash\Gamma_N[\mathcal{X}]\}$$
$$= \{\lambda \in \mathbb{C}: \mathcal{N}(\lambda I - T) \text{ is not complemented}\},$$

$$\vartheta(T) = \vartheta_R(T) \cup \vartheta_N(T)$$
$$= \{\lambda \in \mathbb{C}: \lambda I - T \in \mathcal{B}[\mathcal{X}]\backslash\Gamma_R[\mathcal{X}] \cap \Gamma_N[\mathcal{X}]\}$$
$$= \{\lambda \in \mathbb{C}: \lambda I - T \in \mathcal{B}[\mathcal{X}]\backslash\Gamma[\mathcal{X}]\}$$
$$= \{\lambda \in \mathbb{C}: \mathcal{R}(\lambda I - T)^- \text{ or } \mathcal{N}(\lambda I - T) \text{ is not complemented}\}.$$

According to Theorem A.22, left-right essential spectra and upper-lower semi-Fredholm spectra are related as follows.

Corollary A.29. *Let T be an operator on a Banach space $\mathcal{X}$.*

$$\sigma_{\ell e}(T) = \sigma_{e_+}(T) \cup \vartheta_R(T)$$
$$= \{\lambda \in \mathbb{C}: \mathcal{R}(\lambda I - T) \text{ is not closed, or } \dim \mathcal{N}(\lambda I - T) = \infty,$$
$$\text{or } \mathcal{R}(\lambda I - T) \text{ is not a complemented subspace}\}.$$

$$\sigma_{re}(T) = \sigma_{e_-}(T) \cup \vartheta_N(T)$$
$$= \{\lambda \in \mathbb{C}: \mathcal{R}(\lambda I - T) \text{ is not closed, or } \dim \mathcal{X}/\mathcal{R}(\lambda I - T) = \infty,$$
$$\text{or } \mathcal{N}(\lambda I - T) \text{ is not a complemented subspace}\}.$$

Proof. Straightforward by Definitions A.26, A.27, A.28 and Theorem A.22:

$$\sigma_{\ell e}(T) = \{\lambda \in \mathbb{C}: \lambda I - T \notin \mathcal{F}_\ell[\mathcal{X}]\}$$
$$= \{\lambda \in \mathbb{C}: \lambda I - T \notin (\Phi_+[\mathcal{X}] \cap \Gamma_R[\mathcal{X}])\}$$
$$= \{\lambda \in \mathbb{C}: \lambda I - T \notin \Phi_+[\mathcal{X}] \text{ or } \lambda I - T \notin \Gamma_R[\mathcal{X}]\}$$
$$= \sigma_{e_+}(T) \cup \vartheta_R(T),$$

$$\sigma_{re}(T) = \{\lambda \in \mathbb{C}: \lambda I - T \notin \mathcal{F}_r[\mathcal{X}]\}$$
$$= \{\lambda \in \mathbb{C}: \lambda I - T \notin (\Phi_-[\mathcal{X}] \cap \Gamma_N[\mathcal{X}])\}$$
$$= \{\lambda \in \mathbb{C}: \lambda I - T \notin \Phi_-[\mathcal{X}] \text{ or } \lambda I - T \notin \Gamma_N[\mathcal{X}]\}$$
$$= \sigma_{e_-}(T) \cup \vartheta_N(T). \qquad \square$$

The next proposition is the complement of Proposition A.16.

Proposition A.30. *The following inclusions hold for every $T \in \mathcal{B}[\mathcal{X}]$.*

(a) $\vartheta_N(T) \subseteq \sigma_{o_+}(T)$

(b) $\vartheta_R(T) \subseteq \sigma_{e_-}(T)$.

(c) $\vartheta(T) \subseteq \sigma_e(T)$.

Proof. Since $\Phi_+[\mathcal{X}] \subseteq \Gamma_N[\mathcal{X}]$ and $\Phi_-[\mathcal{X}] \subseteq \Gamma_R[\mathcal{X}]$ (Proposition A.16), we get $\mathcal{B}[\mathcal{X}]\backslash\Gamma_N[\mathcal{X}] \subseteq \mathcal{B}[\mathcal{X}]\backslash\Phi_+[\mathcal{X}]$ and $\mathcal{B}[\mathcal{X}]\backslash\Gamma_R[\mathcal{X}] \subseteq \mathcal{B}[\mathcal{X}]\backslash\Phi_-[\mathcal{X}]$. Then (Definitions A.27 and A.28) $\vartheta_N(T) \subseteq \sigma_{e_+}(T)$ and $\vartheta_R(T) \subseteq \sigma_{e_-}(T)$. $\qquad \square$

Remark A.31. ADDITIONAL PROPERTIES. These are consequences of Corollary A.29 and Proposition A.30, and constitute the complement of Remark A.24.

(a) $\sigma_{\ell e}(T) \cap \sigma_{re}(T) = (\sigma_{e_+}(T) \cap \sigma_{e_-}(T)) \cup \vartheta(T)$,

$\sigma_{\ell e}(T)\backslash\sigma_{re}(T) = (\sigma_{e_+}(T)\backslash\sigma_{e_-}(T))\backslash\vartheta_N(T) = (\sigma_{e_+}(T)\backslash\sigma_{e_-}(T))\backslash\vartheta(T)$,

$\sigma_{re}(T)\backslash\sigma_{\ell e}(T) = (\sigma_{e_-}(T)\backslash\sigma_{e_+}(T))\backslash\vartheta_R(T) = (\sigma_{e_-}(T)\backslash\sigma_{e_+}(T))\backslash\vartheta(T)$.

(b) If $\sigma_{e_-}(T) = \sigma_{e_+}(T)$, then $\sigma_{\ell e}(T) = \sigma_{re}(T) = \sigma_{e_-}(T) = \sigma_{e_+}(T)$.

(c) If $\vartheta_R(T) \subseteq \sigma_{e_+}(T)$ and $\vartheta_N(T) \subseteq \sigma_{e_-}(T)$

$$\text{then } \sigma_{e_+}(T) = \sigma_{\ell e}(T) \text{ and } \sigma_{e_-}(T) = \sigma_{re}(T).$$

In particular,

(d) If $\vartheta_R(T) = \vartheta_N(T)$, then $\sigma_{e_+}(T) = \sigma_{\ell e}(T)$ and $\sigma_{e_-}(T) = \sigma_{re}(T)$.

In this case $\vartheta_R(T)$ and $\vartheta_N(T)$ play no role in Corollary A.29.

Actually, as we will see next, if $\vartheta_R(T)$ and $\vartheta_N(T)$ are open subsets of the complex plane, then they play no role in Corollary A.29 either.

Corollary A.32. *Take $T \in \mathcal{B}[\mathcal{X}]$ on a Banach space $\mathcal{X}$.*

(a) *If $\Gamma_R[\mathcal{X}]$ is open (closed) in $\mathcal{B}[\mathcal{X}]$, then $\vartheta_R(T)$ is closed (open) in $\mathbb{C}$.*
   If $\Gamma_N[\mathcal{X}]$ is open (closed) in $\mathcal{B}[\mathcal{X}]$, then $\vartheta_N(T)$ is closed (open) in $\mathbb{C}$.

(b) *If $\vartheta_R(T)$ is open in $\mathbb{C}$, then $\sigma_{e_+}(T) = \sigma_{\ell e}(T)$.*
   If $\vartheta_N(T)$ is open in $\mathbb{C}$, then $\sigma_{e_-}(T) = \sigma_{re}(T)$.

Proof. Let T be an operator on a Banach space $\mathcal{X}$.

(a) Item (a) follows by the complementary roles played between $\Gamma_R[\mathcal{X}]$ and $\vartheta_R(T)$, and between $\Gamma_N[\mathcal{X}]$ and $\vartheta_N(T)$. Indeed, as $\vartheta_R(T)$ is bounded (Proposition A.30), take an arbitrary $\lambda \in \mathbb{C}\backslash\vartheta_R(T)$. Thus $\lambda I - T \in \Gamma_R[\mathcal{X}]$ (Definition A.28). Suppose $\Gamma_R[\mathcal{X}]$ is nonempty and open in $\mathcal{B}[\mathcal{X}]$. Thus there exists $\varepsilon_\lambda > 0$ for which, if $A \in \mathcal{B}[\mathcal{X}]$ is such that $\|A - (\lambda I - T)\| < \varepsilon_\lambda$, then $A \in \Gamma_R[\mathcal{X}]$. So if $\zeta \in \mathbb{C}$ is such that $|\zeta - \lambda| < \varepsilon_\lambda$, then $\|(\zeta I - T) - (\lambda I - T)\| = |\zeta - \lambda| < \varepsilon_\lambda$, and hence $\zeta I - T \in \Gamma_R(T)$, which means $\zeta \in \mathbb{C}\backslash\vartheta_R(T)$. Therefore $\mathbb{C}\backslash\vartheta_R(T)$ is open (i.e., $\vartheta_R(T)$ is closed). Now suppose $\vartheta_R(T) \neq \varnothing$ (otherwise $\vartheta_R(T)$ is trivially open). Take an arbitrary $\lambda' \in \vartheta_R(T)$. Thus $\lambda' I - T \in \mathcal{B}[\mathcal{X}]\backslash\Gamma_R[\mathcal{X}]$ (Definition A.28). Suppose $\Gamma_R[\mathcal{X}]$ is closed in $\mathcal{B}[\mathcal{X}]$. Then $\mathcal{B}[\mathcal{X}]\backslash\Gamma_R[\mathcal{X}]$ is nonempty and open in $\mathcal{B}[\mathcal{X}]$. So there is an $\varepsilon_{\lambda'} > 0$ for which, if $A \in \mathcal{B}[\mathcal{X}]$ is such that $\|A - (\lambda'I - T)\| < \varepsilon_{\lambda'}$, then $A \in \mathcal{B}[\mathcal{X}]\backslash\Gamma_R[\mathcal{X}]$. Hence if $\zeta' \in \mathbb{C}$ is such that $|\zeta' - \lambda'| < \varepsilon_{\lambda'}$, then $\|(\zeta'I - T) - (\lambda'I - T)\| = |\zeta' - \lambda'| < \varepsilon_{\lambda'}$, which implies $(\zeta'I - T) \in \mathcal{B}[\mathcal{X}]\backslash\Gamma_R(T)$, and so $\zeta' \in \vartheta_R(T)$. Thus $\vartheta_R(T)$ is open. Similarly (applying the same argument), if $\Gamma_N[\mathcal{X}]$ is open (closed) in $\mathcal{B}[\mathcal{X}]$, then $\mathbb{C}\backslash\vartheta_N(T)$ is open (closed) in $\mathbb{C}$.

(b) By Corollary A.29, $\sigma_{\ell e}(T) = \sigma_{e_+}(T)\cup\vartheta_R(T) = \sigma_{e_+}(T)\cup(\vartheta_R(T)\backslash\sigma_{e_+}(T))$. Suppose $\sigma_{\ell e}(T) \neq \sigma_{e_+}(T)$. So $\vartheta_R(T) \not\subseteq \sigma_{e_+}(T)$, which means $\vartheta_R(T)\backslash\sigma_{e_+}(T) \neq \varnothing$. Since $\sigma_{\ell e}(T)$ and $\sigma_{e_+}(T)$ are closed, $\vartheta_R(T)\backslash\sigma_{e_+}(T)$ is not open. If $\vartheta_R(T)$ is open, then $\vartheta_R(T)\backslash\sigma_{e_+}(T) = \vartheta_R(T) \cap (\mathbb{C}\backslash\sigma_{e_+}(T))$ is open (because $\sigma_{e_+}(T)$ is closed), which is a contradiction. Thus $\vartheta_R(T)$ is not open. Similarly, since $\sigma_{re}(T) = \sigma_{e_-}(T)\cup\vartheta_N(T)$ (Corollary A.29) and $\sigma_{re}(T)$ and $\sigma_{e_-}(T)$ are closed, the same argument ensures $\vartheta_N(T)$ is not open whenever $\sigma_{re}(T) \neq \sigma_{e_-}(T)$. This concludes a contrapositive proof for item (b). $\qquad\square$

Remark. If $\Gamma_R[\mathcal{X}]$ and $\Gamma_N[\mathcal{X}]$ are closed, then $\vartheta_R(T)$ and $\vartheta_N(T)$ are open (by Corollary A.32(a)), which implies (by Corollary A.32(b))

$$\sigma_{e_+}(T) = \sigma_{\ell e}(T) \quad \text{and} \quad \sigma_{e_-}(T) = \sigma_{re}(T).$$

The sets $\Gamma_R[\mathcal{X}]$ and $\Gamma_N[\mathcal{X}]$ are not necessarily closed. For instance, there may be sequences of operators T_n in $\Gamma_R[\ell^p]$ converging uniformly to an operator T not in $\Gamma_R[\ell^p]$ for $1 < p \neq 2$ [25, Example 1], and so $\Gamma_R[\mathcal{X}]$ may not be closed. On the other hand, even though the sets $\Gamma_R[\mathcal{X}]$ and $\Gamma_N[\mathcal{X}]$ may not be open, they include the nonempty open sets $\Phi_+[\mathcal{X}]$ and $\Phi_-[\mathcal{X}]$ respectively (Proposition A.16), which in turn include the open group $\mathcal{G}[\mathcal{X}]$.

Consider again the sets $\mathbb{Z}$ of all integers and $\overline{\mathbb{Z}} = \mathbb{Z} \cup \{-\infty\} \cup \{+\infty\}$ of all extended integers. Take an operator $T \in \mathcal{B}[\mathcal{X}]$ on a Banach space $\mathcal{X}$. If $\mathcal{R}(T)$ is a subspace of $\mathcal{X}$ (i.e., if $\mathcal{R}(T)$ is closed in $\mathcal{X}$), and if one of $\dim \mathcal{N}(T)$ or $\operatorname{codim} \mathcal{R}(T)$ is finite, then the *Fredholm index* of T is defined in $\overline{\mathbb{Z}}$ by

$$\operatorname{ind}(T) = \dim \mathcal{N}(T) - \operatorname{codim} \mathcal{R}(T).$$

Thus $\operatorname{ind}(T)$ is well defined if and only if $T \in \Phi_+[\mathcal{X}] \cup \Phi_-[\mathcal{X}]$ (cf. Definition A.15). In particular, $\operatorname{ind}(T)$ is well defined if $T \in \mathcal{F}_\ell[\mathcal{X}] \cup \mathcal{F}_r[\mathcal{X}]$ (cf. Remark A.24(a)). If $\operatorname{ind}(T)$ is finite (i.e., if $\operatorname{ind}(T) \in \mathbb{Z}$), then $T \in \Phi_+[\mathcal{X}] \cap \Phi_-[\mathcal{X}] = \Phi[\mathcal{X}] = \mathcal{F}[\mathcal{X}] = \mathcal{F}_\ell[\mathcal{X}] \cap \mathcal{F}_r[\mathcal{X}]$ (i.e., then T is Fredholm). In particular, a *Weyl operator* is a Fredholm operator with null index, where

$$\mathcal{W}[\mathcal{X}] = \{T \in \Phi[\mathcal{X}] : \operatorname{ind}(T) = 0\}$$

denotes the class of all Weyl operators from $\mathcal{B}[\mathcal{X}]$. Compare with Chapter 5 (see Remarks 5.3(f) and 5.7(b) in Section 5.1). With respect to the essential spectrum $\sigma_{\ell e}(T) \cup \sigma_{re}(T) = \sigma_e(T) = \sigma_{e_+}(T) \cup \sigma_{e_-}(T)$, consider the set

$$
\begin{aligned}
\sigma_w(T) &= \{\lambda \in \mathbb{C} : \lambda I - T \in \mathcal{B}[\mathcal{X}] \backslash \mathcal{W}[\mathcal{X}]\} \\
&= \{\lambda \in \mathbb{C} : \lambda I - T \text{ is not a Fredholm operator of index zero}\} \\
&= \{\lambda \in \mathbb{C} : \lambda I - T \text{ is not Fredholm or is Fredholm of index not zero}\} \\
&= \{\lambda \in \mathbb{C} : \text{either } \lambda \in \sigma_e(T) \text{ or } \lambda I - T \in \Phi[\mathcal{X}] \text{ and } \operatorname{ind}(\lambda I - T) \neq 0\}.
\end{aligned}
$$

This is the *Weyl spectrum* of T. Since $\lambda I - T \in \Phi[\mathcal{X}]$ and $\operatorname{ind}(\lambda I - T) \neq 0$ if and only if $\mathcal{R}(\lambda I - T)$ is closed and $\dim \mathcal{N}(\lambda I - T) \neq \dim \mathcal{N}(\lambda I - T^*)$ with both dimensions finite (cf. Definition A.15 and Remark A.2(b)), then following the argument in the proof of Theorem 5.16(b) we get $\{\lambda \in \mathbb{C} : \lambda I - T \in \Phi[\mathcal{X}]$ and $\operatorname{ind}(\lambda I - T) \neq 0\} \subseteq \sigma(T)$. Hence $\sigma_w(T) \subseteq \sigma(T)$, and therefore

$$\sigma_e(T) \subseteq \sigma_w(T) \subseteq \sigma(T).$$

Let $\sigma_0(T)$ denote the complement of $\sigma_w(T)$ in $\sigma(T)$,

$$
\begin{aligned}
\sigma_0(T) &= \sigma(T) \backslash \sigma_w(T) \\
&= \{\lambda \in \sigma(T) : \lambda I - T \in \mathcal{W}[\mathcal{X}]\} \\
&= \{\lambda \in \sigma(T) : \lambda I - T \text{ is a Fredholm operator of index zero}\} \\
&= \{\lambda \in \sigma(T) : \lambda \notin \sigma_e(T) \text{ and } \operatorname{ind}(\lambda I - T) = 0\}.
\end{aligned}
$$

Moreover, for each nonzero (finite) integer $k \in \mathbb{Z} \backslash \{0\}$, set

$$\sigma_k(T) = \{\lambda \in \mathbb{C} : \lambda I - T \in \Phi[\mathcal{X}] \text{ and } \operatorname{ind}(\lambda I - T) = k \neq 0\},$$

which are open subsets of $\sigma(T)$, being the *holes of the essential spectrum* $\sigma_{e_+}(T) \cup \sigma_{e_-}(T) = \sigma_e(T) = \sigma_{\ell e}(T) \cup \sigma_{re}(T)$ (again by the same argument as in the proof of Theorem 5.16). Also set

$$\sigma_{+\infty}(T) = \sigma_{e_+}(T) \backslash \sigma_{e_-}(T) \quad \text{and} \quad \sigma_{-\infty}(T) = \sigma_{e_-}(T) \backslash \sigma_{e_+}(T).$$

These open sets are holes of $\sigma_{e_-}(T)$ and $\sigma_{e_+}(T)$, respectively (but are not holes of $\sigma_e(T)$), referred to as the *pseudoholes of* $\sigma_e(T)$. So for each nonzero extended integer $k \in \overline{\mathbb{Z}} \backslash \{0\}$ the sets $\sigma_k(T)$ are the holes of $\sigma_+(T) \cap \sigma_-(T)$ (Theorem 5.16 and Fig. § 5.2). The next partition of the spectrum is the *Spectral Picture* analogous to that given in Corollary 5.18. For $T \in \mathcal{B}[\mathcal{X}]$,

$$\sigma(T) = \sigma_e(T) \cup \bigcup\nolimits_{k \in \mathbb{Z}} \sigma_k(T),$$

$$\sigma_e(T) = \big(\sigma_{e_+}(T) \cap \sigma_{e_-}(T)\big) \cup \sigma_{+\infty}(T) \cup \sigma_{-\infty}(T),$$

and the Weyl spectrum is the union of the essential spectrum and all its holes

$$\sigma_w(T) = \sigma(T) \backslash \sigma_0(T) = \sigma_e(T) \cup \bigcup\nolimits_{k \in \mathbb{Z} \backslash \{0\}} \sigma_k(T),$$

as in Theorem 5.24 (Schechter Theorem).

Remark. $\sigma_e(T)$, $\sigma_w(T)$ and $\sigma_k(T)$ for finite k are exactly the same sets as defined in Chapter 5. However, $\sigma_{+\infty}(T)$ and $\sigma_{-\infty}(T)$ as defined above are counterparts of the sets with same notation defined in Chapter 5. In fact, Remark A.31(a) says $\sigma_{\ell e}(T) \backslash \sigma_{re}(T) = (\sigma_{e+}(T) \backslash \sigma_{e-}(T)) \backslash \vartheta(T)$ and $\sigma_{re}(T) \backslash \sigma_{\ell e}(T) = (\sigma_{e-}(T) \backslash \sigma_{e+}(T)) \backslash \vartheta(T)$. If the Banach space $\mathcal{X}$ is such that $\Gamma[\mathcal{X}] = \mathcal{B}[\mathcal{X}]$ (and so $\vartheta(T) = \varnothing$; for instance, if $\mathcal{X}$ is a Hilbert space), then the differences coincide: $\sigma_{\ell e}(T) \backslash \sigma_{re}(T) = \sigma_{e+}(T) \backslash \sigma_{e-}(T)$ and $\sigma_{re}(T) \backslash \sigma_{\ell e}(T) = \sigma_{e-}(T) \backslash \sigma_{e+}(T)$.

Corollary A.33. *The essential spectrum has no pseudoholes if and only if upper and lower semi-Fredholm spectra coincide,*

$$\sigma_{+\infty}(T) = \sigma_{-\infty}(T) = \varnothing \iff \sigma_{e_+}(T) = \sigma_{e_-}(T),$$

which implies that right and left essential spectra coincide,

(a) $\sigma_{e_+}(T) = \sigma_{e_-}(T) \implies \sigma_{\ell e}(T) = \sigma_{re}(T).$

Conversely,

(b) $\sigma_{\ell e}(T) = \sigma_{re}(T) \Rightarrow \sigma_{+\infty}(T) = \vartheta_N(T) \backslash \sigma_{e_-}(T)$ & $\sigma_{-\infty}(T) = \vartheta_R(T) \backslash \sigma_{e_+}(T).$

Certain special cases are of particular interest. Suppose $\sigma_{\ell e}(T) = \sigma_{re}(T)$.

(b$_1$) *If $\vartheta_R(T)$ and $\vartheta_N(T)$ are open, then $\sigma_{+\infty}(T) = \sigma_{-\infty}(T) = \varnothing$.*

(b$_2$) *If $\vartheta_R(T)$ and $\vartheta_N(T)$ are closed, then $\sigma_{+\infty}(T) = \sigma_{-\infty}(T) = \varnothing$.*

(b$_3$) *If $\vartheta_R(T) \cup \vartheta_N(T) \subseteq \sigma_{e_+}(T) \cap \sigma_{e_-}(T)$ then $\sigma_{+\infty}(T) = \sigma_{-\infty}(T) = \varnothing$.*

(b$_4$) *If $\vartheta_R(T) = \vartheta_N(T)$, then $\sigma_{+\infty}(T) = \sigma_{-\infty}(T) = \varnothing$.*

(b$_5$) *If $\vartheta_R(T) \cap \sigma_{e_+}(T) = \vartheta_N(T) \cap \sigma_{e_-}(T) = \varnothing$, then $\sigma_{+\infty}(T) = \sigma_{-\infty}(T) = \varnothing$.*

(b$_6$) *If $\vartheta_R(T) = \vartheta_N(T) = \varnothing$, then $\sigma_{+\infty}(T) = \sigma_{-\infty}(T) = \varnothing$.*

Proof. Let $\sigma_{+\infty}(T) = \sigma_{e_+}(T) \backslash \sigma_{e_-}(T)$ and $\sigma_{-\infty}(T) = \sigma_{e_-}(T) \backslash \sigma_{e_+}(T)$ be the pseudoholes of $\sigma_e(T)$, and consider the following assertions.

(i) $\sigma_{+\infty}(T) = \varnothing$ and $\sigma_{-\infty}(T) = \varnothing$ (i.e., $\sigma_e(T)$ has no pseudoholes).

(ii) $\sigma_{e_+}(T) = \sigma_{e_-}(T)$.

(iii) $\sigma_{\ell e}(T) = \sigma_{re}(T)$.

(iv) $\sigma_{+\infty}(T) = \vartheta_N(T) \backslash \sigma_{e_-}(T)$ and $\sigma_{-\infty}(T) = \vartheta_R(T) \backslash \sigma_{e_+}(T)$.

Assertions (i) and (ii) are equivalent by definition of $\sigma_{+\infty}(T)$ and $\sigma_{-\infty}(T)$.

(a) Assertion (ii) implies (iii) according to Remark A.31(b).

(b) If $\sigma_{\ell e}(T) = \sigma_{re}(T)$, then $\sigma_{\ell e}(T) \backslash \sigma_{re}(T) = \sigma_{re}(T) \backslash \sigma_{\ell e}(T) = \varnothing$. Thus by Remark A.31(a) and Proposition A.30(a,b) we get $\sigma_{e_+}(T) \backslash \sigma_{e_-}(T) \subseteq \vartheta_N(T)$ and $\sigma_{e_-}(T) \backslash \sigma_{e_+}(T) \subseteq \vartheta_R(T)$. Then using Proposition A.30(a,b) again we have $\sigma_{e_+}(T) \backslash \sigma_{e_-}(T) \subseteq \vartheta_N(T) \backslash \sigma_{e_-}(T) \subseteq \sigma_{e_+}(T) \backslash \sigma_{e_-}(T)$ and $\sigma_{e_-}(T) \backslash \sigma_{e_+}(T) \subseteq \vartheta_R(T) \backslash \sigma_{e_+}(T) \subseteq \sigma_{e_-}(T) \backslash \sigma_{e_+}(T)$. Thus (iii) implies (iv).

(b$_1$) This is straightforward from Corollary A.32(b) — so it does not need item (b). Indeed, if $\sigma_{\ell e}(T) = \sigma_{re}(T)$, and if $\vartheta_R(T)$ and $\vartheta_N(T)$ are open, then $\sigma_{e_+}(T) = \sigma_{e_-}(T)$ by Corollary A.32(b) and so $\sigma_{+\infty}(T) = \sigma_{-\infty}(T) = \varnothing$.

(b$_2$) Assumption 1: $\sigma_{\ell e}(T) = \sigma_{re}(T)$. Thus $\sigma_{+\infty}(T) = \vartheta_N(T) \backslash \sigma_{e_-}(T)$ by item (b). Recall: $\sigma_{+\infty}(T)$ is open and bounded. Assumption 2: $\vartheta_N(T)$ is closed. Suppose $\sigma_{+\infty}(T) \neq \varnothing$. If $\vartheta_N(T) \cap \sigma_{e_-}(T) \neq \varnothing$, then $\sigma_{+\infty}(T) = \vartheta_N(T) \backslash \sigma_{e_-}(T) \subseteq \vartheta_N(T)$ is not open (because $\vartheta_N(T)$ and $\sigma_{0_-}(T)$ are closed), which is a contradiction. On the other hand, if $\vartheta_N(T) \cap \sigma_{e_-}(T) = \varnothing$, then $\sigma_{+\infty}(T) = \vartheta_N(T)$, which is another contradiction (since $\sigma_{+\infty}(T)$ is open and bounded and $\vartheta_N(T)$ is closed). Therefore $\sigma_{+\infty}(T) = \varnothing$. Outcome: $\sigma_{\ell e}(T) = \sigma_{re}(T)$ and $\vartheta_N(T)$ closed imply $\sigma_{+\infty}(T) = \varnothing$. Similarly (same argument by using (b) again), $\sigma_{\ell e}(T) = \sigma_{re}(T)$ and $\vartheta_R(T)$ closed imply $\sigma_{-\infty}(T) = \varnothing$.

(b$_3$) By item (b) if $\sigma_{\ell e}(T) = \sigma_{re}(T)$ and if $\vartheta_R(T) \cup \vartheta_N(T) \subseteq \sigma_{e_+}(T) \cap \sigma_{e_-}(T)$, then $\sigma_{+\infty}(T) = \vartheta_N(T) \backslash \sigma_{e_-}(T) = \varnothing$ and $\sigma_{-\infty}(T) = \vartheta_R(T) \backslash \sigma_{e_+}(T) = \varnothing$.

(b$_4$) This follows from Remark A.31(d) — it does not need item (b) either.

(b$_5$) If $\sigma_{\ell e}(T) = \sigma_{re}(T)$ and if $\vartheta_R(T) \cap \sigma_{e_+}(T) = \vartheta_N(T) \cap \sigma_{e_-}(T) = \varnothing$ then item (b) ensures $\sigma_{+\infty}(T) = \vartheta_N(T)$ and $\sigma_{-\infty}(T) = \vartheta_R(T)$. Since $\sigma_{+\infty}(T)$ and $\sigma_{-\infty}(T)$ are open sets, then $\vartheta_N(T)$ and $\vartheta_R(T)$ are open sets, and therefore $\sigma_{+\infty}(T) = \sigma_{-\infty}(T) = \varnothing$ according to item (b$_1$).

(b$_6$) A trivial (or tautological) particular case of any of the above (b$_i$). $\qquad\square$

Remark. The identity $\sigma_{e_+}(T) = \sigma_{e_-}(T)$ implies the identity $\sigma_{\ell e}(T) = \sigma_{re}(T)$ by Corollary A.33(a). The converse holds if $\vartheta_R(T)$ and $\vartheta_N(T)$ satisfy any of the assumptions in Corollary A.33(b_i). The converse, however, might fail if for instance $\vartheta_R(T)$ and $\vartheta_N(T)$ are nonclosed or nonopen. For example, take a compact nonempty subset $\mathbb{D}$ of $\mathbb{C}$ (e.g., the closed unit disk) and two proper subsets D_R and D_N of $\mathbb{D}$, both are neither closed nor open, such that $(\mathbb{D}\backslash D_R)^- \neq (\mathbb{D}\backslash D_N)^-$. Consider the following configuration: $\sigma_{\ell e}(T) = \mathbb{D}$, $\sigma_{re}(T) = \mathbb{D}$, $\vartheta_R(T) = D_R$, $\vartheta_N(T) = D_N$, $\sigma_{e_+}(T) = (\mathbb{D}\backslash D_R)^-$, $\sigma_{e_-}(T) = (\mathbb{D}\backslash D_N)^-$. If such a configuration is possible, then $\sigma_{\ell e}(T) = \sigma_{re}(T)$ and $\sigma_{e_+}(T) \neq \sigma_{e_-}(T)$.

Remark A.34. Let T be an arbitrary operator acting on a Banach space $\mathcal{X}$.

(a) No Holes nor Pseudoholes. The essential spectrum of $T \in \mathcal{B}[\mathcal{X}]$ has no holes if $\sigma_k(T) = \varnothing$ for every finite nonzero integer $k \in \mathbb{Z}\backslash\{0\}$. Equivalently,

$$\sigma_e(T) \text{ has no holes if and only if } \sigma_e(T) = \sigma_w(T)$$

as in Theorem 5.24 (Schechter Theorem). The essential spectrum of T has no pseudoholes if $\sigma_{+\infty}(T) = \sigma_{-\infty}(T) = \varnothing$. Equivalently,

$$\sigma_e(T) \text{ has no pseudoholes if and only if } \sigma_{e_+}(T) = \sigma_{e_-}(T)$$

by Corollary A.33(a) — which trivially implies $\sigma_{e_+}(T) = \sigma_{e_-}(T) = \sigma_e(T)$ (since $\sigma_{e_+}(T) \cup \sigma_{e_-}(T) = \sigma_e(T) = \sigma_{\ell e}(T) \cup \sigma_{re}(T)$). Therefore

$$\sigma_e(T) \text{ has no holes and no pseudoholes}$$

if and only if

$$\sigma_{e_+}(T) = \sigma_{e_-}(T) = \sigma_e(T) = \sigma_w(T), \tag{$*$}$$

which implies

$$\sigma_{\ell e}(T) = \sigma_{re}(T) = \sigma_e(T) = \sigma_w(T) \tag{$**$}$$

according to Corollary A.33(a). Although ($*$) always implies ($**$), the identities in ($**$) will imply the identities in ($*$) if any of the assumptions in Corollary A.33(b_i) are satisfied; in particular, ($**$) implies ($*$) if $\vartheta_R(T) = \vartheta_N(T) = \varnothing$ as in Corollary A.33(b_6), which is the case whenever $\mathcal{X}$ is a Hilbert space.

(b) Compact Operators. Take a compact operator K on an *infinite-dimensional* Banach space $\mathcal{X}$. The Fredholm Alternative as in Remark 5.7(b,ii) says

$$\lambda \neq 0 \quad \Longrightarrow \quad \lambda I - K \in \mathcal{W}[\mathcal{X}].$$

Since $\sigma_w(T) = \{\lambda \in \mathbb{C} \colon \lambda I - T \notin \mathcal{W}[\mathcal{X}]\}$, then $\sigma_w(K) \subseteq \{0\}$. Since $\sigma_w(T) \neq \varnothing$ if $\dim \mathcal{X} = \infty$ (cf. Remark 5.27(a)), then $\sigma_w(K) = \{0\}$. Moreover, $\sigma_e(T) \neq \varnothing$ if $\dim \mathcal{X} = \infty$ (cf. Remark 5.15(a)) and $\sigma_e(T) \subseteq \sigma_w(T)$. Hence

$$\sigma_e(K) = \sigma_w(K) = \{0\}.$$

Since $\mathcal{B}_\infty[\mathcal{X}]$ is an ideal of $\mathcal{B}[\mathcal{X}]$, and since the identity is not compact in an infinite-dimensional space, we get by Definition A.20 (cf. Remark 5.7(b))

$$K \notin \mathcal{F}_\ell[\mathcal{X}] \cup \mathcal{F}_r[\mathcal{X}] = \mathcal{SF}[\mathcal{X}].$$

Thus $0 \in \sigma_{\ell e}(K) \cap \sigma_{re}(K)$ (Definition A.26). Since $\sigma_{\ell e}(T) \cup \sigma_{re}(T) = \sigma_e(T)$ (Definition A.26 again) and $\sigma_e(K) = \{0\}$,

$$\sigma_{\ell e}(K) = \sigma_{re}(K) = \{0\}.$$

According to Corollary A.25(c)

$$\lambda I - K \in \Gamma_N[\mathcal{X}] \cap \Gamma_R[\mathcal{X}] = \Gamma[\mathcal{X}] \quad \text{for} \quad \lambda \neq 0.$$

So by Definition A.28

$$\vartheta_R(K) \cup \vartheta_N(K) = \vartheta(K) \subseteq \{0\}.$$

Hence $\vartheta_R(K)$ and $\vartheta_N(K)$ are both closed. Since $\sigma_{\ell e}(K) = \sigma_{re}(K)$, then

$$\sigma_{e_+}(T) = \sigma_{e_-}(K)$$

by Corollary A.33(b_2). If $\vartheta_R(K)$ and $\vartheta_N(K)$ are both $\{0\}$ or both $\varnothing$, then by Remark A.31(d) $\sigma_{e_+}(K) = \sigma_{\ell e}(K)$ and $\sigma_{e_-}(K) = \sigma_{re}(K)$. On the other hand if $\vartheta_R(K) = \varnothing$ and $\vartheta_N(K) = \{0\}$, then $\sigma_{\ell e}(K) = \sigma_{e_+}(K)$ and $\sigma_{re}(K) = \sigma_{e_-}(K) \cup \{0\}$ by Corollary A.29. Similarly, if $\vartheta_R(K) = \{0\}$ and $\vartheta_N(K) = \varnothing$, then $\sigma_{re}(K) = \sigma_{e_-}(K)$ and $\sigma_{\ell e}(K) = \sigma_{e_+}(K) \cup \{0\}$. Thus in any case

$$\sigma_{\ell e}(K) = \sigma_{re}(K) = \sigma_{e_+}(K) = \sigma_{e_-}(K).$$

Therefore

$$\sigma_{\ell e}(K) = \sigma_{re}(K) = \sigma_{e_+}(K) = \sigma_{e_-}(K) = \sigma_e(K) = \sigma_w(K) = \{0\}.$$

So the identities $(*)$ and $(**)$ of item (a) not only coincide but they do hold for a compact operator on an infinite-dimensional Banach space, and so compact operators have no holes and no pseudoholes. The class of operators with no holes and no pseudoholes (which properly includes the compact operators) is referred to as the class of *biquasitriangular* operators (see [85, Section 6]).

Suggested Readings

Abramovich and Aliprantis [1]
Aiena [3]
Arveson [7]
Barnes, Murphy, Smyth, and West [11]
Caradus, Pfaffenberger, and Yood [26]
Conway [30]
Harte [60]

Heuser [65]
Kato [72]
Laursen and Neumann [87]
Megginson [93]
Müller [95]
Schechter [106]
Taylor and Lay [111]

B

A Glimpse at Multiplicity Theory

It has been written somewhere in the Web, "Multiplicity is a lot of something." Indeed, this matches the common (colloquial) meaning, and perhaps might suggest that the term "multiplicity" has a multiplicity of meanings.

B.1 Meanings of Multiplicity

The first time we met the term in this book was in Section 2.2 with regard to the *multiplicity of an eigenvalue*, defined as the dimension of the respective eigenspace. For an isolated eigenvalue this is sometimes called the *geometric multiplicity* of the isolated eigenvalue, while the *algebraic multiplicity* of an isolated point of the spectrum, as considered in Section 4.5, is the dimension of the range of the Riesz idempotent associated with the isolated point of the spectrum (see also Section 5.3). Related to these definitions, there is the notion of *multiplicity of a point in a disjoint union*, introduced in Section 3.3, which refers to how often an element is repeated in a disjoint union; in particular, in a disjoint union of spectra of direct summands of a normal operator.

Multiplicity of a shift as in Section 2.7 is the common dimension of the shifted spaces. So there is a multiplicity function on the set of shifts (bilateral or unilateral) assigning to each shift a cardinal number: the shift multiplicity.

Multiplicity of an arc was defined in Section 4.3 as a function assigning to each point in the unit interval a cardinal number called the *multiplicity of a point in an arc* indicating how often an arc transverses a given point.

Remark. The term is also applied with respect to multiplicity of prime factors, to multiplicity of roots of polynomials (or of holomorphic — i.e., analytic — complex functions). More generally, there is the notion of multiplicity of zeroes and poles of meromorphic (i.e., analytic up to a discrete set of isolated points) or rational complex functions, among a multiplicity of other instances.

In this appendix, however, we return to Hilbert spaces and will treat multiplicity from the spectral point of view, as recapitulated in the first paragraph

© Springer Nature Switzerland AG 2020

C. S. Kubrusly, *Spectral Theory of Bounded Linear Operators*,

https://doi.org/10.1007/978-3-030-33149-8

of this section. In particular, multiplicity is approached here in connection with the notion of spectral measure, with the central focus being placed upon the spectral theorems of Chapter 3, so that it is revealed as a concept intrinsically linked to normal operators. Indeed, this appendix can be viewed as an extension of (or a complement to) Chapter 3.

B.2 Complete Set of Unitary Invariants

Let $\mathcal{H}$ and $\mathcal{K}$ be Hilbert spaces. Let $U \in \mathcal{B}[\mathcal{H}, \mathcal{K}]$ be a *unitary transformation*, which means an invertible isometry, or equivalently a surjective isometry, or (still equivalently) an isometry and a coisometry. So U is unitary if and only if $U^*U = I$ (identity on $\mathcal{H}$) and $UU^* = I$ (identity on $\mathcal{K}$), and so $U^{-1} = U^*$. Since $U^{**} = U$ and $U^{*-1} = U^{-1*}$, the transformation $U^* \in \mathcal{B}[\mathcal{K}, \mathcal{H}]$ is as unitary as U is. If there is a unitary $U \in \mathcal{B}[\mathcal{H}, \mathcal{K}]$ between Hilbert spaces $\mathcal{H}$ and $\mathcal{K}$, then these are *unitarily equivalent spaces* and we write $\mathcal{H} \cong \mathcal{K}$ (or $\mathcal{K} \cong \mathcal{H}$), where $\cong$ is an equivalence relation on the collection of all Hilbert spaces. *Two Hilbert spaces are unitarily equivalent if and only if they have the same dimension* (i.e., same orthogonal or Hilbert dimension — see, e.g., [78, Theorem 5.49]).

Take $T \in \mathcal{B}[\mathcal{H}]$ and $S \in \mathcal{B}[\mathcal{K}]$. These are *unitarily equivalent operators* if there exists a unitary transformation $U \in \mathcal{B}[\mathcal{H}, \mathcal{K}]$ for which $UT = SU$, or equivalently $T = U^*SU$, or (still equivalently) $S = UTU^*$. In this case we write $T \cong S$ (or $S \cong T$), where $\cong$ is now an equivalence relation on the collection of all Hilbert-space operators. Clearly, $T \cong S$ implies $\mathcal{H} \cong \mathcal{K}$. If two operators are unitarily equivalent, then there is no possible criterion based only on the geometry of Hilbert space by which T and S can be distinguished from each other, and so two unitarily equivalent operators are abstractly the same operator. Given a Hilbert-space operator, consider the equivalence class of all Hilbert-space operators unitarily equivalent to it, which is referred to as a *unitary equivalence class*. If $\mathcal{H}$ is a Hilbert space and $T \in \mathcal{B}[\mathcal{H}]$, then the unitary equivalence class of all Hilbert-space operators unitarily equivalent to T consists of all operators $S \in \mathcal{B}[\mathcal{K}]$ such that $S \cong T$ on an arbitrary $\mathcal{K} \cong \mathcal{H}$.

Throughout this appendix $\cong$ means *unitarily equivalent*.

An *invariant* (property) for a collection of elements is a property common to all elements in the collection. In particular, a *unitary invariant* is a property common to all operators belonging to a unitary equivalence class (i.e., a property shared by all operators in a unitary equivalence class — which is invariant under unitary equivalence — universal to all operators unitarily equivalent to a given operator — trivial example: norm). A *set of unitary invariants* is a set of properties common to all operators in a unitary equivalence class. A *complete set of unitary invariants* is a set of properties (i.e., a set of unitary invariants) such that two operators belong to a unitary equivalence class if and only if they satisfy all properties in the set (i.e., if and only if they posses all unitary invariants in the set). In other words, it is a set of properties sufficient for determining whether or not two operators are unitarily equivalent.

Thus the notion of a complete set of unitary invariants is related (actually, it is the answer) to the following question. When are two Hilbert-space operators unitarily equivalent? More generally, when do a family of Hilbert-space operators belong to the same unitary equivalence class? The question may be faced by trying to attach to each operator in a fixed family of operators a set of attributes such that two operators in the family are unitarily equivalent if and only if the set of attributes attached to them is identical (or equivalent in some sense). Such a set of attributes, which is enough for determining unitary equivalence, is referred to as a *complete set of unitary invariants* for the family of operators. There are a few families of operators for which a reasonable complete set of unitary invariants is known. In this context the term reasonable means it is easier to verify if these attributes are satisfied by every operator in the family than to verify if the operators in the family are unitarily equivalent. The theory of *spectral multiplicity* yields a reasonable complete set of unitary invariants for the family of normal operators.

B.3 Multiplicity Function for Compact Operators

The Spectral Theorem fully describes a compact normal operator $T \in \mathcal{B}[\mathcal{H}]$ by breaking it down into its spectral decomposition

$$T = \sum_{\lambda \in \sigma_P(T)} \lambda E_\lambda,$$

where E_λ is the orthogonal projection with

$$\mathcal{R}(E_\lambda) = \mathcal{N}(\lambda I - T).$$

The family $\{\mathcal{N}(\lambda I - T)\}_{\lambda \in \sigma_P(T)}$ of eigenspaces is the heart of the spectral decomposition for compact normal operators because, besides forming an orthogonal family of subspaces, it spans the space (cf. proof of Theorem 3.3):

$$\left(\sum_{\lambda \in \sigma_P(T)} \mathcal{N}(\lambda I - T) \right)^- = \left(\sum_{\lambda \in \sigma_P(T)} \mathcal{R}(E_\lambda) \right)^- = \mathcal{H}.$$

If T is an operator on any Banach space, then $\mathcal{N}(\lambda I - T) = \{0\}$ if and only if $\lambda \notin \sigma_P(T)$. If T is compact, then $\dim \mathcal{N}(\lambda I - T) < \infty$ whenever $\lambda \neq 0$, and it is just $\dim \mathcal{N}(T)$ that may not be finite. With this in mind, let $\mathbb{N}$ and $\mathbb{N}_0$ be the sets of all positive and nonnegative integers, respectively, and set $\overline{\mathbb{N}}_0 = \mathbb{N}_0 \cup \{+\infty\}$, the *extended nonnegative integers*. By the Fredholm Alternative (Corollary 1.20 and Theorem 2.18), if T is compact, then $\dim \mathcal{N}(\lambda I - T) \in \mathbb{N}$ for every $\lambda \in \sigma(T)\backslash\{0\} = \sigma_P(T)\backslash\{0\}$ and $\dim \mathcal{N}(\lambda I - T) \in \mathbb{N}_0$ for $\lambda \in \mathbb{C}\backslash\{0\}$. However, for the case of $\lambda = 0$, $\dim \mathcal{N}(T)$ may not be finite, and since the actual cardinality of any basis for $\mathcal{N}(T)$ is immaterial in this context, we simply say $\dim \mathcal{N}(T) \in \overline{\mathbb{N}}_0$. Thus, in general, $\dim \mathcal{N}(\lambda I - T) \in \overline{\mathbb{N}}_0$ for every $\lambda \in \mathbb{C}$.

Definition B.1. The *multiplicity* $m_T(\lambda)$ *of a point* $\lambda \in \mathbb{C}$ for a compact operator $T \in \mathcal{B}[\mathcal{H}]$ is the extended nonnegative integer

$$m_T(\lambda) = \dim \mathcal{N}(\lambda I - T).$$

(Equivalently, $m_T(\lambda) = \dim \mathcal{R}(E_\lambda)$.) The *multiplicity function* $m_T \colon \mathbb{C} \to \overline{\mathbb{N}}_0$ for a compact operator T is the extended-nonnegative-integer-valued map on $\mathbb{C}$ assigning $\dim \mathcal{N}(\lambda I - T)$ to each complex number $\lambda \in \mathbb{C}$:

$$\lambda \mapsto \dim \mathcal{N}(\lambda I - T).$$

Theorem B.2. *Two compact normal operators are unitarily equivalent if and only if they have the same multiplicity function.*

Proof. Take $T \in \mathcal{B}[\mathcal{H}]$ and $S \in \mathcal{B}[\mathcal{K}]$ on Hilbert spaces $\mathcal{H}$ and $\mathcal{K}$.

(a) If $T \cong S$, then $\dim \mathcal{N}(\lambda I - T) = \dim \mathcal{N}(\lambda I - S)$ for every $\lambda \in \mathbb{C}$. Indeed, if $W \in \mathcal{G}[\mathcal{H}, \mathcal{K}]$ (i.e., an invertible transformation), then $x \in \mathcal{N}(\lambda I - WTW^{-1})$ if and only if $WTW^{-1}x = \lambda x$, which means $TW^{-1}x = \lambda W^{-1}x$, or equivalently $W^{-1}x \in \mathcal{N}(\lambda I - T)$. This ensures $W(\mathcal{N}(\lambda I - T)) = \mathcal{N}(\lambda I - WTW^{-1})$. Since W is an isomorphism between the linear spaces $\mathcal{H}$ and $\mathcal{K}$, $\dim \mathcal{N}(\lambda I - T) = \dim W(\mathcal{N}(\lambda I - T)) = \dim \mathcal{N}(\lambda I - WTW^{-1})$. Thus $T \cong S \implies m_T = m_S$.

(b) Now consider the family of all Hilbert-space compact normal operators. Let $T \in \mathcal{B}_\infty[\mathcal{H}]$ and $S \in \mathcal{B}_\infty[\mathcal{K}]$ be normal operators on Hilbert spaces $\mathcal{H}$ and $\mathcal{K}$. Conversely, suppose $m_T = m_S$ (i.e., $\dim \mathcal{N}(\lambda I - T) = \dim \mathcal{N}(\lambda I - S)$ for every $\lambda \in \mathbb{C}$). The program is to show the existence of a unitary transformation in $\mathcal{B}[\mathcal{H}, \mathcal{K}]$ which makes $T \cong S$. Applying the Spectral Theorem for compact operators (Theorem 3.3) consider the spectral decompositions of T and S,

$$T = \sum\nolimits_{k \in \mathbb{K}} \lambda_k E_k \quad \text{and} \quad S = \sum\nolimits_{j \in \mathbb{J}} \zeta_j P_j,$$

where $\{\lambda_k\}_{k \in \mathbb{K}} = \sigma_P(T)$ and $\{\zeta_j\}_{j \in \mathbb{J}} = \sigma_P(S)$ are (nonempty countable) sets of all (distinct) eigenvalues of T and S, and $\{E_k\}_{k \in \mathbb{K}}$ and $\{P_j\}_{j \in \mathbb{J}}$ are $\mathcal{B}[\mathcal{H}]$-valued and $\mathcal{B}[\mathcal{K}]$-valued resolutions of the identity. Without loss of generality, assume the countable index sets $\mathbb{K}$ and $\mathbb{J}$ are subsets of the nonnegative integers $\mathbb{N}_0$. The operators $E_k \in \mathcal{B}[\mathcal{H}]$ and $P_j \in \mathcal{B}[\mathcal{K}]$ are orthogonal projections onto the eigenspaces $\mathcal{N}(\lambda_k I - T) = \mathcal{R}(E_k) \subseteq \mathcal{H}$ and $\mathcal{N}(\zeta_j I - S) = \mathcal{R}(P_j) \subseteq \mathcal{K}$. Again with no loss of generality, assume $0 \in \mathbb{K} \cap \mathbb{J}$, set $\lambda_0 = \zeta_0 = 0$, and let E_0 and P_0 be the projections onto $\mathcal{N}(\lambda_0 I - T) = \mathcal{N}(T)$ and $\mathcal{N}(\zeta_0 I - S) = \mathcal{N}(S)$. (Note: without modifying the above sums, zero may or may not be originally included in them, whether or not zero lies in their point spectra.) Take an arbitrary $k \neq 0$. Since $m_T = m_S$ (and $0 \neq \lambda_k \in \sigma_P(T)$),

$$0 < \dim \mathcal{N}(\lambda_k I - T) = \dim \mathcal{N}(\lambda_k I - S).$$

By the unique spectral decomposition of T and S, for each λ_k there is a unique ζ_j such that $\zeta_j = \lambda_k$. So there is an injective function $\theta \colon \mathbb{K} \setminus \{0\} \to \mathbb{J} \setminus \{0\}$ for which $\zeta_{\theta(k)} = \lambda_k$. Similarly,

$$0 < \dim \mathcal{N}(\zeta_j I - S) = \dim \mathcal{N}(\zeta_j I - T),$$

and so for each ζ_j there is a unique λ_k such that $\lambda_k = \zeta_j$. Hence the function $\theta \colon \mathbb{K} \backslash \{0\} \to \mathbb{J} \backslash \{0\}$ is bijective. Set $\theta(0) = 0$ and extend the function θ over $\mathbb{K}$ onto $\mathbb{J}$. The extended $\theta \colon \mathbb{K} \to \mathbb{J}$ remains bijective (i.e., a permutation). Then $\#\mathbb{K} = \#\mathbb{J}$: the cardinalities of the index sets $\mathbb{K}$ and $\mathbb{J}$ coincide. Since the subspaces $\mathcal{R}(E_k) = \mathcal{N}(\lambda_k I - T)$ and $\mathcal{R}(P_{\theta(k)}) = \mathcal{N}(\zeta_{\theta(k)} I - S)$ have the same dimension, they are unitarily equivalent, and so there exists a unitary transformation $U_k \colon \mathcal{R}(E_k) \to \mathcal{R}(P_{\theta(k)})$.

Claim. There exists a unitary $U \in \mathcal{B}[\mathcal{H}, \mathcal{K}]$ for which $UT = SU$.

Proof. Since $\{E_k\}$ and $\{P_{\theta(k)}\}$ are resolutions of the identity, consider the orthogonal topological sums $\mathcal{H} = \left(\sum_k \mathcal{R}(E_k) \right)^-$ and $\mathcal{K} = \left(\sum_k \mathcal{R}(P_{\theta(k)}) \right)^-$. Let $U \colon \left(\sum_k \mathcal{R}(E_k) \right)^- \to \left(\sum_k \mathcal{R}(P_{\theta(k)}) \right)^-$ be defined as follows. Take an arbitrary $x = \sum_k x_k \in \sum_k \mathcal{R}(E_k)$ with $x_k \in \mathcal{R}(E_k)$. Set $Ux = \sum_k U_k x_k$ where $U_k x_k$ lies in $\mathcal{R}(P_{\theta(k)})$. This defines a bounded linear transformation U of $\sum_k \mathcal{R}(E_k)$ onto $\sum_k \mathcal{R}(P_{\theta(k)})$. It may be extended by continuity to a bounded linear transformation, also denoted by U, over $\mathcal{H}$ onto $\mathcal{K}$. As is readily verified, its adjoint $U^* \colon \left(\sum_k \mathcal{R}(P_{\theta(k)}) \right)^- \to \left(\sum_k \mathcal{R}(E_k) \right)^-$ is given by $U^* y = \sum_k U_k^* y_k$ for an arbitrary $y = \sum_k y_k \in \sum_k \mathcal{R}(P_{\theta(k)})$ with $y_k \in \mathcal{R}(P_{\theta(k)})$. Thus $UU^* x = x$ and $U^* U y = y$ for every $x \in \sum_k \mathcal{R}(E_k)$ and $y \in \sum_k \mathcal{R}(P_{\theta(k)})$. So when extended over $\mathcal{H}$ onto $\mathcal{K}$ the transformation $U \in \mathcal{B}[\mathcal{H}, \mathcal{K}]$ is unitary. Hence $\mathcal{H}$ and $\mathcal{K}$ are unitarily equivalent. Moreover, if $x_k \in \mathcal{R}(E_k)$, then $T x_k = \lambda_k x_k$, which implies

$$UT x_k = U \lambda_k x_k = \zeta_{\theta(k)} U x_k = SU x_k$$

because $U x_k$ lies in $\mathcal{R}(P_{\theta(k)})$ and so $SU x_k = \zeta_{\theta(k)} U x_k$. Therefore

$$UTx = UT \sum_k x_k = \sum_k UT x_k = \sum_k SU x_k = SU \sum_k x_k = SUx$$

for every $x \in \sum_k \mathcal{R}(E_k)$. Extending by continuity from $\sum_k \mathcal{R}(E_k)$ to $\mathcal{H} = \left(\sum_k \mathcal{R}(E_k) \right)^-$, $UTx = SUx$ for every $x \in \mathcal{H}$. Thus $m_T = m_S \Longrightarrow T \cong S$.

(c) Outcome: $T \cong S \Longleftrightarrow m_T = m_S$. $\qquad\square$

> *Therefore if two compact normal operators have the same multiplicity function, then they have the same spectrum, and the multiplicity function alone is a complete set of unitary invariants.*

Remark. According to the Fredholm Alternative, zero is the only accumulation point of the spectrum of a compact operator T and the spectrum itself is countable, and so the set of all scalars λ for which the dimension of $\mathcal{N}(\lambda I - T)$ is not zero must be countable, and the dimension of $\mathcal{N}(\lambda I - T)$ is also finite provided λ is not zero (cf. Corollaries 1.20 and 2.20). Under these necessary constraints, *every multiplicity function for a compact normal operator is possible* (see, e.g., [30, Exercise II.8.5]). Although Theorem B.2 holds for compact normal operators, it does not hold for nonnormal, even compact, operators. The classical example is the Volterra operator, which is compact, quasinilpotent (nonnormal with empty point spectrum), and so its multiplicity function is the zero function, but it is not unitarily equivalent to the null operator.

B.4 Multiplicity-Free Normal Operators

Roughly speaking, a normal operator is free of multiplicity if just part (a) in the proof of the first version of the Spectral Theorem (Theorem 3.11) is enough for a full description of the unitary equivalence with a multiplication operator. This means it has a star-cyclic vector or, equivalently, it is a star-cyclic operator. Star-cyclicity for normal operators, however, coincides with cyclicity (Corollary 3.14). Summing up: a normal operator is *multiplicity-free* if it is cyclic (equivalently, star-cyclic). A word on terminology: perhaps counterintuitively, being *multiplicity-free* does not mean "being of multiplicity zero" (but it means "being of multiplicity one").

So by the first version of the Spectral Theorem (cf. proof of Theorem 3.11 part (a)), $T \in \mathcal{B}[\mathcal{H}]$ is a *multiplicity-free normal operator if and only if*

$$T \cong M_\varphi,$$

where the normal operator M_φ in $\mathcal{B}[L^2(\sigma(T), \mu)]$ is a multiplication operator,

$$M_\varphi f = \varphi f \quad \text{for every} \quad f \in L^2(\sigma(T), \mu),$$

with $\mu \colon \mathcal{A}_{\sigma(T)} \to \mathbb{R}$ being a positive finite measure on the σ-algebra $\mathcal{A}_{\sigma(T)}$ of Borel subsets of $\sigma(T) = \sigma(M_\varphi)$, and $\varphi \colon \sigma(T) \to \sigma(T)$ is the identity map (i.e., $\varphi(\lambda) = \lambda$ for $\lambda \in \sigma(T)$) *and hence $\varphi \in L^\infty(\sigma(T), \mu)$ since $\sigma(T)$ is bounded. Accordingly, this multiplication operator $M_\varphi \in \mathcal{B}[L^2(\sigma(T), \mu)]$ is again cyclic and normal and so* multiplicity-free *as well.*

Such a measure μ can be extended (same notation) to a σ-algebra $\mathcal{A}_\Omega$ of Borel subsets of any Borel set Ω including $\sigma(M_\varphi)$ and so its support remains the same: support $(\mu) = \sigma(M_\varphi) \subseteq \Omega \subseteq \mathbb{C} \in \mathcal{A}_\mathbb{C}$ where $\mathcal{A}_\Omega = \mathcal{A}_\mathbb{C} \cap \wp(\Omega)$ is a sub-σ-algebra of the σ-algebra $\mathcal{A}_\mathbb{C}$ of all Borel subsets of $\mathbb{C}$. If in addition Ω is compact (so bounded), then the identity map φ on Ω is bounded (so $\varphi \in L^\infty(\Omega, \mu)$) and support $(\mu) = \text{ess } \Omega = \sigma(M_\varphi) \subseteq \Omega \subset \mathbb{C}$ (Proposition 3.B).

Thus let μ and ν be positive *finite* measures with *compact support* on the σ-algebra $\mathcal{A}_\mathbb{C}$ of Borel subsets of $\mathbb{C}$, consider their restrictions (denoted again by μ and ν) to a σ-algebra $\mathcal{A}_\Omega$ of Borel subsets of any nonempty *compact* (thus Borel) set $\Omega \subset \mathbb{C}$ *including the supports* of μ and ν, let φ be the *identity map* on Ω, and consider two multiplication operators $M_\varphi \in \mathcal{B}[L^2(\Omega, \mu)]$ and $N_\varphi \in \mathcal{B}[L^2(\Omega, \nu)]$ so that (cf. paragraph following Proposition 3.B)

$$\sigma(M_\varphi) = \text{support}\,(\mu) \quad \text{and} \quad \sigma(N_\varphi) = \text{support}\,(\nu).$$

Now recall from Sections 3.6 and 4.2: a measure μ is *absolutely continuous* with respect to a measure ν (both acting on the same σ-algebra $\mathcal{A}_\Omega$) if $\nu(\Lambda) = 0$ implies $\mu(\Lambda) = 0$ for $\Lambda \in \mathcal{A}_\Omega$ (notation: $\mu \ll \nu$). The RADON–NIKODÝM THEOREM says: *if μ and ν are σ-finite measures and if $\mu \ll \nu$, then there exists a unique (ν-almost everywhere unique) positive measurable function ψ on Ω for which $\mu(\Lambda) = \int_\Lambda \psi \, d\nu$ for every $\Lambda \in \mathcal{A}_\Omega$* (see, e.g., [81, Theorem 7.8]). This unique function is called the *Radon–Nikodým derivative* of μ with respect to ν, denoted by $\psi = \frac{d\mu}{d\nu}$. Two measures μ and ν on the same σ-algebra are *equivalent* (notation $\mu \equiv \nu$ or $\nu \equiv \mu$) if they are *mutually absolutely continuous* (i.e., $\mu \ll \nu$ and $\nu \ll \mu$ — one is absolutely continuous with respect to the other and vice versa), and so they share exactly the same sets of measure zero. In other words, $\mu \equiv \nu$ means $\mu(\Lambda) = 0$ if and only if $\nu(\Lambda) = 0$ for $\Lambda \in \mathcal{A}_\Omega$.

If μ and ν have the same sets of measure zero, then by definition their supports coincide: $\mu \equiv \nu$ trivially implies $\mathrm{support}\,(\mu) = \mathrm{support}\,(\nu)$. But the converse fails: take the Lebesgue measure on $[0,1]$ and its sum with a Dirac measure on the singleton $\{\frac{1}{2}\}$. So under the previous assumptions (i.e., finite measures μ and ν on the same σ-algebra $\mathcal{A}_\Omega$ of subsets of the same compact set Ω, and multiplication operators by the identity map φ) we get

$$\mu \equiv \nu \quad \Longrightarrow \quad \sigma(M_\varphi) = \sigma(N_\varphi).$$

Equivalent measures, however, go beyond such an identity of spectra.

Thus as discussed above, let $M_\varphi \in \mathcal{B}[L^2(\Omega, \mu)]$ and $N_\varphi \in \mathcal{B}[L^2(\Omega, \nu)]$ be *multiplicity-free* multiplication operators, where μ and ν are regarded as finite positive measures with compact support on a σ-algebra $\mathcal{A}_\Omega$ of Borel subsets of any compact set $\varnothing \neq \Omega \subset \mathbb{C}$ for which $\mathrm{support}\,(\mu) \cup \mathrm{support}\,(\nu) \subseteq \Omega$, and φ is their common identity map on Ω. This is the setup for the next theorem.

Theorem B.3. $\quad M_\varphi \cong N_\varphi \iff \mu \equiv \nu$.

Proof. Recall: $\sigma(M_\varphi) = \mathrm{support}\,(\mu)$ and $\sigma(N_\varphi) = \mathrm{support}\,(\nu)$.

(a) Suppose $M_\varphi \cong N_\varphi$. This implies $\sigma(M_\varphi) = \sigma(N_\varphi)$. Thus set $\Omega = \sigma(M_\varphi) = \sigma(N_\varphi)$. Also let $U \colon L^2(\Omega, \mu) \to L^2(\Omega, \nu)$ be the unitary transformation for which $UM_\varphi = N_\varphi U$. By a trivial induction $UM_\varphi^k = N_\varphi^k U$ and $UM_\varphi^{*j} = N_\varphi^{*j}U$ for any integers $j, k \geq 0$. Let $P(\Omega)$ be the set of all polynomials $p(\cdot, \bar{\cdot}) \colon \Omega \to \mathbb{C}$ in λ and $\bar{\lambda}$ (i.e., $P(\Omega)$ is a set of all maps $\lambda \mapsto p(\lambda, \bar{\lambda})$ taking each λ in Ω into $p(\lambda, \bar{\lambda}) = \sum_{j,k=0}^n \alpha_{j,k} \lambda^j \bar{\lambda}^k = \sum_{j,k=0}^n \alpha_{j,k} \varphi(\lambda)^j \overline{\varphi}(\lambda)^k$ in $\mathbb{C}$ for an arbitrary nonnegative integer n and an arbitrary set of complex coefficients $\{\alpha_{j,k}\}_{j,k=0}^n$). For each $p \in P(\Omega)$ take the operators $p(M_\varphi, M_\varphi^*) = \sum_{j,k=0}^n \alpha_{j,k} M_\varphi^j M_\varphi^{*k}$ in $\mathcal{B}[L^2(\Omega, \mu)]$ and $p(N_\varphi, N_\varphi^*) = \sum_{j,k=0}^n \alpha_{j,k} N_\varphi^j N_\varphi^{*k}$ in $\mathcal{B}[L^2(\Omega, \nu)]$. Thus

$$Up(M_\varphi, M_\varphi^*) = p(N_\varphi, N_\varphi^*)\, U.$$

Since $M_\varphi f = \varphi f$ and $M_\varphi^* f = \overline{\varphi} f$, and hence $M_\varphi^j M_\varphi^{*k} f = \varphi^j \overline{\varphi}^k f$, we get $pf = \sum_{jk} \alpha_{j,k} \lambda^j \bar{\lambda}^k f = [p(M_\varphi, M_\varphi^*)]f$ and also $\|pf\|_2^2 = \int \left|\sum_{j,k} \alpha_{j,k} \lambda^j \bar{\lambda}^k f(\lambda)\right|^2 d\mu \leq$

$\operatorname{diam}(\Omega)^{4n} \sum_{j,k}^{n} |\alpha_{j,k}|^2 \int |f|^2 d\mu$, for $f \in L^2(\Omega, \mu)$ and $p \in P(\Omega)$. Therefore

$$pf = [p(M_\varphi, M_\varphi^*)]f \in L^2(\Omega, \mu) \quad \text{and} \quad pg = [p(N_\varphi, N_\varphi^*)]g \in L^2(\Omega, \nu)$$

for every $f \in L^2(\Omega, \mu)$ and every $g \in L^2(\Omega, \nu)$. Moreover, for $f \in L^2(\Omega, \mu)$

$$U(pf) = U[p(M_\varphi, M_\varphi^*)](f) = [p(N_\varphi, N_\varphi^*)]U(f) = pU(f).$$

Since U is an isometry, $\|pf\|_2 = \|U(pf)\|_2 = \|pU(f)\|_2$, and so $\int |pf|^2 d\mu = \int |pU(f)|^2 d\nu$. Then, by setting $f = 1 \in L^2(\Omega, \mu)$,

$$\int |p|^2 d\mu = \int |p|^2 |U(1)|^2 d\nu \quad \text{for every } p \in P(\Omega).$$

Recall from the proof of Theorem 3.11 part (a): $P(\Omega)^- = L^2(\Omega, \eta)$ in the L^2-norm topology for every finite measure η defined on the σ-algebra $\mathcal{A}_\Omega$ of Borel subsets of a compact $\Omega \subset \mathbb{C}$. Then $P(\Omega)^- = L^2(\Omega, \mu) \cap L^2(\Omega, \nu)$. Thus (by continuity of the integral) the above identity holds for $p \in P(\Omega)$ replaced with the characteristic function $\chi_\Lambda \in L^2(\Omega, \mu) \cap L^2(\Omega, \nu)$ for any $\Lambda \in \mathcal{A}_\Omega$. So

$$\mu(\Lambda) = \int_\Lambda d\mu = \int_\Lambda |U(1)|^2 d\nu \leq \sup |U(1)|^2 \int_\Lambda d\nu \leq \sup |U(1)|^2 \nu(\Lambda)$$

(where $\sup |U(1)| = \sup_{\lambda \in \Omega} |U(1)(\lambda)| > 0$ because $U(1) \neq 0$ since $\|U(1)\|_2^2 = \|1\|_2^2 = \int d\nu = \int_\Omega d\nu = \nu(\Omega) > 0$ — see, e.g., [81, Remark 10.1]). Therefore

$$\{\nu(\Lambda) = 0 \implies \mu(\Lambda) = 0\} \implies \mu \ll \nu.$$

Similarly, since U is unitary, replace U by U^{-1} which is again an isometry, and apply the same argument to prove $\nu \ll \mu$. Outcome: $M_\varphi \cong N_\varphi \implies \mu \equiv \nu$.

(b) Conversely, first suppose $\mu \ll \nu$. Then the measures μ and ν act on the same σ-algebra $\mathcal{A}_\Omega$. Since these measures are finite (thus σ-finite), take the Radon–Nikodým derivative $\psi = \frac{d\mu}{d\nu}$ of μ with respect to ν (which is a positive measurable $\mathbb{C}$-valued function on Ω). If $g \in L^1(\Omega, \mu)$, then $\int g \, d\mu = \int g \frac{d\mu}{d\nu} d\nu = \int g\psi \, d\nu$ and so $g\psi \in L^1(\Omega, \nu)$. Thus for $f \in L^2(\Omega, \mu)$ we get $f^2 \in L^1(\Omega, \mu)$ and hence $\psi f^2 \in L^1(\Omega, \nu)$. Then $\psi^{\frac{1}{2}} f \in L^2(\Omega, \nu)$ and

$$\|\psi^{\frac{1}{2}} f\|_2^2 = \int |\psi^{\frac{1}{2}} f|^2 d\nu = \int |f|^2 \psi \, d\nu = \int |f|^2 d\mu = \|f\|_2^2.$$

Thus the multiplication operator $U_{\psi^{\frac{1}{2}}} : L^2(\Omega, \mu) \to L^2(\Omega, \nu)$ given by $U_{\psi^{\frac{1}{2}}} f = \psi^{\frac{1}{2}} f$ for every $f \in L^2(\Omega, \mu)$ is an isometry. If in addition $\nu \ll \mu$ (and so $\mu \equiv \nu$) then the Radon–Nikodým derivative of ν with respect to μ is $\frac{d\nu}{d\mu} = \psi^{-1}$ (see, e.g., [81, Problem 7.9(d)]). Using the same argument for any $h \in L^2(\Omega, \nu)$,

$$\|h\|_2^2 = \int |h|^2 d\nu = \int |h|^2 \frac{d\nu}{d\mu} d\mu = \int |h|^2 \psi^{-1} d\mu = \int |\psi^{-\frac{1}{2}} h|^2 d\mu = \|\psi^{-\frac{1}{2}} h\|_2^2.$$

Then the multiplication operator $U_{\psi^{-\frac{1}{2}}} \colon L^2(\Omega, \nu) \to L^2(\Omega, \mu)$ given for each $h \in L^2(\Omega, \nu)$ by $U_{\psi^{-\frac{1}{2}}} h = \psi^{-\frac{1}{2}} h$ is an isometry as well. Since these isometries are the inverses of each other, $U_{\psi^{\frac{1}{2}}} \colon L^2(\Omega, \mu) \to L^2(\Omega, \nu)$ is a unitary transformation. Also, $U_{\psi^{\frac{1}{2}}} M_\varphi f = \psi^{\frac{1}{2}} \varphi f = \varphi \psi^{\frac{1}{2}} f = N_\varphi U_{\psi^{\frac{1}{2}}} f$ for every f in $L^2(\Omega, \mu)$, and so $U_{\psi^{\frac{1}{2}}} M_\varphi = N_\varphi U_{\psi^{\frac{1}{2}}}$. Outcome: $\mu \equiv \nu \implies M_\varphi \cong N_\varphi$. $\qquad\square$

Corollary B.4. *Two multiplicity-free normal operators are unitarily equivalent if and only if their scalar spectral measures are equivalent. Equivalently, if and only if their spectral measures are equivalent.*

Proof. Let $T \in \mathcal{B}[\mathcal{H}]$ and $S \in \mathcal{B}[\mathcal{K}]$ be multiplicity-free normal operators. So $\mathcal{H}$ and $\mathcal{K}$ are separable Hilbert spaces (Corollary 3.14). By Theorem 3.11 (in particular, by part (a) in the proof of Theorem 3.11) they are unitarily equivalent to multiplicity-free multiplication operators $M_\varphi \in L^2(\Omega, \mu)$ and $N_\varphi \in L^2(\Omega, \nu)$. That is, $T \cong M_\varphi$ and $S \cong N_\varphi$. By transitivity $T \cong S$ if and only if $M_\varphi \cong N_\varphi$. Thus $\sigma(T) = \sigma(S) = \sigma(M_\varphi) = \sigma(N_\varphi)$ and Ω denotes these coincident spectra. Then M_φ and N_φ have the same identity function φ on Ω. By Theorem B.3, $M_\varphi \cong N_\varphi$ if and only if $\mu \equiv \nu$. Summing up:

$$T \cong S \quad\Longleftrightarrow\quad M_\varphi \cong N_\varphi \quad\Longleftrightarrow\quad \mu \equiv \nu.$$

Let E and P be the unique spectral measures of the spectral decompositions of $T = \int_{\sigma(T)} \lambda \, dE_\lambda$ and $S = \int_{\sigma(S)} \lambda \, dP_\lambda$ as in Theorem 3.15. Let $\pi_{e,e}$ and $\omega_{f,f}$ be scalar spectral measures for T and S which are positive finite measures equivalent to E and P, respectively (cf. Definition 4.4 and Lemma 4.7). That is,

$$\pi_{e,e} \equiv E \quad \text{and} \quad \omega_{f,f} \equiv P.$$

Thus, by the remark following Lemma 4.7 (since the Hilbert spaces are separable, and since in the multiplicity-free case $\hat{\pi}_{e,e} = \pi_{e,e}$ on $\mathcal{A}_\Omega = \mathcal{A}_{\sigma(T)}$), we get

$$\pi_{e,e} = \mu \quad \text{and} \quad \omega_{f,f} = \nu.$$

All these measures act on the same σ-algebra $\mathcal{A}_\Omega = \mathcal{A}_{\sigma(T)} = \mathcal{A}_{\sigma(S)}$. Therefore

$$\mu \equiv \nu \quad\Longleftrightarrow\quad \pi_{e,e} \equiv \omega_{f,f} \quad\Longleftrightarrow\quad E \equiv P. \qquad\square$$

Spectral measure (up to equivalence) is a complete set of unitary invariants for multiplicity-free normal operators.

B.5 Multiplicity Function for Normal Operators

Again, the hard work has already been done in the proof of the first version of the Spectral Theorem (Theorem 3.11), now referring to part (b) of that proof. Take a nonempty index set Γ. For each $\gamma \in \Gamma$ let μ_γ be a positive finite measure with compact support acting on the σ-algebra $\mathcal{A}_\mathbb{C}$ of all Borel subsets of the complex plane $\mathbb{C}$. Let Ω_γ be a compact set including the support of μ_γ (support $(\mu_\gamma) \subseteq \Omega_\gamma$). So the positive finite measure μ_γ can be regarded as acting on the σ-algebra $\mathcal{A}_{\Omega_\gamma}$ of Borel subsets of the compact subset Ω_γ of $\mathbb{C}$. Let φ_γ be the identity map on the compact set Ω_γ, so that φ_γ is bounded, and consider the multiplicity-free multiplication operator

$$M_{\varphi_\gamma} \in \mathcal{B}[L^2(\Omega_\gamma, \mu_\gamma)].$$

Thus $\sigma(M_{\varphi_\gamma}) = \text{support}\,(\mu_\gamma) \subseteq \Omega_\gamma$. Suppose the ordinary union of $\{\Omega_\gamma\}_{\gamma \in \Gamma}$ is a bounded subset of $\mathbb{C}$. Let $\tilde{\Omega}$ be any compact subset of $\mathbb{C}$ including the ordinary union of $\{\Omega_\gamma\}_{\gamma \in \Gamma}$. (In the proof of Theorem 3.11 part (b) we had $\tilde{\Omega} = \sigma(T)$.) Also let $\tilde{\varphi} \colon \tilde{\Omega} \to \tilde{\Omega}$ be the identity map on $\tilde{\Omega}$ (i.e., $\tilde{\varphi}(\lambda) = \lambda$ for every $\lambda \in \tilde{\Omega}$), which lies in $L^\infty(\tilde{\Omega}, \mu_\gamma)$ for every γ because $\tilde{\Omega}$ is bounded. Consider the proof of Theorem 3.11 part (b). Since support $(\mu_\gamma) \subseteq \Omega_\gamma$ for each γ, then we may replace each multiplicity-free multiplication operator $M_{\varphi_\gamma} \in \mathcal{B}[L^2(\Omega_\gamma, \mu_\gamma)]$ by the multiplicity-free multiplication operator

$$M_{\gamma, \tilde{\varphi}} \in \mathcal{B}[L^2(\tilde{\Omega}, \mu_\gamma)],$$

with a common identity map $\tilde{\varphi}$ on a common set $\tilde{\Omega}$ for every $\gamma \in \Gamma$, for which $\sigma(M_{\gamma, \tilde{\varphi}}) = \text{support}\,(\mu_\gamma) \subseteq \Omega_\gamma \subseteq \tilde{\Omega} \subset \mathbb{C}$.

Now consider the disjoint union $\Omega = \biguplus_{\gamma \in \Gamma} \Omega_\gamma$. Let $\mathcal{A}_\Omega$ be the σ-algebra of Borel subsets of the disjoint union Ω and take the positive measure μ on $\mathcal{A}_\Omega$ as defined in the proof of Theorem 3.11 part (b) (i.e., $\mu(\Lambda) = \sum_{\gamma \in \mathbb{J}} \mu_\gamma(\Lambda_\gamma)$ for $\Lambda = \biguplus_{\gamma \in \mathbb{J}} \Lambda_\gamma \in \mathcal{A}_\Omega$ with $\Lambda_\gamma \in \mathcal{A}_{\Omega_\gamma}$ and $\mathbb{J} \subseteq \Gamma$ countable). Take the *identity map with multiplicity* $\varphi \colon \Omega \to \mathbb{C}$ as defined in the proof of Theorem 3.11 part (b) (i.e., if $\lambda \in \Omega = \biguplus_{\gamma \in \Gamma} \Omega_\gamma$, then $\lambda = \lambda \in \Omega_\beta \subset \mathbb{C}$ for a unique $\beta \in \Gamma$ and a unique $\lambda \in \Omega_\beta$, thus set $\varphi(\lambda) = \varphi_\beta(\lambda) = \lambda \subset \mathbb{C}$). Since the ordinary union of $\{\Omega_\gamma\}_{\gamma \in \Gamma}$ (included in $\tilde{\Omega}$) is bounded in $\mathbb{C}$, then φ is a scalar-valued bounded function on Ω. Consider the multiplication operator

$$M_\varphi \in \mathcal{B}[L^2(\Omega, \mu)].$$

Take multiplication operators $M_{\gamma, \tilde{\varphi}} \in \mathcal{B}[L^2(\tilde{\Omega}, \mu_\gamma)]$ with a common identity function $\tilde{\varphi}$ on $\tilde{\Omega}$ (which as explained above replace multiplication operators $M_{\varphi_\gamma} \in \mathcal{B}[L^2(\Omega_\gamma, \mu_\gamma)]$), and consider the external orthogonal direct sum

$$\bigoplus_{\gamma \in \Gamma} M_{\gamma, \tilde{\varphi}} \in \mathcal{B}\big[\bigoplus_{\gamma \in \Gamma} L^2(\tilde{\Omega}, \mu_\gamma)\big].$$

So (cf. proof of Theorem 3.11 part (b) once again)

$$L^2(\Omega, \mu) \cong \bigoplus_{\gamma \in \Gamma} L^2(\tilde{\Omega}, \mu_\gamma) \quad \text{and} \quad M_\varphi \cong \bigoplus_{\gamma \in \Gamma} M_{\gamma, \tilde{\varphi}}$$

with $\sigma(M_\varphi) = \big(\bigcup_{\gamma \in \Gamma} \sigma(M_{\gamma, \tilde{\varphi}})\big)^- = \varphi\big(\biguplus_{\gamma \in \Gamma} \sigma(M_{\gamma, \tilde{\varphi}})\big)^-$.

Definition B.5. Consider the above setup. The *multiplicity* $m_{M_\varphi}(\lambda)$ of a *point* $\lambda \in \mathbb{C}$ for the multiplication operator M_φ is the cardinality of the inverse image $\varphi^{-1}(\{\lambda\})$ (denoted by λ) of the singleton $\{\lambda\}$ with respect to φ,

$$m_{M_\varphi}(\lambda) = \#\lambda = \#\varphi^{-1}(\{\lambda\}) = \#\biguplus_{\gamma \in \Gamma} \{\{\lambda\} \cap \Omega_\gamma\}.$$

The *multiplicity function* for a multiplication operator M_φ is a cardinal-number-valued map m_M on $\mathbb{C}$ assigning to each $\lambda \in \mathbb{C}$ its multiplicity $m_{M_\varphi}(\lambda)$,

$$m_{M_\varphi} : \lambda \mapsto m_M(\lambda).$$

[Note: $m_{M_\varphi}(\lambda) = 0$ for every $\lambda \notin \mathcal{R}(\varphi) = \varphi(\Omega) = \bigcup_{\gamma \in \Gamma} \Omega_\gamma$ (the ordinary union of $\{\Omega_\gamma\}_{\gamma \in \Gamma}$); in particular, if $\Omega_\gamma = \sigma(M_{\gamma,\tilde{\varphi}})$ so that $\varphi(\Omega)^- = \sigma(M_\varphi)$ (cf. Proposition 2.F(d)), then $m_{M_\varphi}(\lambda) = 0$ for every $\lambda \in \rho(M_\varphi)$.]

Let $M_\varphi \in \mathcal{B}[L^2(\Omega,\mu)]$ and $N_{\varphi'} \in \mathcal{B}[L^2(\Omega',\nu)]$ be multiplication operators where Ω and Ω' are disjoint unions of compact subsets Ω_γ and Ω'_γ of $\mathbb{C}$ whose ordinary unions are bounded, μ and ν are positive measures on the σ-algebras $\mathcal{A}_\Omega$ and $\mathcal{A}_{\Omega'}$ of Borel subsets of Ω and Ω', and φ and φ' are identity maps with multiplicity on Ω and Ω'. Let m_{M_φ} and $m_{N_{\varphi'}}$ be their multiplicity functions. This is the setup of the next theorem.

Theorem B.6. $\quad M_\varphi \cong N_{\varphi'} \iff \mu \equiv \nu \text{ and } m_{M_\varphi} = m_{N_{\varphi'}}.$
$\quad\quad\quad\quad$ (*In this case* $\Omega = \Omega'$ *and so* $\varphi = \varphi'$.)

Proof. As $\tilde{\Omega}$ and $\tilde{\varphi}$ were defined associated with Ω (over Γ) and φ, let $\tilde{\Omega}'$ and $\tilde{\varphi}'$ be analogously defined with respect to Ω' (over Γ') and φ'.

(a) Suppose $M_\varphi \cong N_{\varphi'}$. Then $L^2(\Omega,\mu) \cong L^2(\Omega',\nu)$. Thus by transitivity $\bigoplus_{\gamma \in \Gamma} L^2(\tilde{\Omega},\mu_\gamma) \cong \bigoplus_{\gamma \in \Gamma'} L^2(\tilde{\Omega}',\nu_\gamma)$ and also $\bigoplus_{\gamma \in \Gamma} M_{\gamma,\tilde{\varphi}} \cong \bigoplus_{\gamma \in \Gamma'} N_{\gamma,\tilde{\varphi}'}$. Therefore $M_{\gamma,\tilde{\varphi}} \cong N_{\gamma,\tilde{\varphi}'}$ and $L^2(\tilde{\Omega},\mu_\gamma) \cong L^2(\tilde{\Omega}',\nu_\gamma)$ for every γ, and so Γ' is in a one-to-one correspondence with Γ. Thus we may identify them and set $\Gamma = \Gamma'$. Since $\tilde{\Omega}$ and $\tilde{\Omega}'$ are arbitrary compact subsets of $\mathbb{C}$ including the ordinary unions of $\{\Omega_\gamma\}_{\gamma \in \Gamma}$ and $\{\Omega'_\gamma\}_{\gamma \in \Gamma'}$, then we take $\tilde{\Omega} = \tilde{\Omega}'$ (including both), and hence $\tilde{\varphi} = \tilde{\varphi}'$. In this case $M_{\gamma,\tilde{\varphi}} \cong N_{\gamma,\tilde{\varphi}'}$ implies $\mu_\gamma \equiv \nu_\gamma$ by Theorem B.3, and so $\mu \equiv \nu$ by definition. Moreover, $\mu_\gamma \equiv \nu_\gamma$ also implies support(μ_γ) = support(ν_γ). Thus set $\Omega_\gamma = \Omega'_\gamma$ such that $\Omega_\gamma = \sigma(M_{\gamma,\tilde{\varphi}})$ = support(μ_γ) = support$(\nu_\gamma) = \sigma(N_{\gamma,\tilde{\varphi}'}) = \Omega'_\gamma$. Since $\Omega_\gamma = \Omega'_\gamma$, then $\varphi_\gamma = \varphi'_\gamma$ for every γ in $\Gamma = \Gamma'$ and $\Omega = \bigcup_{\gamma \in \Gamma} \Omega_\gamma = \bigcup_{\gamma \in \Gamma} \Omega'_\gamma = \Omega'$, and therefore $\varphi = \varphi'$. Since $\varphi = \varphi'$ on $\Omega = \Omega'$ we get $\varphi^{-1}(\{\lambda\}) = \varphi'^{-1}(\{\lambda\})$ for every $\lambda \in \mathbb{C}$ (and so $\#\varphi^{-1}(\{\lambda\}) = \#\varphi'^{-1}(\{\lambda\})$ trivially). Thus $m_{M_\varphi} = m_{N_{\varphi'}}$ by Definition B.5.

(b) Conversely, if $\mu \equiv \nu$, then they act on the same σ-algebra and hence $\Omega = \Omega'$. Since $\Omega = \Omega'$ and $m_{M_\varphi} = m_{N_{\varphi'}}$, then $\Omega_\gamma = \Omega'_\gamma$ for every γ and so we may regard $\Gamma = \Gamma'$. Thus $\{\Omega_\gamma\}_{\gamma \in \Gamma} = \{\Omega'_\gamma\}_{\gamma \in \Gamma'}$ is a set of sets whose ordinary union is *a priori* bounded, and hence $\tilde{\varphi} = \tilde{\varphi}'$ (on the closure of their common bounded ordinary union). Moreover, $\Omega_\gamma = \Omega'_\gamma$ and $\mu \equiv \nu$ imply $\mu_\gamma \equiv \nu_\gamma$ for each γ. Since $\mu_\gamma \equiv \nu_\gamma$ and $\tilde{\varphi} = \tilde{\varphi}'$, we get $M_{\gamma,\tilde{\varphi}} \cong N_{\gamma,\tilde{\varphi}'}$ by Theorem B.3. Therefore $M_\varphi \cong \bigoplus_{\gamma \in \Gamma} M_{\gamma,\tilde{\varphi}} \cong \bigoplus_{\gamma \in \Gamma'} N_{\gamma,\tilde{\varphi}'} \cong N_{\varphi'}$. $\quad\square$

Let $T \in \mathcal{B}[\mathcal{H}]$ be a normal operator on a Hilbert space $\mathcal{H}$. Take the unique spectral measure $E : \mathcal{A}_{\sigma(T)} \to \mathcal{B}[\mathcal{H}]$ of the spectral decomposition $T = \int_{\sigma(T)} \lambda \, dE_\lambda$ as in Theorem 3.15. By Theorem 3.11, $T \cong M_\varphi$ with $M_\varphi \in \mathcal{B}[L^2(\Omega,\mu)]$ being a multiplication (thus normal) operator on the Hilbert space $L^2(\Omega, \mu)$ where Ω is a disjoint union included in $\bigcup_{\gamma \in \Gamma} \sigma(T)$, μ is a positive measure on $\mathcal{A}_\Omega$, and $\varphi \colon \Omega \to \mathbb{C}$ is the identity map with multiplicity. Take an arbitrary $\lambda \in \sigma(T)$. As we had seen in the proof of Theorem 3.15 parts (a,b,c),

$$E(\{\lambda\}) = \hat{E}(\varphi^{-1}(\{\lambda\})) \cong E'(\mathbb{\lambda}) = M_{\chi_{\mathbb{\lambda}}}$$

with $\mathbb{\lambda} = \varphi^{-1}(\{\lambda\}) \in \Omega$ and $M_{\chi_{\mathbb{\lambda}}} \in \mathcal{B}[L^2(\Omega, \mu)]$. Let $\mu|_{\mathbb{\lambda}} \colon \mathcal{A}_{\mathbb{\lambda}} \to \mathbb{C}$ be the restriction of the measure $\mu \colon \mathcal{A}_\Omega \to \mathbb{C}$ to the sub-σ-algebra $\mathcal{A}_{\mathbb{\lambda}} = \mathcal{A}_\Omega \cap \wp(\mathbb{\lambda})$. So

$$\mathcal{R}(E(\{\lambda\})) \cong \mathcal{R}(M_{\chi_{\mathbb{\lambda}}}) = L^2(\mathbb{\lambda}, \mu|_{\mathbb{\lambda}}),$$

and hence

$$\dim \mathcal{R}(E(\{\lambda\})) = \dim L^2(\mathbb{\lambda}, \mu|_{\mathbb{\lambda}}).$$

Definition B.7. Let E be the spectral measure on $\mathcal{A}_{\sigma(T)}$ for the spectral decomposition of a normal operator $T \in \mathcal{B}[\mathcal{H}]$ on a Hilbert space $\mathcal{H}$. The *multiplicity* $m_T(\lambda)$ of a point $\lambda \in \mathbb{C}$ for T is defined as follows. If $\lambda \in \sigma(T)$, then

$$m_T(\lambda) = \dim \mathcal{R}(E(\{\lambda\}))$$

(since $\{\lambda\} \in \mathcal{A}_{\sigma(T)}$) and so, according to Proposition 3.G,

$$m_T(\lambda) = \dim \mathcal{R}(E(\{\lambda\})) = \dim \mathcal{N}(\lambda I - T).$$

If $\lambda \in \rho(T))$, then $m_T(\lambda) = 0$. The *multiplicity function* for T is a cardinal-number-valued map m_T on $\mathbb{C}$ assigning to each $\lambda \in \mathbb{C}$ its multiplicity,

$$m_T \colon \lambda \mapsto m_T(\lambda).$$

Corollary B.8. *Two normal operators on separable Hilbert spaces are unitarily equivalent if and only if they have the same multiplicity function and their scalar spectral measures are equivalent (i.e., if and only if they have the same multiplicity function and their spectral measures are equivalent).*

Proof. Let $T \in \mathcal{B}[\mathcal{H}]$ and $S \in \mathcal{B}[\mathcal{K}]$ be normal operators. According to Theorem 3.11, they are unitarily equivalent to multiplication operators $M_\varphi \in L^2(\Omega, \mu)$ and $N_{\varphi'} \in L^2(\Omega', \nu)$. That is, $T \cong M_\varphi$ and $S \cong N_{\varphi'}$. By transitivity $T \cong S$ if and only if $M_\varphi \cong N_{\varphi'}$. By Theorem B.6 $M_\varphi \cong N_{\varphi'}$ if and only if $\mu \equiv \nu$ and $m_{M_\varphi} = m_{N_{\varphi'}}$, and in this case $\Omega = \Omega'$ and $\varphi = \varphi'$. Summing up:

$$T \cong S \quad \Longleftrightarrow \quad M_\varphi \cong N_{\varphi'} \quad \Longleftrightarrow \quad \mu \equiv \nu \ \text{ and } \ m_{M_\varphi} = m_{N_{\varphi'}}.$$

Now assuming $T \cong S$, let E and P be the spectral measures on $\mathcal{A}_{\sigma(T)} = \mathcal{A}_{\sigma(S)}$ of the spectral decompositions of $T = \int_{\sigma(T)} \lambda \, dE_\lambda$ and $S = \int_{\sigma(S)} \lambda \, dP_\lambda$ as in Theorem 3.15, where by unitary equivalence $\sigma(T) = \sigma(M_\varphi) = \sigma(N_{\varphi'}) = \sigma(S)$. Take any $\lambda \in \mathbb{C}$. Set $\mathbb{\lambda} = \varphi^{-1}(\{\lambda\})$ and $\mathbb{\lambda}' = \varphi'^{-1}(\{\lambda\})$. If $\mu \equiv \nu$ and $m_{M_\varphi} = m_{N_{\varphi'}}$, then $\varphi = \varphi'$ on $\Omega = \Omega'$ (cf. Theorem B.6) and so $\mathbb{\lambda} = \mathbb{\lambda}'$. Thus since $\mathbb{\lambda} = \mathbb{\lambda}'$ and $\mu \equiv \nu$, then $\dim L^2(\mathbb{\lambda}, \mu|_{\mathbb{\lambda}}) = \dim L^2(\mathbb{\lambda}', \nu|_{\mathbb{\lambda}'})$ naturally. Conversely, if

$\mu \equiv \nu$ and $\dim L^2(\mathbb{A}, \mu|_{\mathbb{A}}) = \dim L^2(\mathbb{A}', \nu|_{\mathbb{A}'})$, then $\#\mathbb{A} = \#\mathbb{A}'$, which means $m_{M_\varphi} = m_{N_{\varphi'}}$. (Recall: $L^2(\Omega, \mu) \cong L^2(\Omega', \nu)$ are separable once $\mathcal{H} \cong \mathcal{K}$ are, and this implies Γ and Γ' countable, and so are $\mathbb{A}$ and $\mathbb{A}'$.) Therefore since $\dim \mathcal{R}(E(\{\lambda\})) = \dim L^2(\mathbb{A}, \mu|_{\mathbb{A}})$ and $\dim \mathcal{R}(P(\{\lambda\})) = \dim L^2(\mathbb{A}', \nu|_{\mathbb{A}'})$, we get

$$\mu \equiv \nu \quad \text{and} \quad m_{M_\varphi} = m_{N_{\varphi'}} \quad \Longleftrightarrow \quad \mu \equiv \nu \quad \text{and} \quad m_T = m_S$$

by Definition B.7. Since $\mathcal{H}$ and $\mathcal{K}$ are separable, let $\hat{\pi}_{e,e}$ and $\hat{\omega}_{f,f}$ be scalar spectral measures equivalent to the spectral measures $\hat{E}$ and $\hat{P}$ on $\mathcal{A}_\Omega$ of the spectral decompositions of $T = \int_\Omega \lambda \, d\hat{E}_\lambda$ and $S = \int_\Omega \lambda \, d\hat{P}_\lambda$ as in the proof of Theorem 3.15 part (b) — cf. Definition 4.4 and Lemma 4.7. That is,

$$\hat{\pi}_{e,e} \equiv \hat{E} \quad \text{and} \quad \hat{\omega}_{f,f} \equiv \hat{P}.$$

By the remark following Lemma 4.7,

$$\hat{\pi}_{e,e} = \mu \quad \text{and} \quad \hat{\omega}_{f,f} = \nu.$$

All these measures act on the same σ-algebra $\mathcal{A}_\Omega$. Thus

$$\mu \equiv \nu \quad \Longleftrightarrow \quad \hat{\pi}_{e,e} \equiv \hat{\omega}_{f,f} \quad \Longleftrightarrow \quad \hat{E} \equiv \hat{P}.$$

Since $E(\cdot) = \hat{E}(\varphi^{-1}(\cdot))$ and $P(\cdot) = \hat{P}(\varphi'^{-1}(\cdot))$ on $\mathcal{A}_{\sigma(T)} = \mathcal{A}_{\sigma(S)}$ with $\varphi = \varphi'$ on $\Omega = \Omega'$ as in the proof of Theorem 3.15 part (c), we finally get

$$\mu \equiv \nu \quad \Longleftrightarrow \quad \hat{E} \equiv \hat{P} \quad \Longleftrightarrow \quad E \equiv P. \qquad \square$$

Multiplicity function and (equivalence of) spectral measure form a complete set of unitary invariants for normal operators on separable spaces.

Remark. By the argument in the proof of Corollary B.8 and under the separability assumption, Theorem B.6 can be restated as in Corollary B.8 with E and P being the spectral measures of the spectral decompositions of $M_\varphi = \int_{\sigma(M_\varphi)} \lambda \, dE_\lambda$ and $N_{\varphi'} = \int_{\sigma(N_{\varphi'})} \lambda \, dP_\lambda$ with $\sigma(M_\varphi) = \sigma(N_{\varphi'})$ by unitary equivalence (as in Theorem 3.15). Theorem B.2 is also naturally translated along the lines of Corollary B.8 in the case of compact normal operators on separable Hilbert spaces, with $\mathcal{R}(E_\lambda) = \mathcal{R}(E(\{\lambda\})) = \mathcal{N}(\lambda I - T)$ and now with countable spectra (and so the appropriate measures are counting measures).

Suggested Readings

Arveson [6]	Davidson [34]
Brown [21]	Dunford and Schwartz [46]
Conway [30]	Halmos [54]

References

1. Y.A. ABRAMOVICH AND C.D. ALIPRANTIS, *An Invitation to Operator Theory*, Graduate Studies in Mathematics, Vol. 50, Amer. Math. Soc., Providence, 2002.

2. L.V. AHLFORS, *Complex Analysis*, 3rd edn. McGraw-Hill, New York, 1978.

3. P. AIENA, *Fredholm and Local Spectral Theory, with Applications to Multipliers*, Kluwer, Dordrecht, 2004.

4. N.I. AKHIEZER AND I.M. GLAZMAN, *Theory of Linear Operators in Hilbert Space – Volume I*, Pitman, London, 1981; reprinted: Dover, New York, 1993.

5. N.I. AKHIEZER AND I.M. GLAZMAN, *Theory of Linear Operators in Hilbert Space – Volume II*, Pitman, London, 1981; reprinted: Dover, New York, 1993.

6. W. ARVESON, *An Invitation to C*-Algebras*, Springer, New York, 1976.

7. W ARVESON, *A Short Course in Spectral Theory*, Springer, New York, 2002.

8. F.V. ATKINSON, *The normal solvability of linear equations in normed spaces*, Mat. Sbornik **28** (1951), 3–14.

9. G. BACHMAN AND L. NARICI, *Functional Analysis*, Academic Press, New York, 1966; reprinted: Dover, Mineola, 2000.

10. B.A. BARNES, *Riesz points and Weyl's theorem*, Integral Equations Operator Theory **34** (1999), 187–196.

11. B.A. BARNES, G.J. MURPHY, M.R.F. SMYTH, AND T.T. WEST, *Riesz and Fredholm Theory in Banach Algebras*, Pitman, London, 1982.

12. R.G. BARTLE, *The Elements of Integration and Lebesgue Measure*, Wiley, New York, 1995; enlarged 2nd edn. of *The Elements of Integration*, Wiley, New York, 1966.

13. R. BEALS, *Topics in Operator Theory*, The University of Chicago Press, Chicago, 1971.

14. R. BEALS, *Advanced Mathematical Analysis*, Springer, New York, 1973.

15. S.K. BERBERIAN, *Notes on Spectral Theory*, Van Nostrand, New York, 1966.

© Springer Nature Switzerland AG 2020
C. S. Kubrusly, *Spectral Theory of Bounded Linear Operators*,
https://doi.org/10.1007/978-3-030-33149-8

16. S.K. BERBERIAN, *An extension of Weyl's theorem to a class of not necessarily normal operators*, Michigan Math. J. **16** (1969), 273–279.

17. S.K. BERBERIAN, *The Weyl spectrum of an operator*, Indiana Univ. Math. J. **20** (1971), 529–544.

18. S.K. BERBERIAN, *Lectures in Functional Analysis and Operator Theory*, Springer, New York, 1974.

19. S.K. BERBERIAN, *Introduction to Hilbert Space*, 2nd edn. Chelsea, New York, 1976.

20. J. BRAM, *Subnormal operators*, Duke Math. J. **22** (1955), 75–94.

21. A. BROWN, *A version of multiplicity theory*, Topics in Operator Theory, Mathematical Surveys no. 13, Amer. Math. Soc., Providence, 2nd pr. 1979, 129–160.

22. A. BROWN AND C. PEARCY, *Spectra of tensor products of operators*, Proc. Amer. Math. Soc. **17** (1966), 162–166.

23. A. BROWN AND C. PEARCY, *Introduction to Operator Theory I – Elements of Functional Analysis*, Springer, New York, 1977.

24. A. BROWN AND C. PEARCY, *An Introduction to Analysis*, Springer, New York, 1995.

25. S.L. CAMPBELL AND G.D. FAULKNER, *Operators on Banach spaces with complemented ranges*, Acta Math. Acad. Sci. Hungar. **35** (1980), 123–128.

26. S.R. CARADUS, W.E. PFAFFENBERGER, AND B. YOOD, *Calkin Algebras and Algebras of Operators on Banach Spaces*, Lecture Notes in Pure and Applied Mathematics, Vol. 9. Marcel Dekker, New York, 1974.

27. L.A. COBURN, *Weyl's theorem for nonnormal operators*, Michigan Math. J. **13** (1966), 285–288.

28. J.B. CONWAY, *Every spectral picture is possible*, Notices Amer. Math. Soc. **24** (1977), A-431.

29. J.B. CONWAY, *Functions of One Complex Variable*, Springer, New York, 1978.

30. J.B. CONWAY, *A Course in Functional Analysis*, 2nd edn. Springer, New York, 1990.

31. J.B. CONWAY, *The Theory of Subnormal Operators*, Mathematical Surveys and Monographs, Vol. 36, Amer. Math. Soc., Providence, 1991.

32. J.B. CONWAY, *A Course in Operator Theory*, Graduate Studies in Mathematics, Vol. 21, Amer. Math. Soc., Providence, 2000.

33. J.B. CONWAY, *A Course in Abstract Analysis*, Graduate Studies in Mathematics, Vol. 141, Amer. Math. Soc., Providence, 2012.

34. K.R. DAVIDSON, *C*-Algebras by Example*, Fields Institute Monographs, Vol. 6, Amer. Math. Soc., Providence, 1996.

35. J. DIEUDONNÉ, *Foundations of Modern Analysis*, Academic Press, New York, 1969.

36. D.S. DJORDJEVIĆ, *Semi-Browder essential spectra of quasisimilar operators*, Novi Sad J. Math. **31** (2001), 115–123.

37. R.G. DOUGLAS, *On majorization, factorization, and range inclusion of operators on Hilbert space*, Proc. Amer. Math. Soc. **17** (1966), 413–415.

38. R.G. DOUGLAS, *Banach Algebra Techniques in Operator Theory*, Academic Press, New York, 1972; 2nd edn. Springer, New York, 1998.

39. H.R. DOWSON, *Spectral Theory of Linear Operators*, Academic Press, New York, 1978.

40. B.P. DUGGAL, *Weyl's theorem for totally hereditarily normaloid operators*, Rend. Circ. Mat. Palermo **53** (2004), 417–428.

41. B.P. DUGGAL, *Hereditarily normaloid operators*, Extracta Math. **20** (2005), 203–217.

42. B.P. DUGGAL AND S.V. DJORDJEVIĆ, *Generalized Weyl's theorem for a class of operators satisfying a norm condition*, Math. Proc. Royal Irish Acad. **104** (2004), 75–81.

43. B.P. DUGGAL, S.V. DJORDJEVIĆ, AND C.S. KUBRUSLY, *Hereditarily normaloid contractions*, Acta Sci. Math. (Szeged) **71** (2005), 337–352.

44. B.P. DUGGAL AND C.S. KUBRUSLY, *Weyl's theorem for direct sums*, Studia Sci. Math. Hungar. **44** (2007), 275–290.

45. N. DUNFORD AND J.T. SCHWARTZ, *Linear Operators – Part I: General Theory*, Interscience, New York, 1958.

46. N. DUNFORD AND J.T. SCHWARTZ, *Linear Operators – Part II: Spectral Theory – Self Adjoint Operators in Hilbert Space*, Interscience, New York, 1963.

47. N. DUNFORD AND J.T. SCHWARTZ, *Linear Operators – Part III: Spectral Operators*, Interscience, New York, 1971.

48. P.A. FILLMORE, *Notes on Operator Theory*, Van Nostrand, New York, 1970.

49. P.A. FILLMORE, *A User's Guide to Operator Algebras*, Wiley, New York, 1996.

50. M. GONZÁLES AND V.M. ONIEVA, *On Atkinson operators in locally convex spaces*, Math. Z. **190** (1985), 505–517.

51. K. GUSTAFSON, *Necessary and sufficient conditions for Weyl's theorem*, Michigan Math. J. **19** (1972), 71–81.

52. K. GUSTAFSON AND D.K.M. RAO, *Numerical Range*, Springer, New York, 1997.

53. P.R. HALMOS, *Measure Theory*, Van Nostrand, New York, 1950; reprinted: Springer, New York, 1974.

54. P.R. HALMOS, *Introduction to Hilbert Space and the Theory of Spectral Multiplicity*, 2nd edn. Chelsea, New York, 1957; reprinted: AMS Chelsea, Providence, 1998.

55. P.R. HALMOS, *Finite-Dimensional Vector Spaces*, Van Nostrand, New York, 1958; reprinted: Springer, New York, 1974.

56. P.R. HALMOS, *Shifts on Hilbert spaces*, J. Reine Angew. Math. **208** (1961), 102–112.

57. P.R. HALMOS, *What does the spectral theorem say?*, Amer. Math. Monthly **70** (1963), 241–247

58. P.R. HALMOS, *A Hilbert Space Problem Book*, Van Nostrand, New York, 1967; 2nd edn. Springer, New York, 1982.

59. P.R. HALMOS AND V.S. SUNDER, *Bounded Integral Operators on L^2 Spaces*, Springer, Berlin, 1978.

60. R. HARTE, *Invertibility and Singularity for Bounded Linear Operators*, Marcel Dekker, New York, 1988.

61. R. HARTE AND W.Y. LEE, *Another note on Weyl's theorem*, Trans. Amer. Math. Soc. **349** (1997), 2115–2124.

62. W. HEISENBERG, *Physics and Beyond: Encounters and Conversations*, Harper & Row, New York, 1971.

63. G. HELMBERG, *Introduction to Spectral Theory in Hilbert Space*, North-Holland, Amsterdam, 1969.

64. D. HERRERO, *Approximation of Hilbert Space Operators – Volume 1*, 2nd edn. Longman, Harlow, 1989.

65. H.G. HEUSER, *Functional Analysis*, Wiley, Chichester, 1982.

66. E. HILLE AND R.S. PHILLIPS, *Functional Analysis and Semi-Groups*, Colloquium Publications Vol. 31, Amer. Math. Soc., Providence, 1957; reprinted: 1974.

67. J.R. HOLUB, *On perturbation of operators with complemented range*, Acta Math. Hungar. **44** (1984), 269–273.

68. T. ICHINOSE, *Spectral properties of tensor products of linear operators I*, Trans. Amer. Math. Soc. **235** (1978), 75–113.

69. V.I. ISTRĂŢESCU, *Introduction to Linear Operator Theory*, Marcel Dekker, New York, 1981.

70. M.A. KAASHOEK AND D.C. LAY, *Ascent, descent, and commuting perturbations*, Trans. Amer. Math. Soc. **169** (1972), 35–47.

71. N.J. KALTON, *The complemented subspace problem revisited*, Studia Math. **188** (2008), 223–257.

72. T. KATO, *Perturbation Theory for Linear Operators*, 2nd edn. Springer, Berlin, 1980; reprinted: 1995.

73. D. KITSON, R. HARTE, AND C. HERNANDEZ, *Weyl's theorem and tensor products: a counterexample*, J. Math. Anal. Appl. **378** (2011), 128–132.

74. C.S. KUBRUSLY, *An Introduction to Models and Decompositions in Operator Theory*, Birkhäuser, Boston, 1997.

75. C.S. KUBRUSLY, *Hilbert Space Operators: A Problem Solving Approach*, Birkhäuser, Boston, 2003.

76. C.S. KUBRUSLY, *A concise introduction to tensor product*, Far East J. Math. Sci. **22** (2006), 137–174.

77. C.S. KUBRUSLY, *Fredholm theory in Hilbert space – a concise introductory exposition*, Bull. Belg. Math. Soc. – Simon Stevin **15** (2008), 153–177.

78. C.S. KUBRUSLY, *The Elements of Operator Theory*, Birkhäuser-Springer, New York, 2011; enlarged 2nd edn. of *Elements of Operator Theory*, Birkhäuser, Boston, 2001.

79. C.S. KUBRUSLY, *A note on Browder spectrum*, Bull. Belg. Math. Soc. – Simon Stevin **19** (2012), 185–191.

80. C.S. KUBRUSLY, *Spectral Theory of Operators on Hilbert Spaces*, Birkhäuser-Springer, New York, 2012.

81. C.S. KUBRUSLY, *Essentials of Measure Theory*, Springer-Switzerland, Cham, 2015.

82. C.S. KUBRUSLY, *Range-kernel complementation*, Studia Sci. Math. Hungar. **55** (2018), 327–344.

83. C.S. KUBRUSLY AND B.P. DUGGAL, *On Weyl and Browder spectra of tensor products*, Glasgow Math. J. **50** (2008), 289–302.

84. C.S. KUBRUSLY AND B.P. DUGGAL, *On Weyl's theorem of tensor products*, Glasgow Math. J. **55** (2013), 139–144.

85. C.S. KUBRUSLY AND B.P. DUGGAL, *Upper-lower and left-right semi-Fredholmness*, Bull. Belg. Math. Soc. – Simon Stevin **23** (2016), 217–233.

86. S. LANG, *Real and Functional Analysis*, Springer, New York, 1993; enlarged 3rd edn. of *Real Analysis*, 2nd edn. Addison-Wesley, Reading, 1983.

87. K.B. LAURSEN AND M.M. NEUMANN, *Introduction to Local Spectral Theory*, Clarendon Press, Oxford, 2000.

88. D.C. LAY, *Characterizations of the essential spectrum of F.E. Browder*, Bull. Amer. Math. Soc. **74** (1968), 246–248.

89. W.Y. LEE, *Weyl's theorem for operator matrices*, Integral Equations Operator Theory **32** (1998), 319–331.

90. W.Y. LEE, *Weyl spectrum of operator matrices*, Proc. Amer. Math. Soc. **129** (2001), 131–138.

91. J. LINDENSTRAUSS AND L. TZAFRIRI, *On the complemented subspaces problem*, Israel J. Math. **9** (1971), 263–269.

92. P. LOCKHART, *A Mathematician's Lament*, Bellevue Lit. Press, New York, 2009.

93. R. MEGGINSON, *An Introduction to Banach Space Theory*, Springer, New York, 1998.

94. A. MELLO AND C.S. KUBRUSLY, *Residual spectrum of power bounded operators*, Funct. Anal. Approx. Comput. **10**-3 (2018), 205–215.

95. V. MÜLLER, *Spectral Theory of Linear Operators: and Spectral Systems in Banach Algebras*, 2nd edn. Birkhäuser, Basel, 2007.

96. G. MURPHY, *C*-Algebras and Operator Theory*, Academic Press, San Diego, 1990.

97. C.M. PEARCY, *Some Recent Developments in Operator Theory*, CBMS Regional Conference Series in Mathematics No. 36, Amer. Math. Soc., Providence, 1978.

98. H. RADJAVI AND P. ROSENTHAL, *Invariant Subspaces*, Springer, Berlin, 1973; 2nd edn. Dover, New York, 2003.

99. M. REED AND B. SIMON, *Methods of Modern Mathematical Physics I: Functional Analysis*, 2nd ed., Academic Press, New York, 1980.

100. F. RIESZ AND B. SZ.-NAGY, *Functional Analysis*, Frederick Ungar, New York, 1955; reprinted: Dover, New York, 1990.

101. S. ROMAN, *Advanced Linear Algebra*, 3rd edn. Springer, New York, 2008.

102. H.L. ROYDEN, *Real Analysis*, 3rd edn. Macmillan, New York, 1988.

103. W. RUDIN, *Real and Complex Analysis*, 3rd edn. McGraw-Hill, New York, 1987.

104. W. RUDIN, *Functional Analysis*, 2nd edn. McGraw-Hill, New York, 1991.

105. M. SCHECHTER, *On the essential spectrum of an arbitrary operator. I*, J. Math. Anal. Appl. **13** (1966), 205–215.

106. M. SCHECHTER, *Principles of Functional Analysis*, Academic Press, New York, 1971; 2nd edn. Graduate Studies in Mathematics, Vol. 36, Amer. Math. Soc., Providence, 2002.

107. J. SCHWARTZ, *Some results on the spectra and spectral resolutions of a class of singular operators*, Comm. Pure Appl. Math. **15** (1962), 75–90.

108. Y.-M. SONG AND A.-H. KIM, *Weyl's theorem for tensor products*, Glasgow Math. J. **46** (2004), 301–304.

109. V.S. SUNDER, *Functional Analysis – Spectral Theory*, Birkhäuser, Basel, 1998.

110. B. SZ.-NAGY, C. FOIAŞ, H. BERCOVICI, AND L. KÉRCHY, *Harmonic Analysis of Operators on Hilbert Space*, Springer, New York, 2010; enlarged 2nd edn. of B. SZ.-NAGY AND C. FOIAŞ, North-Holland, Amsterdam, 1970.

111. A.E. TAYLOR AND D.C. LAY, *Introduction to Functional Analysis*, Wiley, New York, 1980; reprinted: Krieger, Melbourne, 1986; enlarged 2nd edn. of A.E. TAYLOR, Wiley, New York, 1958.

112. A. UCHIYAMA, *On the isolated points of the spectrum of paranormal operators*, Integral Equations Operator Theory **55** (2006), 145–151.

113. J. WEIDMANN, *Linear Operators in Hilbert Spaces*, Springer, New York, 1980.

114. H. WEYL, *Über beschränkte quadratische Formen, deren Differenz vollstetig ist*, Rend. Circ. Mat. Palermo **27** (1909), 373–392.

List of Symbols

$\varnothing$, 1
$\mathbb{N}_0$, 1
$\mathbb{N}$, 1
$\mathbb{Z}$, 1
$\mathbb{Q}$, 1
$\mathbb{R}$, 1
$\mathbb{C}$, 1
$\mathbb{F}$, 1
$\{x_n\}$, 1
$\langle x, y \rangle$, 1
$\times$, 1
$\|T\|$, 1
$\mathcal{B}[\mathcal{X}, \mathcal{Y}]$, 1
$\mathcal{B}[\mathcal{X}]$, 1
$\{0\}$, 2
I, 2
$\xrightarrow{u}$, 2
$\xrightarrow{s}$, 2
$\xrightarrow{w}$, 2
$\mathcal{M}^-$, 2
$\mathcal{N}(T)$, 2
$\mathcal{R}(T)$, 2
$\oplus$, 3
$\cap$, 3

T^{-1}, 3
$\mathcal{G}[\mathcal{X}, \mathcal{Y}]$, 3
$\mathcal{G}[\mathcal{X}]$, 3
O, 5
$\perp$, 5, 89
$A^\perp$, 5
$\mathcal{M}^{\perp\perp}$, 5
$\simeq$, 6, 23, 73, 191, 194, 220
$\ominus$, 6
$\operatorname{span} A$, 6
$\bigvee A$, 6
$\cup$, 6
$\bigoplus$, 7
$\mathcal{D}(T)$, 9, 194
T^*, 9, 193
T^{**}, 9, 194
$T|_\mathcal{M}$, 10
$\leq$, 12
$<$, 12
$\prec$, 12
$r(T)$, 15, 38
$\mathcal{B}_\infty[\mathcal{X}, \mathcal{Y}]$, 18
$\mathcal{B}_\infty[\mathcal{X}]$, 18
$\dim \mathcal{M}$, 20

© Springer Nature Switzerland AG 2020
C. S. Kubrusly, *Spectral Theory of Bounded Linear Operators*,
https://doi.org/10.1007/978-3-030-33149-8

Index

A

absolutely continuous measure, 89, 96, 225
absolutely continuous unitary, 89
adjoint operator, 9, 11, 193
adjoint transformation, 9
affine space, 190
algebra, 1, 91
algebra homomorphism, 51, 91
algebra isomorphism, 91
algebra with identity, 91
algebraic complements, 3, 166, 189
algebraic isomorphism, 191
algebraic multiplicity, 129, 157, 173, 219
algebraic similarity, 122
analytic function, 29
analytic function on neighborhoods, 110
analytic function on spectra, 110, 111
analytic functional calculus, 102, 111
approximate eigenvalue, 32
approximate point spectrum, 32
approximation spectrum, 32
arc, 103
ascent of an operator, 163
Atkinson operator, 208
Atkinson Theorem, 145
Axiom of Choice, 31, 74

B

backward bilateral shift, 50
backward unilateral shift, 50
Banach algebra, 2, 91
Banach space, 1
Banach–Steinhaus Theorem, 2, 41
bilateral shift, 49, 89
bilateral weighted shift, 50
biquasitriangular operator, 217
Bolzano–Weierstrass Property, 57
Borel measure, 64
Borel sets, 64
Borel σ-algebra, 64
boundary of a set, 30
bounded below, 1
bounded component, 51
bounded inverse, 3
bounded inverse on its range, 3
Bounded Inverse Theorem, 3
bounded linear transformation, 1
bounded measurable function, 65, 95
bounded sequence, 2
bounded variation, 103
Browder operator, 162, 165
Browder spectrum, 162, 170
Browder Theorem, 173, 178, 179

© Springer Nature Switzerland AG 2020
C. S. Kubrusly, *Spectral Theory of Bounded Linear Operators*,
https://doi.org/10.1007/978-3-030-33149-8

Printed in the United States
By Bookmasters